Therapeutic Lipidology

CONTEMPORARY CARDIOLOGY

CHRISTOPHER P. CANNON, MD
SERIES EDITOR

Therapeutic Lipidology, edited by *Michael H. Davidson* MD, FACC, FACP, *Peter P. Toth* MD, PhD, FAAP, FICA, FAHA, FCCP, FACC, *and Kevin C. Maki*, PhD, *2007*

Current Concepts in Cardiology: Cardiac Rehabilitation, edited by *William E. Kraus*, MD, FACC, FACSM, *and Staven J. Keteyian*, PhD, *2007*

Management of Acute Pulmonary Embolism, edited by *Stavros V. Konstantinides*, MD, *2007*

Stem Cells and Myocardial Regeneration, edited by *Marc S. Penn*, MD, PhD, *2007*

Essential Echocardiography: *A Practical Handbook With DVD*, edited by *Scott D. Solomon*, MD, *2006*

Preventive Cardiology: *Insights Into the Prevention and Treatment of Cardiovascular Disease, Second Edition*, edited by *JoAnne Micale Foody*, MD, *2006*

The Art and Science of Cardiac Physical Examination: *With Heart Sounds and Pulse Wave Forms on CD*, by *Narasimhan Ranganathan*, MD, *Vahe Sivaciyan*, MD, *and Franklin B. Saksena*, MD, *2006*

Cardiovascular Biomarkers: *Pathophysiology and Disease Management*, edited by *David A. Morrow*, MD, *2006*

Cardiovascular Disease in the Elderly, edited by *Gary Gerstenblith*, MD, *2005*

Platelet Function: *Assessment, Diagnosis, and Treatment*, edited by *Martin Quinn*, MB BCh BAO, PhD, *and Desmond Fitzgerald*, MD, FRCPI, FESC, APP, *2005*

Diabetes and Cardiovascular Disease, Second Edition, edited by *Michael T. Johnstone*, MD, CM, FRCP(C), *and Aristidis Veves*, MD, DSc, *2005*

Angiogenesis and Direct Myocardial Revascularization, edited by *Roger J. Laham*, MD, *and Donald S. Baim*, MD, *2005*

Interventional Cardiology: *Percutaneous Noncoronary Intervention*, edited by *Howard C. Herrmann*, MD, *2005*

Principles of Molecular Cardiology, edited by *Marschall S. Runge*, MD, *and Cam Patterson*, MD, *2005*

Heart Disease Diagnosis and Therapy: *A Practical Approach, Second Edition*, by *M. Gabriel Khan*, MD, FRCP(LONDON), FRCP(C), FACP, FACC, *2005*

Cardiovascular Genomics: *Gene Mining for Pharmacogenomics and Gene Therapy*, edited by *Mohan K. Raizada*, PhD, *Julian F. R. Paton*, PhD, *Michael J. Katovich*, PhD, *and Sergey Kasparov*, MD, PhD, *2005*

Surgical Management of Congestive Heart Failure, edited by *James C. Fang*, MD *and Gregory S. Couper*, MD, *2005*

Cardiopulmonary Resuscitation, edited by *Joseph P. Ornato*, MD, FACP, FACC, FACEP *and Mary Ann Peberdy*, MD, FACC, *2005*

CT of the Heart: *Principles and Applications,* edited by *U. Joseph Schoepf*, MD, *2005*

Coronary Disease in Women: *Evidence-Based Diagnosis and Treatment*, edited by *Leslee J. Shaw*, PhD *and Rita F. Redberg*, MD, FACC, *2004*

Cardiac Transplantation: *The Columbia University Medical Center/New York-Presbyterian Hospital Manual*, edited by *Niloo M. Edwards*, MD, *Jonathan M. Chen*, MD, *and Pamela A. Mazzeo*, *2004*

Heart Disease and Erectile Dysfunction, edited by *Robert A. Kloner*, MD, PhD, *2004*

Complementary and Alternative Cardiovascular Medicine, edited by *Richard A. Stein*, MD *and Mehmet C. Oz*, MD, *2004*

Nuclear Cardiology, The Basics: *How to Set Up and Maintain a Laboratory*, by *Frans J. Th. Wackers*, MD, PhD, *Wendy Bruni*, BS, CNMT, *and Barry L. Zaret*, MD, *2004*

Minimally Invasive Cardiac Surgery, Second Edition, edited by *Daniel J. Goldstein*, MD, *and Mehmet C. Oz*, MD *2004*

Cardiovascular Health Care Economics, edited by *William S. Weintraub*, MD, *2003*

Platelet Glycoprotein IIb/IIIa Inhibitors in Cardiovascular Disease, Second Edition, edited by *A. Michael Lincoff*, MD, *2003*

Heart Failure: *A Clinician's Guide to Ambulatory Diagnosis and Treatment*, edited by *Mariell L. Jessup*, MD *and Evan Loh*, MD, *2003*

Management of Acute Coronary Syndromes, Second Edition, edited by *Christopher P. Cannon*, MD *2003*

Aging, Heart Disease, and Its Management: *Facts and Controversies*, edited by *Niloo M. Edwards*, MD, *Mathew S. Maurer*, MD, *and Rachel B. Wellner*, MPH, *2003*

Peripheral Arterial Disease: *Diagnosis and Treatment*, edited by *Jay D. Coffman*, MD *and Robert T. Eberhardt*, MD, *2003*

Cardiac Repolarization: *Bridging Basic and Clinical Science*, edited by *Ihor Gussak*, MD, PhD, *Charles Antzelevitch*, PhD, *Stephen C. Hammill*, MD, *Win K. Shen*, MD, *and Preben Bjerregaard*, MD, DMSC, *2003*

THERAPEUTIC LIPIDOLOGY

Edited by

MICHAEL H. DAVIDSON, MD, FACC, FACP

*Director of Preventive Cardiology,
Rush University Medical Center,
Radiant Research,
Chicago, IL*

PETER P. TOTH, MD, PhD, FAAP, FICA, FAHA, FCCP, FACC

*Director of Preventive Cardiology,
Sterling Rock Falls Clinic,
Chief of Medicine,
CGH Medical Center
Clinical Associate Professor,
University of Illinois College of Medicine,
Peoria, IL
Southern Illinois University School of Medicine,
Springfield, IL*

and

KEVIN C. MAKI, PhD

*Chief Science Officer, Provident Clinical Research,
Bloomington, IN*

Foreword by

ANTONIO M. GOTTO, Jr., MD, DPhil

*Weill Cornell Medical College
New York, NY*

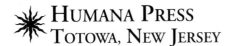

© 2007 Humana Press Inc.
999 Riverview Drive, Suite 208
Totowa, New Jersey 07512
www.humanapress.com

All rights reserved. No part of this book may be reproduced, stored in a retrieval system, or transmitted in any form or by any means, electronic, mechanical, photocopying, microfilming, recording, or otherwise without written permission from the Publisher.

The content and opinions expressed in this book are the sole work of the authors and editors, who have warranted due diligence in the creation and issuance of their work. The publisher, editors, and authors are not responsible for errors or omissions or for any consequences arising from the information or opinions presented in this book and make no warranty, express or implied, with respect to its contents.

Due diligence has been taken by the publishers, editors, and authors of this book to assure the accuracy of the information published and to describe generally accepted practices. The contributors herein have carefully checked to ensure that the drug selections and dosages set forth in this text are accurate and in accord with the standards accepted at the time of publication. Notwithstanding, as new research, changes in government regulations, and knowledge from clinical experience relating to drug therapy and drug reactions constantly occurs, the reader is advised to check the product information provided by the manufacturer of each drug for any change in dosages or for additional warnings and contraindications. This is of utmost importance when the recommended drug herein is a new or infrequently used drug. It is the responsibility of the treating physician to determine dosages and treatment strategies for individual patients. Further it is the responsibility of the health care provider to ascertain the Food and Drug Administration status of each drug or device used in their clinical practice. The publisher, editors, and authors are not responsible for errors or omissions or for any consequences from the application of the information presented in this book and make no warranty, express or implied, with respect to the contents in this publication.

Production Editor: Amy Thau

Cover design by Karen Schulz

Cover Illustration: See Acknowledgments

For additional copies, pricing for bulk purchases, and/or information about other Humana titles, contact Humana at the above address or at any of the following numbers: Tel.: 973-256-1699; Fax: 973-256-8341, E-mail: orders@humanapr.com; or visit our Website: www.humanapress.com

This publication is printed on acid-free paper. ∞

ANSI Z39.48-1984 (American National Standards Institute) Permanence of Paper for Printed Library Materials.

Photocopy Authorization Policy:

Authorization to photocopy items for internal or personal use, or the internal or personal use of specific clients, is granted by Humana Press Inc., provided that the base fee of US $30.00 is paid directly to the Copyright Clearance Center at 222 Rosewood Drive, Danvers, MA 01923. For those organizations that have been granted a photocopy license from the CCC, a separate system of payment has been arranged and is acceptable to Humana Press Inc. The fee code for users of the Transactional Reporting Service is: [978-1-58829-551-4/07 $30.00].

Printed in the United States of America. 10 9 8 7 6 5 4 3 2 1

eISBN 978-1-59745-533-6

Library of Congress Control Number: 2007928341

To the memory of my father, Dr. David M. Davidson, who died at age 47 of a myocardial infarction.

Michael H. Davidson

To RSM (especially MMQ, MMMc, MBA, and MYML) for teaching me the fundamentals of living my life with faith, fortitude, and ferocity.

Peter P. Toth

To the PCRC team, whose commitment to "doing well while doing good" contributes every day to our mission of helping people acheive greater quality of life by preventing diseases that erode good health and function.

Kevin C. Maki

Preface

Lipids constitute a diverse array of molecules with an astounding range of functions in biological systems in both health and disease. Many of the greatest scientific minds of the last century committed their energies to the identification of lipids and sterols and to characterizing how these molecules are synthesized and metabolized by a large number of cell types. This led to the identification of many species of lipids, including phospholipids, sphingolipids, cerebrosides, triglycerides, fatty acids and their metabolites (eicosanoids, leukotrienes), and cholesterol/cholesterol esters and other sterols. Lipids are involved in cell membrane and organelle formation, intracellular and intercellular signaling, cell surface receptor function, inflammation, and immunity. Cholesterol is a modulator of cell membrane fluidity, is a precursor to steroid hormones and bile salts, and, along with many lipid species, is a key modulator of risk for developing atherosclerotic disease. Cholesterol can also be a source of fatty acid in its esterified form. Fifteen Nobel Prizes have been awarded to investigators for work related to the biosynthesis and metabolism of cholesterol, and more are sure to follow. The identification of the enzymes and intermediates of the cholesterol biosynthetic pathway will forever be a milestone in human creativity and investigative ingenuity. The regulatory circuitry of lipid and cholesterol metabolism is understandably of enormous complexity, and many areas of uncertainty and gaps in knowledge remain.

In the clinical arena, the last 30 years have seen tremendous advances in the fields of lipidology and cardiovascular medicine. The Framingham Study and other epidemiologic investigations conducted around the world unequivocally demonstrated strong relationships between lipoproteins and risk for cardiovascular events. Seminal work by Fredrickson and colleagues classified dyslipidemias according to derangements in specific classes of lipoproteins. Goldstein and Brown ushered in the modern era of "therapeutic lipidology" by discovering the low-density lipoprotein (LDL) receptor and subsequently showing that (1) mutations in this receptor were etiologic for familial hypercholesterolemia and (2) inhibitors of the β-hydroxymethylglutaryl coenzyme A reductase upregulated this receptor and facilitated the clearance of LDL from the circulation. Gotto and coworkers subsequently sequenced the amino acid structure of apoprotein B100. Other apoproteins were identified and sequenced and their three-dimensional conformations and cell surface receptors elucidated. Mutations in these apoproteins were correlated with both gain and loss of function, yielding many identifiable and characteristic changes in lipoprotein distributions. More recently, the role of membrane cassette transport proteins in lipid trafficking between the gastrointestinal tract, liver, and blood vessel walls has been characterized. Mutations in these membrane cassette transport proteins have provided mechanistic insights into such lipid disorders as Tangier's disease and β-sitosterolemia. The role of lipolytic enzymes (lipoprotein lipase, hepatic lipase, endothelial lipase, phospholipase A2, and hormone-sensitive lipase) and their various mutant forms in lipid metabolism have also been elucidated and remain foci of intensive investigation. Polymorphisms from around the world continue to be identified and catalogued. The number of nuclear transcription factors (e.g., peroxisomal proliferator-activated receptors, liver X receptor, and farnesoid X receptor) involved in lipid metabolism continues to expand. In the next few years, the areas of lipidomics, isoprenoid metabolism, and vascular biology will continue to provide us with more expansive vistas from which to view the relationship between lipid dynamics and atherogenesis.

Until the mid-1980s, serum LDL cholesterol (LDL-C) was recognized as a critical risk factor for atherosclerotic disease. However, prospective, placebo-controlled data showing that therapeutically reducing the burden of this lipoprotein in serum decreases risk for cardiovascular morbidity and mortality were lacking. The first proof of benefit from this approach came from the Lipid Research Clinics Coronary Primary Prevention Trial. In this trial, the bile acid binding resin cholestyramine was used to treat men at high risk for coronary artery disease. Patients experienced a 19% reduction in major coronary events with a 9% reduction in LDL-C. In its Adult Treatment Panel I, the National Cholesterol Education Program (NCEP) made its initial attempt to provide evidence-based guidelines for dyslipidemia management. The first statin, lovastatin, was approved in 1987. Over the course of the last 20 years, we have observed a remarkable evolution in our understanding of lipid metabolism and have greatly expanded the therapeutic armamentarium with which to decrease the burden of atherogenic lipoproteins. We have a large number of prospective clinical trials which demonstrate that lipid lowering with statins, fibrates, and niacin beneficially impacts risk for cardiovascular morbidity and mortality. Millions of patients have benefited from these therapies.

Despite these great successes, much work remains to be done. Even in the face of very aggressive lipid lowering with statins, the majority of acute cardiovascular events are not prevented. Clinicians must keep pace with rapidly changing guidelines, which call for ever greater reductions in both serum LDL-C and non-high-density lipoprotein cholesterol levels. The percentage of the population that develops the metabolic syndrome and diabetes is increasing, a greater percentage of the population is elderly with increased risk, and more patients are surviving ischemic strokes and acute coronary syndromes such as myocardial infarction and unstable angina. All of these patients warrant aggressive management according to updated NCEP guidelines. Many new drug classes are being developed to help address these challenges. Therapeutic approaches to more reliably increase serum levels of high-density lipoproteins, drugs to reduce appetite and weight gain, and deeper insights into specific dietary and lifestyle approaches to dyslipidemia are actively being developed and evaluated. The atherogenicity of novel pathways are burgeoning areas of exploration. Many new markers of risk have also been introduced into medicine in recent years. These include C-reactive protein, myeloperoxidase, lipoprotein-associated phospholipase A2, and soluble CD36. Evaluation of lipoprotein particle size and number offer valuable insights into lipoprotein metabolism and risk for cardiovascular events. Considerable skill is required to integrate information gleaned from these various modes of patient testing and to tailor and optimize therapeutic approaches. It is these approaches and skills that *Therapeutic Lipidology* addresses.

This book is intended to provide practicing clinicians with a focused and intensive, but useable, source of information on the identification and management of dyslipidemias. The need for a volume such as this has been longstanding. Since the creation of the American Board of Clinical Lipidology, many clinicians have asked us to create an up-to-date, comprehensive reference on lipidology. Subsequent editions are certain to expand in size and scope as more species of atherogenic enzymes, lipids, and inflammatory mediators are identified and novel therapeutic interventions are introduced. The pace of scientific and clinical advances in lipidology is astounding. It is our sincerest wish that this reference will serve as a lifelong stimulus to the reader to continue to learn about the ever-changing and fascinating field of therapeutic lipidology. We also hope that this book will empower our readers to improve and extend the lives of the patients they serve.

Michael H. Davidson, MD, FACC, FACP
Peter P. Toth, MD, PhD, FAAP, FICA, FAHA, FCCP, FACC
Kevin C. Maki, PhD

Contents

Preface vii
Contributors xiii
Acknowledgments xv
Foreword xvii

1. Lipoprotein Metabolism and Vascular Biology 1
   Brian G. Choi, Juan J. Badimon, Pedro R. Moreno, and Valentin Fuster

2. Genetic Disorders of Lipoprotein Metabolism 23
   Marina Cuchel, Atif Qasim, and Daniel J. Rader

3. International Guidelines for Dyslipidemia Management 37
   Cathleen E. Maki

4. Pathophysiology and Management of Dyslipidemias Associated
   with Obesity, Type 2 Diabetes, and Other Insulin-Resistant States 55
   Kevin C. Maki

5. C-Reactive Protein and Other Inflammatory Markers
   in Cardiovascular Disease 69
   Natalie Khuseyinova and Wolfgang Koenig

6. Dietary Prescriptions to Control Dyslipidemias and Coronary Disease Risk 113
   Margo A. Denke

7. Pharmacological Therapy for Cardiovascular Disease:
   Current and Emerging Therapies 121
   Michael H. Davidson

8. Effects of Thiazolidinediones on Serum Lipoproteins 149
   Anjli Maroo and W. H. Wilson Tang

9. High-Density Lipoproteins 159
   Peter P. Toth

10. Management of Hypertriglyceridemia 201
    Reginald Labossiere and Ira J. Goldberg

11. Lowering Low-Density Lipoprotein Cholesterol Levels to Reduce
    Atherosclerotic Coronary Heart Disease Risk..........................221
    Harold E. Bays

12. Lipoprotein(a) as an Emerging Risk Factor for Atherothrombosis:
    Principles from Bench to Bedside..241
    Michael B. Boffa, Santica M. Marcovina, and Marlys L. Koschinsky

13. Familial Hypercholesterolemia and Lipid Apheresis 267
    Patrick M. Moriarty, Cheryl A. Gibson, and Klaus Flechsenhar

14. Phytosterolemia: *Synthesis, Absorption, Trafficking, and Excretion
    of Cholesterol and Noncholesterol Sterols*..................................291
    Thomas Dayspring

15. Utilization of Lipoprotein Subfractions 321
    William C. Cromwell and James D. Otvos

16. Cardiovascular Disease in Women: *The Management of Dyslipidemia* 349
    Emma A. Meagher

17. Management of Lipids in the Elderly.......................................369
    Micah J. Eimer and Neil J. Stone

18. The Clinical Use of Noninvasive Modalities in the Assessment
    of Atherosclerosis 389
    **Atul R. Chugh, Samir N. Patel, Venkataraman Rajaram,
    Rachel Neems, Matt Feinstein, Marshall Goldin,
    and Steven B. Feinstein**

19. Management of Dyslipidemia in Children 409
    Stephen R. Daniels

20. The Allied Health Professional's Role in the Management of Dyslipidemia 425
    Lynne T. Braun and Joan E. Mathien

21. Development and Management of a Lipid Clinic 441
    Carol M. Mason

22. The American Board of Clinical Lipidology Physician Certification Program... 477
    Nicola A. Sirdevan

Index 481

Contributors

JUAN J. BADIMON, PhD • *Cardiovascular Biology Research Laboratory, Zena and Michael A. Wiener Cardiovascular Institute, Mount Sinai School of Medicine, New York, NY*

HAROLD BAYS, MD, FACP • *L-MARC Research Center, Louisville, KY*

MICHAEL B. BOFFA, MSc, PhD • *Department of Biochemistry, Queen's University, Kingston, Ontario, Canada*

LYNNE T. BRAUN, PhD, RN, CNP • *Preventive Cardiology Center, Heart Center for Women, Rush University, College of Nursing, Chicago, IL*

BRIAN G. CHOI, MD, MBA • *Cardiovascular Biology Research Laboratory, Zena and Michael A. Wiener Cardiovascular Institute, Mount Sinai School of Medicine, New York, NY*

ATUL R. CHUGH, MD • *Rush University Medical Center, Chicago, IL*

WILLIAM C. CROMWELL, MD • *Division of Lipoprotein Disorders, Presbyterian Center for Preventive Cardiology, Presbyterian Cardiovascular Institute, Charlotte, NC; and Hypertension and Vascular Disease Center, Wake Forest University School of Medicine, Charlotte, NC*

MARINA CUCHEL, MD, PhD • *654 BRBII/III Labs, University of Pennsylvania School of Medicine, Philadelphia, PA*

STEPHEN R. DANIELS, MD, PhD • *Department of Pediatrics and Environmental Health, Cincinnati Children's Hospital Medical Center and the University of Cincinnati College of Medicine, Cincinnati, OH*

MICHAEL H. DAVIDSON, MD, FACC, FACP • *Davidson, Director of Preventive Cardiology, Rush University Medical Center, Radiant Research, Chicago, IL*

THOMAS DAYSPRING, MD, FACP • *Dayspring and Macaluso, Wayne, NJ*

MARGO A. DENKE, MD • *University of Texas Health Science Center at San Antonio, San Antonio, TX*

MICAH J. EIMER, MD • *Feinberg School of Medicine, Northwestern University, Chicago, IL*

STEVEN B. FEINSTEIN, MD • *Department of Medicine, Section of Cardiology, Rush University Medical Center, Chicago, IL*

MATT FEINSTEIN, BA • *Rush University Medical Center, Chicago, IL*

KLAUS FLECHSENHAR, MD • *Gelita Health Initiative, Heidelberg, Germany*

VALENTIN FUSTER, MD, PhD • *Cardiovascular Biology Research Laboratory, Zena and Michael A. Wiener Cardiovascular Institute, Mount Sinai School of Medicine, New York, NY*

CHERYL A. GIBSON, PhD • *Department of Internal Medicine, University of Kansas School of Medicine, Kansas City, KS*

IRA J. GOLDBERG, MD • *Division of Preventive Medicine and Nutrition, Columbia University School of Medicine, New York, NY*

MARSHALL GOLDIN, MD • Department of Cardiovascular/Vascular Surgery, *Rush University Medical Center, Chicago, IL*
NATALIE KHUSEYINOVA MD • *Department of Internal Medicine II, Cardiology, University of Ulm Medical Center, Ulm, Germany*
WOLFGANG KOENIG, MD, PhD • *Department of Internal Medicine II, Cardiology, University of Ulm Medical Center, Ulm, Germany*
MARLYS L. KOSCHINSKY, PhD • *Department of Biochemistry, Queen's University, Kingston, Ontario, Canada*
REGINALD LABOSSIERE, MD • *Division of Preventive Medicine and Nutrition, Columbia University School of Medicine, New York, NY*
CATHLEEN E. MAKI, RN, MSN, NP • *Provident Clinical Research, Glen Ellyn, IL*
KEVIN C. MAKI, PhD • *Chief Science Officer, Provident Clinical Research, Glen Ellyn, IL*
SANTICA M. MARCOVINA PhD, SCD • *Northwest Lipid Research Laboratories, University of Washington, Seattle, WA*
ANJLI MAROO, MD, RVT • *Department of Cardiovascular Medicine, The Cleveland Clinic, Cleveland, OH*
CAROL M. MASON, ARNP, FAHA • *USF Heart Health, University of South Florida College of Medicine, Tampa, FL*
JOAN E. MATHIEN, BSN • *Preventive Cardiology Center, Chicago, IL*
EMMA A. MEAGHER, MD • *Department of Medicine, University of Pennsylvania School of Medicine, Philadelphia, PA*
PEDRO R. MORENO, MD • *Cardiovascular Biology Research Laboratory, Zena and Michael A. Wiener Cardiovascular Institute, Mount Sinai School of Medicine, New York, NY*
PATRICK M. MORIARTY, MD • *1336 KU Hospital, University of Kansas Medical Center, Kansas City, KS*
RACHEL NEEMS, MD • *Rush University Medical Center, Chicago, IL*
JAMES D. OTVOS, PhD • *LipoScience, Inc, Raleigh, NC; and Department of Biochemistry, North Carolina State University, Raleigh, NC*
SAMIR N. PATEL, MD • *Rush University Medical Center, Chicago, IL*
ATIF QASIM, MD • *654 BRBII/III Labs, University of Pennsylvania School of Medicine, Philadelphia, PA*
DANIEL J. RADER, MD • *654 BRBII/III Labs, University of Pennsylvania School of Medicine, Philadelphia, PA*
VENKATARAMAN RAJARAM, MD • *Rush University Medical Center, Chicago, IL*
NICOLA A. SIRDEVAN, MPA • *American Board of Clinical Lipidology, Inc, Jacksonville, FL*
NEIL J. STONE, MD • *Chicago, Illinois*
W. H. WILSON TANG, MD • *Department of Cardiovascular Medicine, The Cleveland Clinic, Cleveland, OH*
PETER P. TOTH, MD, PhD, FAAP, FICA, FAHA, FCCP, FACC • *Director of Preventive Cardiology, Sterling Rock Falls Clinic, Sterling, IL; and University of Illinois College of Medicine, Peoria, IL*

Acknowledgments

The editors would like to thank Paul Dolgert of the Humana Press and Chrsitopher Cannon, MD, editor of the *Contemporary Cardiology* series, for their unwavering support for developing a textbook devoted to therapeutic lipidology. We also thank Mary Lou Briglio and Theresa Tardi for providing administrative assistance and for obtaining the permissions to reproduce the large number of figures contained within this volume. Michael Davidson thanks Ruth Kleinpell RN, PhD for editorial assistance with chapter 7.

It is with much gratitude that we also acknolwedge colleagues who provided the images for the cover of *Therapeutic Lipidology*. The upper left panel was provided by Zahi Fayad, MD, of the Mount Sinai School of Medicine, New York. This image demonstrates inflamed carotid artery plaque within the red bars by measuring the uptake of 18-fluorine-fluorodeoxyglucose using fused positron emission tomography and computed tomography. Tony DeFranco, MD of the CVCTA Education Center in San Francisco provided the central panel, a color-enhanced 64-slice computed tomography image of the frontal view of a human heart. We also thank Dr. Eva Istvan of Washington University in Saint Louis for the bottom right panel, demonstrating the interaction of simvastatin with the active site of HMG CoA reductase (reproduced from Istvan and Diesenhofer [2001]; Science 292: 1160-4 with permssion).

Foreword

As an investigator of the field of lipid disorders for the last four decades, I have been a first-hand witness to the many successes, controversies, and breakthroughs that have radically advanced the field of atherosclerotic cardiovascular disease prevention. Early skepticism through the mid-1980s that dyslipidemia and other risk factors could account for only about half of all coronary events has given way to a sophisticated understanding of the intricate interplay of cardiovascular risks and the realization that risk factor modification is critical to stemming the tide of cardiovascular disease.

I began in the field in 1967 at the National Institutes of Health (NIH), in the laboratory of Donald Fredrickson and Robert Levy, but the history of lipid research extends much farther back. Cholesterol, cholesteryl ester, and phospholipid circulate in blood in macromolecular complexes called lipoproteins. Michel Macheboeuf, working at the Pasteur Institute in 1929, first described the plasma lipoproteins by using ammonium sulfate fractionation of horse serum to isolate alpha-lipoproteins, what today almost certainly would be recognized as high-density lipoprotein (HDL). In the years surrounding World War II, Edwin J. Cohn and J.L. Oncley utilized fractionation procedures that included precipitation under different acid and salt conditions, as well as electrophoresis, to purify the constituents of blood plasma and serum.

In 1949, John Gofman and his colleagues at the University of California at Berkeley used the newly developed ultracentrifuge to observe that the lipoprotein fraction that corresponds with LDL was associated with increased risk for cardiovascular disease. Around the same time, at the Rockefeller University, E.H. "Pete" Ahrens studied the connection of diet with lipid metabolism and cholesterol homeostasis. At Cornell University Medical College (now Weill Cornell Medical College), Russ, Eder, and Barr used Cohn and Oncley's techniques to argue that higher concentrations of alpha-lipoprotein contributed to the lower cardiovascular disease event rate seen in pre-menopausal women.

At the NIH, Donald Fredrickson took advantage of the preparative ultracentrifuge and paper electrophoresis to characterize the lipoproteins further. In a series of landmark articles in the *New England Journal of Medicine*, Fredrickson, Robert Levy, and Robert Lees described a classification system based on which groups of lipoproteins were elevated. Fredrickson phenotyping has proved to be a popular and enduring principle for describing the dyslipidemias, although it does not distinguish between dyslipidemias with a primary or secondary etiology.

In those early days, two large observational trials were central to establishing the cardiac dangers of excess cholesterol levels. Ancel Keys' Seven Countries study, during the 1950s and 1960s, established the association between dietary fat consumption, dyslipidemia, and coronary risk on an international scale. In those populations that consumed a higher proportion of saturated fat as the total dietary caloric intake, there were higher levels of cholesterol in the blood and a higher incidence of CHD mortality. The other study began in 1948, under the auspices of the National Heart Institute (the precursor to the National Heart, Lung, and Blood Institute). Conducted in a working class community in Massachussetts, the Framingham Heart Study began to collect longitudinal data on 1980 men and 2421 women. In 1961, the Framingham investigators showed that high blood pressure, smoking, and high cholesterol levels were major preventable factors in heart disease, a report that cemented the concept of modifiable and nonmodifiable "risk factors."

The crux of the lipid hypothesis was that reducing cholesterol would reduce coronary events. The epidemiology had made a clear connection between cholesterol and coronary risk, but compelling evidence that treating cholesterol would make a difference was not available until the publication of the Lipid Research Clinic Coronary Primary Prevention trial (LRC-CPPT) in 1984. The trial recruited 3806 middle-aged men with primary hypercholesterolemia, no evidence of coronary disease, and tolerance to the then-available preparation of cholestyramine, which participants compared to sand in terms of texture and palatability. Although the drug was supposed to be administered as 24 g/day, patients could manage an average dosage of around 12 g/day versus placebo. There was a 12% reduction in LDL-C and 9% in total cholesterol, corresponding to a reduction in CHD events of 19% after 7.4 years of follow-up.

While some investigators criticized the trial's design and statistical analysis, many others felt that LRC-CPPT confirmed the fundamental idea that lipid modification could reduce coronary disease. The study raised public awareness of the issue of cholesterol and heart disease and inspired the formation of an NIH committee to put together a National Cholesterol Consensus that was chaired by Dan Steinberg. The National Heart, Lung, and Blood Institute launched a program that turned into the National Cholesterol Education Program. The announcement of the positive results of the primary-prevention Helsinki Heart Study of gemfibrozil shortly thereafter provided additional affirmation of the lipid approach, although the paucity of evidence that treatment could improve survival hindered wider acceptance.

In September 1987, lovastatin became the first statin to be introduced into the market. The launch of the statins was a watershed moment for the field. These drugs lowered LDL-C more effectively than other pharmacotherapies that were available at the time, and early angiographic trials of these agents made a positive case for their use in slowing the progression of atherosclerotic disease. Beginning in 1994 and through the present day, large-scale clinical trials have overwhelmingly demonstrated that statins reduce the risks for clinical cardiovascular events across a broad range of patients. Indeed, three of the landmark statin trials have achieved the "Holy Grail" of a reduction in total mortality with active treatment: the Scandinavian Simvastatin Survival Study (4S); the Long-term Intervention with Pravastatin in Ischaemic Disease (LIPID); and the Heart Protection Study.

More recently, trials have investigated how aggressively we should treat LDL-C. On the whole, the findings suggest that achieving lower levels of LDL-C is associated with greater cardiovascular risk reduction, and guidelines now reflect that even lower LDL-C targets may be desirable in certain higher-risk groups. While no threshold has yet been convincingly identified, it is possible that we may eventually discover a lower bound of LDL-C below which no further coronary benefit may be achieved.

Although targeting LDL-C has taken primacy in the management of the dyslipidemias, other lipid fractions may also contribute to cardiovascular risk, such as HDL cholesterol (HDL-C) and triglyceride, and have garnered increasing attention. The Framingham Heart Study observed that the higher the HDL-C, the lower the risk for an event. Trials with gemfibrozil, such as the Helsinki Heart Study and the Veterans Affairs HDL-C

Intervention Trial (VA-HIT), showed primary HDL-cholesterol raising and triglyceride-lowering effects and reported fewer clinical events with treatment compared with placebo. Improved understanding of the cardioprotective effects of HDL has nurtured interest in potential therapeutic applications, although such efforts were dealt a blow in 2006 by the withdrawal of torcetrapib, a member of a new class of HDL-C-raising compounds, from further study because of cardiovascular toxicity. Nevertheless, the hypothesis that raising HDL-C may prevent cardiovascular disease remains a viable one, and other avenues are being explored.

In summary, atherosclerosis was once thought to be an irreversible, inevitable consequence of aging. The recognition of dyslipidemia as a major modifiable risk factor introduced the possibilities of both treatment and prevention. As time has passed and the body of evidence has grown, investigators and clinicians have increasingly appreciated the multifactorial nature of cardiovascular risk. There is now an emphasis on treating overall global risk for near-term cardiovascular disease, rather than treating individual risk factors per se. At the same time, basic science has further elucidated the complex pathologies that underlie atherosclerosis and has suggested novel therapeutic concepts.

All of this leads to the book that is now in your hands. This excellent text, edited by Michael H. Davidson, Peter P. Toth, and Kevin C. Maki, is an important synthesis of current information on the management of lipid disorders and cardiovascular risk. The authors and editors have produced a book that not only presents a thoughtful, complete, and authoritative guide to contemporary clinical challenges, but also often takes the long view of the field, putting information in the context of a rich five decades of research. Physicians who treat patients at risk for cardiovascular disease will find it an invaluable resource in their daily practice.

Although the advances of the last 50 years have been truly impressive, there is much about the origins and prevention of atherosclerosis that remains undiscovered. Continued exploration is imperative in order to maximize our efforts to reduce the burden of cardiovascular disease in our patients and around the world.

Antonio M. Gotto, Jr., MD, DPhil
Weill Cornell Medical College
New York, NY

1 Lipoprotein Metabolism and Vascular Biology

*Brian G. Choi, MD, MBA,
Juan J. Badimon, PhD, Pedro R. Moreno, MD,
and Valentin Fuster, MD, PhD*

CONTENTS

INTRODUCTION
LIPOPROTEIN CLASSIFICATION
APOLIPOPROTEIN CLASSIFICATION
LIPID METABOLISM
HDL METABOLISM
LIPOPROTEIN METABOLISM REGULATION
LIPIDS IN VASCULAR BIOLOGY AND ATHEROTHROMBOSIS
KEY POINTS
REFERENCES

INTRODUCTION

Cholesterol is vitally important for the maintenance of cellular membranes, production of steroid hormones and bile acids, and a supply of energy for metabolic needs. However, an excess of cholesterol may lead to the development of atherosclerosis and coronary artery disease; imbalances in lipoprotein metabolism are responsible for arterial lipid deposition. As lipids are intricately entwined with the atherothrombotic disease process, therapeutic interventions aimed at favorably altering lipoprotein metabolism have shown success in limiting atherogenesis and its subsequent complications. To understand how these different therapeutic strategies work, one must possess a basic understanding of the metabolic pathways governing lipidology.

Lipoproteins are complex particles that are generally composed of a hydrophobic core of mainly nonpolar cholesterol esters (CEs) and triglycerides (TGs) surrounded by an amphipathic phospholipid monolayer that includes unesterified free cholesterol (FC) and proteins known as apolipoproteins or aproteins. Apolipoproteins play four critical roles: (1) they provide a framework for lipoprotein assembly; (2) they add structural stability to the lipoprotein complex; and (3) they determine the metabolic fate of the lipoprotein by activating or inhibiting key enzymes; and (4) they act as ligands

From: *Contemporary Cardiology: Therapeutic Lipidology*
Edited by: M. H. Davidson, P. P. Toth, and K. C. Maki © Humana Press Inc., Totowa, NJ

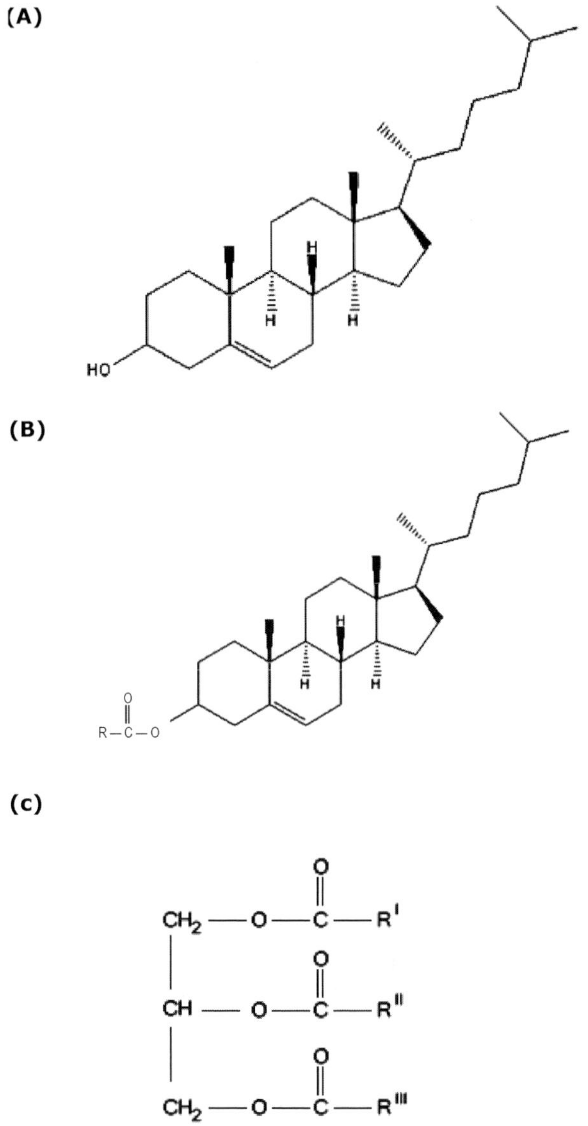

Fig. 1. (**A**) Free cholesterol (FC), (**B**) cholesterol ester (CE), and (**C**) triglyceride (TG). These nonpolar molecules are not water soluble and, therefore, require transport via lipoproteins.

for receptor molecules. Figure 1 details the chemical structures of these components; as these molecules are hydrophobic, they require a carrier to circulate in the aqueous bloodstream. Phospholipids have a polar head containing a glycerol group attached to two fatty acid chains. The hydrophilic outer shell of the glycerol backbones of the phospholipids allows lipoproteins to be water-soluble carriers of the hydrophobic inner contents through the circulating blood (Fig. 2).

LIPOPROTEIN CLASSIFICATION

There are five major classes of lipoproteins, defined by their respective density and electrophoretic mobility (Table 1). Most proteins have densities in the range of 1.3–1.4 g/mL, and lipid aggregate density is approximately 0.8 g/mL; hence,

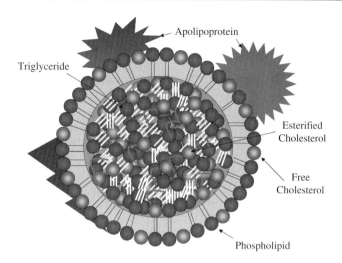

Fig. 2. Lipoprotein structure. A hydrophilic outer surface surrounds an inner hydrophobic core, allowing the transport of triglycerides and cholesterol in circulating blood. The apolipoproteins on the outer surface present to receptors on cell membranes and determine the ultimate destination for the lipoprotein molecule.

Table 1
Lipoprotein Classification and Properties (90,91,94)

Class	Density (g/mL)	Electrophoretic mobility	Diameter (nm)	Molecular weight (Da)
Chylomicrons	0.93	Remain at origin	75–1200	$(50–1000) \times 10^6$
Very low density lipoproteins	0.93–1.006	Pre-β	30–80	$(10–80) \times 10^6$
Intermediate-density lipoproteins	1.006–1.019	Slow pre-β	25–35	$(5–10) \times 10^6$
Low-density lipoproteins	1.019–1.063	β	18–25	$(2–3) \times 10^6$
High-density lipoproteins	1.063–1.21	α	5–12	$(65–386) \times 10^3$

lipoproteins with a higher protein:lipid ratio are denser than those with a lower protein:lipid ratio (Fig. 3). The least dense and largest lipoproteins are the TG-rich chylomicrons, and the densest and smallest lipoproteins are the high-density lipoproteins (HDLs). Between the chylomicrons and HDL are the very low density lipoproteins (VLDLs), intermediate-density lipoproteins (IDLs), and low-density lipoproteins (LDLs), listed in order of increasing density.

APOLIPOPROTEIN CLASSIFICATION

Although lipoproteins are classified based on their density that is determined by their combination of components, their functional uniqueness is dependent on the apolipoproteins that they carry. The apolipoproteins play a critical role in the recognition of lipoproteins as they serve as receptor ligands and enzyme cofactors. The

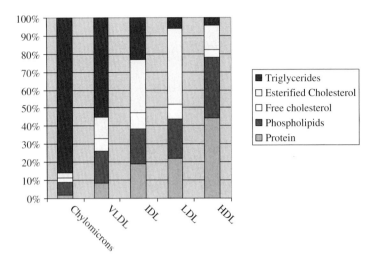

Fig. 3. Lipoprotein composition as percent of mass *(90,91)*.

apolipoproteins are typically classified by an alphabetical designation (A through E) with a Roman numeral suffix defined by the order in which the apolipoprotein emerges from a chromatographic column (Table 2). Apolipoproteins have four major functions: provide a framework for lipoprotein assembly, add structural stability to the lipoprotein complex, activate or inhibit processing enzymes, and signal to receptor molecules for uptake.

- *ApoA-I*: principal structural protein of HDL but also found in chylomicrons and activates lecithin:cholesterol acyltransferase (LCAT).
- *ApoA-II*: another structural protein of HDL also found in chylomicrons and activates hepatic lipase (HL).
- *ApoA-IV*: predominantly found in HDL and activates LCAT and lipoprotein lipase (LPL).
- *ApoB-48*: exclusively found in chylomicrons, derived from the apoB-100 gene, and reduced to 48% of the N-terminal component of B-100 by RNA editing, with no LDL receptor (LDLr) binding domain.
- *ApoB-100*: principal structural protein of LDL, alsofound in VLDL, IDL, and Lp(a); ligand for LDLr.
- *ApoC-I*: primarily found in HDL and chylomicrons and also in IDL and VLDL and activates LCAT.
- *ApoC-II*: protein primarily of VLDL and chylomicrons and also found in HDL and IDL and activates LPL.
- *ApoC-III*: found broadly in HDL, IDL, LDL, VLDL, and chylomicrons but more consistently in VLDL and chylomicrons and inhibits LPL *(1)*.
- *ApoD*: exclusively found in HDL and possibly activates CE transfer protein (CETP) *(2)*.
- *ApoE*: also found broadly in HDL, IDL, LDL, VLDL, and chylomicrons and binds to LDLr with varying affinity dependent on the inherited apoE allele. Three different apoE isoforms exist in humans: apoE2 with lower affinity to LDLr, apoE3 with intermediate binding affinity, and apoE4 with higher affinity *(3)*.
- *Apo(a)*: distinguishing structural protein of Lp(a) that is covalently bound to apoB-100 and inhibits plasminogen activation *(4,5)*.

Table 2
Apolipoprotein Classification (95)

	Predominant lipoprotein	Minor lipoproteins	Plasma concentration (mg/dL)	Role	Molecular weight (kDa)	Chromosome
ApoA-I	HDL	Chylomicrons	90–160	LCAT activation	28.3	11q23
ApoA-II	HDL	Chylomicrons	25–45	HL activation	17	1q21–23
ApoA-IV	HDL		10–20	LCAT and LPL activation	45	11q23
ApoB-48	Chylomicrons		0–100	Structural	241	2q23–24
ApoB-100	LDL, VLDL	IDL, Lp(a)	50–150	LDLr binding	512	2q23–24
ApoC-I	Chylomicrons, HDL	IDL, VLDL	5–6	LCAT activation	6.63	19q13.2
ApoC-II	Chylomicrons, VLDL	HDL, IDL	3–5	LPL activation	8.84	19q13.2
ApoC-III	Chylomicrons, VLDL	HDL, IDL, LDL	10–14	LPL activation	8.76	11q23
ApoD	HDL		4–7	CETP activation	33	3q26.2
ApoE	Chylomicron remnants, IDL	HDL, LDL, VLDL	2–8	LDLr binding	34	19q13.2
Apo(a)	Lp(a)		0–200	Plasminogen activation inhibition	250–800	6q27

CETP, cholesterol ester transfer protein; HDL, high-density lipoprotein; HL, hepatic lipase; IDL, intermediate-density lipoprotein; LCAT, lecithin:cholesterol acyltransferase; LDL, low-density lipoprotein; LDLr, LDL receptor; LPL, lipoprotein lipase; VLDL, very low density lipoprotein.

LIPID METABOLISM

Lipoprotein metabolism serves the two major functions of providing TGs to adipose and muscle for storage or energy substrate and of transporting cholesterol for cellular membrane, steroid hormone, and bile acid syntheses. The majority of carbon and hydrogen that is used for cellular fuel passes through lipid intermediaries prior to ultimate use. Lipid metabolism has two pathways to maintain the movement of lipids from diet to blood to cells: (1) the exogenous pathway (Fig. 4) that starts with intestinal absorption of dietary fat and cholesterol and (2) the endogenous pathway (Fig. 5) that starts with VLDL production from the liver.

Exogenous Pathway

Consumed dietary fat is emulsified in bile salts within the intestinal lumen and then converted to monoglycerides or diglycerides and free fatty acids (FFAs) by pancreatic

EXOGENOUS PATHWAY

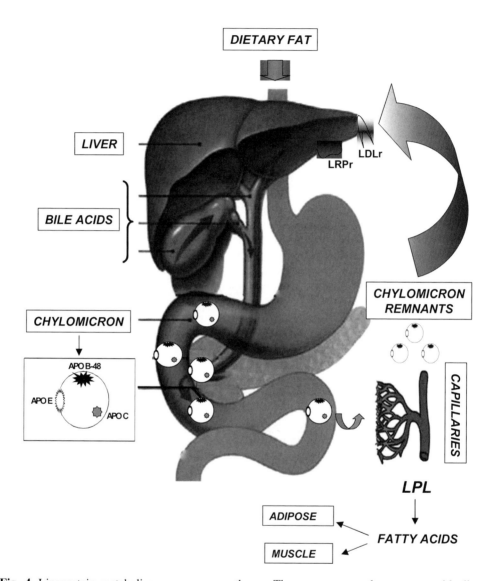

Fig. 4. Lipoprotein metabolism: exogenous pathway. The exogenous pathway starts with dietary consumption of fat that is broken down by pancreatic enzymes and then absorbed by intestinal cells. Here, they are packaged into chylomicrons for export into circulation. Triglycerides carried by the chylomicrons are hydrolyzed into fatty acids by lipoprotein lipase (LPL) for local metabolic needs or for storage in adipose; the remnants return to the liver for removal from circulation.

lipases. Micelles of monoglycerides, FFAs, phospholipids, and bile salts diffuse into intestinal mucosal cells where FFAs are recombined with glycerols to produce TGs, and absorbed dietary cholesterol is esterified by acyl:cholesterol acyltransferase (ACAT) to form CEs *(6)*. Micelle absorption may occur via a Niemann-Pick C1-like 1 protein *(7)*. The TGs and CEs are packaged within chylomicrons intracellularly and then extruded into either the portal circulation or the lymphatic circulation and travel via the thoracic

ENDOGENOUS PATHWAY

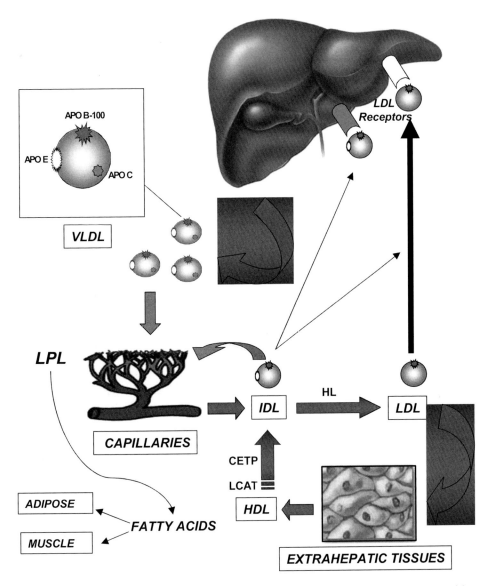

Fig. 5. Lipoprotein metabolism: endogenous pathway. The endogenous pathway starts with very low density lipoprotein (VLDL) production from the liver. The triglycerides carried by the VLDL are hydrolyzed by LPL into fatty acids for local metabolic needs or for storage in adipose; the remaining VLDL remnants, intermediate-density lipoproteins (IDLs), may be absorbed by the liver for removal from circulation or further hydrolyzed by hepatic lipase to form low-density lipoprotein (LDL). LDLs are either returned to the liver or absorbed by nonhepatic tissues to provide cholesterol esters for metabolic needs such as hormone production or cell membrane synthesis.

duct before entering systemic circulation. The chylomicrons initially contain mostly the apolipoproteins apoA-I, apoA-II, and apoB-48 but later acquire increased concentration of apoC-I, apoC-II, apoC-III, and apoE via transfer from HDL and lose their apoA-I with passage to the lymph. By binding to the C-terminal portion of LPL, the apoC-II activates LPL located on luminal side of capillary endothelial cells, which hydrolyzes TG into glycerol (a 3-carbon alcohol) and FFA *(8)*; FFA may then be used for local metabolic needs or stored in adipose. TG hydrolysis is modulated by apoC-III that acts as an LPL inhibitor *(9)*. The remaining glycerol is returned to the liver and intestines via the systemic circulation. ApoE on the remaining chylomicron remnants binds to hepatic LDLr, facilitating the removal of circulating chylomicrons into the liver. Hepatic uptake may also occur via LDL-related protein receptor (LRPr) *(10)*. Thus, the major fate of dietary fat via the exogenous pathway, unless used for immediate metabolic needs, is storage in adipocytes or transfer to the liver.

Endogenous Pathway

The endogenous pathway starts with the production of TG-rich VLDLs in the liver. In situations where dietary fat consumption is insufficient to meet metabolic needs for FFAs (i.e., starvation), the liver can synthesize VLDLs for exportation to other organ systems. Alternatively, cells may generate cholesterol de novo using acetate as a substrate in a process regulated by HMG-CoA reductase and HMG-CoA synthase, but this process is much more metabolically intensive.

Nascent TG-rich VLDLs prior to leaving the liver express apoE and apoB-100 around a core of mostly TGs and some CEs; upon export, nascent VLDLs interact with HDLs, and apoC-I, apoC-II, and apoC-III are added to form mature VLDLs. The apoC-II serves as a cofactor for LPL so that while VLDL is in circulation the TG core of VLDL is hydrolyzed by LPL on capillary endothelial cells forming relatively TG-depleted VLDL remnants, or IDL. The half-life of circulating VLDL is about 30–60 min *(11)*. The hydrolysis of TG by LPL is again moderated by apoC-III that inhibits LPL. The FFA liberated by LPL can again be used for local metabolic needs or stored in adipose as in the endogenous pathway.Because the IDL contains apoB-100 and apoE, these VLDL remnants can be resorbed by the liver via LDLr. Alternatively, approximately half of IDLs may be further hydrolyzed by HL to form LDLs.

Whereas VLDLs are mostly TGs with some CEs, LDLs are mostly CEs with less than 10% TGs. In a TG-rich milieu, LDLs can absorb more TGs, resulting in smaller, denser LDL subfractions that are thought to be more atherogenic. The primary structural apolipoprotein of LDL is apoB-100 that allows LDL to return to the liver for resorption via the LDLr. In the liver, the LDL may be transformed to bile acids and excreted via the biliary tree into the intestinal lumen. From the intestinal lumen, the bile acids may either be removed with the feces (approximately 1 g/day) or reabsorbed via the exogenous pathway in the distal ileum, thus completing the enterohepatic circulatory process.

The physiologic half-life of LDLs in humans is 2–3 days. Approximately half of LDLs, however, are absorbed by nonhepatic tissue. The primary physiologic purpose of LDL is to deliver CEs for critical cellular metabolic needs such as hormone production and cell membrane synthesis. Rapidly dividing cells, for example, because of their need for cellular membrane synthesis exuberantly express LDLr, and the highest level of expression, as expected, occurs in malignant cells. Peripheral cells also express LDLr, which recognize apoE and, to a lesser extent, apoB-100.

HDL METABOLISM

HDLs play a critical role in preventing the deposition of lipids in the periphery. HDL metabolism starts with the synthesis of pre-β lipid-poor apoA-I in the liver and gut, approximately half from each. The discoid-shaped apoA-I is the primary structural protein of HDL. The apoA-I interacts with hepatic and peripheral ATP-binding cassette transporter A1 (ABCA1), which mediates efflux of phospholipids and FC from intracellular storage pools to the structural scaffold provided by apoA-I, thereby forming lipid-laden nascent pre-β-HDL *(12,13)*. LCAT, also activated by apoA-I, esterifies the FC to form CE by transferring a fatty acid chain from lecithin to cholesterol, thereby transforming the nascent pre-β-HDL to the more spherical α-HDL. α-HDL may further accept FC via SRB1 receptors and by passive diffusion *(14)*. SRB1 receptors on hepatic cells may facilitate the selective uptake of α-HDL to the liver *(15)*. The subfractions of HDLs that contain apoE may also undergo hepatic uptake via the LDLr. Thus, HDL, particularly in its HDL_2 subfraction form, serves as the primary vehicle for the return of FC from the periphery back to the liver, a process termed reverse cholesterol transport (Fig. 6). Several epidemiologic studies have postulated an antiatherogenic effect associated with high HDL levels, but the first experimental evidence of the

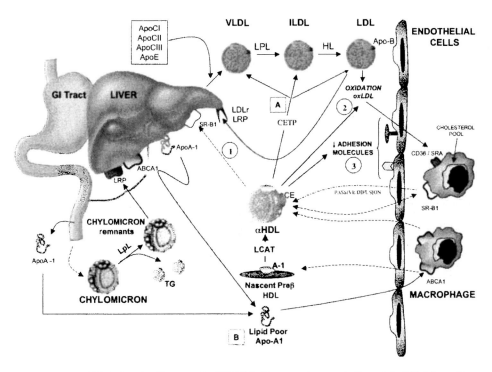

Fig. 6. The role of high-density lipoprotein (HDL) in lipoprotein metabolism. Modified from Brewer *(92)*. The accepted mechanisms behind the potential benefit of HDL are (1) reverse cholesterol transport (RCT) from lipid-laden macrophages in the vessel wall to the liver, (2) shunting of low-density lipoprotein (LDL) from being oxidized, thereby preventing accumulation of lipids within foam cells, and (3) blocking monocyte entry into the vessel wall via downregulation of endothelial cell adhesion molecules (ECAMs). Two sites of action for pharmacologic intervention are identified: (A) the inhibition of cholesterol ester transfer protein (CETP) and (B) augmentation of RCT by increase in lipid-poor apoA-I (i.e., apoA-I Milano infusion).

HDL-mediated reverse cholesterol transport comes from an experimental model by the decrease in lipid deposition in rabbits treated with weekly HDL infusion *(16,17)*. ApoD on α-HDL may also activate CETP, which transfers CEs from α-HDLs to VLDLs, IDLs, and LDLs. HDLs also serve as a reservoir of apoC and apoE for transfer to VLDLs and chylomicrons.

LIPOPROTEIN METABOLISM REGULATION

In the nonpathologic state, consumed dietary fat provides necessary cholesterol for metabolic needs via the exogenous pathway, and in times of fasting, the endogenous pathway supplies cholesterol. The balance between cholesterol deficit and excess is maintained by key regulatory processes.

In the fed state, LPL in adipose is upregulated so that as chylomicrons enter the circulation from the gut, they are hydrolyzed for storage into adipose. This process appears to be insulin mediated. Decreased adipose LPL activity during fasting allows chylomicrons to be used for metabolic demands of other tissues, but also in the fasting state, LPL activity is upregulated in more metabolically active cardiac and skeletal cells to steer the FFA energy substrate toward the organ systems with a greater requirement. Local demands for cholesterol are also regulated by LDLr expression. If cells are in positive cholesterol balance, LDLr is downregulated so that cellular intake of CEs is reduced. Excess cellular cholesterol in the form of 27-hydroxycholesterol also binds to the liver X receptor (LXR) that dimerizes with retinoid X receptor (RXR); this LXR/RXR complex binds to the LXR response element (LXRE) promoter to increase the expression of ABCA1 and SRB1 *(18–20)*. This upregulation of ABCA1 and SRB1 promotes the efflux of FC from the lipid-overloaded cells to HDL for disposal in the liver.

LIPIDS IN VASCULAR BIOLOGY AND ATHEROTHROMBOSIS

In 1852, Virchow first identified the importance of lipid accumulation in the formation of atherosclerotic plaques *(21)*. Perturbations in lipoprotein metabolism, whether by inherited disorders or by environmental factors, may promote atherogenesis. For example, if dietary fat and cholesterol consumption exceed metabolic needs and fecal excretion (i.e., Western diet), then this positive cholesterol balance may result in pathogenic consequences.

Atherothrombosis Phases

As the balance of lipid influx exceeds that of lipid efflux, atherosclerotic plaques progress through a series of five progressive phases (Fig. 7) as described by Fuster et al. *(21)*, based on a classification scheme previously described by the American Heart Association Committee on Vascular Lesions *(22)* and Stary *(23)*. Lipid influx/efflux plays a greater role in atherogenesis in Phases 1–3, whereas Phases 4 and 5 are more lipid independent.

PHASE 1

Early Phase 1 consists of type I lesions that have lipid-laden foam cells, but as lipid accumulates, smooth muscle cells migrate into the vessel intima, and extracellular lipid deposits are formed (type II). Type II lesions can be further divided into the progression-prone type IIa and the progression-resistant type IIb *(24)*. The anatomic

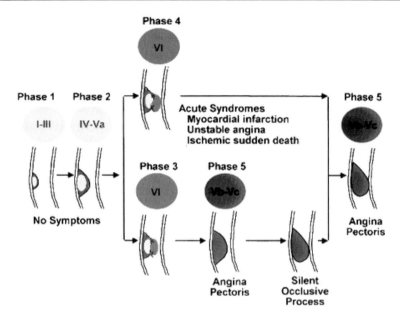

Fig. 7. Clinicopathologic correlation of asymptomatic atherosclerosis leading to symptomatic atherothrombosis. Modified from Corti and Fuster *(93)*.

location of the lesion may determine its fate as IIa or IIb. Departure of blood from laminar flow (e.g., at bifurcation points, curvatures, or stenoses) results in turbulence and areas of relative low shear stress; type IIa lesions are morelikely to be in areas of low shear stress, as LDL particles and monocytes have increased time to migrate into the vessel wall *(25)*. Type III lesions have more smooth muscle cells that have started to form extracellular matrix and have increased extracellular lipid deposition.

PHASE 2

The more advanced Phase 2 lesions have a high lipid content and a thin fibrous cap over the lipid core, making them more prone to rupture, especially in an inflammatory state. Plaques that have high free:esterified cholesterol ratios at the plaque center but high esterified cholesterol level at the plaque edge, possibly reflecting high local macrophage activity, are especially prone to rupture *(26)*. These plaques are categorized morphologically as one of two variants: (1) type IV lesions consist of confluent cellular lesions with a great deal of extracellular lipid intermixed with normal intima, which may predominate as an outer layer or cap; and (2) type Va lesions possess an extracellular lipid core covered by an acquired fibrous cap. The American Heart Association classification falls short of identifying plaque erosion or the thin-cap fibroatheroma (TCFA); new classifications have been proposed including these two categories, as proposed by Virmani et al. *(27)*. Once rupture occurs, the lesions are characterized as Phase 3 (clinically silent) or Phase 4 (symptomatic).

PHASE 3

Phase 3 lesions are the acute "uncomplicated" type VI lesions, originating from ruptured (type IV or Va) or eroded lesions, and leading to mural, non-obstructive thrombosis. This process is clinically silent but occasionally may lead to the onset of angina *(28)*. As LDL is associated with lipid accumulation and HDL with reverse cholesterol transport, plaque rupture was more associated with increased total cholesterol:HDL ratio than with either smoking or hypertension *(29)*. HDL may exert an antiapoptotic effect on vascular endothelial cells, which suggests that a plaque may be less vulnerable to disruption and rupture *(30)*.

PHASE 4

Phase 4 lesions are characterized by acute "complicated" type VI lesions, with fixed or repetitive, occlusive thrombosis. This process becomes clinically apparent in the form of acute coronary syndrome (ACS), although not infrequently it is silent *(31,32)*. In about two-thirds of ACS, occlusive thrombosis occurs on a nonstenotic plaque. In the remaining one-third, the thrombus occurs on the surface of a stenotic plaque *(33)*. In Phases 3 and 4, changes in the geometry of ruptured plaques, as well as organization of the occlusive or mural thrombus by connective tissue, can lead to occlusive or significantly stenotic and fibrotic plaques. Lipids do not play a direct role in the development of the Phase 4 lesion; however, increased lipid deposition within plaques can make Phase 4 more severe by increasing the thrombus size *(34)*. Lipid-rich atheromatous plaque core contains the most concentrated amount of tissue factor; thus, if disruption of the plaque results in exposure of the lipid core, a larger, more occlusive thrombus may develop *(35)*.

PHASE 5

Phase 5 is characterized by type Vb (calcific) or Vc (fibrotic) lesions that may cause angina; however, if preceded by stenosis or occlusion with associated ischemia, the myocardium may be protected by collateral circulation and such lesions may then be silent or clinically inapparent *(36,37)*. Lipid deposition does not play a role in Phase 5 development; Phase 5 may be described as a "healed" lesion, but if it results in a tight stenosis of the vessel lumen without adequate collaterals, these lesions result in significant morbidity.

Pathogenesis of Phase 1: Endothelial Dysfunction

The endothelium is a cellular monolayer that lines the blood vessels, separating the blood from the vascular media, and endothelial dysfunction is the first pathogenic sign in atherogenesis (Fig. 8) and initiates Phase 1 development *(38)*. Endothelial cells perform both autocrine and paracrine functions, and by secreting nitric oxide (NO), they can signal for downregulation of cell adhesion molecules (CAMs), prevent vasoconstriction, shift the balance from thrombosis to fibrinolysis, prevent lipoprotein oxidation, and decrease vessel permeability to lipids *(39)*. Shear stress, caused by blood flow perturbations at bifurcation segments or bends in vessel course, disturbs endothelial function, making the artery at these points more vulnerable to lipid infiltration *(40,41)*. The endothelium responds to shear stress by activating CAMs; these CAMs attract monocytes to enter the vessel wall, which accelerates the atherogenic process. As the monocytes enter and transform into macrophages and become foam cells with the absorption of lipids, they begin to express cytokines that activate the formation of extracellular matrix, fibrosis, and migration of smooth muscle cells into the intima *(42)*.

Hyperlipidemia, in and of itself, also induces endothelial dysfunction; other risk factors for endothelial dysfunction include hypertension (which increases mechanical shear forces), advanced glycation end products (found in diabetics and the elderly),

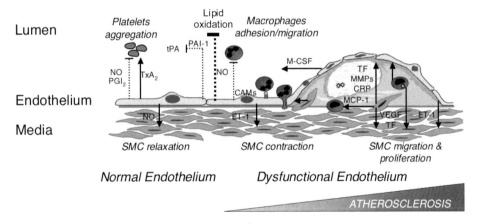

Fig. 8. Hyperlipidemia augments endothelial dysfunction. Dysfunctional endothelium facilitates atherosclerosis development via impaired endothelium-dependent vasodilation, upregulated platelet aggregation and thrombogenicity, increased macrophage adhesion and migration, and smooth muscle cell (SMC) proliferation *(47)*. CAM, cell adhesion molecule; CRP, C-reactive protein; ET, endothelin; MCP, monocyte chemotactic protein; M-CSF, monocyte colony-stimulating factor; MMP, matrix metalloproteinase; NO, nitric oxide; PAI, plasminogen activator inhibitor; PGI_2, prostacyclin; TF, tissue factor; tPA, tissue-type plasminogen activator; TxA2, thromboxane A2; VEGF, vascular endothelial growth factor.

tobacco smoke, vasoactive amines, and immune complexes *(43–46)*. With high lipid levels, NADPH oxidase is upregulated causing nitric oxide (NO) breakdown and superoxide formation *(47)*. Lower NO translates to poorer flow-mediated dilation and increased monocyte and platelet adhesion. Thus, as hyperlipidemia induces endothelial dysfunction, the endothelium loses its ability to prevent the inward migration of lipids and to prevent the oxidation of LDL, which also enhances lipid accumulation, causing a cycle of dysregulation. LDL apoB binds to the extracellular matrix proteins such as proteoglycans, collagen, and fibronectin, leading to the retention of LDL in the vessel wall *(48)*. Once in the wall, LDL, particularly in its oxidized form, may induce inflammation, further increasing plaque vulnerability to rupture *(48)*.

HDL, however, may enhance endothelial function through its anti-inflammatory and antioxidant properties *(49)*, and HDL mediates other mechanisms that may enhance endothelial function (Fig. 9). HDL decreases CAM expression by inhibiting the expression of IL-1α-induced E-selectin *(50,51)*. Also, HDL binds to endothelial SRB1 receptors, thereby activating endothelial NO synthase *(52)*; this mechanism may improve endothelial function by restoring NO bioavailability *(53,54)*. Therefore, by inhibiting endothelial dysfunction, HDL may play a critical role in limiting atherogenesis in addition to its role in reverse cholesterol transport.

Progression into Phase 2: Lipid Influx

If LDL plasma concentrations are high, excess circulating LDLs may be absorbed by macrophages and foam cells. As lipids accumulate, the atherosclerotic lesions progress to Phase 2. Macrophages and foam cells also express LDLr that facilitates the selective uptake of LDL, but these cells also express unregulated scavenger receptors, CD36 or SRA, that absorb oxidized LDL, making oxidized LDL a particularly potent chemoattractant molecule for these cells *(55–57)*. Oxidized LDL contains oxidized

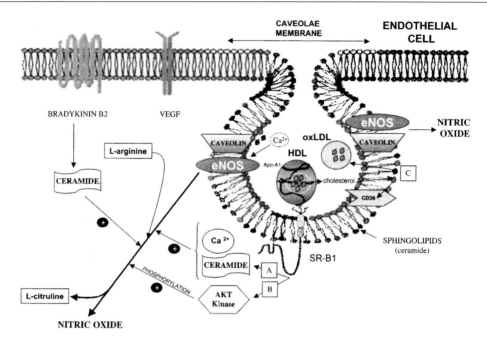

Fig. 9. Nitric oxide dysregulation results in increased lipid accumulation in the vessel wall which may be counterbalanced by high-density lipoprotein (HDL). (A) HDL binding to SRB1, a class B scavenger receptor, can stimulate eNOS by a ceramide-dependent pathway. (B) SRB1–HDL interaction also induces the activation of Akt kinase, which subsequently phosphorylates eNOS and stimulates the enzyme to synthesize nitric oxide. (C) Oxidized low-density lipoprotein (LDL)-initiated CD36-dependent depletion of caveola cholesterol is countered, which allows eNOS to remain associated with caveolae, thereby maintaining the ability of nitric oxide production.

phospholipids that are generated under situations of oxidative stress such as infection and other inflammatory states, and the oxidated state allows binding to these receptors, but native LDL does not *(58)*. These responses to excess oxidized LDLs are adaptive in that they may prevent the LDLs from otherwise causing endothelial injury *(59)*, but as these macrophages continue to absorb LDLs and begin to foam, lipid accumulates within the vessel wall, thereby increasing atherosclerosis *(60)*. Endothelial cells may also selectively uptake oxidized LDLs via the Lox-1 receptor *(61)*. The uptake of oxidized LDLs triggers the activation of transcriptional nuclear factor-kappa-B (NFκB), which upregulates the expression of CAMs that encourage monocyte entry into the vessel wall *(62)*.

In addition to the inhibition of plasminogen, another role that Lp(a) plays is in lipid influx. Lp(a), a modified form of LDL that carries the apo(a), binds with especially high affinity to specific receptor on macrophages that result in rapid accumulation of LDL and foaming *(63)*. Oxidized LDLs may also augment the atherogenicity of Lp(a) *(64)*.

As the vessel thickens from lipid influx and remodeling of the vessel wall, the diffusion of oxygen from blood to the outer vessel wall is impaired, and without neovascularization, oxygen demand may exceed supply at the cellular level, resulting in either anaerobic metabolism or necrosis. Macrophages, however, express cytokines that encourage neovessel formation in the wall to supply oxygenated blood to the thickened large vessels *(65,66)*. Vasa vasorum neovascularization surrounds the adventitia and branch into the outer media of large arteries and are particularly directed to large arterial lipid depositions *(67,68)*. While neovessel content has been associated with vulnerability to plaque rupture *(69)*, atherosclerosis regression is associated with vasa vasorum involution. With lipid-lowering therapy, the concentration of lipids in the vessel wall greatly exceeds the concentration in systemic circulation, so the vasa vasorum may provide a gradient pathway for cholesterol efflux from the vessel wall; as lipids leave the plaque and the vessel wall thins, these microvessels also regress, leaving behind a more stable-type lesion *(21)*.

Phase 2 Reversal: Lipid Efflux

HDL also prevents LDL accumulation in the vessel wall by two principal mechanisms: prevention of LDL oxidation and CAM downregulation. HDL hydrolyzes oxidized LDL, preventing LDL absorption via SRA and CD36 *(70)*, and HDL also downregulates endothelial CAMs that could prevent monocyte entry into the vessel wall *(71)*. Not only does HDL have a high antioxidant content, but it also contains several enzymes that prevent LDL oxidation or degrade bioactive products of oxidized LDLs. HDL carries paraoxonase (PON) that breaks down oxidized LDL phospholipids *(72)*; by lowering oxidized phospholipids, HDL prevents their inflammatory response via cytokine release and subsequent monocyte infiltration *(73)*. ApoA-I may directly exert an antioxidant effect by reducing phospholipids and CE peroxides and removing the "seeding molecules" hydroperoxyeicosatetraenoic acid (HPETE) and hydroperoxyoctadecadienoic acid (HPODE) which induce the oxidation of lipoprotein phospholipids *(74)*.

Not only does HDL play a role in prevention of lipid influx, but it also facilitates active efflux. Excess peripheral lipid accumulation may be counterbalanced by HDL-mediated reverse cholesterol transport (Fig. 6). Lipid-poor apoA-I may accept FC from macrophages via ABCA1, or α-HDL may further accept FC via the macrophage SRB1 receptor. Therefore, as the HDL:LDL ratio is increased, lipid efflux is encouraged and atherosclerosis regression is possible (Fig. 10). Therefore, intervention aimed at increasing HDL may result in reverse remodeling of Phase 2 lesions, thereby preventing their progression to the symptomatic later phases.

Phase 3: Plaque Rupture

The two principal causes of plaque rupture are (1) rheology and (2) intraplaque activity within the vessel. As atherosclerotic plaques grow eccentrically, the shear stress on the vessel wall may increase, and vulnerable points on the plaque (i.e., thin fibrous cap) may rupture. Within the vessel wall, though, numerous processes also additionally contribute to vulnerability to rupture.

Lesions that are heavily infiltrated with macrophages are more likely to rupture *(75)*. These cells express enzymes that aid in the phagocytosis of immunogenic objects; however, the same processes may degrade vessel structure, resulting in rupture. Chiefly among these enzymes are plasminogen activators and matrix metalloproteinases that are a family of collagenases, elastinases, and stromyelysins. In particular, if the internal elastic lamina is degraded, the plaque is particularly prone to rupture *(69)*.

Although more commonly associated with lipid deposition, lipoprotein metabolism affects plaque vulnerability to rupture and subsequent Phase 3 development. Plaque rupture is associated with an increased total cholesterol:HDL ratio *(29)*, and HDL may directly stabilize lesions by preventing the apoptosis of vascular endothelial cells *(30)*.

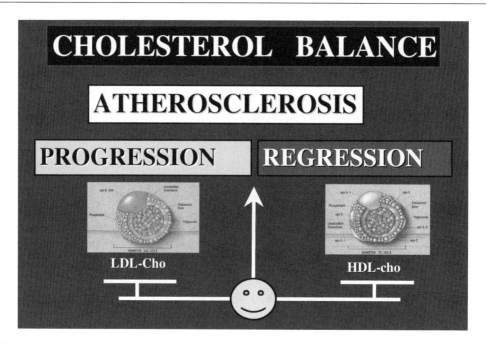

Fig. 10. The high-density lipoprotein to low-density lipoprotein (HDL:LDL) ratio is useful to understand factors that influence atherosclerosis regression versus progression. As HDL increases and LDL decreases, the scale is tipped toward atherosclerosis regression; as LDL increases and HDL decreases, the scale is tipped toward progression.

Exacerbants of Phase 3 and 4 Complications: Vulnerable Blood

Virchow described the triad of thrombogenicity that is appropriate to the atherothrombotic phenomenon (Table 3). When plaques rupture, the severity of their clinical manifestation is increased in those situations with greater blood thrombogenicity. Furthermore, the revealed substrate after plaque rupture also influences the size of the consequent thrombosis. Both blood vulnerability to thrombosis and plaque characteristics that influence thrombogenicity are related to increased lipid content in the blood and the vessel wall; therefore, worse lipid profiles are associated with greater complications of plaque rupture. The magnitude of the thrombus subsequent to rupture modulates the severity of the ACS, and more severe thrombogenicity could make the difference between the asymptomatic Phase 3 and the symptomatic Phase 4 lesion.

Using the Badimon perfusion chamber, various human aortic plaque types were exposed to arterial blood flow, and lipid-rich plaques were found to be the most thrombogenic *(34)*. Blood vulnerability also increases with higher LDL levels. Higher LDL levels result in greater thrombus formation as also determined by perfusion chamber study, and lowering the LDL level by statin therapy reduced thrombus formation *(76,77)*. The mechanism behind this association between LDL and thrombogenicity may occur via the release of tissue factor and thrombin activation, a similar mechanism responsible for the observed increased thrombus formation with smoking and diabetes *(78,79)*.

Whereas LDL is prothrombotic, HDL seems to be antithrombotic. In hypercholesterolemic men, high HDL level predicted decreased platelet thrombus formation within the Badimon chamber *(80)*, and in a mouse model, apoA-I Milano inhibited arterial

Table 3
Virchow's Triad of Thrombogenicity

Vessel wall substrate
   Atherosclerosis
   Inflammation
   Injury (i.e., postintervention)
Rheology
   Shear stress
   Stenosis
   Vasoconstriction
Systemic factors
   Dyslipidemia
   Diabetes mellitus
   Catecholamines
   Hyperreninemia
   Hypercoagulability
   Infection

thrombus formation *(81)*. The antithrombotic effect is further confirmed in an in vitro study which showed that purified HDL-activated proteins C and S significantly inactivated coagulation factor Va *(82)*. In vitro study also showed that the HDL component apolipoproteins apoA-I and apoA-II activate fibrinolysis *(83)*. Furthermore, HDL contains a tissue factor pathway inhibitor (TFPI) that would make the HDL less thrombogenic *(84)*. HDL and apoA-I decrease the procoagulant activity of erythrocytes *(85)*. Another mechanism by which HDL lowers thrombogenicity is by reducing oxidized LDL *(86)*. Even though HDL3 is thought to be less atheroprotective, physiologic levels of HDL3 inhibit thrombin-induced fibrinogen binding and aggregation via inhibition of phosphatidylinositol 4,5-bisphosphate turnover and formation of 1,2-diacylglycerol and phosphatidylinositol 4,5-bisphosphate *(87)*. HDL also increases the bioavailability of prostacyclin (PGI_2), which inhibits platelet aggregation, by stabilizing it *(88)*. HDL also decreases inflammation by inhibiting sphingosine kinase activity, which decreases sphingosine 1-phosphate, a key molecule in mediating TNF-induced adhesion protein expression *(89)*.

KEY POINTS

- Lipoprotein metabolism has two major functions: (1) transport of TGs for storage or for energy substrate for metabolic needs and (2) transport of cholesterol for cellular membrane, steroid hormone, and bile acid syntheses.
- Lipoproteins are complex molecules composed of CE, TGs, phospholipids, FC, and apolipoproteins, and there are five major classes of lipoproteins based on their density: chylomicrons, very low density lipoproteins (VLDLs), intermediate-density lipoproteins (IDLs), low-density lipoproteins (LDLs), and high-density lipoproteins (HDLs).
- Different apolipoproteins distinguish the lipoprotein classes and have four major functions: (1) provide framework for lipoprotein assembly, (2) add structural stability for lipoprotein complex, (3) activate or inhibit key processing enzymes, and (4) act as ligands for receptor molecules.

- Lipoprotein metabolism has two main pathways: (1) the exogenous pathway that starts with intestinal absorption of dietary fat and cholesterol and (2) the endogenous pathway that starts with hepatic VLDL production.
- The exogenous pathway predominates in the well-fed state, whereas the endogenous pathway predominates in fasting.
- HDL metabolism plays a key role in reverse cholesterol transport, in the transfer of key apolipoproteins to other lipoproteins, and as a reservoir for CEs for other lipoproteins.
- Excess of dietary fat or cholesterol, whether by environmental or by heritable factors, contributes to imbalance in lipoprotein metabolism that results in atherogenesis.
- Lipoprotein metabolism plays a greater role in atherogenesis in Phase 1–3 lesions, but complications associated with plaque rupture or erosion in later phase lesions are exacerbated by a worsened lipid profile.
- Hyperlipidemia leads to endothelial cell dysfunction that initiates atherogenesis.
- HDL, however, improves endothelial function, suggesting a beneficial role beyond reverse cholesterol transport.
- Excess LDL results in increased progression of atherosclerosis, but higher HDL may induce regression.
- A similar beneficial effect on blood thrombogenicity is seen with lower thrombotic potential with high HDL and greater with high LDL.
- A high HDL:LDL ratio therapeutically improves atherothrombotic potential.

REFERENCES

1. Ginsberg HN, Le NA, Goldberg IJ, et al. Apolipoprotein B metabolism in subjects with deficiency of apolipoproteins CIII and AI. Evidence that apolipoprotein CIII inhibits catabolism of triglyceride-rich lipoproteins by lipoprotein lipase in vivo. *J Clin Invest* 1986;78:1287–95.
2. Fielding PE, Fielding CJ. A cholesteryl ester transfer complex in human plasma. *Proc Natl Acad Sci USA* 1980;77:3327–30.
3. Davignon J, Gregg RE, Sing CF. Apolipoprotein E polymorphism and atherosclerosis. *Arteriosclerosis* 1988;8:1–21.
4. Danesh J, Collins R, Peto R. Lipoprotein(a) and coronary heart disease. Meta-analysis of prospective studies. *Circulation* 2000;102:1082–5.
5. Palabrica TM, Liu AC, Aronovitz MJ, et al. Antifibrinolytic activity of apolipoprotein(a) in vivo: human apolipoprotein(a) transgenic mice are resistant to tissue plasminogen activator-mediated thrombolysis. *Nat Med* 1995;1:256–9.
6. Buhman KF, Accad M, Farese RV. Mammalian acyl-CoA:cholesterol acyltransferases. *Biochim Biophys Acta* 2000;1529:142–54.
7. Altmann SW, Davis HR, Jr., Zhu LJ, et al. Niemann-Pick C1 Like 1 protein is critical for intestinal cholesterol absorption. *Science* 2004;303:1201–4.
8. Goldberg IJ, Scheraldi CA, Yacoub LK, et al. Lipoprotein ApoC-II activation of lipoprotein lipase. Modulation by apolipoprotein A-IV. *J Biol Chem* 1990;265:4266–72.
9. McConathy WJ, Gesquiere JC, Bass H, et al. Inhibition of lipoprotein lipase activity by synthetic peptides of apolipoprotein C-III. *J Lipid Res* 1992;33:995–1003.
10. Goldstein JL, Brown MS. Regulation of low-density lipoprotein receptors: implications for pathogenesis and therapy of hypercholesterolemia and atherosclerosis. *Circulation* 1987;76:504–7.
11. Stalenhoef AF, Malloy MJ, Kane JP, Havel RJ. Metabolism of apolipoproteins B-48 and B-100 of triglyceride-rich lipoproteins in normal and lipoprotein lipase-deficient humans. *Proc Natl Acad Sci USA* 1984;81:1839–43.
12. Liu L, Bortnick AE, Nickel M, et al. Effects of apolipoprotein A-I on ATP-binding cassette transporter A1-mediated efflux of macrophage phospholipid and cholesterol: formation of nascent high density lipoprotein particles. *J Biol Chem* 2003;278:42976–84.
13. Haghpassand M, Bourassa PA, Francone OL, Aiello RJ. Monocyte/macrophage expression of ABCA1 has minimal contribution to plasma HDL levels. *J Clin Invest* 2001;108:1315–20.
14. Williams DL, Connelly MA, Temel RE, et al. Scavenger receptor BI and cholesterol trafficking. *Curr Opin Lipidol* 1999;10:329–39.

15. Trigatti BL, Krieger M, Rigotti A. Influence of the HDL receptor SR-BI on lipoprotein metabolism and atherosclerosis. *Arterioscler Thromb Vasc Biol* 2003;23:1732–8.
16. Badimon JJ, Badimon L, Fuster V. Regression of atherosclerotic lesions by high density lipoprotein plasma fraction in the cholesterol-fed rabbit. *J Clin Invest* 1990;85:1234–41.
17. Badimon JJ, Badimon L, Galvez A, et al. High density lipoprotein plasma fractions inhibit aortic fatty streaks in cholesterol-fed rabbits. *Lab Invest* 1989;60:455–61.
18. Fu X, Menke JG, Chen Y, et al. 27-Hydroxycholesterol is an endogenous ligand for liver X receptor in cholesterol-loaded cells. *J Biol Chem* 2001;276:38378–87.
19. Malerod L, Juvet LK, Hanssen-Bauer A, et al. Oxysterol-activated LXRalpha/RXR induces hSR-BI-promoter activity in hepatoma cells and preadipocytes. *Biochem Biophys Res Commun* 2002; 299:916–23.
20. Venkateswaran A, Laffitte BA, Joseph SB, et al. Control of cellular cholesterol efflux by the nuclear oxysterol receptor LXR alpha. *Proc Natl Acad Sci USA* 2000;97:12097–102.
21. Fuster V, Moreno PR, Fayad ZA, et al. Atherothrombosis and high-risk plaque part I: evolving concepts. *J Am Coll Cardiol* 2005;46:937–54.
22. Stary HC, Chandler AB, Dinsmore RE, et al. A definition of advanced types of atherosclerotic lesions and a histological classification of atherosclerosis. A report from the Committee on Vascular Lesions of the Council on Arteriosclerosis, American Heart Association. *Circulation* 1995;92:1355–74.
23. Stary HC. Natural history and histological classification of atherosclerotic lesions: an update. *Arterioscler Thromb Vasc Biol* 2000;20:1177–8.
24. Stary HC. Location and development of atherosclerotic lesions in coronary arteries. *Atherosclerosis* 1984;50:237–9.
25. Glagov S, Zarins C, Giddens DP, Ku DN. Hemodynamics and atherosclerosis. Insights and perspectives gained from studies of human arteries. *Arch Pathol Lab Med* 1988;112:1018–31.
26. Felton CV, Crook D, Davies MJ, Oliver MF. Relation of plaque lipid composition and morphology to the stability of human aortic plaques. *Arterioscler Thromb Vasc Biol* 1997;17:1337–45.
27. Virmani R, Kolodgie FD, Burke AP, et al. Lessons from sudden coronary death: a comprehensive morphological classification scheme for atherosclerotic lesions. *Arterioscler Thromb Vasc Biol* 2000;20:1262–75.
28. Davies MJ. Stability and instability: two faces of coronary atherosclerosis. The Paul Dudley White Lecture 1995. *Circulation* 1996;94:2013–20.
29. Burke AP, Farb A, Malcom GT, et al. Coronary risk factors and plaque morphology in men with coronary disease who died suddenly. *N Engl J Med* 1997;336:1276–82.
30. Nofer JR, Levkau B, Wolinska I, et al. Suppression of endothelial cell apoptosis by high density lipoproteins (HDL) and HDL-associated lysosphingolipids. *J Biol Chem* 2001;276:34480–5.
31. Canto JG, Shlipak MG, Rogers WJ, et al. Prevalence, clinical characteristics, and mortality among patients with myocardial infarction presenting without chest pain. *JAMA* 2000;283:3223–9.
32. Sheifer SE, Manolio TA, Gersh BJ. Unrecognized myocardial infarction. *Ann Intern Med* 2001;135:801–11.
33. Falk E, Shah PK, Fuster V. Coronary plaque disruption. *Circulation* 1995;92:657–71.
34. Fernandez-Ortiz A, Badimon JJ, Falk E, et al. Characterization of the relative thrombogenicity of atherosclerotic plaque components: implications for consequences of plaque rupture. *J Am Coll Cardiol* 1994;23:1562–9.
35. Toschi V, Gallo R, Lettino M, et al. Tissue factor modulates the thrombogenicity of human atherosclerotic plaques. *Circulation* 1997;95:594–9.
36. Pohl T, Seiler C, Billinger M, et al. Frequency distribution of collateral flow and factors influencing collateral channel development. Functional collateral channel measurement in 450 patients with coronary artery disease. *J Am Coll Cardiol* 2001;38:1872–8.
37. Werner GS, Ferrari M, Betge S, et al. Collateral function in chronic total coronary occlusions is related to regional myocardial function and duration of occlusion. *Circulation* 2001;104:2784–90.
38. Luscher TF, Tanner FC, Noll G. Lipids and endothelial function: effects of lipid-lowering and other therapeutic interventions. *Curr Opin Lipidol* 1996;7:234–40.
39. Corti R, Fuster V, Badimon JJ. Pathogenetic concepts of acute coronary syndromes. *J Am Coll Cardiol* 2003;41:7S–14S.
40. Ravensbergen J, Ravensbergen JW, Krijger JK, et al. Localizing role of hemodynamics in atherosclerosis in several human vertebrobasilar junction geometries. *Arterioscler Thromb Vasc Biol* 1998;18:708–16.

41. Nerem RM. Vascular fluid mechanics, the arterial wall, and atherosclerosis. *J Biomech Eng* 1992;114:274–82.
42. Libby P, Ridker PM, Maseri A. Inflammation and atherosclerosis. *Circulation* 2002;105:1135–43.
43. Gimbrone MA, Jr., Nagel T, Topper JN. Biomechanical activation: an emerging paradigm in endothelial adhesion biology. *J Clin Invest* 1997;99:1809–13.
44. Traub O, Ishida T, Ishida M, et al. Shear stress-mediated extracellular signal-regulated kinase activation is regulated by sodium in endothelial cells. Potential role for a voltage-dependent sodium channel. *J Biol Chem* 1999;274:20144–50.
45. Kunsch C, Medford RM. Oxidative stress as a regulator of gene expression in the vasculature. *Circ Res* 1999;85:753–66.
46. Cai H, Harrison DG. Endothelial dysfunction in cardiovascular diseases: the role of oxidant stress. *Circ Res* 2000;87:840–4.
47. Viles-Gonzalez JF, Choi BG, Fuster V, Badimon JJ. Peroxisome proliferator-activated receptor ligands in atherosclerosis. *Expert Opin Investig Drugs* 2004;13:1393–403.
48. Khalil MF, Wagner WD, Goldberg IJ. Molecular interactions leading to lipoprotein retention and the initiation of atherosclerosis. *Arterioscler Thromb Vasc Biol* 2004;24:2211–8.
49. Choi BG, Vilahur G, Yadegar D, Viles-Gonzalez JF, Badimon JJ. The role of high-density lipoprotein cholesterol in the prevention and possible treatment of cardiovascular diseases. *Curr Mol Med* (in press).
50. Cockerill GW, Huehns TY, Weerasinghe A, et al. Elevation of plasma high-density lipoprotein concentration reduces interleukin-1-induced expression of E-selectin in an in vivo model of acute inflammation. *Circulation* 2001;103:108–12.
51. Clay MA, Pyle DH, Rye KA, et al. Time sequence of the inhibition of endothelial adhesion molecule expression by reconstituted high density lipoproteins. *Atherosclerosis* 2001;157:23–9.
52. Yuhanna IS, Zhu Y, Cox BE, et al. High-density lipoprotein binding to scavenger receptor-BI activates endothelial nitric oxide synthase. *Nat Med* 2001;7:853–7.
53. Spieker LE, Sudano I, Hurlimann D, et al. High-density lipoprotein restores endothelial function in hypercholesterolemic men. *Circulation* 2002;105:1399–402.
54. Bisoendial RJ, Hovingh GK, Levels JH, et al. Restoration of endothelial function by increasing high-density lipoprotein in subjects with isolated low high-density lipoprotein. *Circulation* 2003;107:2944–8.
55. Leitinger N. Oxidized phospholipids as modulators of inflammation in atherosclerosis. *Curr Opin Lipidol* 2003;14:421–30.
56. Podrez EA, Poliakov E, Shen Z, et al. Identification of a novel family of oxidized phospholipids that serve as ligands for the macrophage scavenger receptor CD36. *J Biol Chem* 2002;277:38503–16.
57. Podrez EA, Febbraio M, Sheibani N, et al. Macrophage scavenger receptor CD36 is the major receptor for LDL modified by monocyte-generated reactive nitrogen species. *J Clin Invest* 2000;105:1095–108.
58. Berliner JA, Watson AD. A role for oxidized phospholipids in atherosclerosis. *N Engl J Med* 2005;353:9–11.
59. Vink H, Constantinescu AA, Spaan JA. Oxidized lipoproteins degrade the endothelial surface layer: implications for platelet-endothelial cell adhesion. *Circulation* 2000;101:1500–2.
60. Podrez EA, Poliakov E, Shen Z, et al. A novel family of atherogenic oxidized phospholipids promotes macrophage foam cell formation via the scavenger receptor CD36 and is enriched in atherosclerotic lesions. *J Biol Chem* 2002;277:38517–23.
61. Sawamura T, Kume N, Aoyama T, et al. An endothelial receptor for oxidized low-density lipoprotein. *Nature* 1997;386:73–7.
62. De Caterina R, Bourcier T, Laufs U, et al. Induction of endothelial-leukocyte interaction by interferon-gamma requires coactivation of nuclear factor-kappaB. *Arterioscler Thromb Vasc Biol* 2001;21:227–32.
63. Zioncheck TF, Powell LM, Rice GC, et al. Interaction of recombinant apolipoprotein(a) and lipoprotein(a) with macrophages. *J Clin Invest* 1991;87:767–71.
64. Tsimikas S, Brilakis ES, Miller BS, et al. Oxidized phospholipids, Lp(a) lipoprotein, and coronary artery disease. *N Engl J Med* 2005;353:46–57.
65. Polverini PJ, Cotran PS, Gimbrone MA, Jr., Unanue ER. Activated macrophages induce vascular proliferation. *Nature* 1977;269:804–6.
66. Lee WS, Jain MK, Arkonac BM, et al. Thy-1, a novel marker for angiogenesis upregulated by inflammatory cytokines. *Circ Res* 1998;82:845–51.

67. Kumamoto M, Nakashima Y, Sueishi K. Intimal neovascularization in human coronary atherosclerosis: its origin and pathophysiological significance. *Hum Pathol* 1995;26:450–6.
68. Fleiner M, Kummer M, Mirlacher M, et al. Arterial neovascularization and inflammation in vulnerable patients: early and late signs of symptomatic atherosclerosis. *Circulation* 2004;110:2843–50.
69. Moreno PR, Purushothaman KR, Fuster V, O'Connor WN. Intimomedial interface damage and adventitial inflammation is increased beneath disrupted atherosclerosis in the aorta: implications for plaque vulnerability. *Circulation* 2002;105:2504–11.
70. Marathe GK, Zimmerman GA, McIntyre TM. Platelet-activating factor acetylhydrolase, and not paraoxonase-1, is the oxidized phospholipid hydrolase of high density lipoprotein particles. *J Biol Chem* 2003;278:3937–47.
71. Barter PJ, Baker PW, Rye KA. Effect of high-density lipoproteins on the expression of adhesion molecules in endothelial cells. *Curr Opin Lipidol* 2002;13:285–8.
72. Aviram M, Hardak E, Vaya J, et al. Human serum paraoxonases (PON1) Q and R selectively decrease lipid peroxides in human coronary and carotid atherosclerotic lesions: PON1 esterase and peroxidase-like activities. *Circulation* 2000;101:2510–7.
73. Navab M, Berliner JA, Subbanagounder G, et al. HDL and the inflammatory response induced by LDL-derived oxidized phospholipids. *Arterioscler Thromb Vasc Biol* 2001;21:481–8.
74. Assmann G, Gotto AM, Jr. HDL cholesterol and protective factors in atherosclerosis. *Circulation* 2004;109:III8–14.
75. Moreno PR, Falk E, Palacios IF, et al. Macrophage infiltration in acute coronary syndromes. Implications for plaque rupture. *Circulation* 1994;90:775–8.
76. Dangas G, Badimon JJ, Smith DA, et al. Pravastatin therapy in hyperlipidemia: effects on thrombus formation and the systemic hemostatic profile. *J Am Coll Cardiol* 1999;33:1294–304.
77. Rauch U, Osende JI, Chesebro JH, et al. Statins and cardiovascular diseases: the multiple effects of lipid-lowering therapy by statins. *Atherosclerosis* 2000;153:181–9.
78. Rauch U, Osende JI, Fuster V, et al. Thrombus formation on atherosclerotic plaques: pathogenesis and clinical consequences. *Ann Intern Med* 2001;134:224–38.
79. Osende JI, Badimon JJ, Fuster V, et al. Blood thrombogenicity in type 2 diabetes mellitus patients is associated with glycemic control. *J Am Coll Cardiol* 2001;38:1307–12.
80. Naqvi TZ, Shah PK, Ivey PA, et al. Evidence that high-density lipoprotein cholesterol is an independent predictor of acute platelet-dependent thrombus formation. *Am J Cardiol* 1999;84:1011–7.
81. Li D, Weng S, Yang B, et al. Inhibition of arterial thrombus formation by ApoA1 Milano. *Arterioscler Thromb Vasc Biol* 1999;19:378–83.
82. Griffin JH, Kojima K, Banka CL, et al. High-density lipoprotein enhancement of anticoagulant activities of plasma protein S and activated protein C. *J Clin Invest* 1999;103:219–27.
83. Saku K, Ahmad M, Glas-Greenwalt P, Kashyap ML. Activation of fibrinolysis by apolipoproteins of high density lipoproteins in man. *Thromb Res* 1985;39:1–8.
84. Lesnik P, Vonica A, Guerin M, et al. Anticoagulant activity of tissue factor pathway inhibitor in human plasma is preferentially associated with dense subspecies of LDL and HDL and with Lp(a). *Arterioscler Thromb* 1993;13:1066–75.
85. Epand RM, Stafford A, Leon B, et al. HDL and apolipoprotein A-I protect erythrocytes against the generation of procoagulant activity. *Arterioscler Thromb* 1994;14:1775–83.
86. Rota S, McWilliam NA, Baglin TP, Byrne CD. Atherogenic lipoproteins support assembly of the prothrombinase complex and thrombin generation: modulation by oxidation and vitamin E. *Blood* 1998;91:508–15.
87. Nofer JR, Walter M, Kehrel B, et al. HDL3-mediated inhibition of thrombin-induced platelet aggregation and fibrinogen binding occurs via decreased production of phosphoinositide-derived second messengers 1,2-diacylglycerol and inositol 1,4,5-tris-phosphate. *Arterioscler Thromb Vasc Biol* 1998;18:861–9.
88. Pirich C, Efthimiou Y, O'Grady J, Sinzinger H. Hyperalphalipoproteinemia and prostaglandin I2 stability. *Thromb Res* 1997;88:41–9.
89. Xia P, Vadas MA, Rye KA, et al. High density lipoproteins (HDL) interrupt the sphingosine kinase signaling pathway. A possible mechanism for protection against atherosclerosis by HDL. *J Biol Chem* 1999;274:33143–7.
90. Havel RJ, Eder HA, Bragdon JH. The distribution and chemical composition of ultracentrifugally separated lipoproteins in human serum. *J Clin Invest* 1955;34:1345–53.

91. Kane JP. Structure and function of the plasma lipoproteins and their receptors. In: Fuster V, Ross R, Topol EJ, eds. *Atherosclerosis and Coronary Artery Disease*. Philadelphia, PA: Lippincott-Raven Publishers, 1996:89–103.
92. Brewer HB, Jr. Increasing HDL cholesterol levels. *N Engl J Med* 2004;350:1491–4.
93. Corti R, Hutter R, Badimon JJ, Fuster V. Evolving concepts in the triad of atherosclerosis, inflammation and thrombosis. *J Thromb Thrombolysis* 2004;17:35–44.
94. Bersot TP, Mahley, RW. Clinical classifications of lipid abnormalities. In: Fuster V, Ross R, Topol EJ, eds. *Atherosclerosis and Coronary Artery Disease*. Philadelphia, PA: Lippincott-Raven Publishers, 1996:163–9.
95. Genest J, Libby P, Gotto AM. Lipoprotein disorders and cardiovascular disease. In: Zipes DP, Libby P, Bonow RO, Braunwald E, eds. *Braunwald's Heart Disease*. Philadelphia, PA: Saunders, 2004:1013–32.

2 Genetic Disorders of Lipoprotein Metabolism

Marina Cuchel, MD, PhD, Atif Qasim, MD, and Daniel J. Rader, MD

CONTENTS

INTRODUCTION
INHERITED DISORDERS OF APOB-CONTAINING
  LIPOPROTEINS
INHERITED DISORDERS OF HDL
REFERENCES

INTRODUCTION

The understanding of genetic disorders affecting lipoprotein metabolism is useful as a guide to accurate diagnosis and effective treatment. The characterization of these disorders has been and will continue to be very important not only to better dissect the complicated pathways of lipid metabolism, but also to guide the identification of new therapeutic targets for the treatment of more common forms of dyslipidemia. Indeed, the study of monogenic conditions such as familial hypercholesterolemia and CETP deficiency led directly to the development of novel therapeutic approaches such as statins and CETP inhibitors. A classification of the genetic disorders of lipoprotein metabolism and their molecular etiology is summarized in the Table 1.

INHERITED DISORDERS OF APOB-CONTAINING LIPOPROTEINS

Lipid Disorders Involving Elevated Triglycerides

FAMILIAL CHYLOMICRONEMIA SYNDROME: LIPOPROTEIN LIPASE DEFICIENCY AND APOLIPOPROTEIN C-II DEFICIENCY

Two different genetic defects can cause the familial hyperchylomicronemia syndrome (FCS): lipoprotein lipase (LPL) deficiency and apoC-II deficiency *(1)*. The hydrolysis of triglycerides in chylomicrons requires the action of LPL in tissue capillary beds, and apoC-II is a required cofactor for the activation of LPL. Mutations in either the *LPL* gene or the *APOC2* gene result in functional deficiency of LPL,

From: *Contemporary Cardiology: Therapeutic Lipidology*
Edited by: M. H. Davidson, P. P. Toth, and K. C. Maki © Humana Press Inc., Totowa, NJ

Table 1
Genetic Disorders of Lipoprotein Metabolism

Genetic disorder	Gene defect	Lipoproteins affected	Clinical findings	Genetic transmission	Estimated incidence
Apo B-containing lipoproteins					
Lipoprotein lipase deficiency	LPL (*LPL*)	↑ Chylomicrons	Eruptive xanthomas, hepatosplenomegaly, pancreatitis	AR	1/1,000,000
Familial apolipoprotein C-II deficiency	ApoC-II (*APOC2*)	↑ Chylomicrons	Eruptive xanthomas, hepatosplenomegaly	AR	<1/1,000,000
Familial hepatic lipase deficiency	Hepatic lipase (*LIPC*)	↑ VLDL remnants	Premature atherosclerosis, pancreatitis	AR	<1/1,000,000
Familial dysbetal-lpoproteinemia	apoE (*APOE*)	↑ Chylomicron, VLDL remnants	Palmar and tuberoeruptive xanthomas, CHD, Peripheral vascular disease	AR AD	1/10,000
Familial hypercholesterolemia	LDL receptor (*LDLR*)	↑ LDL	Tendon xanthomas, CHD	AD	1/500
Familial defective apoB-100	apoB-100 (*APOB*)	↑ LDL	Tendon xanthomas, CHD	AD	1/1000
Autosomal recessive hypercholesterolemia	ARH (*ARH*)	↑ LDL	Tendon xanthomas, CHD	AR	<1/1,000,000
Sitosterolemia	*ABCG5* or *ABCG8*	↑ LDL	Tendon xanthomas, CHD	AR	<1/1,000,000
Autosomal dominant hypercholesterolemia	PCSK9 (*gain of function*)	↑ LDL	Tendon xanthomas, CHD	AD	

Disease	Gene	Lipid change	Clinical features	Inheritance
Cholesteryl esters storage disease (CESD) and Wolman disease	LAL	↑ LDL	Hepatosplenomegalia, hepatic fibrosis	AR
Abetalipoproteinemia	MTP	↓ LDL	Fat malabsorption, spinocerebellar degeneration, retinopathy, possible hepatic steatosis	AR
Hypobetalipo-proteinemia	ApoB	↓ LDL	Fat malabsorption, spinocerebellar degeneration, retinopathy, possible hepatic steatosis	AD
	PCSK9 (loss of function)	↓ LDL		
HDL				
Tangier disease	ABCA1	↓ HDL	Hepatosplenomegaly, enlarged orange tonsils, CHD	AD
LCAT deficiency	LCAT	↓ HDL	Corneal opacification, hemolytic anemia, progressive renal insufficiency	
CETP deficiency	CETP	↑ HDL		

AD, autosomal dominant; AR, autosomal recessive CETP, cholesteryl ester transfer protein; CHD, coronary heart disease; HDL, high-density lipoprotein; LCAT, lecithin: cholesterol acyttransferase; LDL, low-density lipoprotein; VLDL, very low density lipoprotein.

inability to hydrolyze triglycerides in chylomicrons, and consequent massive hyperchylomicronemia. These disorders are autosomal recessive and rare (approximately 1 in 1 million persons). FCS is characterized by presentation in childhood with acute pancreatitis in the setting of triglyceride levels greater than 1000 mg/dL. On physical examination, eruptive xanthomas (small papular lesions that occur in showers on the buttocks and back) are often seen. Lipemia retinalis (a pale appearance of the retinal veins) is a clue to the existence of severe hypertriglyceridemia, and hepatosplenomegaly caused by ingestion of chylomicrons by the reticuloendothelial system is often found. Is not generally a feature of this syndrome.

The diagnosis of FCS is usually made based on the clinical presentation. The plasma is often lactescent, and triglyceride levels are usually greater than 1000 mg/dL, or often much higher. Total cholesterol levels are also elevated because of the presence of cholesterol in chylomicrons. The diagnosis of LPL deficiency can be confirmed at specialized centers by the measurement of LPL activity in the plasma after intravenous heparin injection (postheparin LPL activity). Primary therapy for FCS is restriction of total dietary fat. Caloric supplementation with medium-chain triglycerides, which are absorbed directly into the portal vein and therefore do not promote chylomicron formation, can be useful if necessary. Some patients may respond to a trial of omega-3 fatty acids (fish oils) or fibrates.

FAMILIAL HYPERTRIGLYCERIDEMIA

Familial hypertriglyceridemia (FHTG) is a relatively common inherited disorder of unknown molecular etiology. It is characterized by elevated triglycerides (200–1000 mg/dL) with normal or only mildly increased cholesterol levels (more than 240 mg/dL); low-density lipoprotein cholesterol (LDL-C) is generally not increased in this disorder. FHTG is inherited as an autosomal dominant trait but is not usually expressed until adulthood and occurs in approximately 1 in 500 persons. Both very low density lipoprotein (VLDL) overproduction and reduced VLDL catabolism have been implicated in causing this phenotype. Increased intake of simple carbohydrates, a sedentary lifestyle, obesity, insulin resistance, alcohol use, and estrogens can all exacerbate the hypertriglyceridemia. FHTG is often not associated with a significantly increased risk of atherosclerotic vascular disease.

Therapy for FHTG involves lifestyle management, especially weight control and reduction simple carbohydrates in the diet and of alcohol. Lipid-lowering drug therapy is often needed, especially for patients who have triglycerides greater than 400–600 mg/dL. Statins, fibrates, niacin, and omega-3 fatty acids are all reasonable considerations.

FAMILIAL DYSBETALIPOPROTEINEMIA (TYPE III HYPERLIPOPROTEINEMIA)

Familial dysbetalipoproteinemia (FD) is also known as Type III hyperlipidemia and is caused by mutations in the gene for apolipoprotein E (*APOE*) *(2)*. ApoE is present on chylomicron and VLDL remnants and mediates their removal from the plasma by binding to receptors in the liver. Defective apoE is impaired in its ability to bind to these receptors, resulting in the accumulation of chylomicron and VLDL remnants in the plasma. The apoE gene is polymorphic in humans; the most common form of apoE is known as apoE3. FD is most commonly caused by a variant called apoE2. ApoE2 differs from apoE3 by a single amino acid and does not bind adequately to lipoprotein receptors, resulting in defective removal of chylomicron and VLDL remnants. About 0.5% of individuals are homozygous for apoE2, but many persons with the apoE2/E2 genotype do not have FD, suggesting that a "second hit" is required. Obesity, diabetes mellitus, hypothyroidism, renal disease, and alcohol use are all associated with an increased probability of FD in apoE2/E2 patients, but many patients with FD do not have an obvious second hit.

Patients generally have both hypertriglyceridemia and hypercholesterolemia, often elevated to a relatively similar degree, and the high-density lipoprotein cholesterol (HDL-C) level is often relatively normal. Lipoprotein electrophoresis demonstrates a "broad β band" because of the presence of remnant lipoproteins. "Beta quantification", in which the VLDL cholesterol is directly measured, demonstrates an elevated ratio of VLDL-C to plasma triglycerides of more than 0.3. Advanced lipoprotein analysis usually reveals a substantial increase in intermediate-density lipoprotein (IDL)-sized particles. ApoE genotyping can be performed to confirm the apoE2/E2 genotype. Two types of xanthomas are often seen, uniquely, in patients with FD: (1) tuberoeruptive xanthomas are clusters of small papules on the elbows, knees, or buttocks and (2) palmar xanthomas are orange-yellow discoloration to the creases of the palms. Premature atherosclerotic CHD is often seen in this disorder, making aggressive therapy mandatory. Drug therapy is usually necessary and includes statins, fibrates, niacin, and cholesterol absorption inhibitors.

FAMILIAL COMBINED HYPERLIPIDEMIA

Familial combined hyperlipidemia (FCHL) is characterized by a mixed dyslipidemia usually associated with moderately elevated fasting triglycerides, moderately elevated cholesterol, and reduced HDL-C *(3)*. In most cases, at least one first-degree relative also has hyperlipidemia, and there is often a family history of premature CHD as well. The genetic basis of FCHL is not well understood, and probably several different genes can cause a similar phenotype *(4)*. Hepatic overproduction of VLDL is a common metabolic basis of this condition. FCHL has been estimated to occur in approximately 1 in 200 persons overall, and about one in five patients with CHD under the age of 60. FCHL is suggested by the presence of a mixed hyperlipidemia with fasting triglyceride levels of 200–800 mg/dL, cholesterol levels of 200–400 mg/dL, and decreased HDL-C levels in the absence of secondary causes of hyperlipidemia. A family history of hyperlipidemia and premature coronary disease supports the diagnosis. Patients with FCHL almost always have a significantly elevated plasma level of apoB that is disproportionate to the LDL-C level and is indicative of the presence of small dense LDL particles. Visceral obesity, insulin resistance, and hypertension are sometimes associated with FCHL. Premature atherosclerotic vascular disease is common in patients with FCHL.

Patients with FCHL should be treated aggressively with lifestyle management and drug therapy. Statins, fibrates, niacin, and cholesterol absorption inhibitors are all used, often in different combinations, to achieve adequate control of the lipids.

Inherited Syndromes of Elevated LDL-C

FAMILIAL HYPERCHOLESTEROLEMIA

Familial hypercholesterolemia (FH) is caused by loss-of-function mutations in the LDL receptor *(5)*. FH is autosomal codominant: heterozygotes have high cholesterol and homozygotes (or compound heterozygotes) are even more severely affected.

Heterozygous FH occurs in approximately 1 in 500 persons worldwide, making it one of the most common single-gene disorders. The reduction in functional hepatic LDL receptors leads to reduced clearance of plasma LDL by the liver and substantial elevations in LDL-C (usually 200–400 mg/dL) with normal triglycerides. Tendon xanthomas are common, usually manifest by thickened and irregular Achilles tendons and/or the digit extensor tendons of the metacarpophalangeal (MCP) joints on the dorsum of the hands. A family history of hypercholesterolemia or premature CHD is common. Heterozygous FH is strongly associated with premature atherosclerotic CHD. Patients should be aggressively treated to lower the LDL-C, usually with statins and frequently requiring another drug such as a cholesterol absorption inhibitor. If necessary, a bile acid sequestrant and/or niacin can also be added to the regimen. Even the combination of several drugs sometimes fails to adequately control the cholesterol, in which case LDL apheresis should be considered *(6)*.

Homozygous FH is an even more severe clinical disorder than heterozygous FH and occurs in approximately 1 in 1 million persons. Patients with homozygous FH usually present in childhood with total cholesterol levels greater than 500 mg/dL. Accelerated atherosclerosis often develops first in the aortic root, causing aortic valvular or supravalvular stenosis, and extending into the coronary ostia. Patients with suspected homozygous FH should be referred to a specialized center. Statins and cholesterol absorption inhibitors have only modest effects in reducing cholesterol. Liver transplantation is effective in decreasing LDL-C levels but is associated with substantial risks. LDL apheresis is the therapy of choice at this time, but new therapies are being developed that may hold more promise for this disease.

FAMILIAL DEFECTIVE APOLIPOPROTEIN B-100

Familial defective apoB-100 (FDB) is caused by mutations in the receptor binding region of apoB-100, the ligand for the LDL receptor, which impairs its binding and delays the clearance of LDL *(7)*. Thus, FDB resembles heterozygous FH with substantially elevated LDL-C and normal triglycerides. The most common mutation causing FDB is a substitution of glutamine for arginine at position 3500 in apoB-100; other mutations have also been reported that have a similar effect on apoB binding to the LDL receptor. FDB is a dominantly inherited disorder and occurs in approximately 1 in 700 persons in Europe and North America. FDB is difficult to differentiate from heterozygous FH on clinical grounds but is treated similarly.

AUTOSOMAL RECESSIVE HYPERCHOLESTEROLEMIA

Autosomal recessive hypercholesterolemia (ARH) clinically resembles receptor-defective homozygous FH and is a very rare disorder caused by mutations in the *ARH* gene *(8)*. Although LDL receptor function in cultured fibroblasts is relatively normal, LDL receptor function in the liver is markedly reduced, leading to reduced LDL catabolism. The ARH protein appears to be involved in the regulation of LDL receptor-mediated endocytosis in the liver. In contrast to FH, the condition is recessive and obligate heterozygotes have normal cholesterol levels. Patients sometimes respond partially to treatment with statins but often require LDL apheresis for adequate control of their hypercholesterolemia.

AUTOSOMAL DOMINANT HYPERCHOLESTEROLEMIA

Autosomal dominant hypercholesterolemia (ADH) is another rare disorder caused by apparent gain-of-function mutations in the proprotein convertase subtilisin/kexin type 9 (*PCSK9*) gene *(9)*. The function of PCSK9 and its role in cholesterol metabolism are unclear; however, experimental data suggest that PCSK9 may regulate the density of functional LDL receptors in the liver. Interestingly, loss-of-function mutations in this gene appear to cause low LDL-C levels (see "Familial Hypobetalipoproteinemia" section).

SITOSTEROLEMIA

Sitosterolemia is caused by mutations in one of two members of the adenosine triphosphate (ATP)-binding cassette (ABC) transporter family, *ABCG5* and *ABCG8* *(10)*. These genes are expressed in the intestine and liver where they form a functional complex to limit intestinal absorption and promote biliary excretion of plant-derived and animal-derived neutral sterols. In sitosterolemia, the normally low intestinal absorption of plant sterols is markedly increased and biliary excretion of plant sterols is reduced, resulting in increased plasma levels of sitosterol and other plant sterols. Because the hepatic LDL receptor is downregulated, LDL-C levels tend to be high in this condition. Patients with sitosterolemia often have tendon xanthomas and are at risk for premature CHD.

The hypercholesterolemia in subjects with sitosterolemia is unusually responsive to reductions in dietary cholesterol content, and sitosterolemia should be suspected in patients in whom the plasma LDL-C falls more than 50% on a low cholesterol diet. The diagnosis of sitosterolemia is confirmed by demonstrating an increase in the plasma level of sitosterol. The hypercholesterolemia does not respond all that well to HMG-CoA reductase inhibitors, but cholesterol-absorption inhibitors and bile acid sequestrants are effective in reducing plasma plant sterol levels and LDL-C levels in these patients.

Inherited Syndromes of Low LDL-C

ABETALIPOPROTEINEMIA

Abetalipoproteinemia is a rare autosomal recessive disease caused by mutations in the gene encoding microsomal transfer protein (MTP) *(11)*, a protein that transfers lipids to nascent chylomicrons and VLDL in the intestine and liver, respectively. Plasma levels of cholesterol and triglyceride are extremely low in this disorder, and no chylomicrons, VLDL, LDL, or apoB are detectable in plasma *(12)*. Obligate heterozygotes have normal plasma lipid and apoB levels. Abetalipoproteinemia usually presents in childhood with diarrhea and failure to thrive and is characterized clinically by fat malabsorption, spinocerebellar degeneration, pigmented retinopathy, and acanthocytosis. The initial neurological manifestations are loss of deep tendon reflexes, followed by decreased distal lower extremity vibratory and proprioceptive sense, dysmetria, ataxia, and the development of a spastic gait, often by the third or fourth decade. Patients with abetalipoproteinemia also develop a progressive pigmented retinopathy presenting with decreased night and color vision, followed by reductions in daytime visual acuity and ultimately progressing to near blindness.

Most clinical manifestations of abetalipoproteinemia result from defects in the absorption and transport of fat-soluble vitamins. Vitamin E and retinyl esters are normally transported from enterocytes to the liver by chylomicrons, and vitamin E is dependent on VLDL for transport out of the liver and into the circulation. As a consequence of the inability of these patients to secrete apoB-containing particles, patients with abetalipoproteinemia are markedly deficient in vitamin E and are also mildly to moderately deficient in vitamins A and K. Treatment consists of a low-fat, high-caloric, vitamin-enriched diet accompanied by large supplemental doses of vitamin E. It is imperative that treatment be initiated as soon as possible to obviate the development of the neurological sequelae.

FAMILIAL HYPOBETALIPOPROTEINEMIA

Familial hypobetalipoproteinemia generically refers to low LDL-C levels that have a genetic basis. Historically, it has been used to refer to low LDL-C because of mutations in apoB *(13,14)*. There are a range of missense mutations in apoB that have been shown to reduce secretion and/or accelerate catabolism. Individuals heterozygous for these mutations have LDL-C levels less than 100 mg/dL. They appear to be protected from the development of atherosclerotic vascular disease. There are rare patients who have mutations in both apoB alleles and have plasma lipids similar to those in abetalipoproteinemia, but a less severe neurological phenotype.

More recently, loss-of-function mutations in PCSK9 have been shown to cause low LDL-C levels (usually less than 80 mg/dL) and are therefore another molecular cause of familial hypobetalipoproteinemia *(15,16)*. The mechanism is uncertain but presumably results in the upregulation of the hepatic LDL receptor and increased catabolism of LDL. This condition, which is much more common in people of African descent, provided the opportunity to demonstrate that the effects of lifelong low LDL-C levels are a substantial reduction in with no other adverse consequences *(17)*. This strongly supports the concept that aggressive LDL-C reduction is associated with long-term substantial reduction in cardiovascular risk.

INHERITED DISORDERS OF HDL

There is significant interest in understanding the genetic basis of HDL-C metabolism as HDL-C is inversely associated with CHD, independent of LDL-C levels. Modulating HDL-C metabolism, therefore, has significant potential in the treatment and prevention of coronary disease. Although several environmental factors are known to affect HDL-C levels including cigarette smoking, obesity, physical activity, and diet, it is estimated that at least half of the variation in HDL-C levels in humans is genetic. Yet, to date, we have identified only a few of the genes responsible for inherited syndromes of low or high HDL-C—primarily from familial monogenetic disorders with extremes of HDL-C phenotypes. Nevertheless, this basic understanding has already led to the development of novel pharmacotherapies, such as small-molecule CETP inhibitors. The next steps, however, are to further elucidate the details of HDL-C metabolism and identify the host of other genes that may affect HDL-C levels in the general population. This section will discuss the importance of known monogenetic conditions in our understanding of HDL-C metabolism and CHD, as well the future direction of HDL-C genetics.

Inherited Syndromes of Low HDL-C

APOA-I DEFICIENCY AND STRUCTURAL MUTATIONS

A rare cause of extreme low HDL-C is complete deficiency of apoA-I either from *APOA1* gene deletion or from nonsense mutations that result in virtually absent plasma HDL *(18,19)*. Most these cases are associated with premature CHD. Another relatively rare cause of low HDL-C are missense or nonsense mutations that result in structurally abnormal or truncated apoA-I proteins. The best known of these mutations is apoA-I$_{Milano}$, where a substitution of cysteine for arginine at position 173 *(20)* results in increased turnover of the mutant apoA-I$_{Milano}$ protein, as well of the wild-type apoA-I *(21)*, and a substantial reduction in HDL-C. The low HDL-C levels, however, are not associated with an increased risk of atherosclerosis. Animal studies with intravenous infusion of recombinant apoA-I$_{Milano}$ show less atherosclerosis *(20)*, and a small trial of the intravenous infusion of apoA-I$_{Milano}$–phospholipid complexes in humans demonstrated a reduction from baseline in coronary atheroma volume as measured by intravascular ultrasound *(22)*. There have been several other apoA-I structural mutations described that cause low HDL-C *(23)*, but structural apoA-I mutations are rare, and in the general population, apoA-I mutations are not thought to be a common source of variation in HDL-C levels.

TANGIER DISEASE (ABCA1 DEFICIENCY)

Tangier disease was first described in 1961 as a rare disorder in which individuals had cholesterol accumulation in the reticuloendothelial system causing enlarged orange tonsils, hepatosplenomegaly, intestinal mucosal abnormalities, and peripheral neuropathy, in association with markedly low HDL-C (<5 mg/dL) and apoA-I *(24)*. In 1999, the molecular etiology of Tangier disease was discovered to be due to homozygosity for mutations in encoding for the ATP-binding cassette protein A1 (ABCA1) *(25–27)*. ABCA1 facilitates the efflux of unesterified cholesterol and phospholipids from cells to apoA-I. When ABCA1 is absent or nonfunctional, apoA-I is not appropriately lipidated and is rapidly cleared from the circulation. Furthermore, the impaired cholesterol efflux from tissues results in cholesterol accumulation—especially in macrophages and certain other cell types—leading to many of the typical clinical characteristics.

Heterozygotes for ABCA1 mutations have moderately reduced HDL-C levels (approximately 20–40 mg/dL), and no evidence of cholesterol accumulation in tissues suggesting a gene dosage effect of ABCA1 expression on cellular cholesterol efflux. Both patients with Tangier disease (homozygotes) and obligate heterozygotes are at some increased risk for premature CHD, which may be alleviated by lowering plasma levels of LDL-C. Mutations in ABCA1 are also one possible cause of primary hypoalphalipoproteinemia (low HDL-C levels) in some families *(27,28)*.

Rare private mutations in the *ABCA1* gene may be an important cause of low HDL-C levels in the general population *(29)*. Several large studies also suggest that common genetic variation in the *ABCA1* gene may be an important contributor to the variation in HDL-C levels in the general population and be associated with low HDL-C levels and increased CHD risk *(30,31)*. ABCA1 is now a target for the development of new therapies intended to upregulate ABCA1 expression *(32)*.

LECITHIN:CHOLESTEROL ACYLTRANSFERASE DEFICIENCY

Two genetic forms of lecithin: cholesterol acyltransferase (LCAT) deficiency have been described in humans: complete deficiency known as classic LCAT deficiency and partial deficiency or fish-eye disease *(33)*. Both types are characterized by corneal opacification from deposition of free cholesterol, low HDL-C, and variable hypertriglyceridemia. Although partial deficiency has no clinical sequelae, individuals with complete LCAT deficiency also have low-grade hemolytic anemia and progressive renal insufficiency that eventually leads to end-stage renal disease. Neither is associated with premature coronary disease despite reduced HDL-C and apoA-I levels *(33)*.

Cholesterol effluxes from cells as free cholesterol and is associated in plasma with HDL. LCAT is the enzyme that esterifies the free cholesterol present on HDL to cholesteryl ester and allows the formation of a cholesterol ester core and the maturation of HDL. When LCAT deficiency is present, mature HDL particles are not formed and rapid catabolism of circulating apoA-I and apoA-II occurs *(34)*. HDL-C levels are typically less than 10 mg/dL, and circulating HDLs resemble nascent HDLs. A great proportion of plasma cholesterol is free cholesterol (more than 70% of total plasma cholesterol in many individuals). ApoA-I levels are typically less than 30% of normal. Although LCAT clearly has important effects on HDL metabolism, its relationship to atherosclerosis remains unclear.

PRIMARY OR FAMILIAL HYPOALPHALIPOPROTEINEMIA

Primary or familial hypoalphalipoproteinemia is the most common inherited form of low HDL-C. Individuals who have this condition have HDL-C levels below the 10th percentile, no discernable secondary cause of low HDL-C, and a family history of low HDL-C. Also known as "isolated low HDL," this condition is primarily autosomal dominant, and although its etiology is unknown, accelerated catabolism of apoA-I and HDL-C appears to be the final common pathway *(35)*. Some cases may be caused by ABCA1 mutations, but genetic variation in multiple genes likely contributes to this phenotype.

Some families with hypoalphalipoproteinemia have an increased incidence of CHD; however, other families do not, making the relationship of primary hypoalphalipoproteinemia to premature coronary more dependent on the specific nature of the metabolic cause of the low HDL-C. Nonetheless, although relatively few of the genes that influence HDL-C levels in humans are currently known, ongoing studies may lead to new candidate genes *(36)*.

Inherited Syndromes of Elevated HDL-C Levels

CHOLESTERYL ESTER TRANSFER PROTEIN DEFICIENCY

Cholesterol ester transfer protein (CETP) deficiency in humans is associated with markedly elevated plasma HDL-C levels, usually greater than 120 mg/dL. Multiple different mutations in the gene encoding for CETP have been reported to cause CETP deficiency including a 5′ donor splice site intron 14 G-to-A substitution and a missense mutation in exon 15 (D442G) *(37,38)*. Individuals homozygous for the intron 14 mutation have no measurable CETP mass or activity, but those homozygous for the D442G mutation have only partial CETP deficiency and HDL-C levels that are less markedly elevated *(39)*. Although these are the most frequent CETP mutations in Japan, several other mutations have been identified as well.

Both apoA-I and apoA-II have significantly slower turnover in CETP deficiency *(40)*. LDL-C levels are also low because of the increased catabolism of LDL and apoB with endogenous upregulation of the LDL receptor *(41)*. Heterozygous individuals for CETP deficiency have 60–70% of normal CETP activity, but only a modest increase in HDL-C levels, and otherwise normal LDL-C levels. The relationship of CETP deficiency, elevated HDL-C, and CHD has still not been fully determined *(42)*. Whether homozygous or heterozygous CETP deficiency is associated with increased, decreased, or unchanged cardiovascular risk remains to be resolved.

The relevance to the population at large is another question, as common CETP polymorphisms may also be associated with HDL-C levels and CHD risk. Although intronic and unlikely to be functional, the Taq1B polymorphism has been associated with increased HDL-C and reduced CHD risk *(43)*. The missense I405V SNP is also associated with modestly increased HDL-C levels, but its relationship to cardiovascular outcomes is uncertain *(43)*. Thus, genetic variation in CETP does appear to contribute, modestly, to variation in HDL-C levels in the general population. Because of this, small-molecule inhibitors of CETP, which have been shown to increase HDL-C levels, are currently now in clinical development *(44)*, and trials are underway for several to assess how they modulate CHD risk in combination with statins.

Familial Hyperalphalipoproteinemia

Familial hyperalphalipoproteinemia is defined as an inherited condition of HDL-C level greater than the 90th percentile without a known secondary cause. HDL-C in these families appears to be associated with lower risk for CHD and increased longevity. Other than CETP, additional genes that cause or contribute to this phenotype are not known. An understanding of the genetics of this syndrome would advance the scientific field and provide new targets for the development of therapeutics for HDL.

REFERENCES

1. Santamarina-Fojo, S. 1992. Genetic dyslipoproteinemias: role of lipoprotein lipase and apolipoprotein C-II. *Curr Opin Lipidol* 3:186–195.
2. Mahley, R.W., Huang, Y., and Rall, S.C.J. 1999. Pathogenesis of type III hyperlipoproteinemia (dysbetalipoproteinemia): questions, quandaries, and paradoxes. *J Lipid Res* 40:1933–1949.
3. Grundy, S.M., Chait, A., and Brunzell, J.D. 1987. Familial combined hyperlipidemia workshop. *Arteriosclerosis* 7:203–207.
4. Shoulders, C.C., Jones, E.L., and Naoumova, R.P. 2004. Genetics of familial combined hyperlipidemia and risk of coronary heart disease. *Hum Mol Genet* 13 Spec. No. 1:R149–R160.
5. Rader, D.J., Cohen, J., and Hobbs, H.H. 2003. Monogenic hypercholesterolemia: new insights in pathogenesis and treatment. *J Clin Invest* 111:1795–1803.
6. Sachais, B.S., Katz, J., Ross, J., and Rader, D.J. 2005. Long-term effects of LDL apheresis in patients with severe hypercholesterolemia. *J Clin Apher* 20:252–255.
7. Tybjaerg-Hansen, A., and Humphries, S.E. 1992. Familial defective apolipoprotein B-100: a single mutation that causes hypercholesterolemia and premature coronary artery disease. *Atherosclerosis* 96:91–107.
8. Garcia, C.K., Wilund, K., Arca, M., Zuliani, G., Fellin, R., Maioli, M., Calandra, S., Bertolini, S., Cossu, F., Grishin, N., et al. 2001. Autosomal recessive hypercholesterolemia caused by mutations in a putative LDL receptor adaptor protein. *Science* 292:1394–1398.
9. Abifadel, M., Varret, M., Rabes, J.P., Allard, D., Ouguerram, K., Devillers, M., Cruaud, C., Benjannet, S., Wickham, L., Erlich, D., et al. 2003. Mutations in PCSK9 cause autosomal dominant hypercholesterolemia. *Nat Genet* 34:154–156.
10. Berge, K.E., Tian, H., Graf, G.A., Yu, L., Grishin, N.V., Schultz, J., Kwiterovich, P., Shan, B., Barnes, R., and Hobbs, H.H. 2000. Accumulation of dietary cholesterol in sitosterolemia caused by mutations in adjacent ABC transporters. *Science* 290:1771–1775.
11. Sharp, D., Blinderman, L., Combs, K.A., Kienzle, B., Ricci, B., Wager-Smith, K., Gil, C.M., Turck, C.W., Bouna, M.E., Rader, D.J., et al. 1993. Cloning and gene defects in microsomal triglyceride transfer protein associated with abetalipoproteinemia. *Nature* 365:65–69.
12. Rader, D.J., and Brewer, H.B., Jr. 1993. Abetalipoproteinemia. New insights into lipoprotein assembly and vitamin E metabolism from a rare genetic disease. *JAMA* 270:865–869.
13. Linton, M.F., Farese, R.V., Jr., and Young, S.G. 1993. Familial hypobetalipoproteinemia. *J Lipid Res* 34:521–541.
14. Schonfeld, G., Lin, X., and Yue, P. 2005. Familial hypobetalipoproteinemia: genetics and metabolism. *Cell Mol Life Sci* 62:1372–1378.

15. Cohen, J., Pertsemlidis, A., Kotowski, I.K., Graham, R., Garcia, C.K., and Hobbs, H.H. 2005. Low LDL cholesterol in individuals of African descent resulting from frequent nonsense mutations in PCSK9. *Nat Genet* 37:161–165.
16. Kotowski, I.K., Pertsemlidis, A., Luke, A., Cooper, R.S., Vega, G.L., Cohen, J.C., and Hobbs, H.H. 2006. A spectrum of PCSK9 alleles contributes to plasma levels of low-density lipoprotein cholesterol. *Am J Hum Genet* 78:410–422.
17. Cohen, J.C., Boerwinkle, E., Mosley, T.H., Jr., and Hobbs, H.H. 2006. Sequence variations in PCSK9, low LDL, and protection against coronary heart disease. *N Engl J Med* 354:1264–1272.
18. Norum, R.A., Lakier, J.B., Goldstein, S., Angel, A., Goldberg, R.B., Block, W.D., Noffze, D.K., Dolphin, P.J., Edelglass, J., Bogorad, D.D., et al. 1982. Familial deficiency of apolipoproteins A-I and C-III and precocious coronary-artery disease. *N Engl J Med* 306:1513–1519.
19. Schaefer, E.J., Heaton, W.H., Wetzel, M.G., and Brewer, H.B., Jr. 1982. Plasma apolipoprotein A-1 absence associated with a marked reduction of high density lipoproteins and premature coronary artery disease. *Arteriosclerosis* 2:16–26.
20. Chiesa, G., and Sirtori, C.R. 2003. Apolipoprotein A-IMilano: current perspectives. *Curr Opin Lipidol* 14:159–163.
21. Roma, P., Gregg, R.E., Meng, M.S., Ronan, R., Zech, L.A., Franceschini, G., Sirtori, C.R., and Brewer, H.B., Jr. 1993. In vivo metabolism of a mutant form of apolipoprotein A-I, apo A-IMilano, associated with familial hypoalphalipoproteinemia. *J Clin Invest* 91:1445–1452.
22. Nissen, S.E., Tsunoda, T., Tuzcu, E.M., Schoenhagen, P., Cooper, C.J., Yasin, M., Eaton, G.M., Lauer, M.A., Sheldon, W.S., Grines, C.L., et al. 2003. Effect of recombinant ApoA-I Milano on coronary atherosclerosis in patients with acute coronary syndromes: a randomized controlled trial. *JAMA* 290:2292–2300.
23. von Eckardstein, A. 2006. Differential diagnosis of familial high density lipoprotein deficiency syndromes. *Atherosclerosis* 186:231–239.
24. Fredrickson, D.S. 1964. The inheritance of high density lipoprotein deficiency (Tangier disease). *J lin Invest* 43:228–236.
25. Rust, S., Rosier, M., Funke, H., Real, J., Amoura, Z., Piette, J.C., Deleuze, J.F., Brewer, B., Duverger, N., Denefle, P., et al. 1999. Tangier disease is caused by mutations in the gene encoding ATP-binding cassette transporter 1. *Nat Genet* 22:352–355.
26. Bodzioch, M., Orso, E., Klucken, J., Langmann, T., Bottcher, A., Diederich, W., Drobnik, W., Barlage, S., Buchler, C., Ozcurumez, M.P., et al. 1999. The gene encoding ATP-binding cassette transporter 1 is mutated in Tangier disease. *Nat Genet* 22:347–351.
27. Brooks-Wilson, A., Marcil, M., Clee, S.M., Zhang, L.H., Roomp, K., van Dam, M., Yu, L., Brewer, C., Collins, J.A., Molhuizen, H.O., et al. 1999. Mutations in ABC1 in Tangier disease and familial high-density lipoprotein deficiency. *Nat Genet* 22:336–345.
28. Marcil, M., Brooks-Wilson, A., Clee, S.M., Roomp, K., Zhang, L.H., Yu, L., Collins, J.A., van Dam, M., Molhuizen, H.O., Loubster, O., et al. 1999. Mutations in the ABC1 gene in familial HDL deficiency with defective cholesterol efflux [see comments]. *Lancet* 354:1341–1346.
29. Cohen, J.C., Kiss, R.S., Pertsemlidis, A., Marcel, Y.L., McPherson, R., and Hobbs, H.H. 2004. Multiple rare alleles contribute to low plasma levels of HDL cholesterol. *Science* 305:869–872.
30. Clee, S.M., Zwinderman, A.H., Engert, J.C., Zwarts, K.Y., Molhuizen, H.O., Roomp, K., Jukema, J.W., van Wijland, M., van Dam, M., Hudson, T.J., et al. 2001. Common genetic variation in ABCA1 is associated with altered lipoprotein levels and a modified risk for coronary artery disease. *Circulation* 103:1198–1205.
31. Brousseau, M.E., Bodzioch, M., Schaefer, E.J., Goldkamp, A.L., Kielar, D., Probst, M., Ordovas, J.M., Aslanidis, C., Lackner, K.J., Bloomfield, R.H., et al. 2001. Common variants in the gene encoding ATP-binding cassette transporter 1 in men with low HDL cholesterol levels and coronary heart disease. *Atherosclerosis* 154:607–611.
32. Linsel-Nitschke, P., and Tall, A.R. 2005. HDL as a target in the treatment of atherosclerotic cardiovascular disease. *Nat Rev Drug Discov* 4:193–205.
33. Kuivenhoven, J.A., Pritchard, H., Hill, J., Frohlich, J., Assmann, G., and Kastelein, J. 1997. The molecular pathology of lecithin:cholesterol acyltransferase (LCAT) deficiency syndromes. [Review] [102 refs]. *J Lipid Res* 38:191–205.
34. Rader, D.J., Ikewaki, K., Duverger, N., Schmidt, H., Pritchard, H., Frohlich, J., Dumon, M.F., Fairwell, T., Zech, L., Santamarina-Fojo, S., et al. 1994. Markedly accelerated catabolism of apolipoprotein A-II (ApoA-II) and high density lipoproteins containing ApoA-II in classic lecithin:cholesterol acyltransferase deficiency and fish-eye disease. *J Clin Invest* 93:321–330.

35. Lewis, G.F., and Rader, D.J. 2005. New insights into the regulation of HDL metabolism and reverse cholesterol transport. *Circ Res* 96:1221–1232.
36. Wang, X., and Paigen, B. 2005. Genetics of variation in HDL cholesterol in humans and mice. *Circ Res* 96:27–42.
37. Brown, M.L., Inazu, A., Hesler, C.B., Agellon, L.B., Mann, C., Whitlock, M.E., Marcel, Y.L., Milne, R.W., Koizumi, J., Mabuchi, H., et al. 1989. Molecular basis of lipid transfer protein deficiency in a family with increased high-density lipoproteins. *Nature* 342:448–451.
38. Inazu, A., Brown, M.L., Hesler, C.B., Agellon, L.B., Koizumi, J., Takata, K., Maruhama, Y., Mabuchi, H., and Tall, A.R. 1990. Increased high-density lipoprotein levels caused by a common cholesteryl-ester transfer protein gene mutation. *N Engl J Med* 323:1234–1238.
39. Inazu, A., Jiang, X.C., Haraki, T., Yagi, K., Kamon, N., Koizumi, J., Mabuchi, H., Takeda, R., Takata, K., Moriyama, Y., et al. 1994. Genetic cholesteryl ester transfer protein deficiency caused by two prevalent mutations as a major determinant of increased levels of high density lipoprotein cholesterol. *J Clin Invest* 94:1872–1882.
40. Ikewaki, K., Rader, D.J., Sakamoto, T., Nishiwaki, M., Wakimoto, N., Schaefer, J.R., Ishikawa, T., Fairwell, T., Zech, L.A., and Nakamura, H. 1993. Delayed catabolism of high density lipoprotein apolipoproteins A-I and A-II in human cholesteryl ester transfer protein deficiency. *J Clin Invest* 92:1650–1658.
41. Ikewaki, K., Nishiwaki, M., Sakamoto, T., Ishikawa, T., Fairwell, T., Zech, L.A., Nakamura, H., Brewer, H.B., Jr., and Rader, D.J. 1995. Increased catabolic rate of low density lipoproteins in humans with cholesteryl ester transfer protein deficiency. *J Clin Invest* 96:1573–1581.
42. Rader, D.J. 2004. Inhibition of cholesteryl ester transfer protein activity: a new therapeutic approach to raising high-density lipoprotein. *Curr Atheroscler Rep* 6:398–405.
43. Boekholdt, S.M., and Thompson, J.F. 2003. Natural genetic variation as a tool in understanding the role of CETP in lipid levels and disease. *J Lipid Res* 44:1080–1093.
44. Duffy, D., and Rader, D.J. 2005. Drugs in development: targeting high-density lipoprotein metabolism and reverse cholesterol transport. *Curr Opin Cardiol* 20:301–306.

3 International Guidelines for Dyslipidemia Management

Cathleen E. Maki, RN, MSN, NP

CONTENTS

INTRODUCTION
THE CANADIAN GUIDELINES
THE EUROPEAN GUIDELINES
CONCLUSION
REFERENCES

INTRODUCTION

Cardiovascular disease (CVD) is a leading cause of global mortality and is responsible for approximately 17 million deaths annually *(1)*. In an effort to reduce this number, the World Heart and Stroke Forum (WHSF) Guidelines Task Force of the World Heart Federation (WHF) recommended that every country develop its own policy on CVD prevention. Although the causes of CVD are common to all parts of the world, the approaches to its prevention will differ between countries for cultural, social, medical, and economic reasons. As stated in a World Health Organization (WHO) report from 2002, "Epidemiological theory indicates that, compared with intensive individual treatment of high-risk patients, small improvements in the overall distribution of risk in a population will yield larger gains in disease reduction when the underlying conditions that confer risk are widespread in the population"; therefore, national policy should be developed combining the expertise of governmental, public health, and professional clinical groups.

A number of tools for estimating the risk of coronary heart disease (CHD) have been created including risk-score charts, risk-assessment algorithms, and computer software programs. Many were based on Framingham Heart Study data and others, such as the European Guidelines Systematic Coronary Risk Evaluation (SCORE) system were modified based on results from the second European Action on Secondary Prevention through Intervention to Reduce Events (EUROASPIRE II) *(2)*. The guidelines that describe these tools are widely available from medical societies such as the American Heart Association (AHA), European Society of Cardiology (ESC), the Canadian

From: *Contemporary Cardiology: Therapeutic Lipidology*
Edited by: M. H. Davidson, P. P. Toth, and K. C. Maki © Humana Press Inc., Totowa, NJ

Medical Society (CMS), and the Joint British Societies (JBS). Published guidelines from these sites focus on three strategies *(3)*:

1. A population strategy for altering lifestyle and environmental risk factors as well as social and economic contributions to CHD.
2. A high-risk strategy first to identify those at high risk and then to treat this population.
3. A secondary prevention strategy for those with established coronary artery disease (CAD).

The guidelines agree that risk factors for CHD are multiplicative and that drug therapy must be initiated once a risk threshold is met; however, the guidelines differ on how to measure CV risk and what threshold is best before initiating treatment for dyslipidemia *(4)*. For example, in a study of 100 German men and women, Broedl et al. *(4)* demonstrated how widely several guidelines differ in the identification and quantification of coronary risk as well as when to initiate lipid-lowering drug therapy.

This chapter will review and compare guidelines for the management of dyslipidemia and prevention of CVD from Europe, Canada, and the United Kingdom.

THE CANADIAN GUIDELINES

Introduction

Challenges in the treatment of CVD in Canada, as in all industrialized nations, include aging of the population and increasing prevalence of obesity and diabetes mellitus. When the most recent guidelines were published, more than 20% of Canadians were predicted to be aged over 65 years by 2011 with the fastest growing age group consisting of those aged over 80 years *(5)*, and 31% of Canadian adults were obese (BMI >27 kg/m^2). The authors predicted the incidence of new cases of diabetes to be 60,000 per year with the prevalence increasing from 1.5 million in 1998 to 3 million in 2010. The First Nations population, which comprises about 3% of the total Canadian population, is felt to be at particular risk in the development of diabetes at 3–5 times higher than the general Canadian population *(5)*.

In 1988, the Canadian Consensus Conference of Cholesterol (CCCC) published national guidelines for patient screening and the treatment of dyslipidemia *(6)*. In 1998, the Working Group on Hypercholesterolemia and other Dyslipidemias formed to analyze data and update these Canadian Guidelines. The committee issued updated management recommendations for dyslipidemia in 2000. In 2001, the US National Cholesterol Education Panel, Adult Treatment Panel III (NCEP ATP III) Guidelines were published followed by findings from several large clinical trials on dyslipidemia. Based on these findings, the Canadian Working Group reconvened and published an update in 2003. The guidelines were reviewed by two expert panels that included recognized specialists in the areas of CVD prevention, lipid metabolism, and diabetes along with primary care physicians *(5)*.

Risk Assessment

The 2003 Canadian Update recommend routine screening for all men over age 40 and all women who are postmenopausal or are over age 50. Also, anyone with diabetes, hypertension, abdominal obesity, smoking history, strong family history of premature CVD, manifestations of hyperlipidemia, or evidence of symptomatic or asymptomatic atherosclerosis should be routinely screened *(5)*. The new guidelines attempted to standardize CV risk assessment across North America using the Framingham Heart Study model for estimating the 10-year risk of CVD in men and women without diabetes mellitus or clinical evidence of CVD (Fig. 1). The Framingham CV risk equations were first published by Grundy and colleagues in 1999, and the NCEP ATP III Guidelines (6) adapted the equations based on the estimated 10-year risk of "hard cardiac endpoints" including death from CAD and non-fatal myocardial infarction (MI). Using the revised Framingham risk-stratification algorithm as described in the NCEP ATP III Guidelines (6), the 10-year absolute risk is calculated, and the patient is placed into one of three risk categories. Patients at high ($\geq 20\%$ or a history of diabetes mellitus or other atherosclerotic disease), moderate (11–19%), or low ($\leq 10\%$) risk have established target goals to guide lipid-lowering drug therapy (Table 1). The Canadian Guidelines were further simplified by using two treatment targets [low-density lipoprotein cholesterol (LDL-C) and the total cholesterol to high-density lipoprotein cholesterol ratio (TC: HDL-C)], and both targets must be met. One advantage of this approach is for patients with low HDL levels where more aggressive LDL-C lowering and/or HDL-C raising may be necessary to achieve both targets. To achieve a TC: HDL-C target of <4.0, the LDL-C must often be lowered to <75 mg/dL in those patients with low HDL-C (7). Clinical trials have demonstrated that patients with low HDL-C have a significantly higher residual risk of events on statin therapy. The ratio has also been shown to be a better epidemiologic predictor of CV events than LDL-C and also appears to significantly improve event prediction on statin therapy (14). The Working Group has published the following recommendations to achieve the TC: HDL-C of <4.0.

Lifestyle Therapy

For patients with hypertriglyceridemia, dietary therapy and exercise regimens are intensified with the focus on weight loss and the restriction of refined carbohydrates and alcohol. For patients with low HDL-C levels, increased activity (aerobic exercise), increased intake of monounsaturated fats, and moderate alcohol intake [only with normal triglycerides (TGs)] combined with weight loss are recommended.

Combination Therapy

In patients with combined dyslipidemia and low HDL-C levels, the combination of a statin with niacin is recommended. If a patient is not a candidate for niacin therapy, the group recommends substituting a fibrate. Also, the addition of fish oil in combination with a statin may be beneficial in patients with moderate hypertriglyceridemia.

Specific topics addressed in the 2003 Guidelines include the management of patients that are at high risk of CAD and are already at target for LDL-C (100 mg/dL), the management of patients who have combined dyslipidemia and low HDL-C levels, and the non-invasive assessment of CVD and other risk factors. Factors such as metabolic syndrome, apolipoprotein B, lipoprotein (a), homocysteine, and C-reactive protein are discussed as they relate to their influence on risk assessment.

Metabolic Syndrome

The Working Group defined metabolic syndrome in qualitative terms that included abdominal obesity, insulin resistance, elevated plasma TG levels, low HDL-C levels, and high blood pressure (Table 2). The Working Group used waist circumference

Age (years)	Points Men	Women
20–34	–9	–7
35–39	–4	–3
40–44	0	0
45–49	3	3
50–54	6	6
55–59	8	8
60–64	10	10
65–69	11	12
70–74	12	14
75–79	13	16

Total Cholestral	Points Age 20–39 Men	Women	Age 40–49 Men	Women	Age 50–59 Men	Women	Age 60–69 Men	Women	Age 70–79 Men	Women
>160	0	0	0	0	0	0	0	0	0	0
160–199	4	4	3	3	2	2	1	1	0	1
200–239	7	8	5	6	3	4	1	2	0	1
240–279	9	11	6	8	4	5	2	3	1	2
≥280	11	13	8	10	5	7	3	4	1	2

Smoking Status	Points Age 20–39 Men	Women	Age 40–49 Men	Women	Age 50–59 Men	Women	Age 60–69 Men	Women	Age 70–79 Men	Women
Non-Smoker	0	0	0	0	0	0	0	0	0	0
Smoker	8	9	5	7	3	4	1	2	1	1

HDL (mg/dL)	Points Men	Women
≥60	–1	–1
50–59	0	0
40–49	1	1
<40	2	2

Fig. 1. (*Continued*).

Systolic BP (mm Hg)	If Untreated Men	Women	If Treated Men	Women
<120	0	0	0	0
120–129	0	1	1	3
130–139	1	2	2	4
140–159	1	3	2	5
≥160	2	4	3	6

Total Points: _____. Check next table to determine 10-year CHD risk.

Men Total Points	10-year risk (%)	Women Total Points	10-year risk (%)
≤12	≤10	≤19	≤8
13	12	20	11
14	16	21	14
15	20	22	17
16	25	23	22
≥17	≥30	24	27
		≥25	≥30

10-year risk: _____ %

A 10-year CHD risk >20% is a CHD risk equivalent.

Fig. 1. Modified Framingham Score: estimate of 10-year coronary heart disease (CHD) risk. Reproduced with permission *(6)*.

values as established by the NCEP ATP III Guidelines of 102 cm for men and 88 cm for women as the cut-off values. The guidelines also review the role of visceral fat as being the most strongly related to insulin resistance and dyslipidemia *(5)*.

Apolipoprotein B

The Canadian Guidelines discuss the advantages of measuring apolipoprotein B especially in patients with moderate hypertriglyceridemia and metabolic syndrome. Advantages also include being able to measure a non-fasting blood sample. According to Genest, *(5)* Canadian population values for apolipoprotein B have been established, apolipoprotein B measurement has been standardized, and most Canadian laboratories have the equipment and expertise to measure it. In the Canadian population, an apolipoprotein B level of 0.9 g/L is around the 20th percentile, 1.05 g/L the 50th percentile, and 1.2 g/L the 75th percentile.

Table 1
Risk Categories and Target Lipid Levels

Risk category	Target level	
	LDL-C (mmol/L)	TC: HDL-C
High	<2.5	<4.0
Moderate	<3.5	<5.0
Low	<4.5	<6.0

HDL-C, high-density lipoprotein cholesterol; LDL-C, low-density lipoprotein cholesterol; TC, total cholesterol. Reproduced with permission (5).

Table 2
Clinical Identification of the Metabolic Syndrome

Risk factor	Defining level
Abdominal obesity	
Men	Waist circumference >102 cm
Women	Waist circumference >88 cm
TG level	1.7 mmol/L
HDL-C level	
Men	<1.0 mmol/L
Women	<1.3 mmol/L
Blood pressure	130/85 mmHg
Fasting glucose level	6.2–7.0 mmol/L

HDL-C, high-density lipoprotein cholesterol; TG, triglyceride. Reproduced with permission (5).

Lipoprotein (a)

The Canadian Guidelines suggest that the measurement of lipoprotein (a) although not a routine measurement may be useful in determining CV risk in patients with moderate risk and a family history of early CVD. A lipoprotein level >30 mg/dL in a patient with a TC: HDL-C >5.5 or other risk factors may indicate the need for earlier and more intensive therapy to lower the LDL-C level (5).

Homocysteine

Randomized controlled trials studying the effects of lowering homocysteine levels on CV endpoints are finishing. The position of the Working Group, the Canadian Cardiovascular Society, the AHA, and the Heart and Stroke Foundation of Canada is that there is currently insufficient evidence to warrant homocysteine screening until the results of these trials are known. The group does recommend however that there may be a specific indication for treatment with folic acid and vitamin B_{12} in patient's undergoing percutaneous coronary revascularization (8–10). They also suggest the treatment of patients with homocysteine levels >10 μmol/L using folic acid and vitamins B_{12} and B_6 may be warranted in high-risk patients with renal or CVD *(5)*.

High-Sensitivity C-Reactive Protein

The Canadian Guidelines do not currently recommend the use of C-reactive protein in risk-assessment screening; however, they recognize that the measurement may be useful in identifying people at higher CV risk than that predicted by the global risk-assessment score. At this time, most Canadians would pay out of pocket for the test.

THE EUROPEAN GUIDELINES

Introduction

Cardiovascular disease is the leading cause of death in European adults in their middle and older years. CVD accounts of 49% of all deaths in Europe and 30% of all deaths before the age of 65 *(3)*. In Europe, 1 in 8 men and 1 in 17 women die from CVD before the age of 65 *(3)*. In 2000, CVD accounted for 22% of all disability-adjusted life years lost in Europe *(3)*. There are also large differences between CVD mortality in western, central, and eastern European countries, with the highest rates found in the east. Even though mortality rates are declining, the prevalence of CVD is on the rise. Europe has the oldest population in the world, and projections determine that one in three Europeans will be older than age 65 by the year 2050 *(3)*.

The ESC Committee for Practice Guidelines (CPS) supervises and coordinates the development and subsequent publication of new or updated guidelines produced by expert groups or consensus panels (Task Forces) within the European community. One such Task Force, the Third Joint Task Force of European and other Societies on CVD Prevention in Clinical Practice, formed to represent eight societies across Europe. Historically, the European Atherosclerosis Society, the ESC, and the European Society of Hypertension collaborated and published a set of guidelines in 1994 on the prevention of CHD in clinical practice. The first guideline revision was published in 1998 by the Second Joint Task Force. This revision set lifestyle, risk factor, and therapeutic goals for the prevention of CHD. In the Second Joint Task Force, the European Society of General Practice/Family Medicine, the European Heart Network, and the International Society of Behavioral Medicine joined the original three societies. Two European surveys (EUROASPIRE I and EUROASPIRE II) were done in 1995 and again in 1999–2000. The studies described how secondary prevention was being achieved in clinical practice. Both surveys revealed an unacceptable gap between coronary disease prevention recommendations and what was achieved in the European population. There was little improvement seen between the two surveys *(3)*.

The Third Joint Task Force was joined by the European Association for the Study of Diabetes and the International Diabetes Federation of Europe. The Third Joint Task Force provided the second revision in 2003 which differed from the previous guidelines in four major areas: (i) a shift of focus from CHD prevention to CVD prevention, (ii) the development of the SCORE method of estimating CVD risk, (iii) elaboration on explicit clinical priorities, and (iv) the inclusion of newly published data in preventive cardiology, particularly recent clinical trials. The objective of the revised guidelines was to reduce the incidence of first or recurrent clinical events because of CHD, ischemic stroke, and peripheral artery disease.

In the preparation of these guidelines, the Task Force recognized that these interventions are targeted at populations of highest risk. Therefore, the guidelines suggest the following priorities for CVD prevention in clinical practice:

1. patients with established CHD, PVD, and cerebrovascular atherosclerotic disease;
2. asymptomatic patients with
   a. multiple risk factors,
   b. markedly raised levels of a single risk factor, and
   c. type 2 diabetes mellitus;
3. close relatives of patients with early onset CVD and asymptomatic individuals at particularly high risk; and
4. other individuals encountered in routine clinical practice *(3)*.

Risk Assessment

As in the Canadian and NCEP ATP III Guidelines, the Third Task Force uses a CVD risk-stratification method to guide therapeutic intervention in the European community (Fig. 2). Total risk is estimated using the SCORE system for patients without known vascular disease. SCORE is derived from a large data set of European subjects (EUROASPIRE II) and was designed to predict fatal atherosclerotic endpoints (e.g., fatal CVD events) over a 10-year period *(3)*.

Systematic Coronary Risk Evaluation integrated several modifiable and non-modifiable risk factors including gender, age, smoking, systolic blood pressure, and either TC or HDL-C ratio. Advantages of the SCORE system include the ability to produce risk charts tailored to specific European countries where reliable national mortality data are available, the ability to project estimated CVD risk to age 60, and the ability to estimate relative risk by comparing one risk cell against that of a non-smoking healthy person of the same age/gender *(3)*. Two risk-stratification charts are available, one for high-risk European countries and one for low-risk European countries. The low-risk chart is used in Belgium, France, Greece, Italy, Luxembourg, Spain, Switzerland, and Portugal. The high-risk chart is used in all other European countries. Patients who score ≥5% are candidates for intensive risk-factor intervention.

In addition to SCORE, the SCORECARD system software program provides information on how total risk can be reduced by specific interventions such as lifestyle modification and drug therapy.

Management of CVD Risk Factors

BEHAVIORAL RISK FACTORS

Behavioral risk factors including diet, activity, and smoking are addressed in the new guidelines, which also suggest that recent surveys identified a significant gap between the recommendations provided and the advice given out by physicians in routine clinical practice *(3)*. ESC guideline recommendations in these areas closely follow evidence-based practice guidelines as outlined in all guidelines referenced in this chapter; therefore, these will not be reviewed.

LIPID MANAGEMENT

Prior to the initiation of any therapies, secondary causes of dyslipidemia are ruled out including alcohol, diabetes, hypothyroidism, liver/kidney disease, and concurrent drug therapy. Using the recommendations from the SCORE chart, TC and LDL-C goals are used to guide the decision process and, in general, should be kept <190

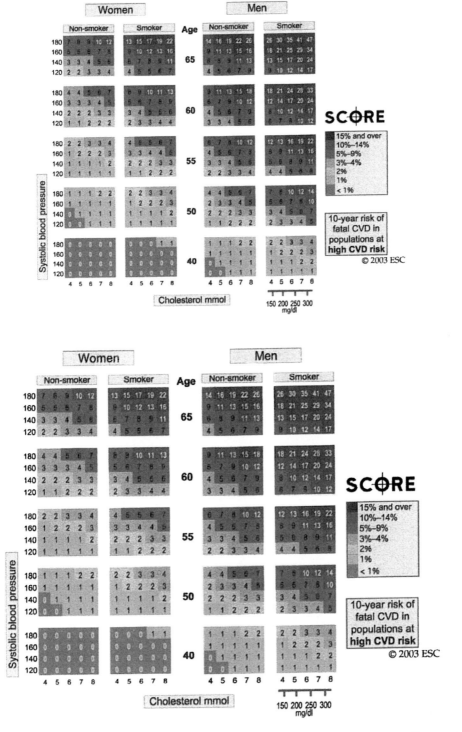

Fig. 2. CVD risk-stratification method to guide therapeutic intervention in the European Community, the SCORE chart. Reproduced with permission (3).

and 115 mg/dL, respectively. Although no treatment goals are defined for TGs and HDL-C, these values are used as risk markers of increased CVD risk. An HDL-C of <40 mg/dL in men and <46 mg/dL in women along with TG of >150 mg/dL increase the patients' risk score. In those asymptomatic patients with high multifactorial risk as defined in the SCORE chart, the guidelines recommend a further reduction of TC to <175 mg/dL and LDL-C to <100 mg/dL using lipid-lowering drug therapy. In asymptomatic patients with a risk score of <5%, the risk assessment is repeated at 5-year intervals. If the risk score is ≥5%, the full lipid panel analysis is completed and interventions are prescribed (including intensive dietary changes exclusive of drug therapies) until the value for LDL-C falls <190 mg/dL and TC falls <155 mg/dL and the total CVD risk falls <5%. The patient is then monitored at yearly intervals. If, however, the total risk remains ≥5%, drug therapy is begun. The goals in this population are TC <175 mg/dL and LDL-C <100 mg/dL.

DIABETES

In general, treatment goals for patients with diabetes are more extensive and are represented in Table 3.

METABOLIC SYNDROME

The European Guidelines use the definition of metabolic syndrome as defined in the NCEP ATP III Guidelines. Although the guidelines suggest that patients with metabolic syndrome are usually at high risk of CVD, no specific guidelines exist beyond those described by the SCORE assessment guidelines.

The European Guidelines also suggest the following in the prevention of CHD in clinical practice:

Aspirin or other platelet-modifying drug therapy in virtually all patients with diagnosed CVD;

Beta-blockers following MI or in patients with left ventricular dysfunction related to CHD;

Table 3
Diabetic Treatment Goals

	Goal
HbA1C	≤6.1%
Venous plasma glucose	
Fasting/pre-prandial	≤6.0 mmol/L
	<110 mg/dL
Self-monitored blood glucose	
Fasting/pre-prandial	4.0–5.0 mmol/L
	79–90 mg/dL
Postprandial	4.0–7.5 mmol/L
	70–135 mg/dL
Blood pressure	<130/80 mmHg
TC	<4.5 (175) mmol/L (mg/dL)
LDL-C	<2.5 (100) mmol/L (mg/dL)

LDL-C, low-density lipoprotein cholesterol; TC, total cholesterol.
Reproduced with permission *(11)*.

Angiotensin-converting enzyme (ACE) inhibitors in patient with symptoms or signs of
  Left ventricular Dysfunction LVD related to CHD and/or arterial hypertension; and
Anti-coagulants in those patients with CHD who are at increased risk of developing a
  thromboembolic event.

Joint British Society Guidelines

Data collected from 1988 to 1991 by the WHO revealed that the death rate from CVD in the United Kingdom was the highest in the world at that time. In the United Kingdom, CHD is the most common cause of premature death *(12)*. In the year 2000, CHD was responsible for one in four male deaths and one in six female deaths totaling around 125,000 deaths. The National Health Service estimated that costs related to CHD are £1.6 billion per year with only 1% spent on primary prevention of CVD. The overall cost to the UK economy is estimated at £10 billion each year *(12)*.

The most recent JBS' guidelines on CVD prevention (JBS2) published in 2005 were developed by nominated representatives from six professional societies including the British Cardiac Society, British Hypertension Society, Diabetes UK, HEART UK, Primary Care Cardiovascular Society, and The Stroke Association. All representatives of the committee contributed to the text and those from the specialist societies of hypertension, lipids, and diabetes were specifically responsible for developing the sections on blood pressure, lipids, and glucose management. All sections of the document represent an evidence-based consensus by all professional societies involved.

The JBS2 guidelines focus on those individuals considered at highest risk from CVD. This includes those patients with any known form of CVD, those at highest risk for the development of CVD ($\geq$20% over the next 10 years), and those with diabetes mellitus *(13)*. The JBS2 guidelines consider a patient at $\geq$20% risk of developing CVD is approximately equivalent to a >15% risk of developing CHD over the next 10 years.

The absolute risk for developing CVD can be estimated from the JBS2 CV risk prediction chart (Fig. 3). The use of these charts is not appropriate for patients with existing CVD or those at higher risk ($\geq$20%) from other medical conditions such as diabetes, familial dyslipidemias, or renal dysfunction. Of note, smoking status is defined as lifetime exposure to tobacco, not only tobacco use only at the time of assessment. The charts are also based on groups of patients with untreated blood pressure, TC, and HDL values. In situations where patients are being treated for one or more of these risk factors, the charts may then only be used as a guide *(13)*. The authors also state that absolute CVD risk is higher than indicated in the charts for those patients with a family history of premature CVD, those with elevated TGs (>1.7 mmol/L), women with premature menopause, and those who are not yet diabetic but demonstrate impaired glucose tolerance.

BEHAVIORAL MANAGEMENT

All patients are encouraged to stop smoking, make healthier food choices, increase aerobic activity, limit alcohol intake to <21 units/week for men and <14 units/week for women, and maintain a BMI of <25 kg/m^2 with no central obesity (waist circumference <102 cm in men and <88 cm in women).

LIPID MANAGEMENT

The decision to initiate lipid-lowering drug therapy is determined by the absolute risk of developing CVD. In general, the guidelines suggest that an absolute risk of $\geq$20%

of developing CVD over the next 10 years is sufficient to consider drug therapy. In patients with existing CVD, the target range for TC is <4.0 mmol/L or a 25% reduction (whichever is lower) and for LDL-C <2.0 mmol/L or a 30% reduction (whichever is lower).

CARDIOPROTECTIVE THERAPY

In addition to the lipid-lowering target ranges, all patients with existing CVD should be considered for cardioprotective therapies that include

Aspirin 75 mg daily,
Blood pressure-lowering therapy as outlined by the British Hypertension Society,
Beta-blockers following MI,
Angiotensin-converting enzyme inhibitors for patients with symptoms of heart failure at the time of MI or those with persistent LVD (ejection fraction <40%),

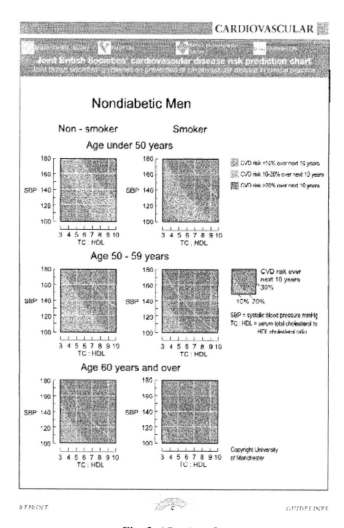

Fig. 3. (*Continued*).

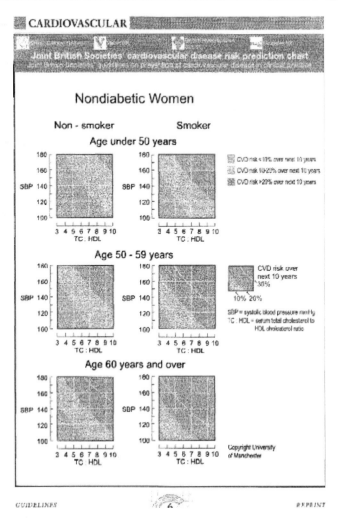

Fig. 3. JBS2 CV risk prediction chart. Reproduced with permission *(12)*.

Consider ACE inhibitor for those with normal LV function if blood pressure targets not achieved,

Use angiotensin II receptor blockers if ACE inhibitor intolerant,

Calcium channel inhibitors in patients with coronary disease when blood pressure targets have not been achieved, and

Anticoagulant therapy for those with CVD patients at high risk for systemic embolization.

GUIDELINES FOR ASYMPTOMATIC PATIENTS AT HIGH RISK

The JBS2 guidelines also discuss management of asymptomatic patients at high risk for developing CVD according to blood pressure management, lipid-lowering drug therapies, cardioprotective drug therapy, and screening of first-degree relatives (Table 4).

Table 4
Guidelines for Asymptomatic People at High Risk of Developing Cardiovascular Disease (CVD)

Category	Parameters	Sub-parameters	Treatment
BP	Healthy subjects		Lifestyle advice drug therapy if sustained
	SBP ≥160 mmHg DBP ≥100 mmHg Healthy subjects	CVD risk ≥20% or target organ damage or diabetes	Lifestyle advice and drug therapy
	SBP = 140–159 mmHg	CVD risk <20% and no target organ damage and no diabetes	Lifestyle advice, monitor BP, and reassess annually
	DBP = 90–99 mmHg	Healthy subjects (BP = 140/85 mmHg)	Lifestyle advice, reassess in 5 years
Cholesterol	Healthy subjects with familial hypercholesterolemia or other inherited dyslipidemias		Lifestyle advice and drug therapy
	Healthy subjects with CVD risk ≥20%	TC >4.0 mmol/L and LDL-C >2.0 mmol/L	Lifestyle advice and drug therapy
		TC >4.0 mmol/L and LDL-C >2.0 mmol/L	Lifestyle advice and reassess annually
	Healthy subjects with CVD risk <20% and no CV complications and no diabetes		Lifestyle advice, reassess in 5 years
Cardioprotective drug therapy Screening of first-degree relatives	Adults aged >50 years	Controlled BP <150/90 mmHg	Aspirin 75 mg daily Screen close relatives if familial hyper-cholesterolemia or other inherited dyslipidemia is suspected

BP, blood pressure; LDL-C, low-density lipoprotein cholesterol; TC, total cholesterol. Reproduced with permission (12).

Table 5
Guideline Comparisons

Guideline	NCEP ATP III	Joint European	Joint British	Canadian
Database Parameters used to calculate absolute risk	Framingham Sex	Eurosphere II Sex	Framingham TC: HDL-C	Framingham TC
	Age SBP Smoking Diabetes TC HDL-C	Tobacco SBP TC Age	SBP Sex Tobacco Age	HDL-C SBP Age Tobacco
Age ranges applicable to guidelines (years)	20–79	40–65	<50	Men >40
			50–59 >60	Women >50 or postmenopausal
Diabetic patients	Considered high risk ≥20% over the next 10 years	Included in the same risk-stratification chart	Aged ≥40 years receive statin and 18–29 based on comorbidities	Considered high risk ≥20% over the next 10 years
Cholesterol management (asymptomatic and known CVD)	Dependent on LDL-C and number of risk factors	High risk ≥5%, drug therapy to decrease TC to <5.0 mmol/L and LDL-C <3 mmol/L	Known CVD or asymptomatic/high risk with TC > 4.0 mmol/L and LDL-C >2.0 mmol/L and prescribe drug therapy	Based on risk level—high, moderate, and low

CVD, cardiovascular disease; HDL-C, high-density lipoprotein cholesterol; LDL-C, low-density lipoprotein cholesterol; NCEP ATP III, National Cholesterol Education Panel, Adult Treatment Panel III; TC, total cholesterol.

DIABETES MELLITUS

The guidelines discuss optimal targets for the following:

HbA1C <6.5%,
Fasting plasma glucose ≤6.0 mmol/L,
Blood pressure <130/80,
TC <4 mmol/L or 25% decrease, and
LDL-C <2.0 mmol/L or 30% decrease.

Angiotensin-converting enzyme inhibitors or angiotensin II receptor blockers are to be prescribed for patients with renal dysfunction and microalbuminuria. The JBS2 guidelines provide specific target goals for diabetic (Type 1 and Type 2) patients regarding cholesterol. All diabetic patients aged ≥40 years are to be prescribed statin therapy. For those aged 18–39, diabetics with at least one of the following conditions are to be prescribed statin therapy:

Retinopathy,
Nephropathy,
HbA1C >9%,
Elevated blood pressure requiring therapy,
TC >6 mmol/L,
Metabolic syndrome, and
Family history of premature CVD.

CONCLUSION

Table 5 summarizes the similarities and differences of the guidelines reviewed. Notable differences remain in whether to utilize the TC: HDL-C or continue to use LDL-C as the primary marker for cholesterol management; yet, all describe what the ideal parameters should be and when to consider initiating drug therapy. Some prefer to include patients with diabetes in the same stratification system as non-diabetics. The Joint European Guidelines are now unique in their use of EUROASPIRE II data versus Framingham Heart Study data. As the population ages and more countries become industrialized, the guidelines and stratification systems will be modified to reflect the changing demographics and will incorporate the latest research data.

REFERENCES

1. Smith SC, Jackson R, Pearson TA, et al. Principles for national and regional guidelines on cardiovascular disease prevention: a scientific statement from the World Heart and Stroke Forum. *Circulation* 2004;109:3112–3121.
2. Brookes L. European guidelines on cardiovascular disease prevention. *Medscape Cardiol* 2003.
3. DeBacker G, Ambrosioni E, Borch-Johnson K, et al. European guidelines on cardiovascular disease prevention in clinical practice (constituted by representatives of eight societies and by invited experts). *Eur J Cardiovasc Prev Rehabil* 2003;10(supplement 1):s1–s78.
4. Broedl UC, Geiss HC, Parhofer KG. Comparison of current guidelines for primary prevention of coronary heart disease. *J Gen Intern Med* 2003;18:190–195.
5. Genest J, Frohlich J, Fodor G, McPherson R. Recommendations for the management of dyslipidemia and the prevention of cardiovascular disease: summary of the 2003 update. *CMAJ* 3004;169(9):921–924.
6. National Cholesterol Education Program Expert Panel, Third Report. Detection, evaluation, and treatment of high cholesterol in adults (Adult Treatment Panel III). NIH Publication 02–5125. Available at http://www.nhlbi.nih.gov/guidelines/cholesterol/atp3_rpt.htm.
7. Davidson, MH. (2004) Emerging Therapeutic Strategies for the Management of Dyslipidomia in Patients with the Metabolic Syndrome. *Am J Cardiol* 2004;93(suppl): 3c–11c.
8. Rossouw JE, Anderson GL, Prentice RL, et al. Writing group for women's health initiative investigators. Risks and benefits of estrogen plus progestin in healthy postmenopausal women: principal results from the Women's Health Initiative randomized controlled trial. *JAMA* 2002;288(3):321–333.
9. Schnyder G, Roffi M, Pin R, et al. Decreased rate of coronary restenosis after lowering of plasma homocysteine levels. *N Engl J Med* 2001;345(22):1593–1600.
10. Schnyder G, Roffi M, Flammer Y, et al. Effect of homocysteine-lowering therapy with folic acid, vitamin B(12), and vitamin B(6) on clinical outcome after percutaneous coronary intervention: the Swiss Heart study: a randomized controlled trial. *JAMA* 2002;288(8):973–979.

11. DeBacke G, Ambrosioni E, Borch-Johnsen K, et al. European guidelines on cardiovascular disease prevention in clinical practice. *Eur Heart J* 2003;24:1601–1610.
12. Poulter N. Global risk of cardiovascular disease. *Heart* 2003;89(supplement II):ii2–ii5.
13. Wood D, Wray R, Poulter N, et al. Joint British Societies Guidelines on prevention of cardiovascular disease in clinical practice. *Heart* 2005;91(supplement 5):v1–v52.
14. Downs JR, Clearfield M, Wers S, Wnitney E, Shapiro DR, Beere DR, Lagendorfer A, Stem EA, Kruyer W, Gotto AM. Primary prevention of acute coronary events with lovastutin in new and women with average cholesterol levels: results of AFCAPS/Tex CAPS. *JAMA* 1998;279:1615–1622.

4 Pathophysiology and Management of Dyslipidemias Associated with Obesity, Type 2 Diabetes, and Other Insulin-Resistant States

Kevin C. Maki, PhD

CONTENTS

INTRODUCTION
OBESITY AND TYPE 2 DIABETES MELLITUS
CARDIOMETABOLIC RISK SYNDROME
LIPID ABNORMALITIES ASSOCIATED WITH IR
FUNCTIONS OF INSULIN
EXCESSIVE PRODUCTION OF VERY LOW-DENSITY LIPOPROTEINS (VLDLs): THE PRIMARY LIPID ABNORMALITY IN THE INSULIN RESISTANT STATE
REASONS FOR ELEVATED FFA LEVELS IN INSULIN-RESISTANT STATES
FORMATION OF SMALL, DENSE LDL PARTICLES
MECHANISMS LINKING IR TO LOW HDL-C
ACTIVITIES OF LIPOPROTEIN AND HEPATIC LIPASES AND THEIR RELATIONSHIPS WITH ATHEROGENIC DYSLIPIDEMIA
LIFESTYLE MANAGEMENT FOR ATHEROGENIC DYSLIPIDEMIA
DRUG THERAPIES FOR ATHEROGENIC DYSLIPIDEMIA
INTENSIFICATION OF EFFORTS TO LOWER LDL-C AS A MEANS OF ACHIEVING THE NON-HDL-C GOAL
TARGETING TG-RICH LIPOPROTEIN REDUCTION AS A MEANS OF ACHIEVING NON-HDL-C GOAL
MANAGEMENT OF DIABETIC DYSLIPIDEMIA
SUMMARY
REFERENCES

From: *Contemporary Cardiology: Therapeutic Lipidology*
Edited by: M. H. Davidson, P. P. Toth, and K. C. Maki © Humana Press Inc., Totowa, NJ

INTRODUCTION

Insulin resistance (IR), defined as an impaired ability of a given circulating concentration of insulin to promote the clearance of glucose from the blood, is common in the United States and other developed countries. It is present in a large majority of people with obesity, polycystic ovary syndrome, impaired glucose tolerance, and type 2 diabetes mellitus *(1)*. In addition, IR appears to be present in approximately half of patients with hypertension *(2)*.

IR has both genetic and environmental determinants. Increased adiposity, sedentary lifestyle, and cigarette smoking all appear to contribute causally to its development *(3)*. However, some individuals have a strong genetic component and may thus manifest IR in the absence of these acquired factors. For example, first-degree relatives of patients with type 2 diabetes mellitus are insulin resistant compared with controls matched for body mass index (BMI) *(1)*. Nevertheless, the degree of IR is further increased when genetic predisposition is combined with lifestyle factors that promote IR.

OBESITY AND TYPE 2 DIABETES MELLITUS

The incidence and prevalence of overweight and obesity have increased dramatically in the United States during the last generation *(4)*. Figure 1 shows the prevalence of obesity among US adult men and women, based on representative samples of the population, from 1960 to 2000. The prevalence of obesity doubled during this time, although the figure makes it clear that the increases have been mainly attributable to the period since 1980. In 2000, it was estimated that approximately 65% of the US population could be classified as overweight (BMI 25.0–29.9 kg/m^2) or obese (BMI $\geq$30 kg/m^2). Obesity is the strongest risk factor for the development of type 2 diabetes mellitus, with the top decile of BMI in the population showing a 40-fold to 50-fold increased risk compared with the lowest decile *(4)*. Therefore, it is not surprising that the incidence of type 2 diabetes mellitus has been rising in concert with that of obesity. Data released by the National Institute of Diabetes and Digestive and Kidney Diseases indicate that approximately 73 million adult Americans, or roughly one-third

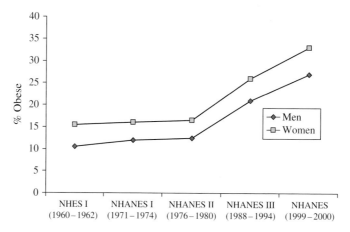

Fig. 1. Prevalence of obesity among US adult men and women, based on representative samples of the population, from 1960 to 2000. Adapted from Manson et al. *(4)*.

of the adult population, have impaired fasting glucose (≥100 mg/dL) or frank diabetes mellitus *(5)*.

CARDIOMETABOLIC RISK SYNDROME

IR is thought to play a central role in the development of a cluster of interrelated metabolic abnormalities that predispose to the development of coronary heart disease (CHD) and occur together more often than would be predicted by chance. Over the years, this clustering of risk factors has been referred to by such terms as the Insulin Resistance Syndrome, Syndrome X, Metabolic Syndrome, and the Cardiometabolic Risk Syndrome. The commonly acknowledged features include increased adiposity, dyslipidemia, glucose intolerance, and hypertension, although more recent evidence indicates that other abnormalities are also associated with this syndrome, including hyperuricemia, low-grade inflammation, and a prothrombotic state *(6)*. The focus of this chapter will be the pathogenesis and management of the characteristic dyslipidemia associated with insulin-resistant states.

LIPID ABNORMALITIES ASSOCIATED WITH IR

Insulin-resistant states are often accompanied by three major disturbances in the lipid profile, which the National Cholesterol Education Program's (NCEP) Third Adult Treatment Panel (ATP III) has termed "atherogenic dyslipidemia" *(7)*. The features of atherogenic dyslipidemia include the following:

1. elevated circulating triglycerides (TGs);
2. reduced high-density lipoprotein cholesterol (HDL-C) concentration; and
3. a predominance of small, dense low-density lipoprotein (LDL) particles.

It should be noted that the term "lipid triad" has also been used in the literature for this group of disturbances, as well as to describe a separate, but related, set of lipid abnormalities (a TG concentration ≥200 mg/dL in combination with an LDL-C/HDL-C ratio >5.0) *(8)*. It is important to distinguish between the two "lipid triads" because elevated LDL-C is not a feature of the dyslipidemia associated with IR per se, although it may be present concurrently.

FUNCTIONS OF INSULIN

Prior to summarizing the influence of IR on lipid metabolism, it is instructive to briefly review some of the functions insulin plays in its role as the "master metabolic hormone." Insulin has a number of actions beyond the promotion of cellular uptake of glucose. It suppresses hepatic gluconeogenesis and shifts the metabolic state of tissues (particularly skeletal muscle) toward the oxidation and storage (as glycogen) of energy from carbohydrate. At the same time, insulin suppresses the activity of hormone-sensitive lipase. This, in turn, reduces the release of free fatty acids (FFAs) from adipose tissues into the circulation and thereby lowers the availability of FFAs as a substrate for oxidation. Insulin also stimulates lipoprotein and hepatic lipases, enhancing the hydrolysis of TGs in circulating lipoproteins and allowing their FFAs to move into cells. In the liver, insulin stimulates the breakdown of apolipoprotein (Apo) B.

Not all of the actions of insulin may be impaired to the same degree. Some of the metabolic abnormalities associated with IR may result from insufficient insulin action.

In patients with IR who do not have diabetes, normal glucose levels are maintained through excessive insulin secretion (compensatory hyperinsulinemia). Thus, some of metabolic abnormalities associated with IR may not result from impaired insulin action, but rather from overstimulation of insulin-mediated processes that are not impaired.

EXCESSIVE PRODUCTION OF VERY LOW-DENSITY LIPOPROTEINS (VLDLs): THE PRIMARY LIPID ABNORMALITY IN THE INSULIN-RESISTANT STATE

The primary metabolic abnormality associated with insulin-resistant states is overproduction of VLDLs. Because newly secreted VLDL particles are TG rich, the result is mild-to-moderate hypertriglyceridemia. If the overproduction of VLDLs is accompanied by other defects, such as Apo C-II deficiency (which reduces lipoprotein lipase activation), or a defect in the hepatic clearance of TG-rich Apo B-containing lipoproteins, the result can be more severe hypertriglyceridemia (Fredrickson Types IV and V) and/or mixed dyslipidemia, involving elevations in TG and LDL-C (Types IIB and III).

Two features of the insulin-resistant state are centrally involved in the pathogenesis of VLDL overproduction: elevated circulating levels of FFAs and hyperinsulinemia. It appears that both must be present to generate overproduction of VLDLs. For example, in subjects with normal insulin sensitivity, a glucose infusion will not only increase the plasma insulin concentration but also lower levels of FFAs and reduce hepatic VLDL secretion *(1)*. In contrast, patients with poorly controlled type 1 diabetes have low insulin levels and high concentrations of FFA but do not have elevated VLDL secretion *(1)*. However, among insulin-resistant subjects with abdominal obesity, fasting and postprandial levels of circulating insulin are elevated, but this hyperinsulinemia does not normally suppress FFA release into the circulation; thus, both insulin and FFA levels are elevated, resulting in VLDL overproduction.

Increased hepatic exposure to FFAs inhibits Apo B degradation and leads to increased VLDL synthesis *(9)*. Insulin promotes lipogenesis, which contributes to the TG pool available for incorporation into VLDL particles. In addition, insulin's ability to enhance Apo B degradation may be impaired in the insulin-resistant patient *(9)*.

REASONS FOR ELEVATED FFA LEVELS IN INSULIN-RESISTANT STATES

The release of FFA into the circulation is directly proportionate to the size of fat cells. Thus, increasing adiposity is accompanied by greater release of FFAs into the circulation. A chronically elevated FFA level is believed to be a cause of IR. This is illustrated by the observation that IR can be induced in normal subjects by the infusion of Intralipid™ (Pfizer, New York, NY) for several hours, which raises the FFA concentration, mimicking the obese state. The mechanisms that are responsible for the effects of a chronically elevated FFA level on insulin sensitivity are beyond the scope of this review, but the interested reader is referred to a recently published review on this topic *(10)*.

FFA turnover in adipose tissues varies according to location. The abdominal visceral fat depots are the most metabolically active and contribute disproportionately to the circulating FFA level. For example, it has been estimated that an abdominally obese male with 20% of his body fat in the visceral stores will have a 50% contribution of these stores to the circulating FFA concentration *(11)*. Upper body subcutaneous fat is less metabolically active than abdominal visceral fat, and lower body subcutaneous fat is least metabolically active. For this reason, an abdominally obese woman is likely to have greater lipid disturbances than a woman of similar BMI with a "gynoid" pattern of obesity, who carries most of her excess adiposity on the hips and thighs. Some ethnic groups (e.g., Asian Indians) tend to have a greater proportion of their body fat carried in the abdominal visceral depots and thus may display IR and other metabolic abnormalities at relatively low BMIs.

It should be noted that an elevated circulating FFA concentration may be present in the absence of obesity *(1)*. In some individuals, the primary metabolic defect responsible for IR may be impaired "fat trapping." When TGs in lipoproteins are hydrolyzed, they enter cells (primarily adipose and muscle) through the action of acylation stimulating protein *(12)*. In some individuals, this mechanism is impaired, resulting in an abnormally large escape of FFAs back into the circulation. Such people may have circulating FFA levels that are much higher than would be predicted by their degree of adiposity, and they may be thought of as being "metabolically obese" *(12)*. In addition, some medications, particularly antiretroviral drugs, may cause peripheral lipodystrophy with a resulting inability of subcutaneous adipose tissues to take up FFAs released by lipoprotein lipase-catalyzed hydrolysis of TGs, resulting in excess return of FFAs liberated to the circulation.

FORMATION OF SMALL, DENSE LDL PARTICLES

Overproduction of VLDL increases the plasma TG concentration. VLDLs compete with chylomicron particles for the available lipoprotein lipase (Fig. 2). The result is a prolongation of the residence time of TG-rich lipoprotein particles in the circulation and increases in the concentrations of partially delipidated remnant particles (chylomicron remnants, small VLDLs, and intermediate-density lipoproteins [IDLs]). Evidence from a variety of sources supports the view that these TG-rich remnant lipoprotein particles are atherogenic *(7)*.

An increase in the circulating TG concentration provides additional substrate for cholesteryl ester transfer protein (CETP), which catalyzes the exchange of TG from TG-rich lipoproteins to LDL and HDL particles in exchange for cholesteryl esters

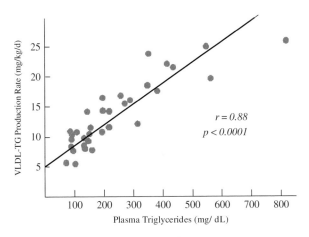

Fig. 2. Relationship between plasma triglyceride concentration and very-low-density lipoprotein triglyceride (VLDL-TG) production rate. Adapted from Olefsky et al. *(39)*.

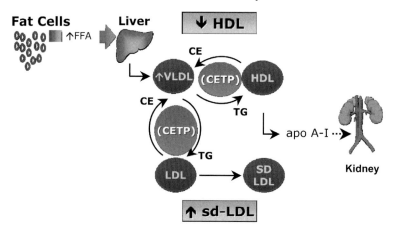

Fig. 3. Generation of the lipid triad. Adapted from Ginsberg *(40)*.

(Fig. 3). CETP activity is also reportedly enhanced by an elevation in FFA concentration *(9,13)*. The result is that the LDL and HDL particles become relatively TG rich and cholesterol poor. The TGs in these particles can be hydrolyzed by hepatic lipase to form smaller, denser particles.

Small, dense LDL particles are believed to have enhanced atherogenicity for several reasons *(14,15)*. They have less affinity for the Apo B receptor, resulting in extended circulation in the plasma before hepatic clearance. Small, dense LDL particles also have greater interactivity with intra-arterial proteoglycans, which can lead to greater residence time within the arterial wall. Moreover, once in the arterial wall, they are more susceptible to oxidative modification. This can lead to unregulated uptake by macrophages, contributing to foam cell formation.

The term LDL subclass pattern A is used to denote a predominance of larger, more buoyant LDL particles, whereas LDL subclass pattern B is a predominance of small, dense LDL particles. The prevalence of LDL subclass pattern B increases progressively with the plasma TG concentration. For example, Austin and colleagues *(16)* found that at a TG concentration of 100 mg/dL the prevalence of pattern B was approximately 18%, whereas at a TG concentration of 250 mg/dL it was approximately 88%. The conversion from pattern A to B appears to be a threshold phenomenon that depends mainly on the circulating TG concentration *(14,15,17)*. Those with a high propensity to form pattern B (e.g., due to high CETP activity) will convert at a lower TG level.

We have recently demonstrated this principle with TG reductions induced by fenofibrate or omega-3 ethyl esters *(17)*. In both studies, substantial TG lowering (30–50%) had no effect on LDL particle size or subclass distribution as long as the on-treatment TG concentration remained above 250 mg/dL. However, at TG concentrations below 250 mg/dL, significant increases in LDL particle size were observed. Furthermore, the proportion of subjects who converted from pattern B to A increased as the on-treatment TG concentration declined below 250 mg/dL.

MECHANISMS LINKING IR TO LOW HDL-C

As discussed earlier, an elevation in the TG concentration increases the CETP-mediated exchange of TG for cholesteryl esters between TG lipoproteins and HDL particles. This results in smaller HDL particles that are relatively enriched with TG and depleted of cholesterol. Apo A-I appears to dissociate from these TG-enriched HDL particles, allowing clearance by the kidney (Fig. 3) *(9,18)*. Thus, insulin-resistant individuals typically have lower levels of both HDL-C and Apo A-I.

ACTIVITIES OF LIPOPROTEIN AND HEPATIC LIPASES AND THEIR RELATIONSHIPS WITH ATHEROGENIC DYSLIPIDEMIA

Although overproduction of VLDLs appears to be the primary lipid disturbance associated with IR and atherogenic dyslipidemia, the TG clearance rate is also generally somewhat reduced. In part, this may relate to an impairment of the ability of insulin to activate lipoprotein lipase. In addition, greater FFA levels in the interstitium may reduce lipoprotein lipase activity through end-product inhibition of lipoprotein lipase activity *(12)*.

LIFESTYLE MANAGEMENT FOR ATHEROGENIC DYSLIPIDEMIA

A twofold approach is recommended for the management of dyslipidemia associated with IR *(7)*. The first is to reduce the underlying lifestyle factors that contribute to the development of the insulin-resistant state, particularly obesity and physical inactivity. The second is to treat the individual lipid and nonlipid risk factors associated with IR.

The NCEP ATP III has emphasized lifestyle changes as the first step in the management of IR and the related risk factors, including atherogenic dyslipidemia. The two cornerstones of lifestyle changes for insulin-resistant patients are increased physical activity and loss of excess body fat. Both exercise and weight loss improve insulin sensitivity and all of the risk factors associated with IR.

Current recommendations for physical activity include at least 30 min of aerobic exercise on most days. It appears that most of the benefits of aerobic exercise can be achieved by lower intensity activities such as walking, which, for most people, is the form of exercise that is most easily incorporated into their daily routine *(19)*. Loss of as little as 5–10% of body weight can produce significant improvements in metabolic risk factors for CHD, including IR and atherogenic dyslipidemia *(7)*.

Cigarette smoking has also been shown to reduce insulin sensitivity and contribute to atherogenic dyslipidemia, adding to the myriad of reasons that smoking cessation should be encouraged. In addition, excessive alcohol intake will increase the TG concentration. Therefore, moderation (or cessation) of alcohol consumption should be encouraged if applicable *(7)*.

The NCEP ATP III altered the dietary recommendations that had been issued in earlier sets of guidelines. The new Therapeutic Lifestyle Changes diet has less emphasis on substitution of carbohydrate for dietary fat. This was done in recognition of the fact that a high-carbohydrate diet will raise the TG concentration, exacerbating atherogenic dyslipidemia. The updated dietary recommendations allow up to 35% of energy from fat, provided that the diet remains low in saturated fats, *trans* fats, and cholesterol *(7)*. Increased intakes of whole grains, nuts, fruits, vegetables, lean meats, and low-fat dairy products should be encouraged *(20)*.

In addition to the lifestyle changes outlined earlier, clinicians should be aware that some drug therapies can worsen the components of atherogenic dyslipidemia and should thus be avoided if possible. These include oral estrogens (transdermal estrogens are a better alternative), retinoids (e.g., Accutane®, Roche Pharmaceuticals, Basel, Switzerland) and thiazide diuretics *(7)*.

DRUG THERAPIES FOR ATHEROGENIC DYSLIPIDEMIA

The primary target for lipid management is LDL-C (Table 1). Once the LDL-C treatment goal has been achieved, the NCEP ATP III guidelines recommend that non-HDL-C (total-C minus HDL-C) is a secondary target for patients with high TGs (200–499 mg/dL). Modifications of circulating levels of TGs and HDL-C are also potential targets for lipid-altering therapies, but the ATP III did not establish specific treatment targets for these lipids. It should be noted that for patients with very high TG ($\geq$500 mg/dL), the initial aim of treatment is to prevent pancreatitis by reducing

Table 1
Treatment Goals for Low-Density Lipoprotein Cholesterol (LDL-C), Non-High-Density Lipoprotein Cholesterol (Non-HDL-C), and Apolipoprotein B According to Coronary Heart Disease (CHD) Risk Status

Risk status	*Primary target: LDL-C (mg/dL)*	*Secondary target: non-HDL-C (mg/dL)*	*Secondary target: apolipoprotein B (mg/dL)*
CHD and CHD risk equivalents[a]	<100	<130	<90
Very high risk subset[b]	<70 (optional)	<100 (optional)	<80
Multiple (2+) risk factors[c]	<130	<160	<110
Moderately high risk subset[d]	<100 (optional)	<130 (optional)	<90
0–1 Risk factor	<160	<190	<130

[a] CHD includes history of myocardial infarction, unstable angina, stable angina, coronary artery procedures (angioplasty or bypass surgery), or evidence of clinically significant myocardial ischemia. CHD risk equivalents include clinical manifestations of noncoronary forms of atherosclerotic disease (peripheral arterial disease, abdominal aortic aneurysm, and carotid artery disease, transient ischemic attacks or stroke of carotid origin or 50% obstruction of a carotid artery), diabetes, and 2+ risk factors with 10-year risk for hard CHD >20%.

[b] The very high risk subset includes established cardiovascular disease and (1) multiple major risk factors (especially diabetes), (2) severe and poorly controlled risk factors (especially continued cigarette smoking), (3) multiple risk factors of the metabolic syndrome (especially high triglycerides $\geq$200 mg/dL and non-HDL-C $\geq$130 mg/dL with low HDL-C [<40 mg/dL]), and (4) patients with acute coronary syndromes.

[c] Risk factors include cigarette smoking, hypertension (BP $\geq$140/90 mmHg or on antihypertensive medication), low HDL cholesterol (<40 mg/dL), family history of premature CHD (CHD in male first-degree relative <55 years of age; CHD in female first-degree relative <65 years of age), and age (men <45 years; women <55 years).

[d] Ten-year CHD event risk 10–20% by the Framingham risk scoring system. Adapted from *(21,22,24)*.

the TG concentration. This situation is covered in detail elsewhere in this book and will not be discussed here.

Apo B may also be given consideration as an alternative secondary target for therapy *(21,22)*. Because VLDL, IDL, LDL, and chylomicron particles each contain one molecule of Apo B (Apo B-100 for those of hepatic origin and Apo B-48 for chylomicron particles), the Apo B concentration is an indicator of the total number of circulating atherogenic particles. Because Apo B and non-HDL-C concentrations are highly correlated, the NCEP ATP III did not recommend the added expense of routine measurement of Apo B. However, some researchers believe that changes in Apo B are superior to those of LDL-C and non-HDL-C for the clinical management of dyslipidemia *(22,23)*.

There are two approaches to achieving non-HDL-C treatment goals in patients with high TG despite having achieved their LDL-C treatment targets. Often such patients will be taking a "standard" dose of a statin drug *(24)*. One approach is to further lower LDL-C by increasing the statin dose, adding a cholesterol absorption inhibitor, or by intensification of nondrug therapies *(21,25)*. The other is to add an agent that lowers VLDL-C and TG-rich remnant lipoproteins (a fibrate, niacin, or omega-3 fatty acids). At present, no clear evidence from clinical trials is available that would allow advocating one or the other of these methods.

INTENSIFICATION OF EFFORTS TO LOWER LDL-C AS A MEANS OF ACHIEVING THE NON-HDL-C GOAL

Statins are recommended as first-line therapy for patients with elevated LDL-C. However, statins also markedly reduce concentrations of both TG and TG-rich lipoproteins in hypertriglyceridemic subjects *(26,27)*. The hypotriglyceridemic effects of statins have been widely underappreciated, largely because few studies of statin therapy had been undertaken in hypertriglyceridemic subjects until recently. TG-lowering with statins is modest in normotriglyceridemic subjects, but much more pronounced in those with hypertriglyceridemia.

Kinetic studies have shown that statins increase the fractional catabolic rate of Apo B-containing lipoproteins (VLDLs, IDLs, and LDLs) while having little effect on VLDL production *(27,28)*. The result is that the circulating levels of TG, TG-rich lipoprotein (VLDL and IDL) cholesterol, and LDL-C all decline. Reductions in TG of 30–50% and non-HDL-C of 50–60% have been reported when hypertriglyceridemic patients are treated with the maximal approved doses of the higher efficacy statins (atorvastatin, rosuvastatin, and simvastatin).

The use of statins is supported by a greater quantity of clinical trial data than any other class of lipid-altering agent. The results from large event trials have consistently shown that statin therapy reduces CHD morbidity and mortality. These effects are present at all levels of baseline lipids studied, as well as in groups that would be expected to be enriched with insulin-resistant individuals, such as subjects with diabetes mellitus, hypertriglyceridemia, and hypertension *(29)*.

Although statin therapy is very safe and well tolerated at lower doses, a large portion (75–85%) of the maximal effects on LDL-C and non-HDL-C are achieved at the usual starting doses. Each additional doubling of the statin dose is associated with further reductions of only 5–7%, whereas the risks of liver and muscle toxicity increase to a larger extent. For this reason, it is not unreasonable to consider using a submaximal dose of statin therapy in combination with another agent rather than using the maximal approved statin doses.

Additional agents that may be considered for further lowering of LDL-C include nondrug therapies (plant sterols/stanols and viscous fibers) or a cholesterol absorption inhibitor. Bile acid sequestrants also reduce LDL-C when added to statin therapy but tend to modestly increase the plasma TG concentration, thus are not ideal for patients with atherogenic dyslipidemia. Plant sterols/stanols and viscous fibers can be expected to provide an additional 5–10% lowering of LDL-C and non-HDL-C when added to statin therapy *(25,30)*. Coadministration of a cholesterol absorption inhibitor (10 mg ezetemibe) will generally lower the LDL-C and non-HDL-C concentrations by a further 10–20% *(31)*.

TARGETING TG-RICH LIPOPROTEIN REDUCTION AS A MEANS OF ACHIEVING NON-HDL-C GOAL

Fibrates, niacin, and omega-3 fatty acids all effectively lower VLDL-C and other TG-rich lipoproteins. Fibrates work by stimulating peroxisome proliferator-activated receptor (PPAR) alpha. This results in enhanced lipoprotein lipase expression, reduced hepatic production of Apo C-III, and enhanced hepatic fat oxidation. In addition, fibrates increase the production rates of Apo A-I and Apo A-II. Studies of lipoprotein kinetics have shown that fibrate therapy increases the clearance rates for VLDL, IDL, and LDL particles, but surprisingly, appears to have little effect on VLDL secretion *(27)*.

At the usual dosages, fibrates typically lower the TG concentration by 30–50% and increase HDL-C by 10–25%. The LDL-C response to fibrate therapy is dependent on both the baseline TG and LDL-C concentrations. In patients with very high TG concentrations ($\geq$500 mg/dL), the LDL-C level may rise. In patients with less severe hypertriglyceridemia, particularly those with concomitantly elevated LDL-C, the LDL-C concentration may decline by as much as 20%.

Clinical outcome trials with fibrate therapy have been generally supportive of a protective effect regarding cardiovascular events *(20,32)*. However, the results are not as robust as those observed with statins. In primary and secondary prevention trials with clofibrate, gemfibrozil, bezafibrate, and fenofibrate, the median relative risk reduction for the primary outcome variable was 21% (range 4–49%), with four of nine trials reaching statistical significance *(20,32)*.

Results from subgroup analyses indicate that the most favorable effects on cardiovascular events have been observed in subjects with elevated TG concentrations at baseline *(33,34)*. For example, the Bezafibrate Infarction Prevention Study had a nonsignificant relative risk reduction of 7.3% in the overall study sample, but a 39.5% relative risk reduction among the quarter of the study sample with a baseline TG concentration $\geq$200 mg/dL *(33)*. Despite the fact that fibrates are most often used in clinical practice to treat hypertriglyceridemia, none of the published event trials have specifically recruited hypertriglyceridemic subjects.

The dramatic effects of niacin on the blood lipid profile were noted more than a half-century ago, but the mechanisms responsible for these have been poorly understood until recently. The effects of niacin on lipid metabolism are due to its ability to suppress FFA release from adipose tissues as well as inhibition of hepatic diacylglycerol acyltransferase (DGAT) *(35)*. The latter is an enzyme involved with TG synthesis and VLDL production. The result of these changes is enhanced hepatic Apo B degradation and reduced VLDL production. Niacin also markedly increases the number of circulating HDL particles by selectively inhibiting the uptake of Apo A-I by hepatocytes, thus reducing the fractional catabolic rate of HDL. At approximately 2 g/day, an extended release niacin will reduce the plasma TG concentration by 20–50%, increase HDL-C by 15–35%, and lower LDL-C by 5–25%.

Although niacin improves all of the features of atherogenic dyslipidemia, two issues raise concerns about its clinical usefulness in such patients. The first is flushing, which is experienced to some degree by most patients. Its intensity is diminished, but not eliminated, by a prescription, extended release preparation (Niaspan®, Kos Pharmaceuticals, Abbott Laboratories, Abbott Park, IL). This side effect is bothersome to many patients, which may limit compliance. The second issue associated with niacin use is the development of IR. One might predict that the reduced FFA release from adipose tissues would lead to improved insulin sensitivity. However, the reverse appears to be true, although the mechanisms responsible for this effect are poorly understood (18). The use of niacin can cause people with mild glucose intolerance to convert to frank diabetes by worsening the degree of IR and therefore increasing demand on the pancreatic beta-cells (18,36). For this reason, niacin should be used with caution in patients with IR or glucose intolerance.

The long-chain omega-3 polyunsaturated fatty acids eicosapentaenoic acid (EPA) and docosahexaenoic acid (DHA), found in high concentrations in the oils of cold water fish, have been known for years to have a hypotriglyceridemic effect when consumed in high doses (1–4 g/day of EPA + DHA). Recently, a prescription preparation of n-3 fatty acid ethyl esters (Omacor®, Reliant Pharmaceuticals, Liberty Corner, NJ) has been made that is more concentrated, thus requiring fewer capsules to be taken to achieve a therapeutic dose of omega-3 fatty acids. A dose of 3.4 g EPA + DHA can be obtained from 4×1 g capsule, which is the equivalent of the EPA + DHA content of 11–12 g of fish oil.

Omega-3 fatty acids reduce VLDL production by inhibiting DGAT, and possibly through a mild stimulatory effect on PPAR alpha, thus stimulating hepatic fat oxidation (28,37). The result is reduced hepatic synthesis and secretion of VLDL, with no apparent effect on hepatic uptake of Apo B-containing particles (28,37). Omega-3 fatty acids therefore reduce circulating levels of TG and VLDL-C (25–50%). They also generally produce a small rise in HDL-C (3–10%). As with fibrates, omega-3 fatty acids sometimes lower LDL-C modestly (5–10%), particularly among subjects with high baseline LDL-C. However, in patients with more severe hypertriglyceridemia, the LDL-C concentration may rise (38). Nevertheless, the reduction in VLDL-C is typically larger than the increase in LDL-C, so the net result is a reduction in cholesterol carried by atherogenic, Apo B-containing lipoproteins (non-HDL-C).

MANAGEMENT OF DIABETIC DYSLIPIDEMIA

"Diabetic dyslipidemia" is essentially atherogenic dyslipidemia that occurs in a person with diabetes mellitus. The fundamental principles of its management do not differ from those for a patient with IR in the absence of diabetes. However, diabetic dyslipidemia may be affected by the degrees of glycemic control and relative insulin deficiency, which can increase the variability in lipoprotein levels.

SUMMARY

IR is common in the United States and other developed countries, being present in the majority of individuals with obesity, impaired glucose tolerance or type 2 diabetes mellitus, and polycystic ovary syndrome. IR also appears to be present in approximately half of those with hypertension.

The combination of elevated circulating FFA and compensatory hyperinsulinemia that are characteristic of insulin-resistant states leads to VLDL overproduction, which is the primary abnormality of lipid metabolism in "atherogenic dyslipidemia" (elevated TG, reduced HDL-C, and a predominance of small, dense LDL particles). VLDL overproduction results in an elevated circulating TG concentration, which drives the CETP-catalyzed exchange of TG from TG-rich lipoprotein particles for cholesteryl ester from LDL and HDL particles. This results in relatively TG-rich and cholesterol-poor LDL and HDL particles.

Apo A-I on TG-enriched HDL particles dissociates and is cleared by the kidney, contributing to low levels of circulating HDL-C. The TG-enriched LDLs undergo delipidation by lipase enzymes (primarily hepatic lipase), forming smaller, more dense particles, which are believed to have increased atherogenicity. LDL subclass pattern B is a predominance of small, dense (more atherogenic) LDL particles. A shift from pattern A (a predominance of larger, more buoyant LDL particles) to pattern B occurs when the TG concentration exceeds a threshold. This threshold varies from person to person but is between 100 and 250 mg/dL for most individuals.

The approach to the management of atherogenic dyslipidemia is twofold: addressing the underlying causes of IR (obesity and physical inactivity) and then treating the specific lipid and nonlipid risk factors associated with the insulin-resistant state. Loss of excess body fat and exercise reduce IR and improve all of the associated CHD risk factors, including atherogenic dyslipidemia. For patients who are unable to adequately control atherogenic dyslipidemia with lifestyle changes, drug therapy may be indicated.

The primary goal of lipid management is to achieve the patient's LDL-C treatment goal. After this, if the TG concentration is still ≥ 200 mg/dL, non-HDL-C (total-C minus HDL-C) becomes a secondary target for intervention. The non-HDL-C goal may be met intensifying efforts to lower LDL-C. This can be accomplished by adding or increasing the dose of statin therapy, adding plant sterols/stanols and viscous dietary fibers to the diet, or using a cholesterol absorption inhibitor. An alternative approach to achieving the non-HDL-C goal is to use an agent (fibrate, niacin, or omega-3 fatty acids) that primarily lowers the levels of TG-rich lipoprotein cholesterol (VLDL-C and IDL-C).

REFERENCES

1. Reaven GM. Syndrome X. Past, present and future. In: Draznrin B, Rizza R, eds. *Clinical Research in Diabetes and Obesity, Vol. II: Diabetes and Obesity*. Totowa, NJ: Humana Press, 1997;357–377.
2. Reaven GM. Insulin resistance/compensatory hyperinsulinemia, essential hypertension, and cardiovascular disease. *J Clin Endocrinol Metab* 2003;88(6):2399–2403.
3. Maki KC. Dietary factors in the prevention of diabetes mellitus and coronary artery disease associated with the metabolic syndrome. *Am J Cardiol* 2004;93(Suppl):12C–17C.
4. Manson JE, Skerrett PJ, Greenland P, VanItallie TB. The escalating pandemics of obesity and sedentary lifestyle. A call for clinicians. *Arch Intern Med* 2004;164:249–258.
5. U.S. Department of Health and Human Services. The National Institute of Diabetes and Digestive and Kidney Diseases (NIDDK). Available at http://www2.niddk.nih.gov/. (accessed October 8, 2006).
6. Reaven GM. The metabolic syndrome: requiescat in pace. *Clin Chem* 2005;51(6):931–938.

7. National Cholesterol Education Program. Third Report of the National Cholesterol Education Program Expert Panel on Detection, Evaluation, and Treatment of High Blood Cholesterol in Adults. NIH Publication No. 02-5215, 2002.
8. Manninen V, Tenkanen L, Koskinen P, et al. Joint effects of serum triglyceride and LDL cholesterol and HDL cholesterol concentrations on coronary heart disease risk in the Helsinki Heart Study: implications for treatment. *Circulation* 1992;85:37–45.
9. Ginsberg HN, Zhang Y, Hernandez-Ono A. Regulation of plasma triglycerides in insulin resistance and diabetes. *Arch Med Res* 2005;36:232–240.
10. Boden G. Fatty acid-induced inflammation and insulin resistance in skeletal muscle and liver. *Curr Diab Rep* 2006;6:177–181.
11. Bjorntorp P. Visceral obesity: a "civilization syndrome". *Obes Res* 1993;1:206–222.
12. Sniderman AD, Cianflone K, Arner P, Summers LKM, Frayn KN. The adipocyte, fatty acid trapping, and atherogenesis. *Arterioscler Thromb Vasc Biol* 1998;18:147–151.
13. Ginsberg HN, Zhang Y, Hernandez-Ono A. Metabolic syndrome: focus on dyslipidemia. *Obesity* 2006;14(Suppl):41S–49S.
14. Packard CJ. Triacylglycerol-rich lipoproteins and the generation of small, dense low-density lipoprotein. *Biochem Soc Trans* 2003;31(Pt 5):1066–1069.
15. Krauss RM. Dense low density lipoproteins and coronary artery disease. *Am J Cardiol* 1995;75:53B–57B.
16. Austin MA, King MC, Vranizan KM, Krauss RM. Atherogenic lipoprotein phenotype. A proposed genetic marker for coronary heart disease risk. *Circulation* 1990;82:495–506.
17. Davidson MH, Bays HE, Stein E, Maki KC, Shalwitz RA, Doyle R. Effects of fenofibrate on atherogenic dyslipidemia in hypertriglyceridemic subjects. *Clin Cardiol* 2006;29:268–273.
18. Ginsberg HN. Niacin in the metabolic syndrome: more risk than benefit? *Nat Clin Pract Endocrinol Metab* 2006;2:300–301.
19. Hu FB, Manson JE. Walking. The best medicine for diabetes? *Arch Intern Med* 2003;163:1397–1398.
20. Maki KC. Fibrates for treatment of the metabolic syndrome. *Curr Atheroscler Rep* 2004;6:45–51.
21. Grundy SM. Low-density lipoprotein, non-high-density lipoprotein, and apolipoprotein, and apolipoprotein B as targets of lipid-lowering therapy. *Circulation* 2002;106:2526–2529.
22. Barter PJ, Ballantyne CM, Carmena R, et al. Apo B versus cholesterol in estimating cardiovascular risk and in guiding therapy: report of the thirty-person/ten country panel. *J Intern Med* 2006;259:247–258.
23. Sniderman AD, Furberg CD, Keech A, et al. Apolipoproteins versus lipids as indices of coronary risk and as targets for statin treatment. *Lancet* 2003;361:777–780.
24. Grundy SM, Cleeman JI, Bairey CN, et al. Implications of recent clinical trials for the National Cholesterol Education Program Adult Treatment Panel III Guidelines. *Circulation* 2004;110:227–239.
25. Maki KC, Galant R, Davidson MH. Non-high-density lipoprotein cholesterol: the forgotten therapeutic target. *Am J Cardiol* 2005;96(Suppl):59K–64K.
26. Isaacsohn J, Hunninghake D, Schrott H, et al. Effects of simvastatin, an HMG-CoA reductase inhibitor, in patients with hypertriglyceridemia. *Clin Cardiol* 2003;26:18–24.
27. Watts GF, Barrett PH, Ji J, et al. Differential regulation of lipoprotein kinetics by atorvastatin and fenofibrate in subjects with the metabolic syndrome. *Diabetes* 2003;52:803–811.
28. Chan DC, Watts GF, Barrett PH, Beilin LJ, Redgrave TG, Mori TA. Regulatory effects of HMG CoA reductase inhibitor and fish oils on apolipoprotein B-100 kinetics in insulin-resistant obese male subjects with dyslipidemia. *Diabetes* 2002;51:2377–2386.
29. Cholesterol Treatment Trialists' Collaboration. Efficacy and safety of cholesterol-lowering treatment: prospective meta-analysis of data from 90,056 participants in 14 randomised trials of statins. *Lancet* 2005;366:1267–1278.
30. Katan MB, Grundy SM, Jones P, et al. Efficacy and safety of plant stanols and sterols in the management of blood cholesterol levels. *Mayo Clin Proc* 2003;78:965–978.
31. Davidson MH, McGarry T, Bettis R, et al. Ezetimibe coadministered with simvastatin in patients with primary hypercholesterolemia. *J Am Coll Cardiol* 2002;40:2125–2134.
32. FIELD Investigators. Effects of long-term fenofibrate therapy on cardiovascular events in 9795 people with type 2 diabetes mellitus (the FIELD study): randomised controlled trial. *Lancet* 2005;366:1849–1861.
33. BIP Study Group. Secondary prevention by raising HDL cholesterol and reducing triglycerides in patients with coronary artery disease. The Bezafibrate Infarction Prevention (BIP) Study. *Circulation* 2000;102:21–27.

34. Robins SJ, Collins D, Wittes JT, et al. Relation of gemfibrozil treatment and lipid levels with major coronary events VA-HIT: a randomized controlled trial. *JAMA* 2001;285:1585–1591.
35. Ganji SH, Kamanna VS, Kashyap ML. Niacin and cholesterol: role in cardiovascular disease (review). *J Nutr Biochem* 2003;14:298–305.
36. Kahn SE, Prigeon RL, Schwartz RS, et al. Obesity, body fat distribution, insulin sensitivity and islet beta-cell function as explanations for metabolic diversity. *J Nutr* 2001;131:354S–360S.
37. Chan DC, Watts GF, Mori TA, Barrett PH, Redgrave TG, Beilin LJ. Randomized controlled trial of the effect of n-3 fatty acid supplementation on the metabolism of apolipoprotein B-100 and chylomicron remnants in men with visceral obesity. *Am J Clin Nutr* 2003;77:300–307.
38. Bays HE. Clinical overview of Omacor: a concentrated formulation of omega-3 polyunsaturated fatty acids. *Am J Cardiol* 2006;98(4A):71i–76i.
39. Olefsky JM, Farquhar JW, Reaven GM. Reappraisal of the role of insulin in hypertriglyceridemia. *Am J Med* 1974;57(4):551–560.
40. Ginsberg HN. Insulin resistance and cardiovascular disease. *J Clin Invest* 2000;106(4):453–458.

5 C-Reactive Protein and Other Inflammatory Markers in Cardiovascular Disease

Natalie Khuseyinova, MD, and Wolfgang Koenig, MD, PhD

CONTENTS

INTRODUCTION
C-REACTIVE PROTEIN
CYTOKINES
MARKERS OF HEMOSTASIS THAT ARE ALSO
   ACUTE-PHASE REACTANTS
WBC COUNT
LIPID-RELATED BIOMARKERS
CONCLUSIONS
KEY POINTS
REFERENCES

INTRODUCTION

During the last ten years, compelling experimental and clinical evidence has demonstrated that both systemic and local inflammation might play a prominent role in the pathogenesis of atherosclerosis and its clinical complications. Because inflammatory processes accompany all stages of atherogenesis, measurement of plasma/serum concentrations of circulating inflammatory biomarkers might aid in identifying individuals at high risk for coronary heart disease (CHD). In particular, such biomarkers might add to the predictive value of the atherogenic lipoprotein phenotype to further improve the assessment of future global cardiovascular (CV) risk. This chapter summarizes our current knowledge based on observations from experimental and clinical studies with emphasis on potential pathophysiological mechanisms of action and on clinical utility of inflammatory biomarkers as predictors of CV risk.

From: *Contemporary Cardiology: Therapeutic Lipidology*
Edited by: M. H. Davidson, P. P. Toth, and K. C. Maki © Humana Press Inc., Totowa, NJ

C-REACTIVE PROTEIN

Among numerous circulating biomarkers of vascular inflammation, C-reactive protein (CRP), the classical acute-phase reactant with a relatively short half-life (~19 h), has been most intensively investigated. More than 25 studies published during the last ten years have provided strong evidence that CRP predicts CV risk in various scenarios not only in initially healthy subjects but also in those with manifest atherosclerosis. Furthermore, a recent AHA/CDC consensus report *(1)* has recommended only the measurement of CRP in asymptomatic subjects at intermediate risk for future coronary events (10-year risk of 10–20%) and in selected patients after an acute coronary syndrome (ACS). This was based on the fact that of all current inflammatory markers only CRP has the analyte and assay characteristics most conducive to use in practice (Class IIa, Level of Evidence B) and that other inflammatory markers should not be measured for determination of CV risk (Class III, Level of Evidence C). However, there is still an ongoing controversy in the scientific community *(2)*.

CRP, a member of the pentraxin family, was first detected in 1930 by Tillett and Francis *(3)*. It received its name because of its ability to precipitate the somatic C-polysaccharide or cell-wall teichoic acid of *Streptococcus pneumonia*. CRP is a nonspecific biochemical marker of inflammation, which, in addition to its involvement in the prediction of CV risk, has also served for many years as an indicator of a variety of pathological processes including infections, tissue damage, autoimmune diseases, or cancer. In healthy individuals, only trace levels of CRP can be detected in circulation. There is no apparent circadian variability with CRP, as has been observed with other acute-phase reactants such interleukin-6 (IL-6) or for plasminogen activator inhibitor-1 (PAI-1) *(4,5)*, and there is also no evidence for seasonal variations as seen with fibrinogen *(6,7)*. Under acute conditions, concentrations of CRP increase during the first 6–8 h and can reach peak levels approaching 300 mg/L after approximately 48 h *(8)*.

CRP is synthesized by hepatocytes with an overall molecular weight of approximately 118 kDa. Its production is under transcriptional control of several cytokines, with IL-6 being a primary stimulus. However, recent evidence has suggested that CRP may be also produced locally in atherosclerotic lesions. For instance, vascular smooth muscle cells (SMCs) within the atherosclerotic plaque have been shown to produce CRP in response to cytokines *(9)*. Jabs et al. *(10)* detected CRP mRNA in venous bypass grafts and in normal vasculature, with CRP expression in diseased vessels being significantly higher. Moreover, Yasojima et al. *(11)* found that CRP expression in macrophages and smooth muscle-like cells in atherosclerotic lesions exceeded CRP expression in the liver or in normal arteries 7.2-fold and 10.2-fold, respectively. It has been also shown that immunoreactivity to CRP in coronary atheromatous plaque was significantly higher in culprit lesions of patients with unstable angina pectoris (AP), compared with that of patients with stable AP *(12)*. In addition, a recent postmortem study further confirmed a pathogenic role of CRP in atheromatous plaque vulnerability *(13)*, demonstrating that higher CRP concentrations strongly correlated with increased numbers of thin cap atheromas. By contrast, Maier et al. *(14)*, measuring the concentrations of inflammatory biomarkers both at the site of the coronary plaque rupture and in the systemic circulation, recently suggested that CRP was not elevated locally at the site of coronary plaque rupture, whereas local concentrations of other inflammatory markers such as IL-6 and serum amyloid A (SAA) were markedly higher than in the systemic circulation. The authors, however, suggested that de concentration at sites of plaque rupture might rather reflect local uptake, or partial catabolism of CRP, because immunohistochemical staining of the ha thrombus had revealed the presence of CRP in the cytoplasm of phagocyting white blood cells (WBCs).

In addition, others sites of local CRP synthesis and possible secretion are being discussed, including alveolar macrophages *(15)*, human neurons *(16)*, and kidney epithelial cells *(17)*. Finally, it has been reported that human adipocytes cultured *in vitro* also produce CRP after treatment with inflammatory cytokines *(18)*.

CRP has initially been considered as an innocent bystander in the atherosclerotic process. Recent evidence, however, suggests that CRP may have direct proinflammatory effects and may contribute to the initiation and progression of atherosclerotic lesions (Fig. 1). Functionally, CRP has several effects that may influence progression of vascular disease, including activation and chemo-attraction of circulating monocytes, mediation of endothelial dysfunction, induction of a prothrombotic state, increase of cytokine release, activation of the complement system, and facilitation of extracellular matrix remodeling as well as lipid-related effects. For instance, CRP opsonizes LDL and facilitates native LDL entry into macrophages *(19)*, primarily through the low-affinity immunoglobulin receptor CD32 *(20,21)*. CRP binds to plasma membranes of damaged cells and activates complement via the classical pathway *(22)*. It stimulates monocytes to release inflammatory cytokines such as IL-1β, IL-6, and tumor necrosis factor-α (TNF-α) *(23,24)* and increases the release of soluble IL-6 receptor by macrophages and foam cells in the neointima *(25)*. Furthermore, CRP itself was found to be chemotactic for freshly isolated human blood monocytes *(26)* and could upregulate the expression of various chemokines such as monocyte chemoattractant protein-1 (MCP-1) *(27)*, or IL-8, as well as CC-chemokine receptor 2, a most dominant

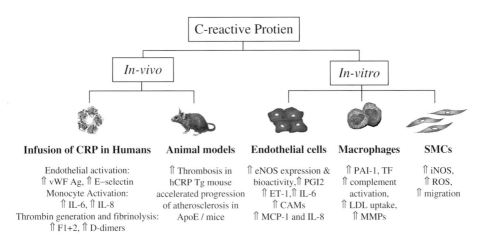

Fig. 1. Potential direct effects of CRP on atherothrombosis. SMCs = smooth muscles cells; vWF Ag = von Willebrand-Factor antigen; IL = interleukin; hCRP Tg = human CRP transgenic; ApoE = apolipoprotein E; eNOS = endothelial nitric oxide synthase; PGI-2 = prostacyclin; ET-1 = endothelin-1; CAMs = cell adhesion molecules; MCP-1 = monocyte chemoattractant protein-1; PAI-1 = plasminogen activaor inhibitor 1; TF = tissue factor; MMPs = matrix metalloproteinases; iNOS = inducible nitric oxide synthase; ROS = reactive oxygen species; ⇑ = increase; ⇓ = decrease.

chemotaxis receptor in monocytes *(28)*. In addition, CRP, at concentrations known to predict CV events, can induce the expression of several cell adhesion molecules (CAMs) such as intercellular adhesion molecule-1 (ICAM-1), vascular cell adhesion molecule-1 (VCAM-1), and E-selectin on human aortic endothelial cells (ECs) *(29)*. It has been shown to facilitate thrombogenesis through stimulation of tissue factor (TF) biosynthesis by mononuclear cells *(30,31)*; however, these data could not be confirmed in a recent study *(32)*. CRP also induces the release of thrombomodulin in human umbilical vein ECs *(33)*. It has been further shown to attenuate the fibrinolytic capacity by decreasing tissue plasminogen activator (t-PA) antigen and activity and inducing the expression of PAI-1, the major endogenous inhibitor of fibrinolysis *(34)*. Finally, several investigators have suggested that CRP may be a direct mediator of endothelial dysfunction because of its ability to modulate endothelial nitric oxide synthase (eNOS) bioactivity. In fact, ECs incubated with CRP decrease eNOS expression and NO release *(35,36)*.

However, several of aforementioned potential proatherogenic properties of CRP should be interpreted with caution, because very recent findings indicated that direct effects of CRP on vasculature, observed in *in vitro* studies, could represent possible artifacts caused by contamination with several additives present in commercial CRP preparations such as bacterial lipopolysaccharide (LPS; endotoxin) or sodium azide that is used as a bacteriostatic preservative *(37–40)*. Two studies, for instance, revealed that the removal of endotoxin contamination from the commercial CRP used abolished CRP-induced secretion of MCP-1, PAI-1, and IL-6, whereas nonpurified CRP did not *(37,40)*. Moreover, van den Berg *(38)* and Taylor *(39)* demonstrated that CRP isolated carefully from human malignant ascites or produced recombinantly in mammalian cells had no proinflammatory effects on ECs and that these effects of the bacterial recombinant product could all be replicated by the addition of either LPS or azide.

Yet, in a recent study, Singh et al. *(41)* using endotoxin-purified, azide-free CRP showed that CRP was able to inhibit t-PA activity via generation of proinflammatory cytokines such as IL-1β and TNF-α. In line with this study are findings by Bisoendial et al. *(42)*, who demonstrated that injection of highly purified recombinant CRP into seven healthy individuals induced a dramatic acute-phase response: CRP treatment has led to an increase in mean CRP level from 1.0 to 23.0 mg/L and to an increased production of inflammatory markers (IL-6 and IL-8, SAA, but not TNF-α) as well as prothrombotic molecules (PAI-1, prothrombin fragments). It is rather unlikely that LPS contamination was responsible for these observations *(43)*.

Animal models have also provided conflicting evidence regarding proatherogenic effects of CRP. This is most probably because mice may not be considered the appropriate model for establishing a functional role of CRP in atherosclerosis, as CRP in rodents is not the major acute-phase reactant and is expressed only at extremely low levels *(44)*. Another alternative approach used the generation of a transgenic mouse that overexpressed nonmouse CRP (i.e., human or rabbit). Two of five hitherto published studies supported the evidence that CRP is prothrombotic and proatherogenic, but the other three studies were negative. In the study by Danenberg et al., the rate of arterial thrombosis, following transluminal wire injury, was faster and higher in human CRP-transgenic mice compared with wild-type mice *(45)*. Transgenic expression of human CRP also caused a 30–50% increase in aortic atherosclerosis in apolipoprotein E (apoE)-deficient male mice *(46)* and was associated with increased complement deposition and elevated expression of angiotensin type-1 (AT-1) receptor, VCAM-1, and collagen within lesions. It is important to keep in mind that CRP concentrations in these CRP-transgenic mice were extremely high (>100 mg/L). By contrast, Hirschfield et al. *(47)* showed that transgenic expression of human CRP had no effect on development, progression, or severity of spontaneous atherosclerosis, nor on morbidity or mortality in male apoE$^{-/-}$hCRP^{+} mice, despite the deposition of human CRP and mouse complement component 3 in plaques. By crossing human CRP-transgenic mice with apolipoprotein E*3-Leiden (E3L) transgenic mice, those mice that exhibited elevated plasma cholesterol and triglyceride levels, corresponding to familial dysbetalipoproteinemia in humans, were found to have moderately increased concentrations of plasma CRP (10.2 ± 6.5 mg/L), but neither the extent nor the severity of atherosclerotic lesion development had been modified *(48)*. Another study, crossing rabbit CRP-transgenic mice into apoE-knockout mice, also demonstrated no marked effect on atherogenesis at 20 weeks and at 52 weeks *(49)*.

Although direct experimental *in vitro* and *in vivo* evidence supporting a causal role for CRP in atherogenesis remains somehow controversial and represents an issue for continuing research, epidemiological studies published during the last ten years, by contrast, have provided strong evidence for CRP to predict future CV risk in a wide variety of clinical settings, including apparently healthy men and women, patients with stable AP, or those with ACS, after myocardial infarction (MI), and in those with the metabolic syndrome.

So far, results from more than 25 different prospective studies have been reported, and the vast majority of these studies clearly demonstrated a significant and independent association between increased concentrations of CRP and future CV events. An earlier meta-analysis summarizing the results of 14 prospective long-term studies with a total of 2557 cases and a mean follow up (FU) period of 8 years *(50)* revealed a summary relative risk (RR) for CHD of 1.9 (95% confidence interval [CI] 1.5–2.3) for the top versus the bottom tertile (T) of the CRP distribution. More recently, however, the results from the Reykjavik study raised some uncertainties regarding the predictive power of CRP by showing a more modest increased risk associated with elevated CRP concentrations with an odds ratio (OR) of 1.45 (95% CI 1.25–1.68) for T3 versus T1 after multivariate adjustment *(2)*. A subsequent meta-analysis of 22 population-based studies, including total of 7068 patients with incident coronary events, showed a similar result *(2)*. However, there are some issues concerning this study, which merit consideration. The Reykjavik study represents a prospective cohort of 18,569 participants, where CRP was measured in approximately 6500 middle-aged men and women without a history of MI at baseline, who were followed for 17.5 years. As compared with previously published studies, the Reykjavik participants had much higher cholesterol levels and lower CRP (upper tertile cutoff point of 2.0 mg/L rather than 3.0 mg/L) as seen in almost all other studies. Thus, underestimation of the true risk associated with elevated CRP is very likely. In addition, the mean FU period of 17 years is extremely long and might also be responsible for the weakening of the association between the risk marker and the disease outcome. Indeed, if we look at the risk estimate at 10 years ("normal" FU time in most prospective studies), the risk estimate was 1.84 (95% CI 1.49–2.28), which is similar to those from other studies, thereby reinforcing the status of CRP as a strong and independent predictor of future CV risk.

More recently, the large EPIC-Norfolk study from the United Kingdom conducted between 1993 and 2003 found that CRP was among the strongest variables predicting risk of CHD; it was the strongest if only fatal cases were analyzed *(51)*, and in contrast to the Reykjavik study, it found that other risk factors had no incremental value.

Additional data in support of a predictive role of CRP came from the Cardiovascular Health Study (CHS), an elderly population without a history of vascular disease at baseline *(52)*. In this study of 3971 men and women aged 65 years or older, elevated CRP levels were associated with an increased 10-year risk of CHD above and beyond traditional risk factors, especially in moderate-risk to high-risk men and in high-risk women.

We also have been able to demonstrate the potential of CRP to modify risk prediction based on the Framingham risk score (FRS) *(53)*. We compared the proportions of incident coronary events within 10 years estimated by the Cox model for the five categories of the FRS alone (Fig. 2, left panel) and for different CRP categories in each category of FRS, adjusted for survey and components of the FRS (Fig. 2, right panel). Probability values of the stratified analyses are given in Fig. 1 (right panel, above each FRS category). Cox regression revealed a considerable modification in coronary event incidence based on CRP concentrations and, more importantly, in categories of FRS associated with a 10 to 20% risk per 10 years, elevated concentrations of CRP were consistently and statistically significantly associated with a further increased risk ($P = 0.03$ and $P = 0.02$, respectively). In contrast, in men with a risk ≤6 to 10% and >20% over 10 years, CRP had no statistically significant additional effect on the prediction of a first coronary event. In ROC analyses, regarding the different AUCs, a remarkable increase was found for the intermediate FRS categories of 11–14% and 15–19% (increase in the AUC from 0.725 to 0.776 and from 0.695 to 0.751, respec-

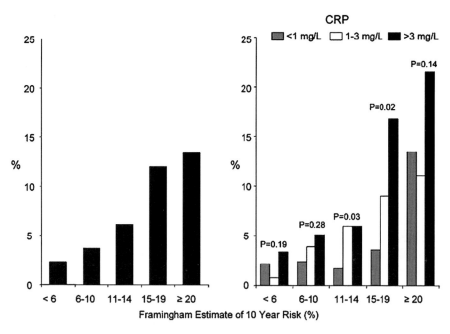

Fig. 2. Occurrence of a first coronary event within 10 years, estimated by Cox proportional hazards models in percentages. Left, Percentage estimated by a model with FRS (5 categories) adjusted for survey. Right, Percentage estimated for each of 5 FRS categories by a model with CRP (3 categories) adjusted for FRS (continuous) and survey. Probability values indicate significance status of CRP in the Cox model.

Source: Koenig *(53)*.

tively). Thus, there is initial evidence that CRP measurements may aid in improving risk prediction in a population for which there is otherwise considerable uncertainty about future prognosis. However, such data have to be confirmed in other populations before general recommendations for the inclusion of CRP in the risk profile can be made.

Whether CRP should be measured to assign statin treatment is a matter of ongoing investigations. Indeed, one such trial, currently in progress in the United States and in Europe, the JUPITER (Justification for the Use of statins in Primary prevention: an Intervention Trial Evaluating Rosuvastatin) study *(54)* is a multicenter, randomized, placebo-controlled trial of rosuvastatin (20 mg/day) among 15,000 individuals with low levels of LDL-C (<130 mg/dL [3.36 mmol/L]) and elevated levels of CRP (≥ 2 mg/L). The study tests whether long-term treatment with rosuvastatin will decrease the rate of major CV events (combined endpoint of CV death, stroke, MI, unstable angina, or arterial revascularization) in this patient population. This trial is expected to provide valuable information concerning the benefits of statin therapy in individuals without overt hyperlipidemia, but at high vascular risk on the basis of an enhanced inflammatory response.

CYTOKINES

Interleukin-6

Weissenbach et al. *(55)* first identified IL-6 in 1980 during the cloning of human fibroblast interferon-β (IFN-β). IL-6 is a 26-kDa single-chain glycoprotein produced by many cell types including activated monocytes/macrophages and ECs *(56,57)* usually as a result of IL-1β, TNF-α, transforming growth factor-β (TGF-β), or LPS stimulation. Recently, much attention has been given to the striking observation that approximately 30% of the IL-6 is produced by adipose tissue *(58)*. IL-6 is involved in diverse biological processes such as final maturation of B cells into plasma cells, T-cell activation, stimulation of growth and differentiation of hematopoietic precursor cells, and proliferation of synovial fibroblasts. However, the most important function of this cytokine is an amplification of the inflammatory cascade. During an acute-phase response, IL-6, acting on hepatocytes, promotes the synthesis of several acute-phase reactants, such as CRP, SAA, and fibrinogen *(59,60)*. Moreover, IL-6 might exert a direct proatherogenic role inside the arterial wall. Possible proatherogenic and prothrombotic mechanisms of IL-6 are illustrated in Fig. 3. Indeed, large amounts of IL-6 have been found in human atherosclerotic plaque *(61,62)*. In addition, a study by Schieffer et al. *(63)* detected IL-6 within the shoulder region of stable and unstable plaques, where it colocalized with the AT-1 receptor. Further investigations showed that IL-6 induced the upregulation of the AT-1 receptor, leading to increased angiotensin II-mediated vasoconstriction, enhanced free oxygen radical production, and the development of endothelial dysfunction *(64)*. IL-6 also stimulates macrophages to secrete MCP-1 *(65)* and participates in the proliferation of SMCs. In addition, ECs, stimulated by IL-6, express ICAM-1 *(66)*. In a murine model of atherosclerosis, injection of excessive amounts of recombinant IL-6 resulted in enhanced fatty lesion development *(67)*. IL-6 is the principal procoagulant cytokine. It has been shown that IL-6 increases mRNA levels of TF and factor VIII. In addition, IL-6 can activate platelets and increase plasma concentrations of fibrinogen and PAI-1. In an animal

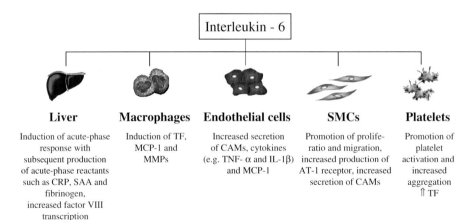

Fig. 3. Potential direct effects of IL-6 on atherothrombosis. CRP = c-reactive protein; SAA = serum amyloid A; TF = tissue factor; MCP-1 = monocyte chemoattractant protein-1; MMPs = matrix metalloproteinases; CAMs = cell adhesion molecules; TNF-α = tumor necrosis factor-α; IL = interleukin; AT-1 = angiotensin II type 1.

model, IL-6 was responsible for the increase in von Willebrand factor (vWF) and the decrease in protein S *(68)*.

During the past few years, a series of clinical and epidemiological studies investigated a predictive value of baseline concentration of IL-6 on future CV events. In patients with unstable angina, elevated levels of IL-6 at 48 h after admission were associated with increased in-hospital morbidity and mortality *(69)*. Furthermore, IL-6 has been measured among 3269 patients of the large prospective, randomized Fragmin and Fast Revascularisation During Instability in Coronary Artery Disease II (FRISC II) trial, which compared invasive versus conservative management in ACS patients. It was found that elevated IL-6 levels (≥ 5 ng/L) in patients with unstable angina were associated with a twofold increased mortality in both the conservative arm of the trial at 6 months and the interventional arm at 12 months, independently of other risk markers, including CRP and Troponin T (TnT) *(70)*. However, no association was found between elevated IL-6 and the composite endpoint of both death and MI at 6 and 12 months. More importantly, this study demonstrated that patients with high IL-6 level might benefit most from an early invasive strategy, which led to a 5.1% absolute or 65% relative reduction in mortality at 1 year in this study group. By contrast, among those without increased concentration of IL-6, there was no significant difference in 12-month mortality between the invasive and conservative strategies *(70)*.

In addition to its predictive value in the ACS, IL-6 might provide even more important information in terms of future prognosis in primary prevention. To date, there are numerous prospective studies, which have consistently shown that baseline levels of IL-6 are a potent predictor of future CV risk in apparently healthy asymptomatic subjects from the general population. For instance, middle-aged men with IL-6 in the top quintile have shown a twofold increased risk for CV events compared with those in the bottom quintile, even after controlling for established risk factors and several inflammatory biomarkers, including CRP *(71)*. This study has also demonstrated that circulating IL-6 values were correlated with CRP ($r = 0.43$, $P < 0.001$), thereby reflecting the fact that IL-6 represents a main stimulus of CRP production by hepatocytes. Within the Women's Health Study (WHS), a large cohort of middle-aged normocholesterolemic women, a baseline IL-6 concentration of 2.86 pg/mL (which corresponds to the median value of the top tertile) was associated with a twofold increased CV risk in a crude model *(72)*. However, IL-6 was not an independent predictor after adjustment for other CVD risk factors in this study. In a further study, median baseline levels of IL-6 were significantly higher among 304 women, who subsequently developed CHD events compared with 304 event-free women, matched by age, smoking status, and ethnicity *(73)*, and elevated levels of IL-6 were significantly associated with future risk of CHD. Finally, increased levels of IL-6 strongly predicted all-cause mortality in older participants of the Iowa 65+ and the Rural Health and Women's Health and Aging Studies *(74,75)*.

Thus, IL-6 might play an important direct pathogenic role in atherogenesis in addition to its role in the amplification of the inflammatory cascade by initiating an acute-phase response. Moreover, circulating or locally produced IL-6 may favor the onset of a prothrombotic state, which could increase the risk of atherosclerotic complications, especially during later stages of atheroma development.

Interleukin-18

IL-18, a pleiotropic proinflammatory cytokine, is widely expressed in cells of hemopoietic and nonhemopoietic lineages, namely, monocytes/macrophages, dendritic cells, Kupffer cells, adipocytes, colon carcinoma cells, keratinocytes, and osteoblasts *(76,77)*. It is synthesized as a 23-kDa biologically inert precursor, which is further cleaved by caspase 1 (or IL-1β-converting enzyme) to yield the mature and active 18.3-kDa glycoprotein *(78)*. IL-18 has been initially discovered as the IFN-γ-inducing factor *(79)*, and this well recognized property of IL-18 is at least in part the result of its synergistic action with IL-12 *(80)*. Beyond induction of IFN-γ with subsequent promotion of Th1 immune response, IL-18 also enhances the expression of matrix metalloproteases (MMPs) *(81–83)*, and these two abilities of IL-18 characterize it as a crucial and potent mediator of atherosclerotic plaque destabilization and vulnerability. Indeed, some experimental studies demonstrated significantly increased expression of IL-18 in human atherosclerotic plaque, especially in lesions prone to rupture, where it is localized mainly in plaque macrophages *(81,84)*. In addition, animal models further support the proatherogenic role of IL-18. It has been shown that the endogenous inhibition of IL-18 by IL-18-binding protein reduces atherosclerotic plaque development and progression in apoE-deficient mice *(85)*. Similarly, IL-18/apo-E double-knockout mice exhibited reduced lesion size *(86)*. In contrast, direct administration of IL-18 enhanced atherogenesis in an IFN-γ-dependent manner *(87)*, even in the absence of T cells *(88)* and induced/promoted a switch to a vulnerable plaque phenotype by decreasing intimal collagen content and cap-to-core ratio *(89)*. Thus, experimental studies on the role of IL-18 in atherogenesis seem to be relatively consistent and promising.

By contrast, the clinical evidence is somehow controversial. Although a number of recent cross-sectional studies favored a role of IL-18 in the development of plaque instability, showing significantly increased levels of this cytokine in patients with ACS, several other investigations were not able to demonstrate any meaningful relationship with CHD *(90–95)*. Furthermore, results from one large prospective study conducted in 1229 patients with angiographically confirmed CHD showed that increased concentrations of IL-18 at baseline were associated with future CV death during a 3.9-year FU, independent of a variety of potential confounders *(96)*. Such association held further true if the prognostic value of increased IL-18 on subsequent CV events was assessed separately in patients with stable and unstable angina on admission. However, when the FU was extended from 3.9 to 5.9 years, IL-18 concentrations were no longer predictive of outcome, thereby questioning its value as an independent marker for future CV risk *(97)*.

To date, only one study assessed the prognostic value of elevated IL-18 for future coronary events in apparently healthy subjects *(98)*. In a nested case–control study within the PRIME (Prospective Epidemiological Study of Myocardial Infarction) cohort, 335 subjects who experienced at least one CHD event (nonfatal MI, coronary death, and AP) during a 5-year FU were compared with 670 age-matched controls. Subjects had been recruited from two European populations (French and Irish populations). Elevated IL-18 concentrations at baseline were associated with a twofold increased risk for subsequent coronary events after multivariable adjustment, if the top tertile of the IL-18 distribution was compared with the bottom tertile. However, such an association was only seen when data from both populations were pooled for analysis. In separate analysis of the French and the Belfast cohort, the statistical significance of the association was completely lost in France, whereas in the Irish populations, it even increased. We tried to replicate these findings and conducted a large case–cohort study in initially healthy, middle-aged men and women from the MONICA/KORA Augsburg populations with a mean FU of 11 years *(99)*. Concentrations of IL-18 were measured in 382 case subjects with incident CHD and in 1980 noncase subjects. In crude and in age-adjusted and survey-adjusted analyses, there was a statistically significant association between increased concentrations of IL-18 and incident CHD in men, whereas no significant association was seen in women. However, in multivariable adjustment for CV risk factors including the total cholesterol/HDL ratio, CRP, and IL-6, this association was attenuated and no longer statistically significant in men and remained nonsignificant in women—Hazard ratio (HR) (95% CI) 1.20 (0.85–1.69) and 1.25 (0.67–2.34), respectively. Thus, elevated concentrations of IL-18 were not statistically significantly associated with the risk of CHD in men and women from an area with moderate absolute risk of CHD. This large population-based case–cohort study therefore suggests that IL-18 may only serve as a marker of future CV events in men with manifest CHD and/or in areas of high absolute risk of CHD.

Tumor Necrosis Factor-α

TNF-α, a pleiotropic cytokine, which is produced by several cells involved in the atherosclerotic process including macrophages, ECs, and SMCs, was originally identified in 1975 as a factor that caused hemorrhagic necrosis in transplanted tumors *(100)*. This cytokine carries potential deleterious effects, such as cell death or immune-modulatory activities when expressed in large amounts. TNF-α is initially secreted as a 26-kDa precursor, anchored to the cell membrane, which, under cleavage with TNF-α-converting enzyme (TACE), transforms to a biologically active mature 17-kDa soluble trimeric protein *(101)*. TNF-α has been identified as an important mediator of vascular dysfunction through its ability to decrease the expression of eNOS *(102)* and, concomitantly, to induce NAD(P)H oxidase activity *(103)*. TNF-α also stimulates ECs and macrophages to synthesize adhesion molecules such as E-selectin and ICAM-1 *(104)*. Moreover, TNF-α has been shown *in vitro* to increase monolayer permeability by direct damaging of ECs *(105,106)*. It triggers the production of other cytokines, for example, IL-6 production by SMCs, and was found to be implicated in atherogenesis by increasing uptake of ox-LDL and by enhancing the expression of its receptor LOX-1 on the surface of ECs *(107)*. This cytokine could also be linked to plaque instability because of its ability to facilitate degradation of collagenous matrix by inducing expression of MMPs *(108–111)*. TNF-α also blocks the production of tissue inhibitors of MMPs (TIMMPs) by fibroblasts *(112)*. In addition, TNF-α promotes macrophage-induced SMC apoptosis that could also contribute to plaque rupture *(113)*. Animal studies also support the proatherogenic properties of TNF-α by demonstrating that the disruption of TNF-α gene reduces the development of atherosclerosis in apoE-deficient mice *(114,115)*. In the clinical setting, although data are somewhat limited, elevated levels of TNF-α were associated with a 3.5-fold increased risk for coronary death in subjects with unstable angina who were followed for 17 months *(116)*. Plasma levels of TNF-α were also measured among patients enrolled in the Cholesterol and Recurrent Events (CARE) trial, a cohort of men in the stable phase after MI *(117)*. Overall, TNF-α levels were significantly higher among patients than in controls. Moreover, elevated TNF-α concentrations (>95th percentile of the control population) were associated with a 2.7-fold increase in the risk of a recurrent nonfatal MI or a fatal CV event independent of other risk factors. A similar strong association for TNF-α was found in 2225 elderly subjects using data from the Health, Aging, and Body Composition (Health ABC) study *(118)*. Among participants aged 70–79 years, TNF-α showed a positive association with CHD risk (HR 1.22 [95% CI 1.04–1.43], per 1 SD increase in TNF-α concentration) during an average FU of 3.6 years. Although the aforementioned studies provide insight into the possible contribution of this proinflammatory cytokine to the development and progression of atherosclerosis, its clinical utility is not clear and further research is warranted.

MARKERS OF HEMOSTASIS THAT ARE ALSO ACUTE-PHASE REACTANTS

Fibrinogen

Fibrinogen, a soluble glycoprotein, which is involved in blood coagulation as the precursor of fibrin, is synthesized in the liver with a molecular mass of 340 kDa. Fibrinogen is a large hexamer, consisting of three paired nonidentical polypeptide chains (Aα, Bβ, γ) linked by disulfide bonds, which are encoded by separate genes *(119)*. Normal concentrations of fibrinogen in the circulation range from 2.0 to 4.5 g/L. Fibrinogen is both a coagulation factor and an acute-phase reactant and therefore is consistently increased during inflammation or tissue necrosis. The primary function of fibrinogen consists in the formation of a fibrin clot through enabling platelet–platelet interactions. Indeed, elevated fibrinogen levels lead to larger thrombi and formation of tight and rigid network structures *(120)*, decrease the deformability of the clot, and render it less amenable to endogenous fibrinolysis *(121)*. Recent data also show that high fibrinogen levels interfere with the binding of plasminogen to its receptor, thus leading to impaired fibrinolysis *(122)*. Apart from its pivotal role in the coagulation cascade as the substrate for thrombin, there is evidence of other mechanisms suggesting that fibrinogen may be involved in both the early and the later stages of the atherothrombotic process (Fig. 4). It has been shown that fibrinogen binding to ICAM-1 causes the release of vasoactive mediators *(123)*. Fibrinogen and fibrinogen degradation products modulate EC permeability, thus enhancing their deposition in the subendothelial space, and further promote EC migration. Fibrinogen has been shown to promote SMC chemotaxis and proliferation and induce monocyte chemotaxis. It provides an adsorptive surface for the extracellular accumulation of LDL *(124)* and

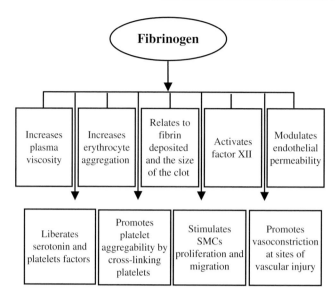

Fig. 4. Fibrinogen and atherothrombosis: possible mechanisms. SMC = smooth muscle cells.

further facilitates cholesterol transfer from platelet to monocytes/macrophages and may therefore play a role in foam cell formation *(125)*. Through all these effects, fibrinogen may be involved in the early stages of plaque formation. Fibrinogen binds to glycoprotein (GP) IIb/IIIa receptors on the platelet membrane and promotes aggregation and formation of platelet-rich thrombi. Elevated plasma fibrinogen levels increase the velocity of platelet aggregation and also increase platelet reactivity *(126)*. Fibrinogen is a major determinant of plasma viscosity (explaining about 50% of its variability), whole blood viscosity, and red blood cell aggregation. Elevated blood and plasma viscosity may lead to impaired microcirculatory flow, endothelial shear stress damage, and predisposition to thrombosis *(127)*.

The fact that fibrinogen is an acute-phase reactant also deserves consideration. Elevated levels of fibrinogen might be an indicator of an underlying low-grade inflammation, the cause of which remains unclear but may be initiated by various stimuli such as oxidized LDL (oxLDL), several cytokines, oxygen free radicals, other factors, and possibly but not very likely by chronic infections. The "inflammation hypothesis" is further supported by the fact that a variety of other systemic markers of inflammation are also related to CHD.

The evidence from prospective studies demonstrating an association between fibrinogen levels at baseline and future CHD (acute nonfatal and fatal MI and coronary death) is unequivocal. The published data are remarkably strong and consistent, despite the diversity of populations studied, the variable length of FU, different definitions of endpoints, and the various analytical methods applied in the absence of an International Standard at the time most of these studies were done. These studies are not discussed here in more details, because they were a subject of recent review *(122)*. In a recent publication from the Fibrinogen Studies Collaboration (FSC) *(128)*, a comprehensive meta-analysis of individual data on 154,211 subjects without known CVD at baseline that come from 31 prospective studies with information on plasma fibrinogen and major disease outcomes has shown a significant and independent association of elevated fibrinogen levels and CV morbidity and mortality. For instance, increase in fibrinogen level by 1 g/L was associated with approximately a twofold increase in risk for CHD (HR 1.93 [95% CI 1.79–2.08], after multivariate adjustment for smoking, total cholesterol, blood pressure, and body mass index). Thus, fibrinogen undoubtedly represents a strong and independent risk factor for future CV events. Despite this well-established association, the proof of a causal involvement of fibrinogen in atherogenesis is still pending. In addition, the lack of adequate standardization of fibrinogen assays limits its clinical utility.

Plasminogen Activator Inhibitor-1

PAIs belong to the superfamily of serpins (i.e., α1-protease inhibitor class of <u>ser</u>ine <u>p</u>rotease <u>in</u>hibitors) *(129)*, and four different proteins have been identified as PAIs. However, the most extensively investigated in relation to atherosclerosis is PAI-1.

PAI-1 is a linear glycoprotein with a molecular weight of approximately 50 kDa. PAI-1, a very unstable molecule with an *in vivo* half-life of approximately 6 min, is synthesized in an active and a latent form *(130)*. PAI-1 is an acute-phase reactant, and various cytokines, such as TNF-α *(131)* and IL-1β *(132)*, are able to induce PAI-1 production. PAI-1 is derived from several sites, including the liver, ECs, and adipose tissue; however, more than 90% of the total circulating PAI-1 is secreted by platelets, where it is stored in α-granules in its inactive form *(133)*. Binding to vitronectin, an abundant cell adhesion protein provides stabilization of PAI-1 that avoids its inactivation and results in nearly doubling of its half-life. Additionally, levels of PAI-1 show a strong circadian variation, with concentration peaks in the morning that correspond to a nadir in fibrinolytic activity and a subsequent fall during the afternoon that occurs in parallel to the activation of endogenous fibrinolysis *(5)*.

PAI-1 represents a pivotal inhibitor of fibrinolysis, the physiological process that degrades blood clots (Fig. 5). When fibrinolysis is initiated, inactive precursor plasminogen is converted by plasminogen activators into the active enzyme plasmin. Plasmin, in turn, degrades the cross-linked insoluble fibrin, resulting in the formation of D-dimers and other fibrin degradation products. PAI-1 is an important component of the plasminogen/plasmin system because it is the main inhibitor of tissue-type and urokinase-type plasminogen activator by forming of a 1:1 stoichiometric complex *(134)*. Thus, under physiological conditions, where a local balance between fibrinolytic proteins and their inhibitors exists, PAI-1 favors the stabilization of fibrin by preventing its degradation, thereby playing a defensive role and protecting the organism against increased risk of bleeding. However, even a light imbalance in this system with excess in PAI-1 levels could cause impaired fibrinolytic function, thus contributing to a prothrombotic state and to atherogenesis. Indeed, large amounts of PAI-1 were found in advanced human atheromatous plaques *(135)*. Genetic deficiency for PAI-1 prolongs the time to occlusive thrombosis following photochemical injury of the carotid atherosclerotic plaque in apoE-deficient (apoE$^{-/-}$) mice *(136)*. Conversely, transgenic mice that overexpress PAI-1 demonstrated development of coronary arterial thrombosis and subendothelial infarction *(137)*. Another study, however, reported a protective role for PAI-1, showing that PAI-1$^{-/-}$/apoE$^{-/-}$ mice had increased atherosclerosis *(138)*.

In the ACS, PAI-1 is released during the first hours of ST-elevation MI and predicted 30-day mortality independently *(139)*. Hamsten et al. were the first to report higher PAI-1 values in young post-MI patients compared with matched controls. The same group also demonstrated an increased risk for re-infarctions associated with higher

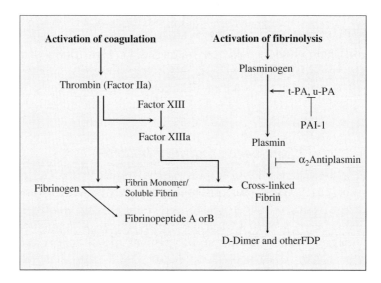

Fig. 5. Schematic overview of coagulation and fibrinolysis cascade. The coagulation cascade, which favours clot formation, is initiated in vivo by tissue factor and factor VIIa (FVIIa) and leads to the conversion of prothrombin to thrombin by the prothrombinase complex (FXa and FVa) (not shown). Exposure of fibrinogen to thrombin (factor IIa) results in rapid proteolysis of fibrinogen and the release of fibrinopeptide A and soluble fibrin. Thrombin also mediates conversion of Factor XIII to Factor XIIIa (also termed transglutaminase), which cross-links fibrin polymers solidifying the clot. Plasmin-mediated fibrinolysis, resulting in fibrin degradation products and clot lysis, occurs following the conversion of the proenzyme, plasminogen to plasmin by tissue-type plasminogen activator (tPA). Inhibition may occur at the level of the plasminogen activators by PAI-1 or at the level of plasmin by forming a 1:1 inhibitory complex with α2-antiplasmin.

levels of PAI-1 activity in survivors of an acute MI during a 3-year FU *(140)*. However, Juhan-Vague et al. *(141)* found that the association between increased PAI-1 levels and subsequent incidence of MI or sudden coronary death (SCD) in stable angina patients of the ECAT study (European Concerted Action on Thrombosis and Disabilities) disappeared after adjustment for components of the insulin resistance syndrome. Moreover, data in initially healthy subjects also do not support an independent relationship between PAI-1 and atherosclerotic complications. Increased PAI-1 concentrations were associated with the development of a first AMI in univariate but not in multivariate analyses when adjusted for traditional risk factors in a small nested case–control study during a mean FU of 15 months *(142)*. The Atherosclerosis Risk in Communities (ARIC) study demonstrated an association between PAI-1 levels and increased incidence of CHD after adjustment for age, race, and gender, but this association was no longer significant after adjustment for other risk factors (ARIC) *(143)*. In a nested case–control study, conducted within CHS, a longitudinal population-based study of CHD in 5000 randomly selected men and women aged 65 years and older *(144)*, PAI-1 did not predict future MI. Consequently, a meta-analysis of five prospective studies with a total number of 833 CHD cases and 3122 controls and a mean weighted FU of 5 years revealed an OR of 0.98 (95% CI 0.53–1.81) across extreme tertiles of baseline PAI-1 concentrations *(145)*. By contrast, in a 13-year FU, increased PAI-1 levels were significantly associated with risk of CVD endpoints in the Caerphilly cohort *(146)*.

D-Dimer

Fibrin D-Dimer is a major fibrin-specific degradation product that detects cross-linked fibrin resulting from endogenous fibrinolysis. Because D-Dimer is a final product of fibrin degradation, it consists of the fragments of two fibrinogen γ-chains connected by two isopeptide bonds, two β-chain fragments, and two α-chain fragments held together by disulfide bonds. The molecular weight of this protein is approximately 180 kDa. D-Dimer can be considered as a global marker for direct measurement of ongoing turnover of cross-linked fibrin and for an activation of the hemostatic system without reflecting the changes in fibrinogen and fibrin status (Fig. 4). In addition, D-Dimer assays are more stable and more practical to detect even small amounts of intravascular clot formation and therefore may be more suitable for routine clinical and epidemiological purposes. Indeed, because of the introduction of ELISAs, D-Dimer is well established in clinical practice as a screening marker of activated coagulation. D-Dimer plays an important role in the detection of hypercoagulable states such as acute symptomatic deep vein thrombosis, pulmonary embolism, or disseminated intravascular coagulation, having a high negative predictive value (147–149).

Taking into account that thrombogenesis is one of the fundamental pathological processes underlying the major complications of atherosclerotic disease, measurement of D-Dimer might be useful to identify patients at high risk for future CV events. Therefore, measuring circulating levels of D-Dimer has gained widespread attention in clinical research as a possible approach to test for increased coagulation in subjects at risk for atherosclerosis (150).

Among 296 initially healthy subjects who developed a first MI over a mean FU period of 60.2 months and 296 event-free participants of the Physicians' Health Study (151), a strong association between future MI and increased D-Dimer concentration was found in multivariate analysis, after controlling for nonlipid CV risk factors and Lp(a). However, if total and HDL-C as well as endogenous t-PA and its primary inhibitor PAI-1 were included in the analysis, this association was attenuated and no longer statistically significant. Future ischemic events were predicted by D-Dimer levels in a study by Lowe et al., but not by other markers of procoagulant activity, such as prothrombin fragment F1 + 2 and thrombin–antithrombin complex (measures of thrombin generation), or by factor VII coagulant activity (152). D-Dimer also predicted future CHD events in apparently healthy middle-aged participants of a large community-based prospective study (153). Comparison of men in the top third with those in the bottom third of baseline fibrin D-Dimer values resulted in an OR for CHD of 1.79 (95% CI 1.36–2.36) after adjustments for smoking, other classic risk factors, and indicators of socioeconomic status. Subsequent meta-analysis, including this and four other population-based cohorts and two studies in patients with stable CVD (CHD or intermittent claudication), yielded almost the same results. A total of 1535 incident CHD events were included in this meta-analysis, and an OR of 1.7 (95% CI 1.3–2.2) was found for subjects with D-Dimer levels in the top tertile versus those in the bottom tertile (153). Further studies were in support of such an association. In the ARIC study, several hemostatic parameters were measured in 326 incident CHD cases, during a mean FU period of 4.3 years and in a randomly stratified, referent cohort of 720 individuals. Among various markers of fibrinolysis and coagulation, such as t-PA, PAI-1, plasminogen, and prothrombin fragment F1 + 2, D-Dimers of subjects in the top quintile compared with those of subjects in the bottom quintile showed the most powerful predictive value for incident CHD with an RR of 4.21

(95% CI 1.9–9.6) *(143)*. Finally, in two more recently published reports, the West of Scotland Coronary Prevention Study (WOSCOPS), a large primary prevention trial of hypercholesterolemic men *(154)*, and the Women's Health Initiative (WHI) Observational Study *(155)*, conducted in apparently healthy postmenopausal women, the strength of the association between elevated levels of D-Dimer and CV events was similar to those previously found in meta-analysis, revealing multivariate-adjusted risk estimates of 1.86 (95% CI 1.24–2.8) and 1.7 (95% CI 1.0–2.9), respectively.

von Willebrand Factor

vWF is a multimeric plasma protein, which plays a key role in primary hemostasis and therefore might be implicated in hemorrhagic as well as in thromboembolic disorders. It is a glycoprotein ranging in size from 600 to 20,000 kDa, which is synthesized by ECs and megakaryocytes *(156)*. ECs secrete vWF in a bipolar manner *(157)*. The luminal secretion, by which endothelial vWF reaches circulation and the abluminal secretion, results in the deposition of this protein in the vascular subendothelial space where it acts as an extracellular matrix protein to bridge circulating platelets. In general, there are two mechanisms of vWF release, namely, its permanent continuous and relatively slow secretion, directly linked to synthesis, and the so-called regulated secretion of vWF, when vWF is released in response to the stimulation by one or more mediators from their specific storage organelles *(158,159)*. Secretion of vWF from its storage granules such as Weibel–Palade bodies in ECs and α-granules of megakaryocytes is triggered by several substances, including histamine, leukotriene D4, platelet-activating factor, vascular permeability factor, the terminal component of complement, adrenaline, fluid mechanical forces, factor VIIa, thrombin, and fibrin *(156)*.

When released, vWF has two major primary roles in hemostasis: mediating platelet adhesion and aggregation, and formation of a noncovalent complex with coagulation factor VIII, thereby stabilizing its procoagulant activity and protecting it from inactivation by activated protein C and factor Xa.

Under high shear stress conditions, vWF in subendothelial matrix undergoes spontaneous and irreversible conformational changes, whereas globular vWF molecules become uncoiled *(160)*. This facilitates the interaction of vWF with GP Ib platelet-surface receptors which further results in platelet tethering to the endothelium. Because vWF due to its multimeric structure has different interaction sites with increased binding potential, simultaneously formation of multiple bonds via separate domains can occur. In addition to interaction with GP Ib, vWF also binds to the collagen receptor GP VI that promotes platelet adhesion and activation. Finally, vWF can mediate platelet–platelet binding, and the epitope I on the surface of vWF, being a GP IIb/IIIa binding site, is responsible for this interaction. This process finally leads to spreading of platelets and their irreversible adhesion and aggregation *(156,160,161)*.

Thus, vWF plays a pivotal role in primary hemostasis, mediating platelet adhesion to damaged arterial surface with subsequent blood clotting and platelet thrombus formation. Because ECs represent the main source of plasma vWF, increased circulating vWF can be regarded as an indicator of current endothelial damage that occurs early in the disease process. However, the published clinical and epidemiological data to date are less convincing.

Several prospective studies have linked higher values of vWF with recurrent coronary events. In the ECAT study, 3043 patients with AP underwent coronary angiography and were followed for 2 years. The multivariable standardized RR for vWF was 1.24 (95% CI 1.00–1.53), which was borderline significant *(162)*. In the Stockholm Heart Epidemiology Program (SHEEP) study, representing a large population-based case–control study, elevated vWF was the strongest predictor of re-infarction in multivariate analysis *(163)*. The relationship of factor VIII activity and vWF with CHD were investigated in 1393 initially healthy men aged 40–64 years at entry to the Northwick Park Heart Study *(164)*. During a 16-year FU, both variables were significantly associated with fatal coronary events: RR for factor VIIIc was 1.28 (95% CI 1.04–1.58) and RR for vWF was 1.34 (95% CI 1.00–1.79) after adjusting for blood group. By contrast, in ARIC *(143)*, as well as in another relatively small study *(142)*, the association was only seen in univariate analysis. Whincup et al. *(165)*, in a nested case–control manner, investigated the prognostic value of vWF in 625 middle-aged men participating in the British Regional Heart Study, who developed a major coronary event during an FU of 16 years, and who were compared with 1266 controls free of incident CHD, frequency-matched for age and town of residence *(165)*. The risk of CHD in men being in the top third of the vWF distribution was increased by 83% compared with that in men in the bottom tertile and remained essentially unchanged after adjustment for risk factors. Subsequent meta-analysis of population-based prospective studies yielded a statistically significant OR of 1.5 (95% CI 1.1–2.0) *(165)*. Moreover, PRIME investigators, using a nested case–control design, also found that vWF independently predicted future CV events among 296 participants who subsequently developed CHD over a 5-year FU compared with 563 control subjects. In a multivariate model, the RR of hard CHD endpoints was 3.40 (95% CI 1.65–7.02) for comparison across extreme quartiles, after adjustment for inflammatory and hemostatic markers such as fibrinogen, CRP, TNF-α, and IL-6 *(166)*. However, when vWF was determined among 2083 patients without evidence of CHD at baseline and 3969 control subjects, who participated in the longitudinal Reykjavik study *(2)*, the association between vWF antigen and CHD risk was no longer statistically significant in multivariate analysis and the comparisons between top and bottom tertiles of the vWF distribution gave a multivariate adjusted OR for CHD of 1.11 (95% CI 0.97–1.27). In line with these findings are the negative results from a more recently published update of the Caerphilly Prospective Study, which represents a cohort of 2208 men who were followed for 13 years *(146)*. Nonetheless, updated meta-analysis *(2)* comprising 3969 subjects yielded an OR of 1.23 (95% CI 1.14–1.33), comparing extreme tertiles, which is however weaker than the previously estimated RR of about 1.5 *(165)*.

Therefore, despite "promising" pathophysiological role of vWF in the atherothrombosis, further studies are needed to evaluate whether increased levels of vWF are a cause of CV disease or merely represent a marker of endothelial damage. Based on the data published to date, there is no role for vWF as a biomarker to predict CHD in clinical practice.

WBC COUNT

An increased WBC count has been suggested as a biomarker of future CHD risk, with the first study being published as early as in 1974 *(167)*.

Because various leukocyte subtypes such as monocytes, lymphocytes, or neutrophils have been found in all stages of atherosclerotic lesion development, WBCs could play a role in atherogenesis through several pathological inflammatory pathways. These possible mechanisms could include increased monocyte adhesiveness with their subsequent infiltration of arterial wall, leukocyte-dependent oxidation of LDL, or direct damaging effect on endothelium by releasing several potent proinflammatory compounds such as cytokines, growth factors, or reactive oxygen species. Vessel plugging caused by abnormal leukocyte aggregation might also facilitate atherothrombosis, in addition to direct prothrombotic effects of WBCs such as induction of TF expression from monocytes and neutrophils *(168,169)*. Histopathological support for the presence of neutrophils in atherosclerotic plaques, as recently reported by Naruko et al. *(170)*, strongly suggests that neutrophils play a role in the destabilization of plaque leading to ACS. Indeed, these cells have been found more frequently and in higher concentrations in the culprit lesion of patients with unstable angina and acute MI than in patients with stable CHD. The release of large amounts of degradative proteases such as elastase *(170)*, as well as of myeloperoxidase (MPO) by neutrophils and macrophages, might represent one particular mechanism implicated in the pathophysiology of WBCs triggering acute CHD events. Another mechanism might be the generation of leukotriene B4 (LTB4) by stimulated neutrophils. Among several LTB4 properties, chemotaxis and adhesion of mononuclear cells to the vascular endothelium are most relevant in this context *(171)*. In addition, together with platelets, aggregated neutrophils may contribute to the plugging of microvessels and thus promote myocardial ischemia/infarction.

Clinically, patients with elevated WBC counts have been shown to be at higher risk of developing acute events in several vascular beds. Indeed, the WBC count was considered an independent predictor of in-hospital acute MI mortality. In the TACTICS–TIMI-18 trial (Treat Angina with Aggrastat Plus Determine Cost of Therapy with an Invasive or Conservative Strategy–Thrombolysis in Myocardial Infarction-18) in patients with unstable angina or non–ST-segment elevation MI *(172)*, leukocytosis at baseline was associated with poorer reperfusion, more extensive CHD, and increased mortality at 6 months. WBC increases with time from symptom onset to presentation with acute MI, suggesting that it is influenced by the inflammatory response to necrosis *(173)*. In meta-analysis, summarizing data from 19 prospective studies, conducted between 1974 and 1996, and involving a total of 7229 cases, a clear and positive association between WBC count and risk of CHD has been shown when individuals in the top tertile were compared with those in the bottom tertile (summary RR of 1.5 [95% CI 1.4–1.6]) even though the majority of these studies made adjustments for smoking and other known risk factors *(174)*.

More recently, Margolis et al. *(175)* from the WHI Research Group reported a higher WBC count to be associated with an increased risk of CV events and mortality among 66,261 postmenopausal women aged 50–79 years, enrolled at 40 centers, who were free of CHD and cancer at baseline. Compared with women with WBC counts in the bottom quartile ($2.5–4.7 \times 10^9$/L), women in the top quartile ($6.7–15.0 \times 10^9$/L) had a more than twofold increased risk for CHD death (HR 2.36 [95% CI 1.51–3.68]), after multivariable adjustment for traditional CV risk factors and aspirin and hormone use. Women in the top quartile of the WBC count also revealed a 40% increase in the risk for nonfatal MI, a 46% increase in the risk for stroke, and a 50% increase in risk for total mortality, independent of other risk factors. In addition, the potential additive value of WBC count to CRP in predicting risk was evaluated in a subgroup of the WHI cohort, showing that women with both higher CRP levels and WBC count were nearly seven times more likely to have an incident CHD event than women being in the lower quartile for both factors. Results from the Bezafibrate Infarction Prevention

Study, a secondary prevention study, revealed that for every 1000/μL increase in WBC count, the risk of total mortality increased by 6% (RR 1.06 [95% CI 1.03–1.10]), whereas the association between increased WBC count and future coronary events was borderline significant *(176)*. The Long-Term Intervention with Pravastatin in Ischemic Disease (LIPID) study, a randomized clinical trial comparing pravastatin (40 mg/day) to placebo in 9014 stable patients with previous unstable angina or MI *(177)*, found that a 1×10^9/L increase in WBC was associated with an 18% greater CHD mortality over 6 years in patients randomized to placebo (HR 1.18 [95% CI 1.12–1.25]) but not in the pravastatin group (HR 1.02 [95% CI 0.96–1.09]). In addition, reduction in CHD mortality observed with pravastatin treatment was greater in patients with a higher baseline WBC than in those with lower values.

The role of total WBC count and WBC subtypes was also evaluated in a high-risk population among 18,558 participants of the Clopidogrel versus Aspirin in Patients at Risk of Ischemic Events (CAPRIE) trial, who were randomized in a double-blind way to receive either clopidogrel (75 mg once daily) or aspirin (325 mg once daily) *(178)*. Entry criteria were ischemic stroke ($n = 6224$), MI ($n = 5974$), and peripheral arterial disease ($n = 6360$). In multivariate analysis, adjusting for vascular risk factors and disease, hematocrit, and study treatment, the risk of recurrent ischemic events was significantly higher in both top quartiles of the WBC counts than in the bottom quartile (RR 1.42 [95% CI 1.25–1.63] for highest versus lowest quartile). Moreover, in this study, neutrophil count most strongly contributed to the increased risk, whereas the impact of monocytes was smaller. In another large cohort of consecutive patients ($n = 3227$) with or at high risk for CHD, who underwent coronary angiography, elevated WBC count had been significantly associated with elevated CV risk (HR 1.4, $P = 0.01$) *(179)*. In addition, in this study, the neutrophil/lymphocyte ratio was the strongest predictor of death/MI with an HR of 2.2 ($P < 0.001$).

Hence, taking into account that various leukocyte subtypes have been considered to play a differential pathogenic role in atherogenesis, it is not entirely clear which leukocyte compartments might be responsible for this increased CV risk. Therefore, numerous studies raised the question of the prognostic ability of a differential WBC count for improved CVD risk prediction *(179–180)*. A recently published meta-analysis has focused on the potential impact of differential leukocyte counts in CHD risk assessment *(180)*. Overall, 1764 incident cases of CHD from seven long-term prospective studies in 30,374 participants with a mean FU time of approximately 12 years were included in the analysis. It could be shown that among all leukocyte compartments, neutrophil counts were the strongest predictors of incident CHD with a risk ratio of 1.32 (95% CI 1.15–1.51, in combined analysis of the top third versus the bottom third) *(180)*. Thus, high WBC counts, in particular increased neutrophil counts, are associated with increased CHD-related morbidity and mortality in various patient populations and clinical settings.

LIPID-RELATED BIOMARKERS

Myeloperoxidase

MPO, a member of the heme peroxidase superfamily, is a leukocyte-derived enzyme, which is stored within the azurophilic granules of polymorphonuclear neutrophils and monocytes and is secreted upon leukocyte activation and degranulation *(181)*. The molecular mass of this hemoprotein is approximately 120–140 kDa. Initially, it has been assumed that the physiological role of this enzyme could be considered as part of the innate immune system and therefore consist in the host defense against infection, taking into account very potent bactericidal and viricidal properties of MPO. Indeed, generation of free radicals and diffusible oxidants by catalyzation of chloride and hydrogen peroxide (H_2O_2) to hypochlorous acid (HOCl) represents a major source of its antimicrobial activity, presumably by oxidizing key functional components of ingested microorganisms *(182)*. MPO is only one enzyme known to generate HOCl in vivo. However, the fact that under certain circumstances MPO-derived reactive oxidizing and chlorinating species can overwhelm local antioxidant defenses and therefore might lead to oxidative damage of the arterial wall have highlighted their possible proatherogenic role.

There are several pathways through which MPO could exert its deleterious effect. MPO together with other enzymes such as lipoxygenase and secretory phospholipase might initiate lipid oxidation in the subendothelial space of the vessel wall. It has been shown that MPO and its oxidation products have been markedly enriched in human atherosclerotic lesion, compared with normal control vessels, where they colocalize with macrophages *(183–185)*. MPO could be also involved in the development of endothelial dysfunction, because MPO utilizes the atheroprotective endothelial-derived NO as a substrate that resulted in reduction of NO bioavailability and in impaired vasodilatation *(186,187)*. Furthermore, there is now strong evidence that apoA-I, the primary protein constituent of HDL, is a selective target for MPO-catalyzed oxidation in vivo *(188)*. As a consequence of this action, a selective inhibition in ABCA1-dependent cholesterol efflux from macrophages occurs *(189)*, thereby impairing the cardioprotective and anti-inflammatory capacity of this antiatherogenic lipoprotein.

In addition, several functional SNPs within the promoter of the MPO gene, which resulted in twofold reduction in MPO expression, appear to be cardioprotective *(190)*. By contrast, animal models did not confirm an involvement of MPO in atherogenesis*(191)*. Moreover, nearly 50% increase in atherosclerotic lesions in the MPO-deficient mice has been shown as compared with the wild-type mice *(191)*. These discrepancies however could be attributed to the fact that murine leukocytes do not secrete active MPO and only low background levels of the enzyme could be detectable in the atherosclerotic lesion in LDLR-knockout mice *(191)*. However, when another murine model was developed, namely, transgenic $LDLR^{-/-}$ mice, which overexpressed a human MPO in macrophages, as driven by the Visna virus promoter, twofold larger atherosclerotic lesion in the aorta of these animals was found compared with that of control mice, thereby again confirming a proatherogenic role of this enzyme *(192)*.

Thus, MPO possesses a large number of potent activities, which could contribute to the development of atherosclerosis. Nonetheless, a most pivotal characteristic of MPO and its endproduct HOCl remains its ability to activate MMPs *(193)* and deactivate inhibitors of MMPs *(194)* that promote the weakening of the fibrous cap and lead to destabilized atherosclerotic plaque.

In line with these findings are the results of two studies in patients with ACS, which evaluated MPO as a predictor of major adverse cardiac events (MACEs). MPO mass concentrations have been measured in the CAPTURE (c7E3 Anti-Platelet Therapy in Unstable REfractory angina) trial in 1090 patients with ACS. Baseline MPO levels predicted an increased risk for adverse CV events, and this effect was even more pronounced (HR 7.48 [95% CI 1.98–28.29]) in patients without myocardial necrosis (negative for cardiac TnT <0.01 µg/L at baseline) *(195)*. In a cohort of 604 patients with chest pain syndromes, a single measurement of MPO on admission to the emergency department, independently predicted acute MI on presentation (OR 3.9 [95% CI 2.2–6.8] for patients in the top MPO quartile [Q] compared with the bottom Q) and demonstrated a 4.7-fold higher likelihood (also for Q4 versus Q1) to have a MACE at 30 days and at 6 months *(196)*. This relationship remained statistically significant even when TnT levels were negligible, whereas high CRP levels only identified high-risk subjects who were positive for TnT, with no prognostic value for TnT-negative subjects. In one cross-sectional study carried out in 158 patients with angiographically confirmed CHD and 175 controls without significant CHD by coronary angiography, blood and leukocyte MPO activity were measured. Increased levels of MPO were significantly associated with the presence of stable CAD with an OR of 11.9 (95% CI 5.5–25.5), and this association remained significant after adjustment for traditional CV risk factors, FRS, and WBC count *(197)*. Serum MPO levels also independently predicted endothelial dysfunction, as monitored by NO-dependent flow-mediated dilatation of the brachial artery. After multivariate adjustment for components of the FRS, prevalence of CVD, and ongoing treatment with CV active compounds, the OR for the presence of endothelial dysfunction was 6.4 (95% CI 2.7–15) for patients in the top versus the bottom quartile of the MPO distribution *(198)*. Thus, MPO might be a very promising prognostic marker for CV events, especially in acute situations. However, further studies are needed to establish whether high MPO levels are predictive of incident CHD in initially healthy subjects.

Oxidized LDL

The oxidative modification hypothesis of atherogenesis suggests that the most significant event in early lesion formation is lipid oxidation, placing oxLDL in a central role for the development of this disease *(199)*. OxLDL has a large number of biological actions and consequences, including injuring ECs, expressing adhesion molecules, recruiting leukocytes and retaining them, as well as the formation of foam cells *(200)*. Proatherogenic properties of oxLDL are summarized in Table 1. Furthermore, elevated oxLDL could play a role in the transition from stable to vulnerable, unstable plaque, and this assumption is supported by recent studies showing that oxLDL stimulates MMP-1 and MMP-9 expressions in human vascular ECs and in monocyte-derived macrophages *(201,202)*. It has also been shown that oxLDL upregulates the expression of MMP-1 and MMP-3 in human coronary ECs, an effect mediated through its endothelial receptor LOX-1 *(203)*. Furthermore, oxLDL triggers the CD40/CD40L signaling pathway, which might also lead to a proinflammatory reaction and induce endothelial injury *(204)*.

To date, a number of cross-sectional studies have examined the involvement of oxidative modification of LDL in subjects with clinical evidence of CV disease. Clinical studies *(205,206)* have demonstrated that patients with both stable CHD and ACS have elevated plasma levels of oxLDL compared with apparently healthy controls. A positive association between oxLDL and the severity of the ACS was found by Ehara et al. *(207)* who reported that oxLDL concentrations were significantly higher in patients with MI than in patients with unstable or stable angina, or age-matched controls. Findings from other studies suggest that plasma levels of oxLDL represent a more sensitive marker for the presence of CHD than the Global Risk Assessment Score (GRAS) *(208)* and that oxLDL also correlates with the extent of CHD in heart

Table 1
Proatherogenic Properties of Oxidized LDL

Biological effect	Possible mechanism
Foam cell formation	Direct uptake of cholesterol by scavenger receptors as well as inhibition of their export from macrophages
Chemoattraction of monocytes, T lymphocytes	Increased expression of MCP-1 and direct chemotactic effect
Macrophage trapping within the intima	Inhibition of the motility of macrophages
Impaired vascular function (vasoconstrictor effect)	Inhibition of nitric oxide release or function
Adhesion of monocytes to endothelium	Increased expression of adhesion molecules
Plaque rupture	Enhanced formation of matrix metalloproteinases
Cell proliferation	Induction of growth factors
Thrombogenesis	Promotion of platelet aggregation and increased tissue factor activity
Increased cellular death	Induction of Fas-mediated apoptosis
Induction of proinflammatory genes	Activation of nuclear factor-kappa B
Increased antigenicity	Induction of autoantibody (IgG) formation

transplant recipients (209). Moreover, circulating oxLDL has been associated with subclinical atherosclerosis in asymptomatic subjects (210,211).

Salonen et al. (212) in 1992 were the first to conduct a prospective, population-based, nested case–control study in which the titer of autoantibodies to malondialdehyde-modified LDL and native LDL in baseline serum samples from 30 Finnish men with accelerated progression of carotid atherosclerosis were compared with those of 30 age-matched controls without progression during an FU of 2 years. They found the titer of autoantibodies to oxLDL to be an independent predictor for the progression of carotid atherosclerosis. Since then, only one small prospective nested case–control study demonstrated that oxLDL concentrations might be associated with acute MI (213). During an FU of 2.6 years, 26 cases and 26 matched controls and a further 26 controls with LDL >5.0 mmol/L were studied. The oxLDL/plasma cholesterol ratio was higher among cases compared with controls and also higher compared with hypercholesterolemic subjects free of an event, suggesting that the high plasma oxLDL/total cholesterol ratio might serve as a possible indicator of increased risk of MI.

More recently, we have reported results from of a nested case–control study, which was conducted within two population-based MONICA/KORA Augsburg surveys (214). The association between plasma oxLDL and risk of future CHD was investigated in 88 middle-aged men with an incident CHD event and 258 age-matched controls during a mean FU of 5.6 years. Baseline oxLDL concentrations were significantly higher in CHD cases compared with those in controls. After multivariable adjustment, plasma oxLDL was the strongest predictor of CHD events compared with a conventional lipoprotein profile, and other traditional risk factors for CHD, with the HR for a future CHD event being 4.25 (95% CI 2.09–8.63) if the top tertile of the oxLDL distribution was compared with the bottom tertile. Furthermore, we also assessed whether the predictive value of oxLDL was additive to other risk factors of CHD and found that oxLDL significantly improved the prediction of incident CHD in addition to the other CV risk factors. Finally, when both oxLDL and CRP were simultaneously assessed in the same model, they still predicted future coronary events, even after multivariable adjustment including the TC/HDL-C ratio, one of the most powerful predictors of risk among conventional lipid variables. However, this study was the first prospective study conducted in apparently healthy men from an area with moderate absolute risk of CHD. Thus, further studies are warranted to establish the clinical relevance of oxLDL measurement in various stages of the atherosclerotic process and to identify the specific pathophysiological mechanisms by which oxLDL exerts its deleterious effects.

Lipoprotein-associated Phospholipase A2

Lipoprotein-associated phospholipase A2 (Lp-PLA2), a 45.4-kDa protein, is a calcium-independent member of the phospholipase A2 family. It is produced mainly by monocytes, macrophages, T lymphocytes, liver, and mast cells *(215–217)*. Lp-PLA2 activity occurs in association with macrophages and has been found to be upregulated in atherosclerotic lesions, especially in complex plaque *(218)*, as well as in the fibrous cap of coronary lesions prone to rupture *(219)*. In the bloodstream, two-thirds of the Lp-PLA2 plasma isoform circulates primarily bound to LDL, and the other one-third is distributed between HDL and very low density lipoproteins (VLDLs) *(220,221)*. Lp-PLA2 seems to play a dual role in atherosclerotic disease *(215)*. The HDL-associated enzyme is considered to be protective against atherosclerosis, whereas Lp-PLA2 bound to LDL is probably proatherogenic. Indeed, Lp-PLA2 was initially denoted as the platelet-activating factor acetylhydrolase (PAF-AH), reflecting its antiatherogenic activity: to catalyze the degradation of PAF and related oxidized phospholipids, thereby preventing LDL from further oxidative modification *(222–224)*. In addition, the HDL-associated Lp-PLA2 fraction may also inhibit foam cell formation and enhance cholesterol efflux in macrophages. On the contrary, Lp-PLA2 may promote oxidation of LDL and subsequent investigations indeed favored the proatherogenic properties of this enzyme *(215,225)*. After LDL oxidation within the arterial wall, a short acyl group at the sn-2 position of phospholipids becomes susceptible to the hydrolytic action of Lp-PLA2 that cleaves an oxidized phosphatidylcholine component of the lipoprotein particle generating two potent proinflammatory and proatherogenic mediators: namely, lysophosphatidylcholine (LysoPC) and oxidized fatty acid (oxFA) (Fig. 6) *(226)*. Of particular note, Lp-PLA2 acts only on oxidatively modified LDLs, and hydrolysis of oxLDL can be carried out solely by Lp-PLA2. Proinflammatory actions of LysoPC, as well as those of oxFA, trigger a cascade of events, which might directly promote atherogenesis. LysoPC is a potent chemoattractant for T cells and monocytes, promotes EC dysfunction, stimulates macrophage proliferation, and induces apoptosis in SMCs and macrophages, as depicted in Fig. 7 *(227–229)*. Thus, Lp-PLA2 may represent an important "missing link" between the oxidative modification of LDL in the intimal layer of the arterial wall and local inflammatory processes within the atherosclerotic plaque.

Hitherto published results from several prospective studies conducted in initially healthy subjects from various populations revealed an independent association between increased concentrations of Lp-PLA2 and future CV events. Initial evidence for such an association came from the WOSCOPS, in which the RR associated with a 1 SD increase in Lp-PLA2 was 1.18 (95% CI 1.05–1.33) after controlling for traditional risk

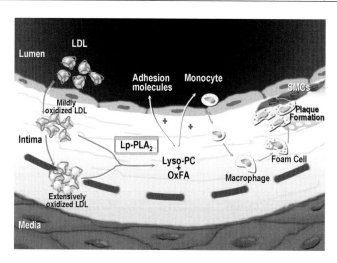

Fig. 6. Schematic representation of action of Lp-PLA2 in the vascular wall. Lp-PLA2 = lipoprotein-associated phospholipase A_2; LysoPC = lysophosphatidylcholine; oxFA = oxidized fatty acids; SMC = smooth muscle cell.

factors, and was independent of various other biomarkers such as CRP, WBC count, and fibrinogen (230). Almost identical results have been found in the MONICA (231) and Rotterdam studies (232), with approximately 20% elevated coronary risk per 1 SD increase in Lp-PLA2. The predictive role of Lp-PLA2 has also been assessed within the WHS (233); however, elevated levels of Lp-PLA2 did not predict subsequent CV events in this low-risk population for CVD. This lack of association could be attributed to existing gender differences for Lp-PLA2 and the low power of this study. Results from the ARIC study (234) demonstrated only a borderline association in the

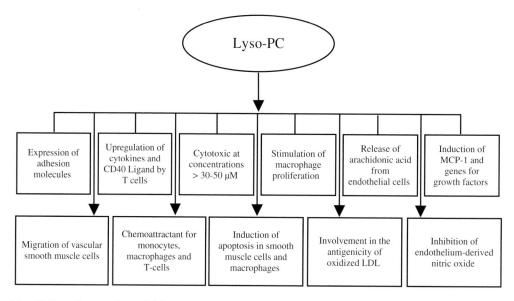

Fig. 7. Pro-atherogenic activities of lysophosphatidylcholine (Lyso-PC) MCP-1, Monocyte chemoattractant protein; LDL, Low density lipoprotein protein-1.

total case–cohort; however, in subjects with low LDL-C, Lp-PLA2 significantly and independently predicted CHD, thereby suggesting that it might be a useful marker for identifying high-risk patients with relatively normal levels of LDL-C, a group in whom additional markers of risk are clearly needed.

There are two studies to date in patients with CHD in which the prognostic value of Lp-PLA2 has been studied. In the first study from the Mayo Clinic (235), 466 consecutive patients scheduled for coronary angiography were followed for a median of 4 years. The RR for a future event for a 1 SD increase in Lp-PLA2 mass was 1.30 after multivariable adjustments and thus comparable with data from WOSCOPS (230) and MONICA (231). In the second study (236), plasma concentrations and activity of Lp-PLA2 were determined at baseline in a cohort of 1051 patients aged 30–70 years with CHD. Secondary CVD events as the outcome were determined during a mean FU of 48.7 months. In multivariable analyses, after controlling for traditional risk factors, severity of CHD, statin treatment, CRP, cystatin C, and N-terminal proBNP, Lp-PLA2 mass was still strongly and significantly associated with CVD events, whereas activity lost its statistical significance. The HR for recurrent events across extreme tertiles was 2.09 (95% CI 1.10–3.96) for mass and 1.81 (95% CI 0.94–3.49) for activity. Thus, increased concentrations of Lp-PLA2 mass predicted future CVD events in patients with manifest CHD, independent of a variety of potential risk factors including markers of inflammation, renal function, and hemodynamic stress. In several studies (231,234,237), the potential additive value of Lp-PLA2 to CRP in predicting risk was evaluated for CHD and stroke. For this purpose, high CRP was defined according to a recent AHA/CDC consensus document as >3.0 mg/L, and for Lp-PLA2, the upper tertile cutpoint was used (422 μg/L in ARIC and 290.8 ng/mL in MONICA). In ARIC, individuals with high Lp-PLA2 and high CRP exhibited the greatest risk for CHD (HR 2.95 [95% CI 1.47–5.94]) (234). The MONICA Augsburg study demonstrated that the combination of elevated Lp-PLA2 and elevated CRP was consistently associated with a statistically significantly increased risk for future coronary events and was superior to either marker alone in predicting risk (HR of 1.93 [95% CI 1.09–3.40]) compared with both markers not being increased (referent) in the fully adjusted model) (231). Taken together, these data suggest that Lp-PLA2 and CRP may be complementary (or additive) in identifying high-risk subjects, and therefore, the combination of both markers may further improve risk assessment.

These intriguingly consistent results have generated much interest in better defining the relationship between Lp-PLA2 and CV risk; data from studies published so far, as well as preliminary data from several subsequent studies, are summarized in Table 2.

In summary, a growing body of evidence from large prospective epidemiological, population-based, and clinical studies suggests that Lp-PLA2 may be an independent and clinically relevant marker for CHD and probably also for stroke. Moreover, because specific inhibitors of Lp-PLA2 are currently under evaluation in clinical trials, lowering Lp-PLA2 in plasma and/or the vessel wall might represent a promising novel strategy for the treatment of atherosclerosis via direct targeting vascular inflammation and thus may open a new avenue to combat this widespread disease.

Adiponectin

Contrary to numerous emerging proinflammatory and proatherogenic biomarkers, adiponectin seems to be a novel antiatherogenic marker. Adiponectin is a 244-amino acid collagen-like protein, a member of a family of obesity-related hormones.

Table 2
Lp-PLA2 and Cardiovascular Disease: Overview of Studies

Author	Study	Design	Population	Outcome variable	FU	N	Relative Risk/Odds Ratio	Reference
Prospective studies (CHD)								
Packard et al. (230)[e]	WOSCOPS	nested case-control	Hyperlipidemic men	Coronary events	4.9	580/1160	1.18 (1.05–1.33)[a]	NEJM. 2000;343(16):1148–55.
Blake et al. (233)[e]	WHS	nested case-control	Healthy women	Coronary events	3	123/123	1.17 (0.45–3.05)[c]	JACC. 2001;38(5):1302–6.
Ballantyne et al. (234)[e]	ARIC	nested case-cohort	Healthy M/F	CV events	6–8	608/740	1.15 (0.81–1.63)[b]; LDL-C <130 mg/dL 2.08(1.20–3.62)[b]	Circulation. 2004;109(7):837–42
Koenig et al. (253)[e]	MONICA	complete cohort	Healthy men	Coronary events	14	934	1.21 (1.01–1.45)[a]	Circulation. 2004;110(14):1903–8
Oei et al. (232)[d]	Rotterdam	nested case-cohort	Elderly (55+)	Coronary events	10	377/1822	1.76 (1.09–2.85)[c]	Circulation. 2005;111(5):570–5.
Brilakis et al. (235)[e]	Mayo	complete cohort	CHD patients	CV events	4	504	1.30 (1.06–1.59)[a]	Eur Heart J. 2005;26(2):137–44.
Koenig et al. (236)[de]	KAROLA	complete cohort	CHD patients	CV events	4	1051	2.09 (1.10–3.96)[eb]; 1.81 (0.94–3.49)[db]	JACC 2005; 45 (Suppl A):371 A
Persson et al. (256)[d]	Malmö	complete cohort	Healthy M/F	CV events	9.4	145/4938	1.75 (1.01–3.08)[b]	Circulation 2005;112(Abstract Suppl):II-802
Prospective studies (Stroke)								
Oei et al. (232)[d]	Rotterdam	nested case-cohort	Elderly (55+)	Stroke	12	200/1822	1.77(1.19–2.64)[c]	Circulation. 2005;111(5):570–5.
Ballantyne et al. (237)[e]	ARIC	nested case-cohort	Healthy M/F	Stroke	4.4	194/766	1.93 (1.14–3.27)[b]	Arch Intern Med. 2005;165(21)2479–84.
Persson et al. (256)[d]	Malmö	complete cohort	Healthy M/F	Stroke	9.4	112/4938	1.69 (0.90–3.18)[b]	Circulation 2005;112(Abstract Suppl):II-802

Cross-section studies (CHD)

Study	Location	Design	Groups	Outcome		N (case/control)	OR (95% CI)	Reference
Caslake et al. (220)[e]	Glasgow	case-control	CHD/Control	Presence of CHD	–	94/54	NA	Atherosclerosis. 2000;150(2):413–9.
Shohet et al. (259)[d]	Dallas	case-control	CHD/Control	Presence of CHD	–	72/72	NA	Am J Cardiol. 1999;83(1):109–11, A8–9.
Blankenberg et al. (258)[d]	Mainz	case-control	CHD/Control	Presence of CHD	–	496/477	1.8 (1.01–3.20)[c]	J Lipid Res. 2003;44(7):1381–6.
Winkler et al. (259)[d]	Freiburg	case-control	T2DM/Control	Presence of CHD	–	42/47	2.09 (1.02–4.29)[c]	J Clin Endocrinol Metab. 2004;89(3):1153–9.
Khuseyinova et al. (260)[e]	Ulm	case-control	CHD/Control	Presence of CHD	–	312/479	1.91 (1.12–3.28)[c]	Atherosclerosis. 2005;182(1):181–8.
Winkler et al. (261)[d]	LURIC	case-control	CHD/Control	Presence of CHD	–	2454/694	1.85 (1.23–2.78)[c]	Circulation. 2005;111(8):980–7.

Cross-sectional studies (CAC)

Study	Location	Design	Groups	Outcome		N	OR (95% CI)	Reference
Iribarren et al. (262)[de]	CARDIA	nested case-control	Young adults	CAC	–	266/266	1.28(1.03–1.66)[da]; 1.09 (0.84–1.42)[da]	ATVB. 2005;25(1):216–21.

[a] Increase of Lp-PLA2 per 1 standard deviation (SD) in multivariable analyses including CRP
[b] Tertile analysis, multivariable including CRP;
[c] Quartile analysis, multivariable adjustment
[d] Measurement of Lp-PLA2 activity;
[e] Measurement of Lp-PLA2 mass

CV = cardiovascular, CHD = coronary heart disease, CAC = coronary artery calcification, NA = not assessed; T2DM = Type 2 diabetes mellitus

Adiponectin is exclusively produced by white adipose tissue. Structurally, it consists of a C-terminal collagenous tail and a complement (C1q)-like globular head, which form trimer–dimers and are found as a high-molecular-weight complex in the circulation in relatively high concentrations (2–30 mg/L) *(238,239)*.

The mechanisms whereby adiponectin exerts its physiological activities are not completely elucidated. Apart from its role as an insulin-sensitizing agent, and its implication in metabolic disorders, adiponectin might also be involved in the regulation of inflammatory processes that contribute to atherosclerosis by, for instance, the inhibition of TNF-α induced expression of adhesion molecules, thereby preventing the attachment of monocytes to the endothelial surface *(240)*, or by the inhibition of class A macrophage scavenger receptor (SR-A) that reduces cholesterol ester accumulation and decreases oxLDL uptake, thereby diminishing the transformation of macrophages into foam cells *(241)* (Fig. 8). Low adiponectin levels are associated with reduced expression of NO, and increased expression of angiotensin II from the endothelium *(239)*. Data in adiponectin-deficient mice have shown a twofold increase in neointimal thickening and increased proliferation of vascular SMCs in arteries after mechanical endothelial injury than wild-type mice *(242,243)*. In addition, adiponectin-knockout mice showed high levels of TNF-α mRNA in adipose tissue *(244)*. By contrast, transfection of human adiponectin gene to apoE-knockout mice leads to an overexpression of adiponectin and, as a result, to the suppression of atherosclerotic plaques *(245)*.

All these properties of adiponectin are in support of a promising role of this molecule as an anti-inflammatory and antiatherogenic biomarker for the prediction of future CV events. Indeed, several cross-sectional clinical studies have reported lower serum levels in patients with CHD than in controls *(246–250)*.

The predictive value of adiponectin for future coronary events in apparently healthy, nondiabetic subjects has been assessed for the first time by Pischon and colleagues in US Health Professionals *(251)*. Of 18,225 male participants of the Health Professionals' Follow-up Study, 266 men who subsequently developed coronary events during a 6-year FU were compared with 532 event-free controls, matched for age, date of blood draw, and smoking status. The authors found a significantly reduced risk of subsequent AMI associated with higher levels of adiponectin in serum at baseline; notably, this

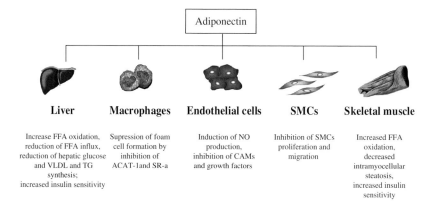

Fig. 8. Anti-inflammatory and anti-atherogenic effects of adiponectin. FFA = free fatty acids; VLDL = very -low-density lipoprotein; TG = triglyceride; ACAT-1 = acyl-coenzyme A: cholesterol acyltransferase-1; SR-a = scavenger receptor a; NO = nitric oxide; CAMs = cell adhesion molecules; SMCs = smooth muscle cells.

association was also reduced after adjustment for covariates but persisted after inclusion of lipids (LDL-C and HDL-C) in the model (RR 0.56 [95% CI 0.32–0.99]; P for trend $= 0.02$). Further adjustment for glycemic status and CRP did not appreciably affect the results.

A second prospective study *(252)* was conducted in elderly women, aged 60–79 years, who participated in the British Women's Heart and Health Study, representing a cohort of 4286 women, randomly selected from 23 British towns between 1999 and 2001. Using nested case–control design, 167 women with incident CHD over 4 years were compared with 334 subjects, who remained free of disease, and the RR for a doubling of adiponectin was found to be 0.99 (95% CI 0.82–1.19) after multivariate adjustment including lipoproteins and CRP. The authors suggested that gender differences in body fat distribution and higher levels of adiponectin in women could reflect a relative resistance to the effect of adiponectin in female participants, thereby explaining the negative findings of this study. Thus, gender differences of adiponectin and their relationship to CHD should be a matter of further research.

Within the MONICA/KORA Augsburg platform, we also studied the risk for future CHD events associated with reduced levels of adiponectin in 938 apparently healthy men, aged 45–64 years, who were followed for 14 years *(253)*. Higher adiponectin levels were associated with considerably lower risk of about 40–50% for a first ever acute CHD event, and this association was independent of a variety of potential confounders. In a Cox model, the HR of a future coronary event, comparing the top tertile of the adiponectin distribution to the bottom tertile was 0.54 (95% CI 0.31–0.93, after adjustment for CV risk factors). Only the introduction of HDL-C in the model resulted in a moderate attenuation of the association, which then became borderline significant. Thus, these data suggest an antiatherogenic role of increased concentrations of adiponectin and that hypoadiponectinemia therefore may be associated with increased risk of atherosclerotic disease. The protective effect of high serum concentrations might, at least in part, be explained through its strong positive correlation with HDL-C. Thus, the molecular mechanisms of this relationship are of particular interest. In a recent study, von Eynatten et al. *(254)* reported data from 206 nondiabetic men and 110 patients with type 2 diabetes in whom they found an association between decreased postheparin lipoprotein lipase (LPL) activity and low plasma adiponectin, independent of systemic inflammation and insulin resistance. Conversely, dramatically raised levels of LPL activity have been found with increased plasma adiponectin in an animal model. Thus, adiponectin may directly stimulate the expression of LPL, which will then result in increased HDL-C levels.

However, there is still more work needed to establish the role of this new biomarker in clinical medicine.

CONCLUSIONS

Because inflammation accompanies all stages of atherogenesis, measurement of plasma/serum concentrations of circulating inflammatory biomarkers might aid in identifying individuals at high risk for CHD. Some of these biomarkers might be involved in the initiation of atherogenesis. However, elevated concentrations of the vast majority of them must be considered as a response to a variety of stimuli, but they may facilitate the progression of atherothrombosis and be involved in other proatherothrombotic mechanisms. Thus, several biomarkers might be used as predictors of early stages

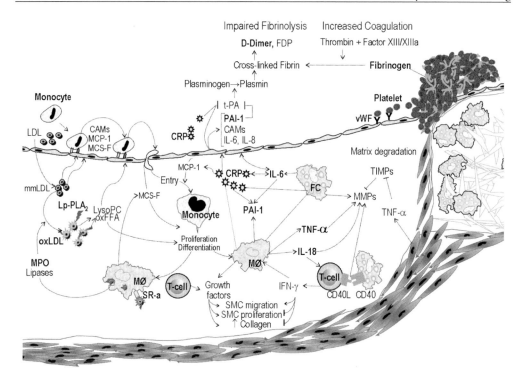

Fig. 9. Schematic representation of plaque development. LDL = low-density lipoprotein; mmLDL = minimally modified low-density lipoprotein; oxLDL = oxidized low-density lipoprotein; Lp-PLA2 = lipoprotein-associated phospholipase A_2; LysoPC = lysophosphatidylcholine; oxFFA = oxidized free fatty acids; MPO = myeloperoxidase; SR-a = scanvenger receptor a; MØ = macrophage; FC = foam cell; SMC = smooth muscle cell; MCS-F = macrophage colony stimulating factor; MCP-1 = monocyte chemoattractant protein-1; CAMs = cell adhesion molecules; CRP = c-reactive protein; t-PA = tissue-type plasminogen activator; PAI-1 = plasminogen activator inhibitor 1; IL = interleukin; vWF = von Willebrand factor; FDP = fibrin degradation products; TNF-α = tumor necrosis factor-α; MMPs = matrix metalloproteinases; IMPs = tissue inhibitors of matrix metalloproteinases; ↑= induction; ⊤ = inhibition.

of atherosclerotic disease, whereas others might represent powerful prognostic markers of advanced disease or even its clinical complications (Fig. 9).

Although an increasing number of emerging biomarkers of CV risk have been identified during the last years, assessment of the clinical utility of markers of systemic inflammation is only at the beginning. In order to be translated into clinical practice, systemic biomarkers should possess certain criteria *(255)*. First, they must provide independent information on risk or prognosis in large prospective population-based studies conducted in apparently healthy men and women. In addition, laboratory characteristics such as sensitivity and precision of biomarkers should be known. Furthermore, appropriate cutoff points must be evaluated for proper use of biomarkers, with emphasis on which cutpoints should be used in primary and in secondary prevention. Most importantly, standardization of assays must be attempted to ensure adequate reproducibility of measurements. Low costs and ease of performance are other characteristics that facilitate the utility of biomarkers in routine laboratory practice.

Thus, among all biomarkers discussed in this chapter, only CRP and partly Lp-PLA2 have been investigated in several large prospective studies with consistent results, fulfill most of these requirements, and might therefore be added to the clinical armamentarium for improved prediction of CV disease in selected populations.

In the future, we probably will see a biomarker profile that covers various aspects of the complex pathophysiology of atherothrombotic disease; we will focus more on biological patterns or systems that may become dysbalanced during the atherosclerotic process rather than on single markers only, and finally, testing the "inflammation hypothesis" in large clinical trials will represent an important goal for clinical research of atherosclerosis.

KEY POINTS

1. The inflammatory response represents an important contributor to atherothrombosis. Screening for low-grade inflammation using several novel biomarkers might provide an important tool for identifying individuals at increased risk who would benefit most from targeted preventive interventions.
2. To be implemented into clinical practice, these markers, however, should fulfill certain requirements: provide independent information on risk prediction in addition to global risk assessment, be reliable and easily reproducible, and show high sensitivity and specificity. Finally, simple and robust assays should be commercially available.
3. In the future, simultaneous assessment of several biomarkers, a so-called multimarker approach, might allow to reveal in more detail the complex and multifactorial origin of atherothrombosis, thereby opening a new avenue to combat this still widespread disease.

REFERENCES

1. Pearson TA, Mensah GA, Alexander RW, Anderson JL, Cannon RO III, Criqui M, Fadl YY, Fortmann SP, Hong Y, Myers GL, Rifai N, Smith SC Jr, Taubert K, Tracy RP, Vinicor F; Centers for Disease Control and Prevention; American Heart Association. Markers of inflammation and cardiovascular disease: application to clinical and public health practice: a statement for healthcare professionals from the Centers for Disease Control and Prevention and the American Heart Association. *Circulation* 2003;107:499–511.
2. Danesh J, Wheeler JG, Hirschfield GM, Eda S, Eiriksdottir G, Rumley A, Lowe GD, Pepys MB, Gudnason V. C-reactive protein and other circulating markers of inflammation in the prediction of coronary heart disease. *N Engl J Med* 2004;350:1387–1397.
3. Tillett WS, Francis TJ. Serological reactions in pneumonia with a nonprotein somatic fraction of pneumococcus. *Exp Med* 1930;52:561–571.
4. Meier-Ewert HK, Ridker PM, Rifai N, Price M, Dinges DF, Mullington JM. Absence of diurnal variation of C-reactive protein concentrations in healthy subjects. *Clin Chem* 2001;47:426–430.
5. Angleton P, Chandler WL, Schmer G. Diurnal variation of tissue-type plasminogen activator and its rapid inhibitor (PAI-1). *Circulation* 1989;79:101–106.
6. Fröhlich M, Sund M, Thorand B, Hutchinson WL, Pepys MB, Koenig W. Lack of seasonal variation in C-reactive protein. *Clin Chem* 2002;48:575–577.
7. Frohlich M, Sund M, Russ S, Hoffmeister A, Fischer HG, Hombach V, Koenig W. Seasonal variations of rheological and hemostatic parameters, and acute phase reactants in young, healthy subjects. *Arterioscler Thromb Vasc Biol* 1997;17:2692–2697.
8. Kushner I, Broder ML, Karp D. Control of the acute phase response. Serum C-reactive protein kinetics after acute myocardial infarction. *J Clin Invest* 1978;61:235–242.
9. Calabro P, Willerson JT, Yeh ET. Inflammatory cytokines stimulated C-reactive protein production by human coronary artery smooth muscle cells. *Circulation* 2003;108:1930.
10. Jabs WJ, Theissing E, Nitschke M, Bechtel JF, Duchrow M, Mohamed S, Jahrbeck B, Sievers HH, Steinhoff J, Bartels C. Local generation of C-reactive protein in diseased coronary artery venous bypass grafts and normal vascular tissue. *Circulation* 2003;108:1428–1431.
11. Yasojima K, Schwab C, McGeer EG, McGeer PL. Generation of C-reactive protein and complement components in atherosclerotic plaques. *Am J Pathol* 2001;158:1039–1051.

12. Ishikawa T, Hatakeyama K, Imamura T, Date H, Shibata Y, Hikichi Y, Asada Y, Eto T. Involvement of C-reactive protein obtained by directional coronary atherectomy in plaque instability and developing restenosis in patients with stable or unstable angina pectoris. *Am J Cardiol* 2003;91: 287–292.
13. Burke AP, Tracy RP, Kolodgie F, Malcom GT, Zieske A, Kutys R, Pestaner J, Smialek J, Virmani R. Elevated C-reactive protein values and atherosclerosis in sudden coronary death: association with different pathologies. *Circulation* 2002;105:2019–2023.
14. Maier W, Altwegg LA, Corti R, Gay S, Hersberger M, Maly FE, Sutsch G, Roffi M, Neidhart M, Eberli FR, Tanner FC, Gobbi S, von Eckardstein A, Luscher TF. Inflammatory markers at the site of ruptured plaque in acute myocardial infarction: locally increased interleukin-6 and serum amyloid A but decreased C-reactive protein. *Circulation* 2005;111:1355–1361.
15. Dong Q, Wright JR. Expression of C-reactive protein by alveolar macrophages. *J Immunol* 1996;156:4815–4820.
16. Yasojima K, Schwab C, McGeer EG, McGeer PL. Human neurons generate C-reactive protein and amyloid P: upregulation in Alzheimer's disease. *Brain Res* 2000;887:80–89.
17. Jabs WJ, Logering BA, Gerke P, Kreft B, Wolber EM, Klinger MH, Fricke L, Steinhoff J. The kidney as a second site of human C-reactive protein formation in vivo. *Eur J Immunol* 2003;33: 152–161.
18. Calabro P, Chang DW, Willerson JT, Yeh ET. Release of C-reactive protein in response to inflammatory cytokines by human adipocytes: linking obesity to vascular inflammation. *J Am Coll Cardiol* 2005;46:1112–1113.
19. Zwaka TP, Hombach V, Torzewski J. C-reactive protein-mediated low density lipoprotein uptake by macrophages: implications for atherosclerosis. *Circulation* 2001;103:1194–1197.
20. Volanakis JE. Complement activation by C-reactive protein complexes. *Ann N Y Acad Sci* 1982;389:235–250.
21. Bharadwaj D, Stein MP, Volzer M, Mold C, Du Clos TW. The major receptor for C-reactive protein on leukocytes is FCγ receptor II. *J Exp Med* 1999;190:585–590.
22. Torzewski J, Torzewski M, Bowyer DE, Frohlich M, Koenig W, Waltenberger J, Fitzsimmons C, Hombach V. C-reactive protein frequently colocalizes with the terminal complement complex in the intima of early atherosclerotic lesions of human coronary arteries. *Arterioscler Thromb Vasc Biol* 1998;18:1386–1392.
23. Verma S, Li SH, Badiwala MV, Weisel RD, Fedak PW, Li RK, Dhillon B, Mickle DA. Endothelin antagonism and interleukin-6 inhibition attenuate the proatherogenic effects of C-reactive protein. *Circulation* 2002;105:1890–1896.
24. Ballou SP, Lozanski G. Induction of inflammatory cytokine release from cultured human monocytes by C-reactive protein. *Cytokine* 1992;4:361–368.
25. Jones SA, Novick D, Horiuchi S, Yamamoto N, Szalai AJ, Fuller GM. C-reactive protein: a physiological activator of interleukin 6 receptor shedding. *J Exp Med* 1999;189:599–604.
26. Torzewski M, Rist C, Mortensen RF, Zwaka TP, Bienek M, Waltenberger J, Koenig W, Schmitz G, Hombach V, Torzewski J. C-reactive protein in the arterial intima: role of C-reactive protein receptor-dependent monocyte recruitment in atherogenesis. *Arterioscler Thromb Vasc Biol* 2000;20:2094–2099.
27. Pasceri V, Chang J, Willerson JT, Yeh ETH. Modulation of C-reactive protein-mediated monocyte chemoattractant protein-1 induction in human endothelial cells by anti-atherosclerosis drugs. *Circulation* 2001;103:2531–2534.
28. Han KH, Hong KH, Park JH, Ko J, Kang DH, Choi KJ, Hong MK, Park SW, Park SJ. C-reactive protein promotes monocyte chemoattractant protein-1—mediated chemotaxis through upregulating CC chemokine receptor 2 expression in human monocytes. *Circulation* 2004;109: 2566–2571.
29. Pasceri V, Willerson JT, Yeh ET. Direct proinflammatory effect of C-reactive protein on human endothelial cells. *Circulation* 2000;102:2165–2168.
30. Nakagomi A, Freedman SB, Geczy CL. Interferon-gamma and lipopolysaccharide potentiate monocyte tissue factor induction by C-reactive protein: relationship with age, sex, and hormone replacement treatment. *Circulation* 2000;101:1785–1791.
31. Cermak J, Key NS, Bach RR, Balla J, Jacob HS, Vercellotti GM. CRP induces human peripheral blood monocytes to synthesize tissue factor. *Blood* 1993;82:513–520.
32. Paffen E, Vos HL, Bertina RM. C-reactive protein does not directly induce tissue factor in human monocytes. *Arterioscler Thromb Vasc Biol* 2004;24:975–981.

33. Blann AD, Lip GY. Effects of CRP on the release of von Willebrand factor, E-selectin, thrombomodulin and intercellular adhesion molecule-1 from HUVEC. *Blood Coagul Fibrinolysis* 2003;14:335–340.
34. Devaraj S, Xu DY, Jialal I. C-reactive protein increases plasminogen activator inhibitor-1 expression and activity in human aortic endothelial cells: implications for the metabolic syndrome and atherothrombosis. *Circulation* 2003;107:398–404.
35. Verma S, Wang CH, Li SH, Dumont AS, Fedak PW, Badiwala MV, Dhillon B, Weisel RD, Li RK, Mickle DA, Stewart DJ. A self-fulfilling prophecy: C-reactive protein attenuates nitric oxide production and inhibits angiogenesis. *Circulation* 2002;106:913–919.
36. Venugopal SK, Devaraj S, Yuhanna I, Shaul P, Jialal I. Demonstration that C-reactive protein decreases eNOS expression and bioactivity in human aortic endothelial cells. *Circulation* 2002;106:1439–1441.
37. Nagoshi Y, Kuwasako K, Cao YN, Kitamura K, Eto T. Effects of C-reactive protein on atherogenic mediators and adrenomedullin in human coronary artery endothelial and smooth muscle cells. *Biochem Biophys Res Commun* 2004;314:1057–1063.
38. van den Berg CW, Taylor KE, Lang D. C-reactive protein-induced in vitro vasorelaxation is an artefact caused by the presence of sodium azide in commercial preparations. *Arterioscler Thromb Vasc Biol* 2004;24:e168–e171.
39. Taylor KE, Giddings JC, van den Berg CW. C-reactive protein-induced in vitro endothelial cell activation is an artefact caused by azide and lipopolysaccharide. *Arterioscler Thromb Vasc Biol* 2005;25:1225–1230.
40. Nerurkar SS, McDevitt PJ, Scott GF, Johanson KO, Willette RN, Yue TL. Lipopolysaccharide (LPS) contamination plays the real role in C-reactive protein-induced IL-6 secretion from human endothelial cells in vitro. *Arterioscler Thromb Vasc Biol* 2005;25:e136.
41. Singh U, Devaraj S, Jialal I. C-reactive protein decreases tissue plasminogen activator activity in human aortic endothelial cells: evidence that C-reactive protein is a procoagulant. *Arterioscler Thromb Vasc Biol* 2005;25:2216–2221.
42. Bisoendial RJ, Kastelein JJ, Levels JH, Zwaginga JJ, van den Bogaard B, Reitsma PH, Meijers JC, Hartman D, Levi M, Stroes ES. Activation of inflammation and coagulation after infusion of C-reactive protein in humans. *Circ Res* 2005;96:714–716.
43. Bisoendial R, Kastelein J, Stroes E. In response to van den Berg et al: on the direct actions of CRP in humans. *Circ Res* 2005;97:e71.
44. Szalai AJ, McCrory MA. Varied biologic functions of C-reactive protein: lessons learned from transgenic mice. *Immunol Res* 2002;26:279–287.
45. Danenberg HD, Szalai AJ, Swaminathan RV, Peng L, Chen Z, Seifert P, Fay WP, Simon DI, Edelman ER. Increased thrombosis after arterial injury in human C-reactive protein-transgenic mice. *Circulation* 2003;108:512–515.
46. Paul A, Ko KW, Li L, Yechoor V, McCrory MA, Szalai AJ, Chan L. C-reactive protein accelerates the progression of atherosclerosis in apolipoprotein E-deficient mice. *Circulation* 2004;109: 647–655.
47. Hirschfield GM, Gallimore JR, Kahan MC, Hutchinson WL, Sabin CA, Benson GM, Dhillon AP, Tennent GA, Pepys MB. Transgenic human C-reactive protein is not proatherogenic in apolipoprotein E-deficient mice. *Proc Natl Acad Sci USA* 2005;102:8309–8314.
48. Trion A, de Maat MP, Jukema JW, van der Laarse A, Maas MC, Offerman EH, Havekes LM, Szalai AJ, Princen HM, Emeis JJ. No effect of C-reactive protein on early atherosclerosis development in apolipoprotein E*3-leiden/human C-reactive protein transgenic mice. *Arterioscler Thromb Vasc Biol* 2005;25:1635–1640.
49. Reifenberg K, Lehr HA, Baskal D, Wiese E, Schaefer SC, Black S, Samols D, Torzewski M, Lackner KJ, Husmann M, Blettner M, Bhakdi S. Role of C-reactive protein in atherogenesis: can the apolipoprotein E knockout mouse provide the answer? *Arterioscler Thromb Vasc Biol* 2005;25: 1641–1646.
50. Danesh J, Whincup P, Walker M, Lennon L, Thomson A, Appleby P, Gallimore JR, Pepys MB. Low grade inflammation and coronary heart disease: prospective study and updated meta-analyses. *BMJ* 2000;321:199–204.
51. Boekholdt SM, Hack CE, Sandhu MS, Luben R, Bingham SA, Wareham NJ, Peters RJ, Jukema JW, Day NE, Kastelein JJ, Khaw KT. C-reactive protein levels and coronary artery disease incidence and mortality in apparently healthy men and women: The EPIC-Norfolk prospective population study 1993–2003. *Atherosclerosis* 2006;187(2):415–422 [2005 Oct 27; Epub ahead of print].

52. Cushman M, Arnold AM, Psaty BM, Manolio TA, Kuller LH, Burke GL, Polak JF, Tracy RP. C-reactive protein and the 10-year incidence of coronary heart disease in older men and women: the Cardiovascular Health Study. *Circulation* 2005;112:25–31.
53. Koenig W, Lowel H, Baumert J, Meisinger C. C-reactive protein modulates risk prediction based on the Framingham Score: implications for future risk assessment: results from a large cohort study in southern Germany. *Circulation* 2004;109:1349–1353.
54. Ridker PM; JUPITER Study Group. Rosuvastatin in the primary prevention of cardiovascular disease among patients with low levels of low-density lipoprotein cholesterol and elevated high-sensitivity C-reactive protein: rationale and design of the JUPITER trial. *Circulation* 2003; 108:2292–2297.
55. Weissenbach J, Chernajovsky Y, Zeevi M, Shulman L, Soreq H, Nir U, Wallach D, Perricaudet M, Tiollais P, Revel M. Two interferon mRNAs in human fibroblasts: in vitro translation and Escherichia coli cloning studies. *Proc Natl Acad Sci USA* 1980;77:7152–7156.
56. Aarden LA, De Groot ER, Schaap OL, Lansdorp PM. Production of hybridoma growth factor by human monocytes. *Eur J Immunol* 1987;17:1411–1416.
57. Jirik FR, Podor TJ, Hirano T, Kishimoto T, Loskutoff DJ, Carson DA, Lotz M. Bacterial lipopolysaccharide and inflammatory mediators augment IL-6 secretion by human endothelial cells. *J Immunol* 1989;142:144–147.
58. Mohamed-Ali V, Goodrick S, Rawesh A, Katz DR, Miles JM, Yudkin JS, Klein S, Coppack SW. Subcutaneous adipose tissue releases interleukin-6, but not tumor necrosis factor-alpha, in vivo. *J Clin Endocrinol Metab* 1997;82:4196–4200.
59. Heinrich PC, Castell JV, Andus T. Interleukin-6 and the acute phase response. *Biochem J* 1990;265:621–636.
60. Amrani DL. Regulation of fibrinogen biosynthesis: glucocorticoid and interleukin-6 control. *Blood Coagul Fibrinolysis* 1990;1:443–446.
61. Seino Y, Ikeda U, Ikeda M, Yamamoto K, Misawa Y, Hasegawa T, Kano S, Shimada K. Interleukin-6 gene transcripts are expressed in human atherosclerotic lesions. *Cytokine* 1994; 6:87–91.
62. Rus HG, Vlaicu R, Niculescu F. Interleukin-6 and interleukin-8 protein and gene expression in human arterial atherosclerotic wall. *Atherosclerosis* 1996;127:263–271.
63. Schieffer B, Schieffer E, Hilfiker-Kleiner D, Hilfiker A, Kovanen PT, Kaartinen M, Nussberger J, Harringer W, Drexler H. Expression of angiotensin II and interleukin-6 in human coronary atherosclerotic plaques: potential implications for inflammation and plaque instability. *Circulation* 2000;101:1372–1378.
64. Schieffer B, Selle T, Hilfiker A, Hilfiker-Kleiner D, Grote K, Tietge UJ, Trautwein C, Luchtefeld M, Schmittkamp C, Heeneman S, Daemen MJ, Drexler H. Impact of interleukin-6 on plaque development and morphology in experimental atherosclerosis. *Circulation* 2004;110:3493–3500.
65. Biswas P, Delfanti F, Bernasconi S, Mengozzi M, Cota M, Polentarutti N, Mantovani A, Lazzarin A, Sozzani S, Poli G. Interleukin-6 induces monocyte chemotactic protein-1 in peripheral blood mononuclear cells and in the U937 cell line. *Blood* 1998;91:258–265.
66. Pigott R, Dillon LP, Hemingway IH, Gearing AJ. Soluble forms of E-selectin, ICAM-1 and VCAM-1 are present in the supernatants of cytokine activated cultured endothelial cells. *Biochem Biophys Res Commun* 1992;187:584–589.
67. Huber SA, Sakkinen P, Conze D, Hardin N, Tracy R. Interleukin-6 exacerbates early atherosclerosis in mice. *Arterioscler Thromb Vasc Biol* 1999;19:2364–2367.
68. Kerr R, Stirling D, Ludlam CA. Interleukin 6 and haemostasis. *Br J Haematol* 2001;115:3–12.
69. Biasucci LM, Liuzzo G, Fantuzzi G, Caligiuri G, Rebuzzi AG, Ginnetti F, Dinarello CA, Maseri A. Increasing levels of interleukin (IL)-1Ra and IL-6 during the first 2 days of hospitalization in unstable angina are associated with increased risk of in-hospital coronary events. *Circulation* 1999;99:2079–2084.
70. Lindmark E, Diderholm E, Wallentin L, Siegbahn A. Relationship between interleukin-6 and mortality in patients with unstable coronary artery disease: effects of an early invasive or noninvasive strategy. *JAMA* 2001;286:2107–2113.
71. Ridker PM, Rifai N, Stampfer MJ, Hennekens CH. Plasma concentration of interleukin-6 and the risk of future myocardial infarction among apparently healthy men. *Circulation* 2000;101: 1767–1772.
72. Ridker PM, Hennekens CH, Buring JE, Rifai N. C-reactive protein and other markers of inflammation in the prediction of cardiovascular disease in women. *N Engl J Med* 2000;342:836–843.

73. Pradhan AD, Manson JE, Rossouw JE, Siscovick DS, Mouton CP, Rifai N, Wallace RB, Jackson RD, Pettinger MB, Ridker PM. Inflammatory biomarkers, hormone replacement therapy, and incident coronary heart disease: prospective analysis from the Women's Health Initiative observational study. *JAMA* 2002;288:980–987.
74. Harris TB, Ferrucci L, Tracy RP, Corti MC, Wacholder S, Ettinger WH Jr, Heimovitz H, Cohen HJ, Wallace R. Associations of elevated interleukin-6 and C-reactive protein levels with mortality in the elderly. *Am J Med* 1999;106:506–512.
75. Volpato S, Guralnik JM, Ferrucci L, Balfour J, Chaves P, Fried LP, Harris TB. Cardiovascular disease, interleukin-6, and risk of mortality in older women: the women's health and aging study. *Circulation* 2001;103:947–953.
76. Gracie JA, Robertson SE, McInnes IB. Interleukin-18. *J Leukoc Biol* 2003;73:213–224.
77. Skurk T, Kolb H, Müller-Scholze S, Röhrig K, Hauner H, Herder C. The proatherogenic cytokine interleukin-18 is secreted by human adipocytes. *Eur J Endocrinol* 2005;152:871–876.
78. Gu Y, Kuida K, Tsutsui H, Ku G, Hsiao K, Fleming MA, Hayashi N, Higashino K, Okamura H, Nakanishi K, Kurimoto M, Tanimoto T, Flavell RA, Sato V, Harding MW, Livingston DJ, Su MS. Activation of interferon-γ inducing factor mediated by interleukin-1β converting enzyme. *Science* 1997;275:206–209.
79. Okamura H, Tsutsi H, Komatsu T, Yutsudo M, Hakura A, Tanimoto T, Torigoe K, Okura T, Nukada Y, Hattori K, Akita K, Namba M, Tanabe F, Konishi K, Fukuda S, Kurimoto M. Cloning of a new cytokine that induces IFN-γ production by T cells. *Nature* 1995;378:88–91.
80. Micallef MJ, Ohtsuki T, Kohno K, Tanabe F, Ushio S, Namba M, Tanimoto T, Torigoe K, Fujii M, Ikeda M, Fukuda S, Kurimoto M. Interferon-γ-inducing factor enhances T helper 1 cytokine production by stimulated human T cells: synergism with interleukin-12 for interferon-γ production. *Eur J Immunol* 1996;26:1647–1651.
81. Gerdes N, Sukhova GK, Libby P, Reynolds RS, Young JL, Schonbeck U. Expression of interleukin (IL)-18 and functional IL-18 receptor on human vascular endothelial cells, smooth muscle cells, and macrophages: implications for atherogenesis. *J Exp Med* 2002;195:245–257.
82. Nold M, Goede A, Eberhardt W, Pfeilschifter J, Muehl H. IL-18 initiates release of matrix metalloproteinase-9 from peripheral blood mononuclear cells without affecting tissue inhibitor of matrix metalloproteinases-1: suppression by TNF-α blockage and modulation by IL-10. *Naunyn Schmiedebergs Arch Pharmacol* 2003;367:68–75.
83. Ishida Y, Migita K, Izumi Y, Nakao K, Ida H, Kawakami A, Abiru S, Ishibashi H, Eguchi K, Ishii N. The role of IL-18 in the modulation of matrix metalloproteinases and migration of human natural killer (NK) cells. *FEBS Lett* 2004;569:156–160.
84. Mallat Z, Corbaz A, Scoazec A, Besnard S, Leseche G, Chvatchko Y, Tedgui A. Expression of interleukin-18 in human atherosclerotic plaques and relation to plaque instability. *Circulation* 2001;104:1598–1603.
85. Mallat Z, Corbaz A, Scoazec A, Graber P, Alouani S, Esposito B, Humbert Y, Chvatchko Y, Tedgui A. Interleukin-18/interleukin-18 binding protein signaling modulates atherosclerotic lesion development and stability. *Circ Res* 2001;89:E41–E45.
86. Elhage R, Jawien J, Rudling M, Ljunggren HG, Takeda K, Akira S, Bayard F, Hansson GK. Reduced atherosclerosis in interleukin-18 deficient apolipoprotein E-knockout mice. *Cardiovasc Res* 2003;59:234–240.
87. Whitman SC, Ravisankar P, Daugherty A. Interleukin-18 enhances atherosclerosis in Apolipoprotein E−/− mice through release of interferon-γ. *Circ Res* 2002;90:e34–e38.
88. Tenger C, Sundborger A, Jawien J, Zhou X. IL-18 accelerates atherosclerosis accompanied by elevation of IFN-γ and CXCL16 expression independently of T cells. *Arterioscler Thromb Vasc Biol* 2005;25:791–796.
89. de Nooijer R, von der Thusen JH, Verkleij CJ, Kuiper J, Jukema JW, van der Wall EE, van Berkel JC, Biessen EA. Overexpression of IL-18 decreases intimal collagen content and promotes a vulnerable plaque phenotype in apolipoprotein-E-deficient mice. *Arterioscler Thromb Vasc Biol* 2004;24:2313–2319.
90. Mallat Z, Henry P, Fressonnet R, Alouani S, Scoazec A, Beaufils P, Chvatchko Y, Tedgui A. Increased plasma concentrations of interleukin-18 in acute coronary syndromes. *Heart* 2002;88:467–469.
91. Rosso R, Roth A, Herz I, Miller H, Keren G, George J. Serum levels of interleukin-18 in patients with stable and unstable angina pectoris. *Int J Cardiol* 2005;98:45–48.

92. Kawasaki D, Tsujino T, Morimoto S, Fujioka Y, Naito Y, Okumura T, Masutani M, Shimizu H, Yuba M, Ueda A, Ohyanagi M, Kashiwamura S, Okamura H, Iwasaki T. Usefulness of circulating interleukin-18 concentration in acute myocardial infarction as a risk factor for late restenosis after emergency coronary angioplasty. *Am J Cardiol* 2003;91:1258–1261.
93. Yamashita H, Shimada K, Seki E, Mokuno H, Daida H. Concentrations of interleukins, interferon, and C-reactive protein in stable and unstable angina pectoris. *Am J Cardiol* 2003;91:133–136.
94. Narins CR, Lin DA, Burton PB, Jin Z-G, Berk BC. Interleukin-18 and interleukin-18 binding protein levels before and after percutaneous coronary intervention in patients with and without recent myocardial infarction. *Am J Cardiol* 2004;94:1285–1287.
95. Chalikias GK, Tziakas DN, Kaski JC, Hatzinikolaou EI, Stakos DA, Tentes IK, Kortsaris A, Hatseras DI. Interleukin-18: Interleukin-10 ratio and in-hospital adverse events in patients with acute coronary syndrome. *Atherosclerosis* 2005;182:135–143.
96. Blankenberg S, Tiret L, Bickel C, Peetz D, Cambien F, Meyer J, Rupprecht HJ; AtheroGene Investigators. Interleukin-18 is a strong predictor of cardiovascular death in stable and unstable angina. *Circulation* 2002;106:24–30.
97. Tiret L, Godefroy T, Lubos E, Nicaud V, Tregouet DA, Barbaux S, Schnabel R, Bickel C, Espinola-Klein C, Poirier O, Perret C, Munzel T, Rupprecht HJ, Lackner K, Cambien F, Blankenberg S; AtheroGene Investigators. Genetic analysis of the interleukin-18 system highlights the role of the interleukin-18 gene in cardiovascular disease. *Circulation* 2005;112:643–650.
98. Blankenberg S, Luc G, Ducimetiere P, Arveiler D, Ferrieres J, Amouyel P, Evans A, Cambien F, Tiret L; PRIME Study Group. Interleukin-18 and the risk of coronary heart disease in European men: the Prospective Epidemiological Study of Myocardial Infarction (PRIME). *Circulation* 2003;108:2453–2459.
99. Koenig W, Khuseyinova N, Baumert J, Thorand B, Loewel H, Chambless L, Meisinger C, Schneider A, Martin S, Kolb H, Herder C. Increased concentrations of C-reactive protein and IL-6 but not IL-18 are independently associated with incident coronary events in middle-aged men and women: Results from the MONICA/KORA Augsburg Case-Cohort Study, 1984–2002. 2006;26:2745–2751.
100. Carswell EA, Old LJ, Kassel RL, Green S, Fiore N, Williamson B. An endotoxin-induced serum factor that causes necrosis of tumors. *Proc Natl Acad Sci USA* 1975;72:3666–3670.
101. Black RA, Rauch CT, Kozlosky CJ, Peschon JJ, Slack JL, Wolfson MF, Castner BJ, Stocking KL, Reddy P, Srinivasan S, Nelson N, Boiani N, Schooley KA, Gerhart M, Davis R, Fitzner JN, Johnson RS, Paxton RJ, March CJ, Cerretti DP. A metalloproteinase disintegrin that releases tumour-necrosis factor-alpha from cells. *Nature* 1997;385:729–733.
102. Anderson HD, Rahmutula D, Gardner DG. Tumor necrosis factor-alpha inhibits endothelial nitric-oxide synthase gene promoter activity in bovine aortic endothelial cells. *J Biol Chem* 2004;279: 963–969.
103. Gertzberg N, Neumann P, Rizzo V, Johnson A. NAD(P)H oxidase mediates the endothelial barrier dysfunction induced by TNF-alpha. *Am J Physiol Lung Cell Mol Physiol* 2004;286:L37–L48.
104. Couffinhal T, Duplaa C, Labat L, Lamaziere JM, Moreau C, Printseva O, Bonnet J. Tumor necrosis factor-alpha stimulates ICAM-1 expression in human vascular smooth muscle cells. *Arterioscler Thromb* 1993;13:407–414.
105. Schuger L, Varani J, Marks RM, Kunkel SL, Johnson KJ, Ward PA. Cytotoxicity of tumor necrosis factor-alpha for human umbilical vein endothelial cells. *Lab Invest* 1989;61:62–68.
106. Goldblum SE, Sun WL. Tumor necrosis factor-alpha augments pulmonary arterial transendothelial albumin flux in vitro. *Am J Physiol* 1990;258:L57–L67.
107. Kume N, Murase T, Moriwaki H, Aoyama T, Sawamura T, Masaki T, Kita T. Inducible expression of lectin-like oxidized LDL receptor-1 in vascular endothelial cells. *Circ Res* 1998;83:322–327.
108. Porter KE, Turner NA, O'Regan DJ, Ball SG. Tumor necrosis factor alpha induces human atrial myofibroblast proliferation, invasion and MMP-9 secretion: inhibition by simvastatin. *Cardiovasc Res* 2004;64:507–515.
109. Migita K, Eguchi K, Kawabe Y, Ichinose Y, Tsukada T, Aoyagi T, Nakamura H, Nagataki S. TNF-alpha-mediated expression of membrane-type matrix metalloproteinase in rheumatoid synovial fibroblasts. *Immunology* 1996;89:553–557.
110. Lee WH, Kim SH, Lee Y, Lee BB, Kwon B, Song H, Kwon BS, Park JE. Tumor necrosis factor receptor superfamily 14 is involved in atherogenesis by inducing proinflammatory cytokines and matrix metalloproteinases. *Arterioscler Thromb Vasc Biol* 2001;21:2004–2010.

111. Creemers EE, Cleutjens JP, Smits JF, Daemen MJ. Matrix metalloproteinase inhibition after myocardial infarction: a new approach to prevent heart failure? *Circ Res* 2001;89:201–210.
112. Shingu M, Nagai Y, Isayama T, Naono T, Nobunaga M, Nagai Y. The effects of cytokines on metalloproteinase inhibitors (TIMP) and collagenase production by human chondrocytes and TIMP production by synovial cells and endothelial cells. *Clin Exp Immunol* 199;94:145–149.
113. Boyle JJ, Weissberg PL, Bennett MR. Tumor necrosis factor-alpha promotes macrophage-induced vascular smooth muscle cell apoptosis by direct and autocrine mechanisms. *Arterioscler Thromb Vasc Biol* 2003;23:1553–1558.
114. Branen L, Hovgaard L, Nitulescu M, Bengtsson E, Nilsson J, Jovinge S. Inhibition of tumor necrosis factor-alpha reduces atherosclerosis in apolipoprotein E knockout mice. *Arterioscler Thromb Vasc Biol* 2004;24:2137–2142.
115. Ohta H, Wada H, Niwa T, Kirii H, Iwamoto N, Fujii H, Saito K, Sekikawa K, Seishima M. Disruption of tumor necrosis factor-alpha gene diminishes the development of atherosclerosis in ApoE-deficient mice. *Atherosclerosis* 2005;180:11–17.
116. Koukkunen H, Penttila K, Kemppainen A, Halinen M, Penttila I, Rantanen T, Pyorala K. C-reactive protein, fibrinogen, interleukin-6 and tumour necrosis factor-alpha in the prognostic classification of unstable angina pectoris. *Ann Med* 2001;33:37–47.
117. Ridker PM, Rifai N, Pfeffer M, Sacks F, Lepage S, Braunwald E. Elevation of tumor necrosis factor-alpha and increased risk of recurrent coronary events after myocardial infarction. *Circulation* 2000;101:2149–2153.
118. Cesari M, Penninx BW, Newman AB, Kritchevsky SB, Nicklas BJ, Sutton-Tyrrell K, Rubin SM, Ding J, Simonsick EM, Harris TB, Pahor M. Inflammatory markers and onset of cardiovascular events: results from the Health ABC study. *Circulation* 2003;108:2317–2322.
119. Henschen A, McDonagh J. Fibrinogen, fibrin and factor XIII. In: Zwaal RFA, Henker HC, eds. *Blood Coagulation*. Amsterdam: Elsevier, 1986:171–242.
120. Scrutton MC, Ross-Murphy SB, Bennett GM, Stirling Y, Meade TW. Changes in clot deformability—a possible explanation for the epidemiological association between plasma fibrinogen concentration and myocardial infarction. *Blood Coagul Fibrinolysis* 1994;5:719–723.
121. McDonagh J, Lee MH. How does hyperfibrinogenemia lead to thrombosis? *Fibrinol Proteol* 1997;11(Suppl 1):13–17.
122. Koenig W. Fibrin(ogen) in cardiovascular disease: an update. *Thromb Haemost* 2003;89:601–609.
123. Retzinger GS, DeAnglis AP, Patuto SJ. Adsorption of fibrinogen to droplets of liquid hydrophobic phases. Functionality of the bound protein and biological implications. *Arterioscler Thromb Vasc Biol* 1998;18:1948–1957.
124. Rabbani LE, Loscalzo J. Recent observations on the role of hemostatic determinants in the development of the atherothrombotic plaque. *Atherosclerosis* 1994;105:1–7.
125. Schneider DJ, Taatjes DJ, Howard DB, Sobel BE. Increased reactivity of platelets induced by fibrinogen independent of its bindings to the IIb-IIIa surface glycoprotein: a potential contributor to cardiovascular risk. *J Am Coll Cardiol* 1999;33:261–266.
126. Fatah K, Hamsten A, Blomback B, Blomback M. Fibrin gel network characteristics and coronary heart disease: relations to plasma fibrinogen concentration, acute phase protein, serum lipoproteins and coronary atherosclerosis. *Thromb Haemost* 1992;68:130–135.
127. Koenig W, Ernst E: The possible role of hemorheology in atherothrombogenesis. *Atherosclerosis* 1992;94:93–107.
128. Fibrinogen Studies Collaboration. Plasma fibrinogen level and the risk of major cardiovascular diseases and nonvascular mortality: an individual participant meta-analysis. *JAMA* 2005;294:1799–1809.
129. Kruithof EK. Plasminogen activator inhibitors—a review. *Enzyme* 1988;40:113–121.
130. Vaughan DE. PAI-1 and atherothrombosis. *J Thromb Haemost* 2005;3:1879–1883.
131. Sawdey MS, Loskutoff DJ. Regulation of murine type 1 plasminogen activator inhibitor gene expression in vivo. Tissue specificity and induction by lipopolysaccharide, tumor necrosis factor-alpha, and transforming growth factor-beta. *J Clin Invest* 1991;88:1346–1353.
132. Emeis JJ, Kooistra T. Interleukin 1 and lipopolysaccharide induce an inhibitor of tissue-type plasminogen activator in vivo and in cultured endothelial cells. *J Exp Med* 1986;163:1260–1266.
133. Brogren H, Karlsson L, Andersson M, Wang L, Erlinge D, Jern S. Platelets synthesize large amounts of active plasminogen activator inhibitor 1. *Blood* 2004;104:3943–3948.
134. Lindahl TL, Ohlsson PI, Wiman B. The mechanism of the reaction between human plasminogen-activator inhibitor 1 and tissue plasminogen activator. *Biochem J* 1990;265:109–113.

135. Falkenberg M, Tjarnstrom J, Ortenwall P, Olausson M, Risberg B. Localization of fibrinolytic activators and inhibitors in normal and atherosclerotic vessels. *Thromb Haemost* 1996;75:933–938.
136. Eitzman DT, Westrick RJ, Xu Z, Tyson J, Ginsburg D. Plasminogen activator inhibitor-1 deficiency protects against atherosclerosis progression in the mouse carotid artery. *Blood* 2000;96:4212–4215.
137. Eren M, Painter CA, Atkinson JB, Declerck PJ, Vaughan DE. Age-dependent spontaneous coronary arterial thrombosis in transgenic mice that express a stable form of human plasminogen activator inhibitor-1. *Circulation* 2002;106:491–496.
138. Luttun A, Lupu F, Storkebaum E, Hoylaerts MF, Moons L, Crawley J, Bono F, Poole AR, Tipping P, Herbert JM, Collen D, Carmeliet P. Lack of plasminogen activator inhibitor-1 promotes growth and abnormal matrix remodeling of advanced atherosclerotic plaques in apolipoprotein E-deficient mice. *Arterioscler Thromb Vasc Biol* 2002;22:499–505.
139. Collet JP, Montalescot G, Vicaut E, Ankri A, Walylo F, Lesty C, Choussat R, Beygui F, Borentain M, Vignolles N, Thomas D. Acute release of plasminogen activator inhibitor-1 in ST-segment elevation myocardial infarction predicts mortality. *Circulation* 2003;108:391–394.
140. Hamsten A, Wiman B, de Faire U, Blomback M. Increased plasma levels of a rapid inhibitor of tissue plasminogen activator in young survivors of myocardial infarction. *N Engl J Med* 1985;313: 1557–1563.
141. Juhan-Vague I, Pyke SD, Alessi MC, Jespersen J, Haverkate F, Thompson SG. Fibrinolytic factors and the risk of myocardial infarction or sudden death in patients with angina pectoris. ECAT Study Group. European Concerted Action on Thrombosis and Disabilities. *Circulation* 1996;94: 2057–2063.
142. Thogersen AM, Jansson JH, Boman K, Nilsson TK, Weinehall L, Huhtasaari F, Hallmans G. High plasminogen activator inhibitor and tissue plasminogen activator levels in plasma precede a first acute myocardial infarction in both men and women: evidence for the fibrinolytic system as an independent primary risk factor. *Circulation* 1998;98:2241–2247.
143. Folsom AR, Aleksic N, Park E, Salomaa V, Juneja H, Wu KK. Prospective study of fibrinolytic factors and incident coronary heart disease: the Atherosclerosis Risk in Communities (ARIC) Study. *Arterioscler Thromb Vasc Biol* 2001;21:611–617.
144. Cushman M, Lemaitre RN, Kuller LH, Psaty BM, Macy EM, Sharrett AR, Tracy RP. Fibrinolytic activation markers predict myocardial infarction in the elderly. The Cardiovascular Health Study. *Arterioscler Thromb Vasc Biol* 1999;19:493–498.
145. Lowe GD, Danesh J, Lewington S, Walker M, Lennon L, Thomson A, Rumley A, Whincup PH. Tissue plasminogen activator antigen and coronary heart disease. Prospective study and meta-analysis. *Eur Heart J* 2004;25:252–259.
146. Smith A, Patterson C, Yarnell J, Rumley A, Ben-Shlomo Y, Lowe G. Which hemostatic markers add to the predictive value of conventional risk factors for coronary heart disease and ischemic stroke? The Caerphilly Study. *Circulation* 2005;112:3080–3087.
147. Fancher TL, White RH, Kravitz RL. Combined use of rapid D-Dimer testing and estimation of clinical probability in the diagnosis of deep vein thrombosis: systematic review. *BMJ* 2004; 329:821–828.
148. Stein PD, Hull RD, Patel KC, Olson RE, Ghali WA, Brant R, Biel RK, Bharadia V, Kalra NK. D-Dimer for the exclusion of acute venous thrombosis and pulmonary embolism: a systematic review. *Ann Intern Med* 2004;140:589–602.
149. Wakai A, Gleeson A, Winter D. Role of fibrin D-Dimer testing in emergency medicine. *Emerg Med J* 2003;20:319–325.
150. Lowe GD. Fibrin D-Dimer and cardiovascular risk. *Semin Vasc Med* 2005;5:387–398.
151. Ridker PM, Hennekens CH, Cerskus A, Stampfer MJ. Plasma concentration of cross-linked fibrin degradation product (D-dimer) and the risk of future myocardial infarction among apparently healthy men. *Circulation* 1994;90:2236–2240.
152. Lowe GD, Rumley A, Sweetnam PM, Yarnell JW, Rumley J. Fibrin D-dimer, markers of coagulation activation and the risk of major ischaemic heart disease in the caerphilly study. *Thromb Haemost* 2001;86:822–827.
153. Danesh J, Whincup P, Walker M, Lennon L, Thomson A, Appleby P, Rumley A, Lowe GD. Fibrin D-dimer and coronary heart disease: prospective study and meta-analysis. *Circulation* 2001;103: 2323–2327.
154. Lowe GD, Rumley A, McMahon AD, Ford I, O'Reilly DS, Packard CJ; West of Scotland Coronary Prevention Study Group. Interleukin-6, fibrin D-dimer, and coagulation factors VII and XIIa in prediction of coronary heart disease. *Arterioscler Thromb Vasc Biol* 2004;24:1529–1534.

155. Pradhan AD, LaCroix AZ, Langer RD, Trevisan M, Lewis CE, Hsia JA, Oberman A, Kotchen JM, Ridker PM. Tissue plasminogen activator antigen and D-dimer as markers for atherothrombotic risk among healthy postmenopausal women. *Circulation* 2004;110:292–300.
156. Ruggeri ZM. Von Willebrand factor. *Curr Opin Hematol* 2003;10:142–149.
157. Wagner DD. Cell biology of von Willebrand factor. *Annu Rev Cell Biol* 1990;6:217–246.
158. Mayadas TN, Wagner DD. von Willebrand factor biosynthesis and processing. *Ann N Y Acad Sci* 1991;614:153–166.
159. van Mourik JA, Romani de Wit T, Voorberg J. Biogenesis and exocytosis of Weibel-Palade bodies. *Histochem Cell Biol* 2002;117:113–122.
160. Mannucci PM. Treatment of von Willebrand's disease. *N Engl J Med* 2004;351:683–694.
161. Mannucci PM. von Willebrand factor: a marker of endothelial damage? *Arterioscler Thromb Vasc Biol* 1998;18:1359–1362.
162. Thompson SG, Kienast J, Pyke SD, Haverkate F, van de Loo JC. Hemostatic factors and the risk of myocardial infarction or sudden death in patients with angina pectoris. European Concerted Action on Thrombosis and Disabilities Angina Pectoris Study Group. *N Engl J Med* 1995;332:635–641.
163. Wiman B, Andersson T, Hallqvist J, Reuterwall C, Ahlbom A, deFaire U. Plasma levels of tissue plasminogen activator/plasminogen activator inhibitor-1 complex and von Willebrand factor are significant risk markers for recurrent myocardial infarction in the Stockholm Heart Epidemiology Program (SHEEP) study. *Arterioscler Thromb Vasc Biol* 2000;20:2019–2023.
164. Meade TW, Cooper JA, Stirling Y, Howarth DJ, Ruddock V, Miller GJ. Factor VIII, ABO blood group and the incidence of ischaemic heart disease. *Br J Haematol* 1994;88:601–607.
165. Whincup PH, Danesh J, Walker M, Lennon L, Thomson A, Appleby P, Rumley A, Lowe GD. von Willebrand factor and coronary heart disease: prospective study and meta-analysis. *Eur Heart J* 200;23:1764–1770.
166. Morange PE, Simon C, Alessi MC, Luc G, Arveiler D, Ferrieres J, Amouyel P, Evans A, Ducimetiere P, Juhan-Vague I; PRIME Study Group. Endothelial cell markers and the risk of coronary heart disease: the Prospective Epidemiological Study of Myocardial Infarction (PRIME) study. *Circulation* 2004;109:1343–1348.
167. Friedman GD, Klatsky AL, Siegelaub AB. The leukocyte count as a predictor of myocardial infarction. *N Engl J Med* 1974;290:1275–1278.
168. Madjid M, Awan I, Willerson JT, Casscells SW. Leukocyte count and coronary heart disease: implications for risk assessment. *J Am Coll Cardiol* 2004;44:1945–1956.
169. Coller BS. Leukocytosis and ischemic vascular disease morbidity and mortality: is it time to intervene? *Arterioscler Thromb Vasc Biol* 2005;25:658–670.
170. Naruko T, Ueda M, Haze K, van der Wal AC, van der Loos CM, Itoh A, Komatsu R, Ikura Y, Ogami M, Shimada Y, Ehara S, Yoshiyama M, Takeuchi K, Yoshikawa J, Becker AE. Neutrophil infiltration of culprit lesions in acute coronary syndromes. *Circulation* 2002;106:2894–2900.
171. Hoover RL, Karnovsky MJ, Austen KF, Corey EJ, Lewis RA. Leukotriene B4 action on endothelium mediates augmented neutrophil/endothelial adhesion. *Proc Natl Acad Sci USA* 1984;81:2191–2193.
172. Sabatine MS, Morrow DA, Cannon CP, Murphy SA, Demopoulos LA, DiBattiste PM, McCabe CH, Braunwald E, Gibson CM. Relationship between baseline white blood cell count and degree of coronary artery disease and mortality in patients with acute coronary syndromes: a TACTICS-TIMI 18 (Treat Angina with Aggrastat and determine Cost of Therapy with an Invasive or Conservative Strategy–Thrombolysis in Myocardial Infarction 18 trial) substudy. *J Am Coll Cardiol* 2002;40:1761–1768.
173. Wong CK, French JK, Gao W, White HD. Relationship between initial white blood cell counts, stage of acute myocardial infarction evolution at presentation, and incidence of Thrombolysis in Myocardial Infarction-3 flow after streptokinase. *Am Heart J* 2003;145:95–102.
174. Danesh J, Collins R, Appleby P, Peto R. Association of fibrinogen, C-reactive protein, albumin, or leukocyte count with coronary heart disease: meta-analyses of prospective studies. *JAMA* 1998;279:1477–1482.
175. Margolis KL, Manson JE, Greenland P, Rodabough RJ, Bray PF, Safford M, Grimm RH Jr, Howard BV, Assaf AR, Prentice R; Women's Health Initiative Research Group. Leukocyte count as a predictor of cardiovascular events and mortality in postmenopausal women: the Women's Health Initiative Observational Study. *Arch Intern Med* 2005;165:500–508.

176. Haim M, Boyko V, Goldbourt U, Battler A, Behar S. Predictive value of elevated white blood cell count in patients with preexisting coronary heart disease: the Bezafibrate Infarction Prevention Study. *Arch Intern Med* 2004;164:433–439.
177. Stewart RA, White HD, Kirby AC, Heritier SR, Simes RJ, Nestel PJ, West MJ, Colquhoun DM, Tonkin AM; Long-Term Intervention with Pravastatin in Ischemic Disease (LIPID) Study Investigators. White blood cell count predicts reduction in coronary heart disease mortality with pravastatin. *Circulation* 2005;111:1756–1762.
178. Grau AJ, Boddy AW, Dukovic DA, Buggle F, Lichy C, Brandt T, Hacke W; CAPRIE Investigators. Leukocyte count as an independent predictor of recurrent ischemic events. *Stroke* 2004;35: 1147–1152.
179. Horne BD, Anderson JL, John JM, Weaver A, Bair TL, Jensen KR, Renlund DG, Muhlestein JB; Intermountain Heart Collaborative Study Group. Which white blood cell subtypes predict increased cardiovascular risk? *J Am Coll Cardiol* 2005;45:1638–1643.
180. Wheeler JG, Mussolino ME, Gillum RF, Danesh J. Associations between differential leukocyte count and incident coronary heart disease: 1764 incident cases from seven prospective studies of 30,374 individuals. *Eur Heart J* 2004;25:1287–1292.
181. Nambi V. The use of myeloperoxidase as a risk marker for atherosclerosis. *Curr Atheroscler Rep* 2005;7:127–131.
182. Klebanoff SJ. Myeloperoxidase: friend and foe. *J Leukoc Biol* 2005;77:598–625.
183. Daugherty A, Dunn JL, Rateri DL, Heinecke JW. Myeloperoxidase, a catalyst for lipoprotein oxidation, is expressed in human atherosclerotic lesions. *J Clin Invest* 1994;94:437–444.
184. Hazen SL, Heinecke JW. 3-Chlorotyrosine, a specific marker of myeloperoxidase-catalyzed oxidation, is markedly elevated in low density lipoprotein isolated from human atherosclerotic intima. *J Clin Invest* 1997;99:1–7.
185. Malle E, Waeg G, Schreiber R, Grone EF, Sattler W, Grone HJ. Immunohistochemical evidence for the myeloperoxidase/H2O2/halide system in human atherosclerotic lesions: colocalization of myeloperoxidase and hypochlorite-modified proteins. *Eur J Biochem* 2000;267:4495–4503.
186. Eiserich JP, Baldus S, Brennan ML, Ma W, Zhang C, Tousson A, Castro L, Lusis AJ, Nauseef WM, White CR, Freeman BA. Myeloperoxidase, a leukocyte-derived vascular NO oxidase. *Science* 2002;296:2391–2394.
187. Abu-Soud HM, Hazen SL. Nitric oxide is a physiological substrate for mammalian peroxidases. *J Biol Chem* 2000;275:37524–37532.
188. Zheng L, Nukuna B, Brennan ML, Sun M, Goormastic M, Settle M, Schmitt D, Fu X, Thomson L, Fox PL, Ischiropoulos H, Smith JD, Kinter M, Hazen SL. Apolipoprotein A-I is a selective target for myeloperoxidase-catalyzed oxidation and functional impairment in subjects with cardiovascular disease. *J Clin Invest* 2004;114:529–541.
189. Bergt C, Pennathur S, Fu X, Byun J, O'Brien K, McDonald TO, Singh P, Anantharamaiah GM, Chait A, Brunzell J, Geary RL, Oram JF, Heinecke JW. The myeloperoxidase product hypochlorous acid oxidizes HDL in the human artery wall and impairs ABCA1-dependent cholesterol transport. *Proc Natl Acad Sci USA* 2004;101:13032–13037.
190. Kutter D, Devaquet P, Vanderstocken G, Paulus JM, Marchal V, Gothot A. Consequences of total and subtotal myeloperoxidase deficiency: risk or benefit? *Acta Haematol* 2000;104:10–15.
191. Brennan ML, Anderson MM, Shih DM, Qu XD, Wang X, Mehta AC, Lim LL, Shi W, Hazen SL, Jacob JS, Crowley JR, Heinecke JW, Lusis AJ. Increased atherosclerosis in myeloperoxidase-deficient mice. *J Clin Invest* 2001;107:419–430.
192. McMillen TS, Heinecke JW, LeBoeuf RC. Expression of human myeloperoxidase by macrophages promotes atherosclerosis in mice. *Circulation* 2005;111:2798–2804.
193. Fu X, Kassim SY, Parks WC, Heinecke JW. Hypochlorous acid oxygenates the cysteine switch domain of pro-matrilysin (MMP-7). A mechanism for matrix metalloproteinase activation and atherosclerotic plaque rupture by myeloperoxidase. *J Biol Chem* 2001;276:41279–41287.
194. Shabani F, McNeil J, Tippett L. The oxidative inactivation of tissue inhibitor of metalloproteinase-1 (TIMP-1) by hypochlorous acid (HOCl) is suppressed by anti-rheumatic drugs. *Free Radic Res* 1998;28:115–123.
195. Baldus S, Heeschen C, Meinertz T, Zeiher AM, Eiserich JP, Munzel T, Simoons ML, Hamm CW; CAPTURE Investigators. Myeloperoxidase serum levels predict risk in patients with acute coronary syndromes. *Circulation* 2003;108:1440–1445.
196. Brennan ML, Penn MS, Van Lente F, Nambi V, Shishehbor MH, Aviles RJ, Goormastic M, Pepoy ML, McErlean ES, Topol EJ, Nissen SE, Hazen SL. Prognostic value of myeloperoxidase in patients with chest pain. *N Engl J Med* 2003;349:1595–1604.

197. Zhang R, Brennan ML, Fu X, Aviles RJ, Pearce GL, Penn MS, Topol EJ, Sprecher DL, Hazen SL. Association between myeloperoxidase levels and risk of coronary artery disease. *JAMA* 2001;286:2136–2142.
198. Vita JA, Brennan ML, Gokce N, Mann SA, Goormastic M, Shishehbor MH, Penn MS, Keaney JF Jr, Hazen SL. Serum myeloperoxidase levels independently predict endothelial dysfunction in humans. *Circulation* 2004;110:1134–1139.
199. Steinberg D. Low density lipoprotein oxidation and its pathobiological significance. *J Biol Chem* 1997;272:20963–20966.
200. Jessup W, Kritharides L, Stocker R. Lipid oxidation in atherogenesis: an overview. *Biochem Soc Trans* 2004;32:134–138.
201. Huang Y, Mironova M, Lopes-Virella MF. Oxidized LDL stimulates matrix metalloproteinase-1 expression in human vascular endothelial cells. *Arterioscler Thromb Vasc Biol* 1999;19:2640–2647.
202. Xu XP, Meisel SR, Ong JM, Kaul S, Cercek B, Rajavashisth TB, Sharifi B, Shah PK. Oxidized low-density lipoprotein regulates matrix metalloproteinase-9 and its tissue inhibitor in human monocyte-derived macrophages. *Circulation* 1999;99:993–998.
203. Li D, Liu L, Chen H, Sawamura T, Ranganathan S, Mehta JL. LOX-1 mediates oxidized low-density lipoprotein-induced expression of matrix metalloproteinases in human coronary artery endothelial cells. *Circulation* 2003;107:612–617.
204. Li D, Liu L, Chen H, Sawamura T, Mehta JL. LOX-1, an oxidized LDL endothelial receptor, induces CD40/CD40L signaling in human coronary artery endothelial cells. *Arterioscler Thromb Vasc Biol* 2003;23:816–821.
205. Holvoet P, Vanhaecke J, Janssens S, Van de Werf F, Collen D. Oxidized LDL and malondialdehyde-modified LDL in patients with acute coronary syndromes and stable coronary artery disease. *Circulation* 1998;98:1487–1494.
206. Toshima S, Hasegawa A, Kurabayashi M, Itabe H, Takano T, Sugano J, Shimamura K, Kimura J, Michishita I, Suzuki T, Nagai R. Circulating oxidized low density lipoprotein levels. A biochemical risk marker for coronary heart disease. *Arterioscler Thromb Vasc Biol* 2000;20:2243–2247.
207. Ehara S, Ueda M, Naruko T, Haze K, Itoh A, Otsuka M, Komatsu R, Matsuo T, Itabe H, Takano T, Tsukamoto Y, Yoshiyama M, Takeuchi K, Yoshikawa J, Becker AE. Elevated levels of oxidized low density lipoprotein show a positive relationship with the severity of acute coronary syndromes. *Circulation* 2001;103:1955–1960.
208. Holvoet P, Mertens A, Verhamme P, Bogaerts K, Beyens G, Verhaeghe R, Collen D, Muls E, Van de Werf F. Circulating oxidized LDL is a useful marker for identifying patients with coronary artery disease. *Arterioscler Thromb Vasc Biol* 2001;21:844–848.
209. Holvoet P, Stassen JM, Van Cleemput J, Collen D, Vanhaecke J. Oxidized low density lipoproteins in patients with transplant-associated coronary artery disease. *Arterioscler Thromb Vasc Biol* 1998;18:100–107.
210. Liu ML, Ylitalo K, Salonen R, Salonen JT, Taskinen MR. Circulating oxidized low-density lipoprotein and its association with carotid intima-media thickness in asymptomatic members of familial combined hyperlipidemia families. *Arterioscler Thromb Vasc Biol* 2004;24:1492–1497.
211. Hulthe J, Fagerberg B. Circulating oxidized LDL is associated with subclinical atherosclerosis development and inflammatory cytokines (AIR Study). *Arterioscler Thromb Vasc Biol* 2002;22:1162–1167.
212. Salonen JT, Yla-Herttuala S, Yamamoto R, Butler S, Korpela H, Salonen R, Nyyssonen K, Palinski W, Witztum JL. Autoantibody against oxidised LDL and progression of carotid atherosclerosis. *Lancet* 1992;339:883–887.
213. Nordin Fredrikson G, Hedblad B, Berglund G, Nilsson J. Plasma oxidized LDL: a predictor for acute myocardial infarction? *J Intern Med* 2003;253:425–429.
214. Meisinger C, Baumert J, Khuseyinova N, Loewel H, Koenig W. Plasma oxidized low-density lipoprotein, a strong predictor for acute coronary heart disease events in apparently healthy, middle-aged men from the general population. *Circulation* 2005;112:651–657.
215. Caslake MJ, Packard CJ. Lipoprotein-associated phospholipase A_2 as a biomarker for coronary disease and stroke. *Nat Clin Pract Cardiovasc Med* 2005;2:529–535.
216. Asano K, Okamoto S, Fukunaga K, Shiomi T, Mori T, Iwata M, Ikeda Y, Yamaguchi K. Cellular source(s) of platelet-activating-factor acetylhydrolase activity in plasma. *Biochem Biophys Res Commun* 1999;261:511–514.
217. Tarbet EB, Stafforini DM, Elstad MR, Zimmerman GA, McIntyre TM, Prescott SM. Liver cells secrete the plasma form of platelet-activating factor acetylhydrolase. *J Biol Chem* 1991;266:16667–16673.

218. Hakkinen T, Luoma JS, Hiltunen MO, Macphee CH, Milliner KJ, Patel L, Rice SQ, Tew DG, Karkola K, Yla-Herttuala S. Lipoprotein-associated phospholipase A_2, platelet-activating factor acetylhydrolase, is expressed by macrophages in human and rabbit atherosclerotic lesions. *Arterioscler Thromb Vasc Biol* 1999;19:2909–2917.
219. Kolodgie F, Burke A, Taye A, Liu W, Sudhir K, Virmani R. Lipoprotein-associated phospholipase A2 is highly expressed in macrophages of coronary lesions prone to rupture. *Circulation* 2004;110 (Suppl III):III-246. Abstract.
220. Caslake MJ, Packard CJ, Suckling KE, Holmes SD, Chamberlain P, Macphee CH. Lipoprotein-associated phospholipase A_2, platelet-activating factor acetylhydrolase: a potential new risk factor for coronary artery disease. *Atherosclerosis* 2000;150:413–419.
221. Tsimihodimos V, Karabina SA, Tambaki AP, Bairaktari E, Miltiadous G, Goudevenos JA, Cariolou MA, Chapman MJ, Tselepis AD, Elisaf M. Altered distribution of platelet-activating factor-acetylhydrolase activity between LDL and HDL as a function of the severity of hypercholesterolemia. *J Lipid Res* 2002;43:256–263.
222. Tjoelker LW, Wilder C, Eberhardt C, Stafforini DM, Dietsch G, Schimpf B, Hooper S, Le Trong H, Cousens LS, Zimmerman GA, Yamadat O, Mclntyre TM, Prescott SM, Gray PW. Anti-inflammatory properties of a platelet-activating factor acetylhydrolase. *Nature* 1995;374: 549–553.
223. Tjoelker LW, Stafforini DM. Platelet-activating factor acetylhydrolases in health and disease. *Biochim Biophys Acta* 2000;1488:102–123.
224. Quarck R, De Geest B, Stengel D, Mertens A, Lox M, Theilmeier G, Michiels C, Raes M, Bult H, Collen D, Van Veldhoven P, Ninio E, Holvoet P. Adenovirus-mediated gene transfer of human platelet-activating factor-acetylhydrolase prevents injury-induced neointima formation and reduces spontaneous atherosclerosis in apolipoprotein E-deficient mice. *Circulation* 2001;103: 2495–2500.
225. Zalewski A, Macphee C. Role of lipoprotein-associated phospholipase A_2 in atherosclerosis: biology, epidemiology, and possible therapeutic target. *Arterioscler Thromb Vasc Biol* 2005;25: 923–931.
226. Macphee CH, Moores KE, Boyd HF, Dhanak D, Ife RJ, Leach CA, Leake DS, Milliner KJ, Patterson RA, Suckling KE, Tew DG, Hickey DM. Lipoprotein-associated phospholipase A_2, platelet-activating factor acetylhydrolase, generates two bioactive products during the oxidation of low-density lipoprotein: use of a novel inhibitor. *Biochem J* 1999;338:479–487.
227. Kume N, Cybulsky MI, Gimbrone MA Jr. Lysophosphatidylcholine, a component of atherogenic lipoproteins, induces mononuclear leukocyte adhesion molecules in cultured human and rabbit arterial endothelial cells. *J Clin Invest* 1992;90:1138–1144.
228. Quinn MT, Parthasarathy S, Steinberg D. Lysophosphatidylcholine: a chemotactic factor for human monocytes and its potential role in atherogenesis. *Proc Natl Acad Sci USA* 1988;85:2805–2809.
229. Macphee C, Benson GM, Shi Y, Zalewski A. Lipoprotein-associated phospholipase A_2: a novel marker of cardiovascular risk and potential therapeutic target. *Expert Opin Investig Drugs* 2005;14:671–679.
230. Packard CJ, O'Reilly DS, Caslake MJ, McMahon AD, Ford I, Cooney J, Macphee CH, Suckling KE, Krishna M, Wilkinson FE, Rumley A, Lowe GD. Lipoprotein-associated phospholipase A_2 as an independent predictor of coronary heart disease. West of Scotland Coronary Prevention Study Group. *N Engl J Med* 2000;343:1148–1155.
231. Koenig W, Khuseyinova N, Lowel H, Trischler G, Meisinger C. Lipoprotein-associated phospholipase A_2 adds to risk prediction of incident coronary events by C-reactive protein in apparently healthy middle-aged men from the general population: results from the 14-year follow-up of a large cohort from southern Germany. *Circulation* 2004;110:1903–1908.
232. Oei HH, van der Meer IM, Hofman A, Koudstaal PJ, Stijnen T, Breteler MMB, Witteman JCM. Lipoprotein-associated phospholipase A_2 activity is associated with risk of coronary heart disease and ischemic stroke: the Rotterdam Study. *Circulation* 2005;111:570–575.
233. Blake GJ, Dada N, Fox JC, Manson JE, Ridker PM. A prospective evaluation of lipoprotein-associated phospholipase A_2 levels and the risk of future cardiovascular events in women. *J Am Coll Cardiol* 2001;38:1302–1306.
234. Ballantyne CM, Hoogeveen RC, Bang H, Coresh J, Folsom AR, Heiss G, Sharrett AR. Lipoprotein-associated phospholipase A_2, high-sensitivity C-reactive protein, and risk for incident coronary heart disease in middle-aged men and women in the Atherosclerosis Risk in Communities (ARIC) study. *Circulation* 2004;109:837–842.

235. Brilakis ES, McConnell JP, Lennon RJ, Elesber AA, Meyer JG, Berger PB. Association of lipoprotein-associated phospholipase A_2 levels with coronary artery disease risk factors, angiographic coronary artery disease, and major adverse events at follow-up. *Eur Heart J* 2005;26: 137–144.
236. Koenig W, Twardella D, Brenner H, Rothenbacher D. Lipoprotein-associated phospholipase A_2 predicts future cardiovascular events in patients with coronary heart disease independently of traditional risk factors, markers of inflammation, renal function, and hemodynamic stress. Arterioscler Thromb Vasc Biol 2006;26: 1586–1593.
237. Ballantyne CM, Hoogeveen RC, Bang H, Coresh J, Folsom AR, Chambless LE, Myerson M, Wu KK, Sharrett AR, Boerwinkle E. Lipoprotein-associated phospholipase A_2, high-sensitivity C-reactive protein, and risk for incident ischemic stroke in middle-aged men and women in the Atherosclerosis Risk in Communities (ARIC) study. *Arch Intern Med* 2005;165:2479–2484.
238. Shimada K, Miyazaki T, Daida H. Adiponectin and atherosclerotic disease. *Clin Chim Acta* 2004;344:1–12.
239. Lau DC, Dhillon B, Yan H, Szmitko PE, Verma S. Adipokines: molecular links between obesity and atheroslcerosis. *Am J Physiol Heart Circ Physiol* 2005;288: H2031–H2041.
240. Ouchi N, Kihara S, Arita Y, Maeda K, Kuriyama H, Okamoto Y, Hotta K, Nishida M, Takahashi M, Nakamura T, Yamashita S, Funahashi T, Matsuzawa Y. Novel modulator for endothelial adhesion molecules: adipocyte-derived plasma protein adiponectin. *Circulation* 1999;100:2473–2476.
241. Ouchi N, Kihara S, Arita Y, Nishida M, Matsuyama A, Okamoto Y, Ishigami M, Kuriyama H, Kishida K, Nishizawa H, Hotta K, Muraguchi M, Ohmoto Y, Yamashita S, Funahashi T, Matsuzawa Y. Adipocyte-derived plasma protein, adiponectin, suppresses lipid accumulation and class A scavenger receptor expression in human monocyte-derived macrophages. *Circulation* 2001;103:1057–1063.
242. Matsuda M, Shimomura I, Sata M, Arita Y, Nishida M, Maeda N, Kumada M, Okamoto Y, Nagaretani H, Nishizawa H, Kishida K, Komuro R, Ouchi N, Kihara S, Nagai R, Funahashi T, Matsuzawa Y. Role of adiponectin in preventing vascular stenosis. The missing link of adipo-vascular axis. *J Biol Chem* 2002;277:37487–37491.
243. Kubota N, Terauchi Y, Yamauchi T, Kubota T, Moroi M, Matsui J, Eto K, Yamashita T, Kamon J, Satoh H, Yano W, Froguel P, Nagai R, Kimura S, Kadowaki T, Noda T. Disruption of adiponectin causes insulin resistance and neointimal formation. *J Biol Chem* 2002;277:25863–25866.
244. Maeda N, Shimomura I, Kishida K, Nishizawa H, Matsuda M, Nagaretani H, Furuyama N, Kondo H, Takahashi M, Arita Y, Komuro R, Ouchi N, Kihara S, Tochino Y, Okutomi K, Horie M, Takeda S, Aoyama T, Funahashi T, Matsuzawa Y. Diet-induced insulin resistance in mice lacking adiponectin/ACRP30. *Nat Med* 2002;8:731–737.
245. Okamoto Y, Kihara S, Ouchi N, Nishida M, Arita Y, Kumada M, Ohashi K, Sakai N, Shimomura I, Kobayashi H, Terasaka N, Inaba T, Funahashi T, Matsuzawa Y. Adiponectin reduces atherosclerosis in apolipoprotein E-deficient mice. *Circulation* 2002;106:2767–2770.
246. Kumada M, Kihara S, Sumitsuji S, Kawamoto T, Matsumoto S, Ouchi N, Arita Y, Okamoto Y, Shimomura I, Hiraoka H, Nakamura T, Funahashi T, Matsuzawa Y; Osaka CAD Study Group. Coronary artery disease. Association of hypoadiponectinemia with coronary artery disease in men. *Arterioscler Thromb Vasc Biol* 2003;23:85–89.
247. Hotta K, Funahashi T, Arita Y, Takahashi M, Matsuda M, Okamoto Y, Iwahashi H, Kuriyama H, Ouchi N, Maeda K, Nishida M, Kihara S, Sakai N, Nakajima T, Hasegawa K, Muraguchi M, Ohmoto Y, Nakamura T, Yamashita S, Hanafusa T, Matsuzawa Y. Plasma concentrations of a novel, adipose-specific protein, adiponectin, in type 2 diabetic patients. *Arterioscler Thromb Vasc Biol* 2000;20:1595–1599.
248. Rothenbacher D, Brenner H, März W, Koenig W. Adiponectin, risk of coronary heart disease and correlations with cardiovascular risk markers. *Eur Heart J* 2005;26:1640–1646.
249. Nakamura Y, Shimada K, Fukuda D, Shimada Y, Ehara S, Hirose M, Kataoka T, Kamimori K, Shimodozono S, Kobayashi Y, Yoshiyama M, Takeuchi K, Yoshikawa J. Implications of plasma concentrations of adiponectin in patients with coronary artery disease. *Heart* 2004;90:528–533.
250. Kojima S, Funahashi T, Sakamoto T, Miyamoto S, Soejima H, Hokamaki J, Kajiwara I, Sugiyama S, Yoshimura M, Fujimoto K, Miyao Y, Suefuji H, Kitagawa A, Ouchi N, Kihara S, Matsuzawa Y, Ogawa H. The variation of plasma concentrations of a novel, adipocyte derived protein, adiponectin, in patients with acute myocardial infarction. *Heart* 2003;89:667.
251. Pischon T, Girman CJ, Hotamisligil GS, Rifai N, Hu FB, Rimm EB. Plasma adiponectin levels and risk of myocardial infarction in men. *JAMA* 2004;291:1730–1737.

252. Lawlor DA, Davey Smith G, Ebrahim S, Thompson C, Sattar N. Plasma adiponectin levels are associated with insulin resistance, but do not predict future risk of coronary heart disease in women. *J Clin Endocrinol Metab* 2005;90:5677–5683.
253. Koenig W, Khuseyinova N, Baumert J, Meisinger C, Löwel H. Serum concentrations of adiponectin and risk of type 2 diabetes mellitus and coronary heart disease in apparently healthy middle-aged men: results from the 18-years follow-up of a large cohort from southern Germany. *J Am Coll Cardiol* 2006;48:1369–1377.
254. von Eynatten M, Schneider JG, Humpert PM, Rudofsky G, Schmidt N, Barosch P, Hamann A, Morcos M, Kreuzer J, Bierhaus A, Nawroth PP, Dugi KA. Decreased plasma lipoprotein lipase in hypoadiponectinemia: an association independent of systemic inflammation and insulin resistance. *Diabetes Care* 2004;27:2925–2929.
255. Manolio T. Novel risk markers and clinical practice. *N Engl J Med* 2003;349:1587–1589.
256. Persson Nilsson J, Hedblad B, Nelson JJ, Bergland G. Lipoprotein-associated phospholipase A2 predicts cardiovascular events. *Circulation* 2005;112(Abstract Suppl):II-802.
257. Shohet RV, Anwar A, Johnston JM, Cohen JC. Plasma platelet-activating factor acetylhydrolase activity is not associated with premature coronary atherosclerosis. *Am J Cardiol* 1999;83(1):109–111, A8–A9.
258. Blankenberg S, Stengel D, Rupprecht HJ, Bickel C, Meyer J, Cambien F, Tiret L, Ninio E. Plasma PAF-acetylhydrolase in patients with coronary artery disease: results of a cross-sectional analysis. *J Lipid Res* 2003;44(7):1381–1386.
259. Winkler K, Abletshauser C, Friedrich I, Hoffmann MM, Wieland H, Marz W. Fluvastatin slow-release lowers platelet-activating factor acetyl hydrolase activity: a placebo-controlled trial in patients with type 2 diabetes. *J Clin Endocrinol Metab* 2004;89(3):1153–1159.
260. Khuseyinova N, Imhof A, Rothenbacher D, Trischler G, Kuelb S, Scharnagl H, Maerz W, Brenner H, Koenig W. Association between Lp-PLA2 and coronary artery disease: focus on its relationship with lipoproteins and markers of inflammation and hemostasis. *Atherosclerosis* 2005;182(1): 181–188.
261. Winkler K, Winkelmann BR, Scharnagl H, Hoffmann MM, Grawitz AB, Nauck M, Bohm BO, Marz W. Platelet-activating factor acetylhydrolase activity indicates angiographic coronary artery disease independently of systemic inflammation and other risk factors: the Ludwigshafen Risk and Cardiovascular Health Study. *Circulation* 2005;111(8):980–987.
262. Iribarren C, Gross MD, Darbinian JA, Jacobs DR Jr, Sidney S, Loria CM. Association of lipoprotein-associated phospholipase A2 mass and activity with calcified coronary plaque in young adults: the CARDIA study. *Arterioscler Thromb Vasc Biol* 2005;25(1):216–221.

6 Dietary Prescriptions to Control Dyslipidemias and Coronary Disease Risk

Margo A. Denke, MD

CONTENTS

INTRODUCTION
DIET-ENTHUSIAST'S SUMMARY OF THE EVOLVING ROLE
  OF DIET
PRESCRIPTION PAD 1: DIETARY THERAPY FOR HIGH
  LDL LEVELS
PRESCRIPTION PAD 2: DIETARY THERAPY
  FOR HYPERTRIGLYCERIDEMIA
PRESCRIPTION PAD 3: DIETARY
  THERAPY FOR COMBINED ELEVATIONS
  IN TRIGLYCERIDES AND LDL CHOLESTEROL
  OR NON-HDL CHOLESTEROL LEVELS
  (COMBINED HYPERLIPIDEMIA)
PRESCRIPTION PAD 4: DIETARY THERAPY FOR LOW
  HDL CHOLESTEROL
REFERENCES

INTRODUCTION

Dietary therapy remains the first step in the treatment of dyslipidemia [1], even when considering high-risk patients, where prompt initiation of drug therapy has proven benefits. The lipid-modifying effects of diet only partially accounts for diet's benefit. For most clinicians, prescription of drug therapy is a faster step than prescription of dietary therapy. This should not be the case, as oftentimes simple adjustments in diet can be as rapidly recommended as drug therapy. To guide dietary advice given by a pharmaceutically educated clinician this chapter will compartmentalize the elements of diet according to its therapeutic indications, and dietary factors within each indication are separated into a general class of intervention. Prior to this more restricted view of dietary therapy, a historical review of the role of diet in cholesterol lowering and coronary disease prevention is presented.

From: *Contemporary Cardiology: Therapeutic Lipidology*
Edited by: M. H. Davidson, P. P. Toth, and K. C. Maki © Humana Press Inc., Totowa, NJ

DIET-ENTHUSIAST'S SUMMARY OF THE EVOLVING ROLE OF DIET

Pre-statin Cholesterol Lowering (1960–1987)

By the 1960s, Ancel Keys and Mark Hegstead had nearly completed their work identifying the major dietary factors that controlled serum cholesterol levels. Saturated fatty acids and dietary cholesterol were shown to raise serum cholesterol levels, and polyunsaturated fatty acids appeared to lower them. The effects of diet were reproducible and predictable.

The prediction of diet's contribution to cholesterol levels of the population came at a time when the food supply was rich with full-fat dairy products, lard leavened baked goods, and marbled meats with $1/2$" trim. The typical intake of saturated fatty acids was 17–22% of total calories. The expected cholesterol-lowering effects of restricting saturated fat intake to 10% of calories from saturated fat led easily to at least a 10% reduction in total cholesterol *(2)*. Niacin, clofibrate, phytosterols, and neomycin were the only drugs available to treat hypercholesterolemia, each drug capable of achieving nearly a 10% reduction in total cholesterol *(3)*. The choice for clinicians was between two equipotent therapies—diet that appeared safe versus drugs that had side effects and risks.

The popularity of diet began to wane as clinical trials focused on disease endpoints. The efficacy and feasibility of dietary modification was called into question when the large-scale Multiple Risk Factor Intervention Trial (MRFIT) achieved only a 2% reduction in total cholesterol levels and no reduction in coronary events *(4)*. Improved trial designs employed in the Lipid Research Clinics Coronary Primary Prevention Trial (LRC-CPPT) using cholestyramine (9% reduction in total cholesterol) *(5)* and Helsinki Heart Study using gemfibrozil (10% reduction in total cholesterol) *(6)* proved that lipid-lowering drug therapy could prevent coronary events. Neither trial was powered to prove mortality benefit, and slight increases in "accidental deaths" in both trials fed lingering doubts that an overall benefit from lipid-modifying drug therapy could be achieved. Even if diet was not as effective, at least it was safe.

Early Statin Years: Focus Shifts to LDL Lowering (1987–1994)

The year 1987 marked significant changes in both diet and drug therapies. Consumer demand created dedicated shelf space for low-fat diary products. Monounsaturated fat was found to be as effective as polyunsaturated fat for cholesterol lowering, clarifying the focus from P to S ratio of the diet to percent calories from saturated fat intake. US Department of Agriculture (USDA) renamed the beef grade "good" (30% less marbling than "prime" and 15% less than "choice" grade) to a more appealing name "select." Marketplace trimming of fat from cuts of beef fell to 1/16". These changes collectively achieved reductions in the typical intake of saturated fat. This is turn reduced the expected benefits of further saturated fat restriction from 10% reduction in serum to cholesterol 3–7% reduction. On the drug front, the first statin (lovastatin) was approved by the FDA. Lovastatin achieved a reduction in total cholesterol of 16–24%, dwarfing the cholesterol-lowering effects of diet. The introduction of six additional statin compounds followed. In 1994, National Cholesterol Education Program (NCEP) II identified low-density lipoprotein (LDL) as the target of therapy.

Landmark Statin Trials, LDL Lowering, and Growth of the Metabolic Syndrome (1994–2004)

In 1994, the 4S study was published showing simvastatin therapy achieved a reduction in all-cause mortality in men with existing coronary disease *(7)*. This year also brought results from the Lyon Diet Heart Study, showing reduction in cardiovascular events and mortality from a Mediterranean dietary intervention that did not improve the lipid profile *(8)*. Over the next 10 years, additional statin trials extended and embellished the benefits of statin therapy, redefining optimum target LDL levels as well as risk categories for drug therapy. More powerful statins, and statins in combination with cholesterol absorption inhibitors, could achieve a 50% reduction in LDL.

Population tracking of health indices showed significant increases in body weight; these were attributed to widespread changes in lifestyle habits *(9)*. NCEP III defined the metabolic syndrome to bring obesity and body weight into consideration of dyslipidemia, but insufficient data were available to bring the metabolic syndrome in as a risk factor benefiting from LDL reduction *(1)*. In 2002, another dietary trial with CHD endpoints emphasized the benefits of lifestyle in reducing events *(10)*; however, the ability of this trial to add to the evidence supporting dietary therapy has been tempered because of scientific concerns regarding the veracity of the trial *(11)*.

Lipid Therapy Beyond LDL and CHD Prevention Dietary Therapies Beyond Lipids (2004 to Present)

Statin therapy has been shown to consistently reduce clinical event rates by 25–35%, meaning that the majority of statin-treated patients continue to have events. Search for alternative, complimentary, or additive therapies has been ongoing. Several new and potent high-density lipoprotein (HDL)-raising drugs are under investigation. A PPAR-alpha and PPAR-gamma agonist that has been shown to improve lipids and insulin resistance was disappointedly associated with greater coronary events *(12)*. In the Veterans Affairs HDL Intervention Trial (VA-HIT), the presence of insulin resistance predicted benefit from gemfibrozil therapy *(13)*. The presence of an elevated high-sensitivity C-reactive protein (hsCRP) has been shown to predict events among statin-treated patients who still achieved their LDL treatment goal; attempts to reduce inflammation through antioxidant vitamins *(14)* and antibiotic therapy *(15)* have been universally unsuccessful. A new theory linking diet with coronary disease suggests that postprandial peaks in glucose and lipids create oxidative stress, promoting inflammation *(16)*. In a long-term diet trial, the Mediterranean style diet was found to reduce markers of inflammation in patients with the metabolic syndrome *(17)*. Whether simple improvement in dietary patterns (more whole grains, fiber, fish, and less concentrated sweets and saturated fats) will result in reductions in coronary heart disease among statin-treated patients remains to be established. The hypothesis is attractive. Beyond the Lyon Diet Heart Trial, lifestyle interventions in the Diabetes Prevention Program *(18)*, and Dietary Approaches to Stop Hypertension (DASH) trials *(19)*, lend support to this theory.

The role of diet has evolved from an adversary to cholesterol lowering drug therapy, with the expectations that both therapies will lead to reduced coronary events. There are far fewer data on diet than on drug therapies, but the data available support prudent recommendations for dietary pattern changes and increased physical activity for all of our patients, irrespective of their lipid profile. Improvements in the lipid profile, could be obtained by targeted advice for the patient's particular lipid disorder. Most of the published data supporting these measures are reviewed in detail elsewhere *(2)*; only new information is referenced directly.

PRESCRIPTION PAD 1: DIETARY THERAPY FOR HIGH LDL LEVELS

There are four basic manuvers to lower LDL. All are complementary to each other; reductions in saturated fat and body weight are likely to achieve greater reductions in LDL cholesterol than reductions in dietary cholesterol intake. Other popular measures (e.g., soy protein, fish oil, and increases in physical activity) have not been shown to be reliably effective in reducing LDL *(20)*.

1. Reduce saturated fat intake. Ingestion of saturated fatty acids depresses hepatic LDL receptor activity. The primary focus of dietary therapy is a reduction in saturated fatty acid intake. Resurrection of the Adkin's diet for weight loss and carbohydrate restriction to control early glucose intolerance have led some patients to select an egg, bacon, and sausage breakfast over cereal for a breakfast meal, with equally poor choices for other meals. Physician can begin the clinical instruction to the patient by reviewing the frequency and portion size of common sources of saturated fatty acids (full-fat dairy products, butter, cheese, ice cream, bacon, sausage, ribs, fatty meats, and donuts). A registered dietician can be invaluable in helping the patient find lower saturated fat substitutes for these foods.
2. Reduce excessive endogenous cholesterol production. Individuals who are overweight have an additional metabolic abnormality that is amenable to dietary therapy: obesity leads to an increased production of very low density lipoproteins by the liver. Reduction of body weight will reduce the production rate of these LDL precursors. The physician can start the weight loss process by recommending an appropriate level of caloric restriction (Table 1).
3. Reduce dietary cholesterol intake. Dietary cholesterol also depresses LDL receptor activity and can contribute to elevated cholesterol levels. Dietary cholesterol comes from a few cholesterol-rich foods (organ meats and egg yolk), and it is fairly simple for a patient to restrict cholesterol intake from these foods by eliminating organ meats from the diet and restricting egg yolks to no more than 3 per week. A more "hidden" source of cholesterol is meat and dairy products. Every ounce of meat has approximately 25 mg of dietary cholesterol, and most dairy products contain 20–40 mg of cholesterol/cup. Physicians can be very effective in counseling the portion size of meat intake; a piece

Table 1
Simplified Equations to Calculate Caloric Requirements

Men = $[900 + 4.5 \times (\text{wt in lb})] \times$ activity factor
Women = $[800 + 3.2 \times (\text{wt in lb})] \times$ activity factor

where activity factor low = 1.2 (e.g., sedentary job),
moderate = 1.4, high = 1.6 (e.g., manual labor; daily exercise program)

For a weight loss of 1–2 lb/week, a daily caloric deficit of 500 cal is required.

Source: Owen et al. *(23,24)*.

of meat the size of a deck of cards is 3 oz and only two portions are recommended each day.
4. Increase water-soluble fiber intake. By an unknown mechanism, water-soluble fiber intake of 3 g/day has been shown to reproducibly lower LDL 3%.

Expected effect: The magnitude of reduction depends on the patient's initial diet, their initial body weight, and the patient's ability to alter diet and weight. In general, LDL reduction of 5–11% can be achieved. Greater LDL reductions are expected from individuals who have an initial diet rich in saturated fatty acids and dietary cholesterol. Greater LDL reductions are expected from individuals who are able to achieve even modest reductions in body weight.

PRESCRIPTION PAD 2: DIETARY THERAPY FOR HYPERTRIGLYCERIDEMIA

There are five primary maneuvers to lower triglyceride levels. As with LDL dietary interventions, all are complementary to each other. Before initiating these prescriptions, note that if the fasting triglycerides are greater than 1000 mg/dL, restricting total fat to less than 5 g/day is indicated to reduce chylomicron triglyceride input and reduce risk for pancreatitis. Within 48–72 h of this diet, fasting triglycerides are expected to fall below 1000, and the dietary measures that can be prescribed are as follows:

1. Restrict excess alcohol intake. Determine alcohol intake as percent of calories; if more than 7%, alcohol is contributing to the hypertriglyceridemia. Patients who make their own drinks should be advised to use a jigger to measure alcohol and a measuring cup to measure wine intake. Limit hard liquor intake to no more than $1\frac{1}{2}$ oz/day, beer intake to 12 oz, and wine intake to 5 oz (2/3 cup).
2. Restrict excessive calorie intake. Evaluate body weight using body mass index (BMI) and waist circumference. Hypocaloric diets, even without significant weight loss (e.g., restricting caloric intake by only 200 cal/day), will improve fasting triglyceride levels.
3. Improve insulin resistance. Evaluate fasting blood sugar. Individuals with glucose intolerance (e.g., fasting glucose > 110 mg/dL) may have an associated derangement in triglyceride metabolism. Weight reduction, spacing out calories with three meals and two snacks, and regular physical activity can improve lipid levels by both reducing hepatic production of triglycerides and reducing input of dietary triglycerides to the liver.
4. Avoid carbohydrate-induced hypertriglyceridemia. Evaluate the percent calories from fat versus carbohydrate. Individuals with stringent, very low fat diets (e.g., Pritikin and Ornish) may have simply a carbohydrate-induced mild hypertriglyceridemia, and adding fats back that are low in saturated and trans fatty acids (e.g., vegetable oils, nuts, and avocados) can reduce the triglycerides and raise the HDL cholesterol level.
5. Consider the use of omega-3 fatty acid supplementation. Fish oil supplementation has been shown in two clinical trials to reduce coronary event rates *(21,22)*. High dose of fish oil (3–12 g/day) can be effective therapy for hypertriglyceridemia.

Expected effect: Many patients with excess body weight can normalize their triglycerides when they lose weight. Similarly, those who consume excess calories from alcohol can improve triglycerides with alcohol restriction. Patients who are lean, do not drink, and do not have glucose intolerance are unlikely to respond to dietary maneuvers unless the cause of their high triglycerides is a very low fat diet. A regular exercise program will reduce fasting triglycerides, but it is unlikely that exercise alone will normalize levels unless significant weight loss occurs.

PRESCRIPTION PAD 3: DIETARY THERAPY FOR COMBINED ELEVATIONS IN TRIGLYCERIDES AND LDL CHOLESTEROL OR NON-HDL CHOLESTEROL LEVELS (COMBINED HYPERLIPIDEMIA)

Mixed dyslipidemia is difficult to treat, and if it is amenable to lifestyle therapy, the dyslipidemia is usually because of insulin resistance.

1. Improve insulin resistance. Evaluate BMI and waist circumference. Weight reduction may be effective in lowering triglyceride levels. Regular physical activity also can lower triglyceride levels. If triglycerides can be reduced to below 250 mg/dL, a single agent may be effective for drug therapy.
2. Restrict excess alcohol. Quantify alcohol intake. As with disorders of simple hypertriglyceridemia, occasional patients will have a contribution of their hypertriglyceridemia from alcohol.
3. Employ all LDL reduction strategies listed earlier. Reductions in saturated fatty acids and dietary cholesterol, as in simple hypercholesterolemia, have also been shown to be effective in producing LDL reductions in these patients.

Expected effect: Dietary therapy will be most effective in patients with Type III hyperlipidemia and in patients who are overweight.

PRESCRIPTION PAD 4: DIETARY THERAPY FOR LOW HDL CHOLESTEROL

Goals of dietary and lifestyle therapy are as follows:

1. Improve insulin resistance. Weight loss and increased physical activity have been shown to independently raise HDL.
2. Smoking cessation. Smoking cessation will increase HDL cholesterol levels and reduce the risk for coronary disease in other ways. When the clinician discusses smoking cessation, he/she should also discuss that the patient may gain weight in the process. By openly discussing weight gain, the physician will help prioritize the relative risk of these two factors. Although both may lower HDL cholesterol levels, smoking is a much more powerful cardiovascular risk factor than obesity. The small increase in cardiovascular risk that may occur from weight gain will be more than offset by the immediate risk reduction from smoking cessation.
3. Liberalize fat intake. In patients who are consuming less than 25% of calories from total fat, an increase in dietary fat intake may raise HDL cholesterol levels. Because increases in saturated fat intake will also raise LDL cholesterol levels, the extra dietary fat should be low in saturated fat. Because fat is more calorically dense than carbohydrate, when recommending high fat intake, it is important to caution the patient that concomitant reductions in intake of other food stuffs must take place. An easy way to add fat back to a diet that is otherwise healthy is to add oil as salad dressing or to add $1/4$ cup of nuts to the diet. Any type of nuts (except macadamia nuts) will do; all are low in saturated fatty acids. Raw or roasted nuts can be used; even if nuts are roasted in coconut oil, little of the roasting fat ends up on the roasted nut.

Expected effect: The magnitude of increase in HDL using the aforementioned measures depends on how much of the patient's HDL level is genetically determined and how much is modifiable by environment. Anecdotally, I have seen individuals whose HDL level rose from 23–30 mg/dL range to 42–48 mg/dL by changing the fat content of the diet or from smoking cessation, but these patients are clearly the exception rather than the rule. On average, a change of 4–6 mg/dL in HDL from weight loss, smoking cessation, or changes in dietary consumption can be expected.

REFERENCES

1. Expert Panel on Detection, Evaluation, and Treatment of High Blood Cholesterol in Adults. Executive summary of the third report of the National Cholesterol Education Program (NCEP) Expert Panel on detection, evaluation and treatment of high blood cholesterol in adults (Adult Treatment Panel III). *JAMA* 2001; 285: 2486–2497.
2. Denke MA. Diet and lifestyle modification and its relationship to atherosclerosis. *Med Clin North Am* 1994; 78: 197–223.
3. Stamler J. The Coronary Drug Project: findings with regard to estrogen, dextrothyroxine, clofibrate and niacin. *Adv Exp Med Biol* 1977; 82: 52–75.
4. Multiple Risk Factor Intervention Trial Research Group. Multiple Risk Factor Intervention Trial: risk factor changes and mortality results. *JAMA* 1982; 248: 1465–1477.
5. Lipid Research Clinics Program. The Lipid Research Clinics Coronary Primary Prevention Trial results. I: Reduction in incidence of coronary heart disease. *JAMA* 1984; 251: 351–364.
6. Manninen V, Elo MO, Frick MH, et al. Lipid alternations and decline in the incidence of coronary heart disease in the Helsinki Heart Study. *JAMA* 1988; 260: 641–651.
7. Scandianvian Simvastatin Survival Study Group. Randomised trial of cholesterol lowering in 4444 patients with coronary heart disease: the Scandinavian Simvastatin Survival Study (4S). *Lancet* 1994; 344: 1383–1389.
8. de Lorgeril M, Renaud S, Mamelle N, et al. Mediterranean alpha-linolenic acid-rich diet in secondary prevention of coronary heart disease. *Lancet* 1994; 343: 1454–1459.
9. Grundy SM, Hanse B, Smith SC Jr, et al. Clinical management of metabolic syndrome: Report of the American Heart Association/National Heart, Lung, and Blood Institute/American Diabetes Association conference on scientific issues related to management. *Circulation* 2004; 109: 551–556.
10. Singh RB, Dubnov G, Niaz MA, et al. Effect of an Indo-Mediterranean diet on progression of coronary artery disease in high risk patients (Indo-Mediterranean Diet Heart Study): a randomized single-blind trial. *Lancet* 2002; 360: 1455–1461.
11. Horton R. Expression of concern: Indo-Mediterranean Diet Heart Study. *Lancet* 2005; 366: 354–356.
12. Nissen SE, Wolski K, Topol EJ. Effect of Muraglitazar on death and major adverse cardiovascular events in patients with type 2 diabetes mellitus. *JAMA* 2005; 294: 2581–2586 (doi:10.1001/jama.294.20.joc50147).
13. Robins SR, Rubins HB, Faas FA, et al. Insulin resistance and cardiovascular events with low HDL cholesterol: The Veterans Affairs HDL Intervention Trial (VA-HIT). *Diabetes Care* 2003; 26: 1513–1517.
14. Kris-Etherton PM, Lichtenstein AH, Howard BV, et al. for the Nutrition Committee of the American Heart Association Council on Nutrition, Physical Activity, and Metabolism. Antioxidant vitamin supplements and cardiovascular disease. *Circulation* 2004; 110: 637–641.
15. Cannon CP, Braunwald E, McCabe CH, et al. Pravastatin or atorvastatin evaluation and infection therapy-thrombolysis in myocardial infarction 22 investigators. Intensive versus moderate lipid lowering with statins after acute coronary syndromes. *N Engl J Med* 2004; 350: 1495–1504.
16. Ceriello A, Taboga C, Tonutti L, et al. Evidence for an independent and cumulative effect of postprandial hypertriglyceridemia and hyperglycemia on endothelial dysfunction and oxidative stress generation. *Circulation* 2002; 106: 1211–1218.
17. Esposito K, Marfella R, Ciotola M, et al. Effects of a mediterranean-style diet on endothelial dysfunction and markers of vascular inflammation in the metabolic syndrome: a randomized trial. *JAMA* 2004; 292: 1440–1446.
18. Haffner S, Temprosa M, Crandall J, et al. Diabetes Prevention Program Research Group Intensive lifestyle intervention or metformin on inflammation and coagulation in participants with impaired glucose tolerance. *Diabetes* 2005; 54(5): 1566–1572.
19. Lopes HF, Martin KL, Nashar K, et al. DASH diet lowers blood pressure and lipid-induced oxidative stress in obesity. *Hypertension* 2003; 41: 422–430.
20. Kerckhoffs DA, Brouns F, Hornstra G, Mensink RP. Effects on the human serum lipoprotein profile of beta-glucan, soy protein and isoflavones, plant sterols and stanols, garlic and tocotrienols. *J Nutr* 2002; 132: 2494–2505.

21. Burr ML, Fehily AM, Gilbert JF, et al. Effects of changes in fat, fish and fibre intakes on death and myocardial reinfarction: Diet and Reinfarction Trial (DART). *Lancet* 1989; 2: 757–761.
22. GISSI-Prevenzione Investigators. Dietary supplementation with n-3 polyunsaturated fatty acids and vitamin E after myocardial infarction: results of the GISSI-Prevenzione trial. *Lancet* 1999; 354: 447–455.
23. Owen OE, Holup JL, D'Alessio DA, et al. A reappraisal of the caloric requirements of men. *Am J Clin Nutr* 1987; 46: 875–885.
24. Owen OE, Kavle E, Owen RS, et al. A reappraisal of caloric requirements in healthy women. *Am J Clin Nutr* 1986; 44: 1–19.

7 Pharmacological Therapy for Cardiovascular Disease
Current and Emerging Therapies

Michael H. Davidson, MD, FACC, FACP

CONTENTS

> INTRODUCTION
> STATINS
> EZETIMIBE
> BILE ACID SEQUESTRANTS
> NIACIN
> FIBRATES
> OMEGA-3 FATTY ACIDS
> EMERGING DRUGS
> SUMMARY
> REFERENCES

INTRODUCTION

The morbidity and mortality associated with coronary heart disease (CHD) are significant, with CHD accounting for 37.3% of all deaths in 2003 (or 1 of every 2.7 deaths) in the United States *(1)*. Current estimates indicate that more than 13,200,000 Americans have CHD *(1)*. As a result, lipid-lowering drugs are the most prescribed medications in the world with more than 20 million people being prescribed this pharmacological class of drug. Despite increasing evidence for the benefits of lipid-lowering therapy on patient outcomes, recent surveys demonstrate that only 18% of very high risk patients are at the optimal low-density lipoprotein (LDL) goal of less than 70 mg/dL *(2)*. As pharmacological options for CHD treatment are increasing, awareness of the benefits of single-agent as well as combination therapies is essential for ensuring the achievement of LDL goal. This chapter discusses current and emerging agents targeting pharmacological regulation of lipid metabolism in patients with dyslipidemia.

STATINS

Cholesterol-lowering therapy with 3-hydroxy-3-methylglutaryl coenzyme A (HMG-CoA) reductase inhibitors (statins) has been established as an effective method of reducing death and myocardial infarction (MI) among patients with CHD. Numerous clinical trials have demonstrated that statins can improve the lipid profile of patients with dyslipidemia, bring large numbers of these patients within treatment guidelines, and lower their risk of cardiovascular-related morbidity and mortality *(3)*. In fact, all of the major statin trials have had a consistent linear correlation between cholesterol lowering and cardiovascular risk reduction (Fig. 1).

Statin outcome trials have proven conclusively that lowering LDL-C results in significant improvement in cardiovascular morbidity or mortality (Table 1). Additionally, primary and secondary prevention studies using statins have established the safety and efficacy of this class of pharmacological agents.

Mechanism

Statins are structural inhibitors of HMG-CoA reductase, the rate-limiting enzyme for hepatic cholesterol biosynthesis resulting in the upregulation of the LDL receptor and the lowering of serum LDL-C (Fig. 2). All statins share a similar structure (the *pharmaphore*) that inhibits HMG-CoA reductase. The fungally derived statins (lovastatin, simvastatin, and pravastatin) have other structural similarities, whereas the synthetic statins (cerivastatin, fluvastatin, atorvastatin, rosuvastatin, and pravastatin) also have common clinical structures. As a result, classifying statins as fungal metabolites (natural statins) or synthetic statins is an acknowledged way to differentiate the two types. Another classification scheme involves identifying statins by their solubility in octanol (lipophilicity) or water (hydrophilicity). Pravastatin, rosuvastatin, and, to a much lesser degree, fluvastatin are considered hydrophilic statins, whereas the other statins are considered lipophilic (Fig. 3).

Statins have clinically relevant differences in efficacy, pharmacokinetics, and safety profiles, yet differences in the statin class of drugs have focused on the pleiotropic or nonlipid-mediated effects. Almost all cells possess the mevalonate pathway that

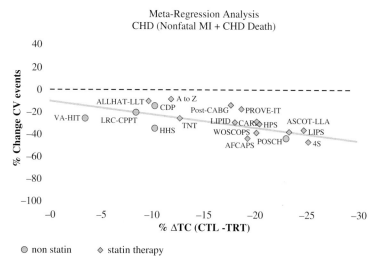

Fig. 1. Meta-regression analysis CHD (nonfatal MI and CHD death). Relation of total cholesterol differential in active treatment versus control group to % change in CV events versus placebo. CHD, coronary heart disease; CTL, control group; CV, cardiovascular; MI, myocardial infarction; TC, total cholesterol; TRT: treatment group.
Reprinted with permission from Davidson and Robinson (83).

Table 1
Outcomes Studies of Statin Medications

Study	Design	Patient characteristics	Treatment Groups	Results
Primary prevention				
Air Force/Texas Coronary Atherosclerosis Prevention Study (AFCAPS/TexCAPS)	Multicenter, double blind (DB), placebo controlled (PC), 5.6-year follow-up	6605 men and women; LDL 151 mg/dL; age 58 years	Diet with placebo or lovastatin 20–40 mg/day	36% reduction in first major coronary event; 33% reduction in percutaneous transluminal coronary angioplasty (PTCA) and coronary artery bypass graft (CABG)
West of Scotland Coronary Prevention Study (WOSCOPS)	Multicenter, double blind (DB), placebo controlled (PC), 5-year follow-up	6595 men; No history of MI; LDL 192 mg/dL; age 55 years	Diet with placebo or pravastatin 40 mg/day	30% reduction in nonfatal MI or CHD death; 22% reduction in death from any cause
Secondary Prevention				
Scandinavian Simvastatin Survival Study (4S)	Multicenter, double blind (DB), placebo controlled (PC), 5.4-year follow-up	4444 men and women; angina or previous MI; cholesterol 261 mg/dL; LDL 187 mg/dL; age 60 years	Diet with placebo or simvastatin 40 mg/day	Reduction in total mortality; 34% reduction in fatal/nonfatal MI and sudden cardiac death; 37% reduction in any coronary event
Anglo-Scandinavian Cardiac Outcomes Trial—Lipid Lowering Arm (ASCOT-LLA)	Multicenter, double blind (DB), placebo controlled (PC), 3.36-year follow-up	10,305 men and women; ≥ 3 additional CV risk factors, no h/o CHD; median LDL 131 mg/dL; median age 63 years	Diet with placebo or atorvastatin 10 mg/day	36% reduction in nonfatal MI and fatal coronary heart disease; 27% reduction in fatal and nonfatal stroke; 21% reduction in total cardiovascular events; 29% reduction in total coronary events
Cholesterol and Recurrent Events (CARE)	Multicenter, double blind (DB), placebo controlled (PC), 5-year follow-up	4159 men and women; previous MI; mean LDL 139 mg/dL; age 59 years	Diet with placebo or pravastatin 40 mg/day	

(Continued)

Table 1
(*Continued*)

Study	Design	Patient characteristics	Treatment Groups	Results
Heart Protection Study (HPS)	Multicenter, double blind (DB), placebo controlled (PC), 5-year follow-up	20,536 adults; coronary disease, other occlusive arterial disease, diabetes, or hypertension	Diet with placebo or simvastatin 40 mg/day	
Long-term Intervention with Pravastatin in Ischaemic Disease (LIPID)	Multicenter, double blind (DB), placebo controlled (PC), 6.1-year follow-up	9014 men and women; angina or previous MI; median LDL 150 mg/dL; age 62 years	Diet with placebo or pravastatin 40 mg/day	
Treat to New Target (TNT)	Multicenter, double blind (DB), placebo controlled (PC), 4.9-year follow-up	10,001 men and women with CHD; age 61 years; median LDL 77 versus 101 mg/dL	Atorvastatin 10 versus 80 mg/day	22% reduction in composite CVD events

CHD, coronary heart disease; CVD, cardiovascular disease; DB, double blind; LDL, low-density lipoprotein; MI, myocardial infarction; PC, placebo controlled.
Source: Adapted from Davidson and Robinson *(71)*. Reprinted with permission.

is affected by statin therapy; thus, the influence of statins may not be limited to the cardiovascular system. In addition to cholesterol, the mevalonate pathway leads to the formation of dolichols, ubiquinones involved in electron transport, and isoprenoids, which are involved in posttranslational modification of many proteins, including those

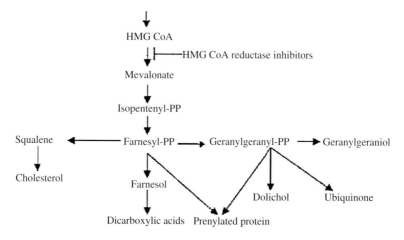

Fig. 2. Biosynthesis pathway and its inhibitors.

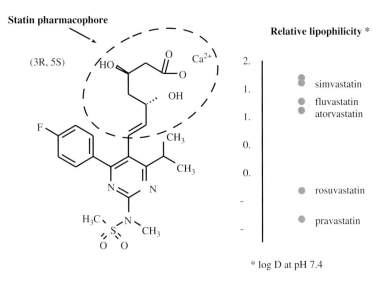

Fig. 3. Properties of rousavastatin.

necessary for cell proliferation, and has a key role in lipoprotein synthesis. A wide range of pleiotropic effects including vasodilation, antithrombosis, antioxidant, antiproliferative, and anti-inflammatory have been reported for statins, yet it becomes difficult to determine which have clinical and practical significance beyond those associated with lipoprotein changes (Fig. 4). Additionally, little is known as to whether these effects are dose dependent, which part of the molecular structure they can be attributed to, and to what extent they contribute to clinical outcomes independent of reduction in the known coronary artery disease risk factors. Supporting the pleiotropic effects for statins that may impact outcomes includes clinical trial evidence supporting early separation of the survival curves—the placebo and treatment groups in the Scandinavian Simvastatin Survival Study (4S), West of Scotland Coronary Prevention Study (WOSCOPS), Air Force/Texas Coronary Atherosclerosis Prevention Study (AFCAPS/TexCAPS), and Heart Protection Study (HPS) *(4–7)*. Post hoc analysis of WOSCOPS demonstrated that patients who had achieved similar on-treatment LDL-C levels with pravastatin had a lower CHD event rate than those who received placebo (6.3 versus 9.6%, a 36% difference). This suggests pleiotropic effects or added effects of statins that may reduce events. The Framingham equation does not include triglycerides (TGs) or use apolipoprotein B (Apo B); therefore, the possibility exists that part of the added benefit may be because of the effects of statins on lowering TGs. Although the main benefits of statins on CHD risk reduction are because of their effects on LDL, the effects on TGs and high-density lipoprotein (HDL) may also contribute to the risk reduction. As Apo B is contained in LDL, intermediate-density lipoprotein, and very low density lipoprotein (VLDL), the impact of statins on the TG-containing intermediate-density lipoprotein and VLDL may additionally contribute to risk-modifying benefits. Additional support is seen in the AFCAPS/TexCAPS trial which demonstrated that the best overall predictor of events on treatment was Apo B/Apo AI, which supports the concept that raising

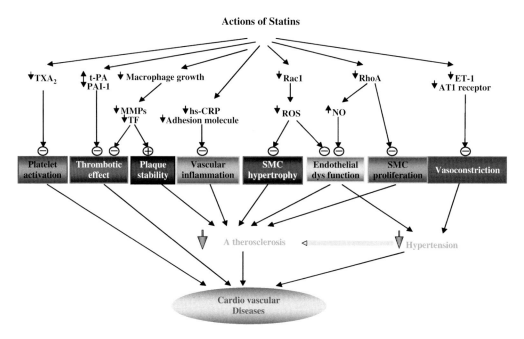

Fig. 4. Actions of statins.

Source: Takemoto and Liao *(84)*.

HDL (the Apo AI-containing particles) also contributes to risk reduction. Of additional significance is that all the major statin trials are remarkably consistent in demonstrating that LDL reduction correlates with event reduction in a linear fashion (Fig. 1). The pleiotropic effects do not appear to differentiate the statins within the classifications, as both fungal metabolites and synthetic or hydrophilic and lipophilic statins have been shown to reduce CHD events in a similar manner based on their efficacy in modifying lipoproteins.

Pharmacokinetics

Table 2 outlines the comparative pharmacokinetics of the statins. The pharmacokinetic profiles of the statins vary, with differences in bioavailability, absorption, major metabolic enzyme effect, systemic active metabolites, hepatic first-pass metabolism, and renal excretion.

LOVASTATIN

Lovastatin is a short-half-life prodrug that is lipophilic and metabolized by the cytochrome P450 3A4 (CYP3A4) pathway. Lovastatin is available in immediate-release formulation (10–80 mg) and extended-release formulations (Altocor® 10–40 mg, or Altoprev® 10–60 mg). Primary prevention benefits of Lovastatin were demonstrated in the AFCAPS/TexCAPS trial which confirmed that Lovastatin (20–40 mg daily) reduced LDL-C by 25% to 2.96 mmol/L (115 mg/dL) and increased HDL-C by 6% to 1.02 mmol/L (39 mg/dL) and significantly reduced the incidence of first acute major coronary events, MI, unstable angina, coronary revascularization procedures, coronary events, and cardiovascular events *(6)*.

Table 2
Summary of the Comparative Pharmacokinetics of Statins in Healthy Volunteers

Variable	Atorvastatin	Fluvastatin	Lovastatin	Pravastatin	Rosuvastatin	Simvastatin
Prodrug	No	No	Yes	No	No	Yes
Lipophilicity (log P)	4.06	3.24	4.30	−0.23	−0.33	4.68
Affinity for Pgp transporter	Yes	No	Yes	Yes	No	Yes
T_{max} (h)	1.0–2.0	0.5–1.0	2.0–4.0	1.0–1.5	3.0–5.0	1.3–3.0
Absorption (%)	30	98	30–31	34	40–60	60–80
Hepatic first-pass metabolism (%)	20–30	40–70	40–70	50–70	50–70	50–80
Bioavailability (%)	12–14	29	<5	18	20	<5
Protein binding (%)	>98	>98	>95	43–54	88	95
Major metabolic enzyme	CYP3A4	CYP2C9	CYP3A4	Minimal CYP450	Minimal CYP4502C9	CYP3A4
Systemic active metabolites (number)	Yes (2)	No	Yes (3)	No	Minimal	Yes (3)
Renal excretion (%)	≤2	<6	≥10	20	10	13
$T_{\frac{1}{2}}$ (h)	14–15	3.0	2.0	2.0	20	1.4–3.0

CYP450, cytochrome P450; NA, not available; NR, not reported; Pgp, P-glycoprotein; T_{max}, time of maximum circulating concentration; $T_{1/2}$, half-life.
Source: Adapted from Davidson and Robinson (71). Reprinted with permission.

PRAVASTATIN

Pravastatin is a short-half-life hydrophilic statin that is not affected by food intake. Pravastatin is available at doses of 10–80 mg and provides LDL reductions ranging from 25 to 40%. Pravastatin has demonstrated the best profile among the statins in outcome trials with established cardiovascular benefits in patients with coronary disease (Cholesterol and Recurrent Events [CARE] trial and in the Long-term Intervention with Pravastatin in Ischemic Disease [LIPID] trial), in high-risk primary prevention (WOSCOPS) and in the elderly (Prospective Study of Pravastatin in the Elderly at Risk [PROSPER]) (5,8–10).

SIMVASTATIN

Simvastatin is a prodrug with a short half-life of 3 h. It is lipophilic, hepatically metabolized by CYP3A4, and not affected by food intake. Simvastatin is available at doses of 5–80 mg and provides LDL reductions of 25–40%. Simvastatin is the only statin to show a reduction in total mortality in both the 4S (4) and HPS trials (4,7). For patients on potent 3A4 inhibitors such as gemfibrozil or cyclosporine, the dosage of simvastatin should not exceed 10 mg/day. For patients on verapamil or amiodarone (less potent 3A4 inhibitors), the dosage of simvastatin should not exceed 20 mg/day (11).

FLUVASTATIN

Fluvastatin is a synthetic, racemic mixture with a short half-life of 2.5 h. It is hepatically metabolized by cytochrome 2C9. A modified release formulation of fluvastatin 80 mg, fluvastatin XL, provides enhanced efficacy and safety compared with the immediate-release dose. Fluvastatin 80 mg XL lowers LDL by approximately 35%. Drug interactions with fluvastatin are less than with the other statins because of metabolism by CYP2C9. Fluvastatin is the only statin that has been shown not to be affected by gemfibrozil (Spence et al. 1995) *(87)* and therefore is probably the statin of choice to combine with this fibrate. Drug interactions can occur with fluconazole (a 2C9 inhibitor) and warfarin. Although the pharmacokinetics of fluvastatin is affected by concomitant cyclosporine, the Assessment of LEscol in Renal Transplantation (ALERT) study of post-renal transplant patients demonstrated reduced incidence of cardiovascular events on fluvastatin and no cases of rhabdomyolysis *(12)*. Findings from the ALERT trial indicate that for patients post renal transplant on cyclosporine, fluvastatin is probably the statin of choice.

ATORVASTATIN

Atorvastatin is a synthetic compound with a long half-life of approximately 14 h. Atorvastatin is a lipophilic statin metabolized by CYP3A4. Atorvastatin is available at doses of 10–80 mg. LDL reduction with the use of atorvastatin is 38–55%. Drug interactions exist for atorvastatin similar to other 3A4 metabolized statin; however, in clinical trials, atorvastatin has an extremely low rate of myopathy *(13)*. Clinical trials with atorvastatin in almost 10,000 patients have not demonstrated any cases of drug-associated myopathy (CPK > 10 times the upper limit of normal [ULN] with symptoms). Atorvastatin 80 mg has an incidence of liver enzyme three times the ULN of approximately 2.5%, which is the highest of all statins, but a low incidence of 0.5% for all the other doses *(11)*. The benefits of atorvastatin have been demonstrated in the Anglo-Scandinavian Cardiac Outcomes Trial (ASCOT) with atorvastain 10 mg in hypertensive patients *(14)* and the Collaborative Atorvastatin Diabetes Study (CARDS) demonstrated a benefit in diabetic patients *(15)*. The clinical benefit of atorvastatin 80 mg compared with 10 mg of atorvastatin was supported by the Treating to New Targets (TNT) *(16)*. Additionally, the benefit of atorvastatin 80 mg compared with 20–40 mg of simvastatin was established in the Incremental Decrease in Endpoints through Aggressive Lipid lowering (IDEAL) trial *(17)*.

Efficacy

The STELLAR (Statin Therapies for Elevated Lipid Levels Compared Across Doses to Rosuvastatin) trial compared the efficacy of the various statins across the dose range *(18)* (Fig. 5). More recent trials have compared the efficacy of ezetimibe/simvastatin and other statins. The relationship between LDL-C and CHD event appears to be linear with considerable evidence supporting the "lower is better" hypothesis. However, even with low LDL levels, the residual risk for subsets of high-risk patients continues to be elevated. In the TNT trial, patients on atorvastatin 80 mg (with a mean LDL-C of 77 mg/dL) had a total of 28% cardiovascular events compared with a 33% event rate for patients on 10 mg atorvastatin with mean LDL-C of 101 mg/dL (a 22% relative risk reduction) *(16)*. As a result, approximately 70% of the events were not avoided despite significant LDL-C reduction. Therefore, a significant number of individuals who are receiving statin therapy continue to have high residual risk. Additional research has demonstrated that large proportions of dyslipidemic patients receiving lipid-lowering therapy are not achieving National Cholesterol Education Program (NCEP) LDL-C target levels *(2)*.

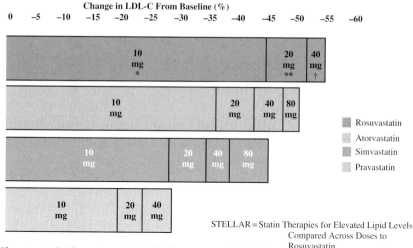

Fig. 5. Comparative efficacy of statins based on the Statin Therapies for Elevated Lipid Levels Compared Across Doses to Rosuvastatin (STELLAR) Trial.

Source: Adapted from Jones et al. *(18)*.

Safety and Drug Interaction

Lovastatin, simvastatin, and atorvastatin are metabolized via the CYP3A4 pathway. Fluvastatin is metabolized by the CYP2C9 pathway, and cerivastatin is metabolized by dual CYP2C9 (or 2C8) and 3A4 pathways. Pravastatin and rosuvastatin are not significantly metabolized by the CYP pathway. A number of clinically relevant statin drug interactions may occur with pharmacological agents that are metabolized by the CYP3A4 pathways. In addition to cyclosporine, drugs that significantly inhibit the CYP34A pathway are of most concern with combination therapy with lovastatin, simvastatin, and atorvastatin. Commonly used CYP3A4 inhibitors include erythromycin, ketoconazole, and nefazodone. A number of case reports have described the development of rhabdomyolysis when lovastatin was combined with erythromycin. As erythromycin can cause QT prolongation, special precaution should be used when combining a CYP3A4 statin with erythromycin in patients with a history of arrhythmias. As erythromycin also affects the metabolism of calcium channel blockers that use the CYP3A4 pathway (i.e., felodipine, nisoldipine, and nifedipine), patients should be carefully monitored and dosages of calcium channel blockers decreased if necessary when erythromycin is added. Erythromycin does not appear to affect the pharmacokinetics of pravastatin, fluvastatin, or rosuvastatin. However, drug interactions may exist with ketoconazole and itraconazole, and there have been reports of myopathy when these antifungals were combined with lovastatin. Fluconazole does not inhibit the CYP3A4 pathway and can be a safer alternative to combine with lovastatin, simvastatin, or atorvastatin.

The antidepressant drugs fluoxetine, fluvoxamine, nefazodone, and sertraline also inhibit CYP3A4 statins and should be used with caution. Paroxetine and venlafaxine, which do not inhibit the CYP3A4 pathway, may be used as alternatives for the treatment of depression.

The protease inhibitors (indinavir, nelfinavir, ritonavir, and saquinavir) inhibit CYP3A4 metabolism and may also induce dyslipidemia. As there are little data regarding the coadministration of these drugs with CYP3A4 statins, caution is indicated with use. Currently, pravastatin, rosuvastatin, and fluvastatin are the preferred statins in patients taking protease inhibitors. Of the protease inhibitors, indinavir appears to be a less potent inhibitor of CYP3A4 and may be more of an option to consider when initiating a protease inhibitor to patients taking a CTP3A4 statin. Another potent CYP3A4 inhibitor is amiodarone and should therefore be used with caution in patients taking these statins. For patients taking amiodarone, the dose of simvastatin should not exceed 20 mg as a myopathy rate of 6% has been demonstrated with simvastatin 80 mg.

Calcium channel blockers are substrates and therefore weak inhibitors of CYP3A4 metabolism. Yet, despite the effects of calcium channel blockers on CYP3A4 statin pharmacokinetics, there is little evidence of clinically significant drug interactions. Concomitant use of diltiazem and simvastatin does result in modest increases in serum concentrations of both drugs. However, pravastatin does not impact diltiazem pharmacokinetics. Clinical trial data also do not indicate risk of significant drug interactions with calcium channel blockers and statins. In the 4S and HPS trials, there were only rare cases of myopathy noted, although there were large numbers of patients taking calcium channel blockers. In the HPS, there were more than 3000 patients taking calcium channel blockers at randomization and more than 10,000 patients in the simvastatin-treated cohort. In the two cases of myopathy with simvastatin 40 mg, neither was associated with calcium channel blocker use. The safety of the use of CYP3A4 statins with up to the 40 mg dose of simvastatin is supported; however, in the Study of the Effectiveness of Additional Reductions in Cholesterol and Homocysteine (SEARCH) trial with 80 mg simvastatin, 0.6% of patients taking verapamil developed myopathy. Therefore, caution is advised with the use of calcium channel blockers and 80 mg of a CYP3A4-metabolized statin.

Fresh or frozen grapefruit inhibits intestinal CYP3A4 but appears to have limited effects on hepatic CYP3A4. Of significance is that grapefruit juice (at least 200 mL) can increase serum concentrations of CYP3A4 substrate drugs that undergo intestinal first-pass metabolism via this enzyme. Although published reports of myopathy because of grapefruit juice and CYP3A4 statins are lacking, it is advisable to separate the dosing of CYP3A4 statins and ingestion of grapefruit juice by 2 h. Little is known about the effects of other citrus fruits on CYP enzymes, but it is known that orange juice, which lacks 6,7-dihydroxybergamuttin, does not inhibit CYP3A4.

All statins are listed in the *Physicians' Desk Reference* as agents that may affect the international normalized ratio for patients on warfarin. The protein binding of the various statins may affect warfarin levels by mutual displacement from plasma protein-binding sites, but these effects are not mediated by the CYP enzymes and are generally not clinically significant.

Cyclosporine is an inhibitor of both the CYP3A4 metabolic pathway and the intestinal P-glycoprotein (Pgp) drug efflux pump system. As a result, cyclosporine may markedly increase statin levels. Cyclosporine appears to significantly increase the area under the curve (AUC) for all statins including pravastatin and rosuvastatin (Fig. 6). This suggests another mechanism by which cyclosporine significantly increases statin serum levels. Inhibition of the hepatic influx organic anion transporter C by cyclosporine may be instrumental to the cyclosporine–statin interaction. Myopathy has been reported with cyclosporine and statin combination therapy, and the frequency

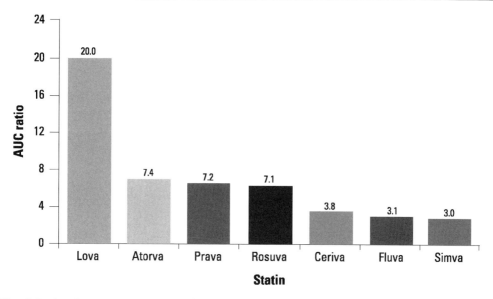

Fig. 6. Ratio of area under the curve (AUC) in cyclosporine-treated patients to AUC in historical control patients.
Source: From Davidson and Toth *(86)* with permission.

is perhaps higher than anticipated based on CYP metabolism alone. The safety of combining low dosages of simvastatin (10 mg/day) and pravastatin with cyclosporine has been documented in clinical trials, but they may underestimate the propensity for significant adverse drug reactions. In the ALERT trial *(19)*, none of the approximately 1000 patients taking concomitant fluvastatin and cyclosporine developed significant myositis. Therefore, fluvastatin may be the preferred statin in posttransplant patients.

In regard to statins interacting with fibrates, there appear to be significant differences between gemfibrozil and fenofibrate. Gemfibrozil inhibits the glucuronidation of statins by uridine diphosphate-glucuronosyltransferase (UGT 1A1, UGT A3), which promotes clearance or lactonization to inactive forms. In comparison, fenofibrate appears to be a weaker inhibitor of glucuronidation of statins (Fig. 7) and may explain the statin interaction profile of gemfibrozil versus fenofibrate. Although gemfibrozil increases the AUC of all the statins evaluated (atorvastatin has not been evaluated), fenofibrate does not significantly increase the AUG of simvastatin, cerivastatin, pravastatin, or rosuvastatin. Table 3 outlines clinically relevant statin drug interactions.

EZETIMIBE

Ezetimibe is classified in a new class of lipid-lowering drugs known as cholesterol absorption inhibitors *(20)*. Ezetimibe inhibits the absorption of dietary and biliary cholesterol without affecting the absorption of TGs or fat-soluble vitamins (Fig. 8) *(21–23)*.

Mechanism

Ezetimibe localizes to the brush border of small intestinal enterocyte and selectively inhibits enterocyte cholesterol uptake and absorption by binding to the NPL-1 sterol transporter *(24)*.

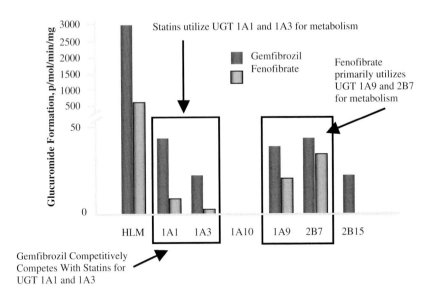

Fig. 7. The glucuronidation of the fibrate fenofibrate utilizes a different family of enzymes than the glucuronidation of the fibrate gemfibrozil. Gemfibrozil utilizes the same enzymes as most statins.

Source: Prueksaritanont et al. *(80)*.

Table 3
Clinically Relevant Statin Drug Interactions

Drug	Atorvastatin	Fluvastatin	Lovastatin	Pravastatin	Simvastatin	Rosuvastatin
Azole antifungals	+	−	+	−	+	−
CCBs	−	−	+	−	+	−
Cyclosporine	+	−	+	+	+	+
Erythromycin	+	−	+	−	+	−
Gemfibrozil	+	−	+	+	+	+
Fenofibrate	−	−	NA	−	−	−
HIV PIs	+	−	+	−	+	−
Warfarin	+	+	+	+	+	+

+, interaction reported; −, no interaction reported; CCB, calcium channel blocker (i.e., diltiazem, verapamil); NA, not available; PI, protease inhibitor.
Source: Adapted from Davidson and Robinson *(71)*. Reprinted with permission.

Pharmacokinetics

Following oral administration, ezetimibe is rapidly glucuronidated in the intestines and undergoes enterohepatic recirculation, promoting drug delivery to its site of action and minimizing systemic exposure *(21)*. The enterohepatic recirculation potentially

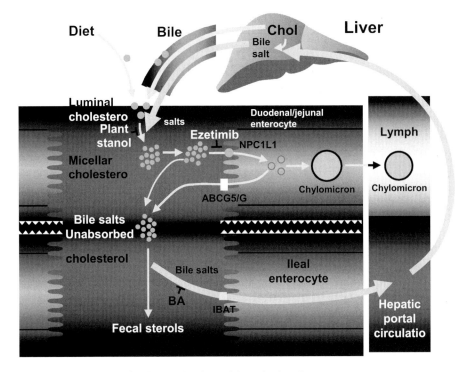

Fig. 8. Mechanism of intestinal-acting agents.

explains the long duration of action of ezetimibe and its long half-life (22 h) permits once daily dosing. Food intake does not affect its bioavailability (25).

Monotherapy with ezetimibe has been shown to effectively reduce LDL-C in patients with hypercholesterolemia (23,26). Ezetimibe does not interact with drugs metabolized by CYP1A2, 2D6, 2C8, 2C9, or 3A4, indicating that it has a low potential for drug–drug interactions (27). Ezetimibe does not interact with statins, including atorvastatin, simvastatin, lovastatin, or fluvastatin, and can therefore be taken concurrently with statins for combination therapy (28–30).

Efficacy

Several randomized clinical trials with ezetimibe 10 mg have demonstrated decreases in LDL-C of 17–18.2%, decreases in total cholesterol of 12.4–12.5%, decreases in Apo B of 15.4–15.5%, and increases of HDL cholesterol of 1.0–1.3% (23,26).

Coadministration of ezetimibe (10 mg) with a statin has been investigated in four randomized clinical trials with simvastatin, atorvastatin, lovastatin, and pravastatin (31–34). Combination therapy with ezetimibe 10 mg plus simvastatin 10 mg reduced LDL-C equivalent to simvastatin 80 mg alone (46–45%) (31). Similarly, ezetimibe plus atorvastatin 10 mg was as effective as atorvastatin 80 mg along (53 versus 54%) (32), and ezetimibe plus lovastatin 10 mg was as effective as lovastatin 40 mg alone (34 versus 31%) (33), and ezetimibe plus pravastatin 10 mg was as effective as pravastatin 40 mg (34 versus 29%) (34). Ezetimibe/simvastatin (Vytorin) is a cholesterol-lowering therapy that inhibits the intestinal absorption (ezetimibe) and synthesis (simvastatin) of cholesterol. A review of muscle-related adverse event (AE) data from 17 randomized blinded clinical trials in which ezetimibe and simvastatin were either coadministered as separate agents or given as combination tablet to 4558 patients did not find an increase in the incidence of muscle-related clinical or laboratory AEs *(35)*.

Safety and Drug Interaction

The overall safety profile of ezetimibe when coadministered with statin therapy has demonstrated similarities to statin monotherapy *(31–34)*. Incidence of elevations in ALT or AST three times or more the ULN ranged from 0 to 0.8% with statin monotherapy compared with 0 to 2.2% with ezetimibe plus statin coadministration. Coadminstration of ezetimibe with caffeine, dextromethorphan, midazolam, tolbutamide, antacid, cemetidine, oral contraceptives, warfarin, digoxin, or glipizide did not result in clinically significant interactions *(27,29,36–41)*.

BILE ACID SEQUESTRANTS

Bile acid sequestrants are nonabsorbed, lipid-lowering polymers that bind bile acids in the intestine, impeding their reabsorption (PDR 2006). Cholestyramine (Questran®), colestipol (Colestid®), and colesevelam (Welchol®) are the three main bile acid sequestrants currently available.

Mechanism and Pharmacokinetics

Bile acid sequestrants decrease intrahepatic cholesterol by interrupting enterohepatic circulation of bile acids, resulting in an increase in the synthesis of bile acids from cholesterol in the liver. LDL receptor activity is upregulated, and LDL clearance from the blood is increased.

Bile acid sequestrants are hydrophilic, water-insoluble polymers that are not hydrolyzed by digestive enzymes and are not absorbed. In binding with bile acids in the intestine, bile acid sequestrants cause an increase in hepatic synthesis of bile acids from cholesterol. This depletion of hepatic cholesterol increases hepatic LDL receptor activity, which removes LDL-C from the plasma (Fig. 8) *(11)*.

Efficacy

Bile acid sequestrants have been demonstrated to reduce total cholesterol, LDL-C, and Apo B and increase HDL-C when administered either alone or in combination with an HMG-CoA reductase inhibitor in patients with primary hypercholesterolemia. The clinical efficacy of bile acid sequestrants has been demonstrated in several clinical trials including the Lipid Research Clinics–Coronary Primary Prevention Trial (LRC-CPPT) *(42)* and the Familial Atherosclerosis Treatment Study *(43)*.

Safety and Drug Interaction

Adherence to bile acid sequestrants is often an issue with their use because of poor palatability and occurrence of gastrointestinal adverse effects, particularly constipation. Compliance with therapy has been estimated to be as low as 50%. Colesevelam, a novel bile acid sequestrant, lacks the constipating effect seen with typical bile acid sequestrants and has become the preferred drug of this class. It has been shown to be a safe and effective cholesterol-lowering agent with a nonsystemic mechanism of action, good tolerability, and minimal side effects *(44)*. Combination therapy of colesevelam with statins, including lovastatin, simvastatin, and atorvastatin, produces an additional reduction (8–16%) in LDL-C levels above that seen with the statin alone *(44)*.

NIACIN

Niacin is a member of the B-vitamin family and is sometimes referred to as vitamin B_3 *(11)*. Niacin or nicotinic acid is a soluble B vitamin that impacts all major lipid subfractions (Table 4) but is not widely used because of the associated side effects. Niacin-induced flushing, which is a result of vasodilation, usually occurs within 20 min following ingestion and may last for up to 1 h *(11)*.

Mechanism

Niacin has antihyperlipidemic activity and may have antiatherogenic activity *(11)*. Niacin reduces total cholesterol, LDL-C, and TGs; increases HDL-C; and decreases levels of Apo B-100 *(11)*. Niacin reduces the mobilization of free fatty acids from the periphery probably by inhibiting hormone-sensitive lipase, resulting in reduced secretion of VLDL that is a precursor of LDL. Niacin appears to increase HDL by decreasing the holo particle hepatic uptake of Apo AI, resulting in delayed catabolism (Fig. 9) *(45)*.

Pharmacokinetics

The pharmacokinetic profile of niacin is complicated because of rapid dose rate-specific and extensive first-pass metabolism *(11)*, cholesterol (HDL-C), and decreases in serum levels of lipoprotein (a) [Lp(a)] and Apolipoprotein B-100 (Apo B). The mechanism of the antihyperlipidemic action of niacin is not well understood but is thought to decrease the release of free fatty acids from adipose tissue, decrease the influx of free fatty acids into the liver, and decrease the rate of production of hepatic VLDL *(46)*. The time to reach peak serum concentrations of the immediate release of niacin is approximately 45 min after ingestion, whereas for the extended-release form (Niaspan), the time is 4–5 h after ingestion *(11)*. Food maximizes the availability of niacin.

Table 4
Effects of Niacin

Niacin decreases	*Niacin increases*
Total cholesterol	HDL cholesterol
Total triglycerides	HDL_2 cholesterol
VLDL-C	HDL_3 cholesterol (less than HDL_2)
LDL-C	Apolipoproteins AI, AII
Small dense LDL	LP AI
Lp(a)	LP AI + AII (less than LP AI)
Apo B	LDL particle size
Total cholesterol/HDL-C	
LDL-C/HDL-C	
Apo B/AI	

Source: Adapted from Davidson and Robinson *(71)*. Reprinted with permission.

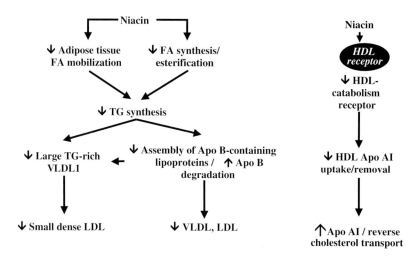

Fig. 9. Mechanism of action of niacin on hepatic Apo B and Apo AI metabolism and plasma lipids. *Source*: Adapted with permission from Kamanna et al. *(46)*.

Efficacy

Niacin is the only known approved lipid-altering drug (with the exception of estrogen) that lowers Lp(a) and is the most potent drug to raise HDL. On average, niacin lowers LDL by 10–20%, TGs by 20–40%, and Lp(a) by 10–30%, and it raises HDL by 15–30% *(47)*. Niacin has demonstrated CHD outcome benefits either as monotherapy or in combination with other lipid-lowering agents (Table 5). In the Coronary Drug Project Trial, 6 years of niacin monotherapy was associated with significant decreased risk of nonfatal MI *(48–50)*. Other clinical trials including the Stockholm Ischemia Heart Study *(51)*, Familial Atherosclerosis Treatment Study *(44)*, and HDL Atherosclerosis Treatment Study (HATS) *(52)* have established the benefits of niacin on CHD endpoints including reduction in definite nonfatal MI, CHD events, CHD deaths, and total mortality.

Safety and Drug Interaction

Niacin has been associated with abnormal liver tests and can cause significant hepatotoxicity. Niacin should be discontinued if liver enzymes (SGOT and SGPT) exceed three times the ULN. The use of over-the-counter niacin should be discouraged, as many niacin supplements may not be labeled as sustained release and, when combined with a fiber such as oat or rice bran, can cause a sustained release and adversely affect the liver enzymes *(53)*. Niacin can also increase uric acid levels, aggravating gout. Niacin-induced flushing is an adverse reaction that is the most bothersome and is the principal reason for compliance issues with the use of high-dose nicotinic acid for the treatment of hyperlipidemia *(11)*. Other side effects of niacin include pruritus, gastrointestinal problems including worsening of esophageal reflux, or peptic ulcers, and headache.

Niacin is often added to a statin in patients with combined hyperlipidemia, especially if the HDL is low of Lp(a) is high. Although statins have demonstrated an approximate reduction in CHD events by 30%, combination therapy with statins and niacin has resulted in reductions in CHD events of 75%. This significant reduction in CHD events suggests that other effects of niacin, such as lowering TG and Lp(a) and raising HDL,

Table 5
Niacin Coronary Heart Disease (CHD) Endpoints Including Reduction in Definite Nonfatal Myocardial Infarction (MI), CHD events, CHD deaths, and total mortality

Study	Population	Results
Coronary Drug Project	8341 post-MI men Baseline TC 250 mg/dL: 9.9% on trial Baseline TG 177 mg/dL: 26.1% on trial	27% reduction in definite nonfatal MI — 24% reduction in CVA. Total mortality decrease 10.6% at 15 years
Stockholm Ischemia Heart Study	555 consecutive MI survivors <70 years old; open label clofibrate/niacin or placebo; Baseline TC 245 mg/dL: 13% on trial Baseline TG 208 mg/dL: 19% on trial	36% reduction in CHD deaths
Familial Atherosclerosis Treatment Study	146 men with Apo B ≥ 125 mg/dL with CHD. Conventional therapy versus niacin/colestipol or lovastatin/colestipol	73% reduction in CHD events in patients who received intensive lipid-lowering therapy
HDL Atherosclerosis Treatment Study (HATS)	160 CHD patients with HDL <35 mg/dL, LDL <145 mg/dL. Treated 3 years with niacin/simvastatin or placebo with or without antioxidants	Reduction of CHD events by 60% (on antioxidants) to 90% (off antioxidants)

Source: Adapted from Davidson and Robinson (71). Reprinted with permission.

also contribute to the benefits. Studies evaluating the combination of a statin with niacin have indicated that the incidence of liver function abnormalities is similar to that in niacin monotherapy (54). Myopathy has been reported with statins in combination with niacin because of niacin-induced hepatotoxicity and reduction in statin catabolism, causing markedly elevated statin levels.

A major drawback of niacin use in diabetic patients and/or the metabolic syndrome is its tendency to increase blood glucose levels (55), thus requiring careful monitoring of glycemic status. However, the Assessment of Diabetes Control and Evaluation of the Efficacy of Niaspan Trial (ADVENT) (56) demonstrated that extended-release niacin produced a significant on both TGs and HDL-C levels, with low dosages (1000 mg/day) not significantly increasing hemoglobin A1C levels. Low doses of extended-release niacin are therefore an option for the treatment of dyslipidemia in patients who have glucose-controlled type 2 diabetes.

FIBRATES

There are five fibrates currently used in human therapy: clofibrate, gemfibrozil, fenofibrate, bezafibrate, and ciprofibrate. However, only gemfibrozil and fenofibrate are available in the United States. Fibrates are peroxisome proliferator-activated receptor (PPAR-alpha) ligands.

Mechanism

Fibrates produce reductions in total TGs and VLDL-C and increase HDL and apoproteins Apo AI and Apo AII (11). Fibrate activation of PPAR-alpha results in increased lipoprotein lipase expression and decreasing Apo CIII expression, causing enhanced catabolism of TG-rich particles. Fibrates also increase the expression of Apo AI and AII (Fruchart et al. 2001) *(88)*. Fibrate therapy culminates in decreased hypertriglyceridemia and an increase in HDL-C. LDL levels may also decrease, most likely because of a reduction of dense LDL, which is more atherogenic than buoyant LDL and has poor affinity for the LDL receptor. However, LDL may also increase in patients with hypertriglyceridemia with fibrate therapy, but these are usually the less atherogenic buoyant LDL particles. As a result, fibrates are used almost exclusively for patients with hypertriglyceridemia and low HDL, but fibrates may lower LDL by 10–15% in patients without hypertriglyceridemia. The fibrates result in similar lipoprotein changes; however, fenofibrate may more potently decrease LDL (Fig. 10).

Pharmacokinetics

Although gemfibrozil and fenofibrate are both fibrates and activate PPAR-alpha, resulting in similar lipoprotein changes, they have different pharmacokinetic profiles. Gemfibrozil is a potent inhibitor of the CYP2C8 metabolic pathway, organic anion transport protein 2 (AOTP2), and statin glucoronidation *(57)*. In comparison, fenofibrate is activated to fenofibric acid that is a substrate for the CYP2C9 metabolic pathway but is not an inhibitor and is also a mild inhibitor of Pgp. As a result of these metabolic pathways, fenofibrate appears to have little known drug interactions.

Efficacy

Both primary and secondary outcome studies have demonstrated efficacy of fibrate therapy on fatal and nonfatal MI, total MI cardiovascular events, sudden death, and CHD death (Table 6). In patients with hypertriglyceridemia and/or low HDL, fibrates reduced CHD events or the angiographic progression of atherosclerosis.

The Veterans Affairs High-Density Lipoprotein Cholesterol Intervention Trial (VA-HIT) study *(58)* showed that gemfibrozil compared favorably with statin therapy in patients with CHD and low HDL. The Benzafibrate Infarction Prevention (BIP) trial

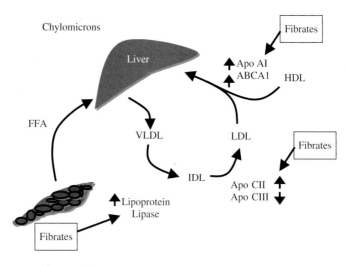

Fig. 10. Metabolic action of fibrates.

Table 6
Fibrate Outcome Studies

Trial	Duration years	N	Treatment	Primary endpoint	RRR	ARR	NNT
Primary prevention							
Helsinki Heart Study (HHS)	5	4081	Gemfibrozil 600 mg BID	Fatal, nonfatal MI, and cardiac death	34%	1.4%	71
Fenofibrate Intervention and Event Lowering in Diabetes (FIELD)	5	9795	Fenofibrate 200 mg	Fatal and nonfatal MI	11%	1.3%	77
FIELD subgroup without CVD	5	7664	Fenofibrate 200 mg	Total MI cardiovascular events	19%	2.0%	50
Secondary prevention							
Bezafibrate Infarction Prevention Study (BIP)	6.2	3090	Bezafibrate 400 mg qd	Fatal, nonfatal MI, sudden death	9% (NS)	NA	NA
BIP subgroup TG > 200	6.2	459	Bezafibrate 400 mg qd	Fatal, nonfatal MI, and sudden death	39%	7.7%	13
Veterans Affairs High Density Lipoprotein Intervention Trial (VA-HIT)	5.1	2531	Gemfibrozil 1200 mg qd	Nonfatal MI and CHD death	22%	4.4%	23

APR, absolute risk reduction; MI, myocardial infarction; N, number of participants; NNT, number needed to treat; RRR, relative risk reduction.

Source: Adapted from Davidson and Robinson *(71)*. Reprinted with permission.

demonstrated a reduction in the cumulative probably of the primary endpoint (fatal or nonfatal MI or sudden death) by 39.5% for patients with TGs greater than 200 mg/dL *(59)*. Similar results were reported in the Helsinki Heart Study (HHS) that demonstrated beneficial results especially in overweight hypertriglyceridemic individuals with low HDL-C levels, with a 78% reduction in CHD risk *(60)*.

The Fenofibrate Intervention and Event Lowering in Diabetes (FIELD) trial *(61)* did not reduce nonfatal and fatal myocardial infarctions significantly (11% reduction, $P = 0.16$), but a secondary endpoint that included fatal CVD events (nonfatal and fatal myocardial infarctions, coronary revascularization, and stroke) was reduced significantly by 11% ($P = 0.035$). The greatest absolute reduction in CHD events (2.5%) was demonstrated in the patients with "dyslipidemia" (high TGs and low HDL).

Safety and Drug Interaction

The fibrates are generally well tolerated. Common side effects are upper gastrointestinal disturbances, headache, myalgias, and loss of libido. Use of fibrates is contraindicated in hepatic or several renal dysfunction and preexisting gallbladder disease. Fibrates should also not be used in pregnant or nursing mothers.

Gemfibrozil is a potent inhibitor of the CYP2C8 metabolic pathway and the glucuronidation of statins. All statins require varying degrees of glucuronidation for metabolism, and the AUCs for statins evaluated, except for fluvastatin, are increased when combined with gemfibrozil (Table 7). As a result, gemfibrozil has a greater propensity to increase the risk of rhabdomyolysis when combined with statins compared with fenofibrate. As CYP2C8 is a common metabolic pathway for many of the diabetic agents such as glimepiride, rosiglitazone, and repaglinide, gemfibrozil raises the AUC for all these drugs significantly. As a result, gemfibrozil should be used cautiously in combination with CYP2C8-metabolized drugs (Table 8). Fenofibrate can be used as a substitute for gemfibrozil without adversely affecting the AUC of CYP2C8-metabolized drugs.

Table 7
Statin/Fibrate Combination Therapy: Pharmacokinetic Interactions

	Gemfibrozil	*Fenofibrate*
Atorvastatin	↑ in C_{max} by 1.2-2.7-fold	No effect
Simvastatin	↑ in C_{max} by 2-fold	No effect
Pravastatin	↑ in C_{max} by 2-fold	No effect
Rosuvastatin	↑ in C_{max} by 2-fold	No effect
Fluvastatin	No effect	No effect
Lovastatin	↑ in C_{max} by 2.8-fold	No effect
Cerivastatin	↑ in C_{max} by 2-3-fold	No effect

Source: Backman et al. *(72,76,77)*, TriCor *(73)*, Kyrklund *(74)*, Pan *(75)*, Abbott Laboratories *(78)*, Davidson *(79)*, Prueksaritanont *(80)*, Martin (81), and Bergman *(82)*.

Table 8
Cytochrome CYP2C8-Metabolized Drugs

CYP2C8 substrates		*CYP2C8 inhibitors*	*CYP2C8 inducers*
Amiodarone	Repaglinide	Anastrozole	Carbamazepine
Benzphetamine	Retinoic acid	Gemfibrozil	Phenobarbital
Carbamazepine	Retinal	Nicardipine	Rifabutin
Docetaxel	Rosiglitazone	Quercetin	Rifampicin
Fluvastatin	Tolbutamide	Sulfaphenazole	Rifampin
Isotretinoin	Tretinoin	Sulfinpyrazone	
Paclitaxel	Verapamil	Trimethoprim	
Phenytoin	Warfarin		
Pioglitazone	Zopiclone		

Although both gemfibrozil and fenofibrate are fibrates and activate PPAR-alpha and result in similar lipoprotein changes, they have very different pharmacokinetic profiles. Gemfibrozil is a competitive inhibitor of the CYP2C8 metabolic pathway, organic anion transport protein 2, and statin glucuronidation. Fenofibrate is a substrate for the CYP2C9 metabolic pathway and is also a mild inhibitor of Pgp, pathways that appear to result in little known drug interactions. The gemfibrozil–statin interaction is most problematic as frequently the addition of a fibrate to a statin is indicated to improve the lipoprotein profile. Fenofibrate and gemfibrozil, although less commonly, may increase serum creatinine by an unknown mechanism. Although the clinical significance is uncertain, homocysteine levels are increased by both fenofibrate and gemfibrozil. As a result, folic acid supplementation is indicated based on a correlation between increased homocysteine levels and CHD risk.

The usual dosage of gemfibrozil is 1200 mg at two divided doses 30 min before the morning and evening meals. Micronized fenofibrate 140 mg is given with a meal as the pharmacokinetics is markedly affected by food. A new formulation of micronized fenofibrate is available that does not require intake with food. The recommended dose of bezafibrate is 400 mg daily as a sustained release tablet and ciprofibrate is given at 100 g daily.

OMEGA-3 FATTY ACIDS

Omega-3 fatty acids or fish oils are a special class of essential fatty acids (EFAs). There are three omega-3 fatty acids: eicosapentaenoic acid (EPA), docosahexaenoic acid (DHA), and alpha-linolenic acid. EPA and DHA are found in fish oils.

Mechanism and Pharmacokinetics

EFAs are thought to inhibit VLDL and TG synthesis in the liver *(62)*. The mechanisms for other TG-lowering therapies, such as fibrates or niacin, are fairly well established; however, for omega-3 fatty acids, TG-lowering effect is not well understood.

Omega-3 fatty acids regulate at least four metabolic nuclear receptors that result in the repartitioning of fatty acids away from TG storage and toward oxidation. As RG synthesis is reduced, fatty acid oxidation in the hepatocyte is increased, and there is decreased substrate for VLDL synthesis and secretion. In addition, omega-3 fatty acid, as an unsaturated fat, may undergo peroxidation that may stimulate Apo B degradation, which also results in the reduction of VLDL secretion. As a result, postprandial chylomicron clearance is enhanced. Omega-3 fatty acids appear to also directly stimulate lipoprotein lipase activity (Fig. 11) *(63)*.

Efficacy

Omega-3 fatty acids lower TGs (by 25–35%) with modest increases in LDL of 5–10% and no apparent effect on serum HDL *(64)*. A dose–response relationship exists between omega-3 fatty acid intake and TG lowering; however, even small intakes can produce significant reductions in TG *(65,66)*.

Safety and Drug Interaction

Omega-3 fatty acids appear to be well absorbed from the oral route from dietary fish, capsule supplements, or in microencapsulated fish-oil-enriched foods. Prescription

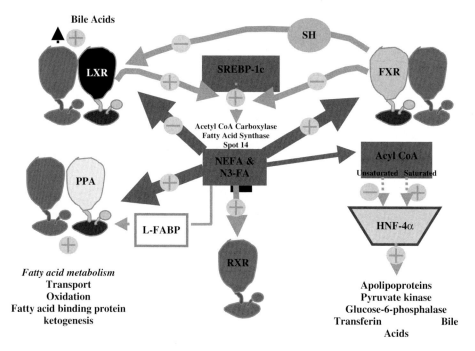

Fig. 11. Schematic overview of the regulation of genes by nonesterified fatty acid (NEFA). (NEFA) affect at least four metabolic nuclear receptors: PPAR, LXR, FXR, and HNF-4alpha. Omega-3 fatty acids (N-3FA) on a carbon-for-carbon basis are more potent regulators of these genes compared with other polyunsaturated fatty acids (PUFA). The net result of the regulation of these genes by omega-3 fatty acids (N-3FA) is the repartitioning of metabolic fatty acids away from triglyceride storage and toward oxidation.

Source: Adapted from Pegorier et al. *(86)*.

omega-3 fatty marine oils are indicated for the treatment of hypertriglyceridemia, and clinical trial evidence has indicated that reductions in serum TG by 19–55% can occur. Omega-3 fatty acids are well tolerated with few side effects other than mild gastrointestinal-related symptoms. No adverse effects of marine oil have been reported regarding elevations in muscle or liver enzymes, hyperglycemia, abnormal bleeding, or kidney or nerve function.

EMERGING DRUGS

CETP Inhibitors

The cholesteryl ester transfer protein (CETP) plays a pivotal role in cholesterol metabolism in exchanging cholesteryl esters (CEs) and TGs between lipoproteins *(35,67)*. CETP transfers CEs from HDL to VLDL and LDL in exchange for TGs. In addition, CETP transfers TGs from VLDL to LDL and HDL in exchange for CEs *(67)*. As a result, CETP activity is potentially proatherogenic, decreasing the CE content of atheroprotective HDL and increasing the CE content of atherogenic VLDL and LDL *(67)*. By exchanging TGs in VLDL for CEs in HDL and LDL, CETP additionally promotes the formation of small dense LDL and HDL by increasing the remodeling of TG-enriched LDL and HDL particles by TG lipases *(67)*. It is well established that small dense LDL is particularly atherogenic. CETP-driven TG enrichment of HDL and LDL particles may be especially important in creating an atherogenic lipid profile, especially in individuals with elevated TG levels as the VLDL pool is typically enlarged. CETP inhibition is therefore a strategy for elevating HDL-C for treating CHD. Two CETP inhibitors, JTT-705 and torcetrapib, are in clinical development.

Mechanism

Torcetrapib binds to CETP with 1:1 stoichiometry and blocks both neutral lipid and phospholipid transfer activities. Because most CETP appears to be bound to HDL, torcetrapib binds to the CETP and prevents all of the major lipid transfer functions of CETP by inducing a nonproductive complex between the transfer protein and HDL (Fig. 12).

Efficacy

In a multicenter randomized double-blind study of 162 subjects with below-average HDL, torcetrapib 10, 30, 60, or 90 mg/day for 8 weeks compared with placebo resulted in dose-dependent changes in HDL-C levels *(35)*. Percent changes in HDL-C from baseline to week 8 ranged from 9.0 to 54.5% with torcetrapib 10 to 90 mg/day. Differences were significant at dosages of 30, 60, and 90 mg/day ($P \leq 0.0001$). Increases in HDL-C were accompanied by dose-dependent increases in the levels of Apo AI (dominant change) and Apo AII. HDL subclass changes determined by nuclear magnetic resonance (NMR) spectroscopy were consistent with findings from ultracentrifugation/precipitation analysis showing that torcetrapib produced dose-dependent increases in levels of the large HDL-C subclasses *(35)*. At week 8, moderate decreases in LDL-C levels from baseline compared with placebo were observed with torcetrapib 60 mg (-8.1%) and 90 mg (-16.5%; $P < 0.01$). Concurrently, LDL-C decreases were accompanied by significant decreases in Apo B-100 levels ($P < 0.01$), suggesting a reduction in the concentration of circulating LDL particles *(35)*. NMR analysis demonstrated a trend to reduction in the concentration of the small LDL-C subclass. NMR spectroscopy also demonstrated that LDL particle size was increased in a dose-dependent manner. Torcetrapib 60 mg and 90 mg increased mean LDL particle size from 20.4 (± 0.7) to 21.2 nm (± 0.6) and 20.4 (± 0.7) to 21.3 nm (± 0.6), respectively

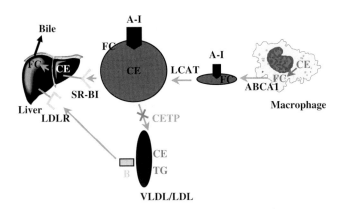

Fig. 12. CETP deficiency is associated with markedly increased HDL-C levels.

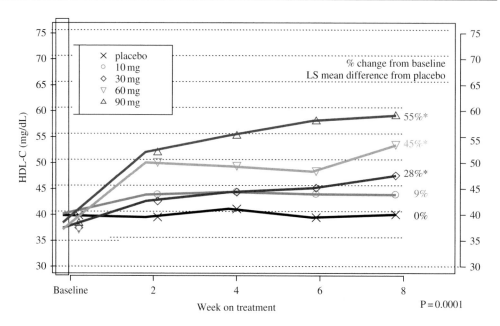

Fig. 13. Mean HDL-C – all subjects. Torcetrapib alone: rapid sustained increase in HDL-C.

($P \leq 0.0001$ for both) (Fig. 13). In other preliminary trials, torcetrapib has also been shown to result in substantial elevations in HDL-C, modest decreases in LDL-C, and increases in lipid particle size. The 60 mg torcetrapib was demonstrated to increase HDL by 45% and lower LDL by 8% *(68)*.

Safety and Drug Interaction

Generally, torcetrapib is well tolerated. Dose-related increase in blood pressure has been reported *(69,70)*. Due to the interim safety findings from the ILLUMINATE study, torcetrapib was discontinued from clinical use due to excess mortality. It is not known whether the increased mortality was the result of torcetrapib's effect on blood pressure or due to the mechanisms by which CETP inhibitors increase HDL. Clarification of this important issue will require further investigation with CETP inhibitors that do not increase blood pressure.

SUMMARY

Despite accumulating evidence of the benefits of (LDL) cholesterol lowering, initiation of treatment and long-term adherence to therapy remain suboptimal *(3)*. As many options exist for the pharmacological management of CHD, determining optimal treatment requires an understanding of the pharmacological properties of commonly prescribed as well as emerging agents. The use of combination therapy can improve the lipid profile for high-risk patients who have not yet achieved optional therapeutic targets. Lipid-altering agents such as bile acid sequestrants, extended-release niacin, and ezetimibe do not have a pharmacokinetic interaction with statins and appear to have a low risk for increasing statin-related side effects. Increased options for pharmacological therapy and combination therapy for CHD hold promise in improving the lipid profile and in reducing CHD risk.

REFERENCES

1. Thom T, Haase N, Rosamond W, et al. for the Statistics Committee and Stroke Statistics Subcommittee. Heart disease and stroke statistics—2006 update. *Circulation* 2006;113:e85–e151.
2. Davidson MH, Maki KC, Pearson TA, et al. Results of the National Cholesterol Education (NCEP) Program Utilizing Novel E-Technology (NEPTUNE) II survey and implications for treatment under the recent NCEP writing group recommendations. *Am J Cardiol* 2005;96:556–563.
3. National Cholesterol Education Program (NCEP) Expert Panel on Detection, Evaluation, and Treatment of High Blood Cholesterol in Adults (Adult Treatment Panel III). Third Report of the National Cholesterol Education Program (NCEP) Expert Panel on Detection, Evaluation, and Treatment of High Blood Cholesterol in Adults (Adult Treatment Panel III) final report. *Circulation* 2002;106:3143–3421.
4. Scandinavian Simvastatin Survival Study Group. Randomised trial of cholesterol lowering in 4444 patients with coronary heart disease: the Scandinavian Simvastatin Survival Study. *Lancet* 1994;344:1383–1389.
5. Packard CJ, Shepherd J, Cobbe SM, et al. Influence of pravastatin and plasma lipids on clinical events in the West of Scotland Coronary Prevention Study (WOSCPOS). *Circulation* 1998;97:1440–1445.
6. Downs JR, Clearfield M, Weiss S, et al. for the AFCAPS/TexCAPS Research Group. Primary prevention of acute coronary events with lovastatin in men and women with average cholesterol levels. *JAMA* 1998;279:1615–1622.
7. Heart Protection Study Collaborative Group. MRC/BHF Heart Protection Study of cholesterol lowering with simvastatin in 20,536 high-risk individuals: a randomized placebo-controlled trial. *Lancet* 2002;360:7–22.
8. Sacks FM, Pfeffer MA, Moye LA, et al. The effect of pravastatin on coronary events after myocardial infarction in patients with average cholesterol levels. *N Engl J Med* 1996;335:1001–1009.
9. The Long-Term Intervention with Pravastatin in Ischemic Disease (LIPID) Study Group. Prevention of cardiovascular events and death with pravastatin in patients with coronary heart disease and a broad range of initial cholesterol levels. *N Engl J Med* 1998;339:1349–1357.
10. Shepherd J, Blauw GJ, Murphy MD, et al. Pravastatin in elderly individuals at risk of vascular disease (PROSPER): a randomized controlled trial. *Lancet* 2002;360:1623–1630.
11. PDR 2006. http://www.pdr.net (accessed September 17, 2006).
12. Holdaas H, Fellstrom B, Jardine AG, et al. for the Assessment of LEscol in Renal Transplantation (ALERT) Study Investigators. Effect of fluvastatin on cardiac outcomes in renal transplant recipients: a multicentre, randomised, placebo-controlled trial. *Lancet* 2003;361:2024–2031.
13. Newman CG, Palmer G, Silbershatz H, Szarek M. Safety of atorvastatin derived from analysis of 44 completed trials in 9,416 patients. *Am J Cardiol* 2003;92:670–676.
14. Sever PS, Dahlof B, Poulter NR, et al. Prevention of coronary and stroke events with atorvastatin in hypertensive patients who have average or lower-than-average cholesterol concentrations, in the Anglo-Scandinavian Cardiac Outcomes Trial—Lipid Lowering Arm (ASCOT-LLA): a multicenter randomized controlled trial. *Lancet* 2003;361:1149–1158.
15. Colhoun HM, Thomason MJ, Mackness MI, et al. Design of the Collaborative AtoRvastatin Diabetes Study (CARDS) in patients with type 2 diabetes. *Diabet Med* 2002;19:201–211.
16. LaRosa JC, Grundy SM, Waters DD, et al. for the Treating to New Targets (TNT) Investigators. Intensive lipid lowering with atorvastatin in patients with stable coronary disease. *N Engl J Med* 2005;352:1425–1435.
17. Pedersen TR, Faergeman O, Kastelein JJ, et al. Incremental Decrease in End Points Through Aggressive Lipid Lowering (IDEAL) Study Group. High-dose atorvastatin vs usual-dose simvastatin for secondary prevention after myocardial infarction: The IDEAL study: a randomized control trial. *JAMA* 2005;294:2437–2445.
18. Jones PH, Davidson MH, Stein EA, et al. for STELLAR Study Group. Comparison of the efficacy and safety of rosuvastatin versus atorvastatin, simvastatin, and pravastatin across doses (STELLAR* Trial). *Am J Cardiol* 2003;92:152–160.
19. Fellstrom B, Holdaas H, Jardine AG, et al. for the Assessment of Lescol in Renal Transplantation Study Investigators. Effect of fluvastatin on renal end points in the Assessment of Lescol in Renal Transplant (ALERT) trial. *Kidney Int* 2004;66:1549–1555.
20. Catapano AL. Ezitimbe: a selective inhibitor of cholesterol absorption: *Eur Heart J* 1002;Suppl 3 (Suppl E):E6–E10.

21. van Heek M, France DF, Compton DS, et al. In vivo metabolism-based discovery of a potent cholesterol absorption inhibitor, SCH58235, in the rat and rhesus monkey through the identification of the active metabolites of SCH48461. *J Pharmacol Exp Ther* 1997;283:157–163.
22. van Heek M, Farley C, Compton DS, et al. Comparison of the activity and disposition of the novel cholesterol absorption inhibitor, SCH58235, and its glucuronide, SCH60663. *Br J Pharmacol* 2000;129:1748–1754.
23. Knopp RH, Bays H, Manion CV, et al. Effect of ezetimibe on serum concentrations of lipid-soluble vitamins [abstract]. *Atherosclerosis* 2001;2(Suppl):90.
24. Davis HR, Compton DS, Hoos L, et al. Ezetimibe (SCH58235) localizes to the brush border of small intestinal enterocyte and inhibits enterocyte cholesterol uptake and absorption. *Eur Heart J* 2000;21(Suppl):636 [Abstract P3500].
25. Bays HE, Moore PB, Drehobl MA, et al. Effectiveness and tolerability of ezetimibe in patients with primary hypercholesterolemia: pooled analysis of two phase II studies. *Clin Ther* 2001;23:1209–1230.
26. Dujovne CA, Ettinger MP, McNeer JF, et al. Efficacy and safety of a potent new selective cholesterol absorption inhibitor, ezetimibe, in patients with primary hypercholesterolemia. *Am J Cardiol* 2002;90,1092–1097.
27. Zhu Y, Statkevich P, Kosoglou T, et al. Effect of SCH 58235 on the activity of drug metabolizing enzymes in vivo [abstract]. *Clin Pharmacol Ther* 2000;67:152 [Abstract PIII-43].
28. Zhu Y, Statkevich P, Kosoglou T, et al. Lack of a pharmacokinetic interaction between ezetimibe and atorvastatin [abstract]. *Clin Pharmacol Ther* 2001;69:68 [Abstract PIII-15].
29. Kosoglou T, Meyer I, Veltri EP, et al. Pharmacodynamic interaction between the new selective cholesterol absorption inhibitor ezetimibe and simvastatin. *Br J Clin Pharmacol* 2002;54:309–319.
30. Reyderman L, Kosoglou T, Statkevich P, et al. No pharmacokinetic drug interaction between ezetimibe and lovastatin [abstract]. *Clin Pharmacol Ther* 2001;69:66 [Abstract PIII-8].
31. Davidson MH, McGarry T, Bettis R, et al. Ezetimibe coadministered with simvastatin in patients with primary. *J Am Coll Cardiol* 2002;40:2125–2134.
32. Ballantyne C, Houri J, Notarbartolo A, et al. Effect of ezetimibe coadministered with atorvastatin in 628 patients with primary hypercholesterolemia: a prospective, randomized, double-blind trial. *Circulation* 2003;107:2409–2415.
33. Kerzner B, Corbelli J, Sharp S, et al. Efficacy and safety of ezetimibe coadministered with lovastatin in primary hypercholesterolemia. *Am J Cardiol* 2003;91:418–424.
34. Melani L, Mills R, Hassman D, et al. Efficacy and safety of ezetimibe coadministered with pravastatin in patients with primary hypercholesterolemia: a prospective, randomized, double-blind trial. *Eur Heart J* 2003;24:717–728.
35. Davidson MH, McKenney JM, Shear CL, Revkin JH. Efficacy and safety of torcetrapib, a novel cholesteryl ester transfer protein inhibitor, in individuals with below-average high-density lipoprotein cholesterol levels. *J Am Coll Cardiol* in press.
36. Statkevich P, Reyderman L, Kosoglou T, et al. Ezetimibe does not affect the pharmacokinetics and pharmacodynamics of glipizide [abstract]. *Clin Pharmacol Ther* 2001;69:67 [Abstract PIII-12].
37. Kosoglou T, Statkevich P, Bauer KS, et al. Ezetimibe does not affect the pharmacokinetics and pharmacodynamics of digoxin [abstract]. *AAPS Pharm Sci* 2.
38. Keung AC, Kosoglou T, Statkevich P, et al. Ezetimibe does not affect the pharmacokinetics of oral contraceptives [abstract]. *Clin Pharmacol Ther* 2001;69:55. Abstract PII-89001;3(Suppl 3):[serial online].
39. Bauer KS, Kosoglou T, Statkevich P, et al. Ezetimibe does not affect the pharmacokinetics or pharmacodynamics of warfarin [abstract]. *Clin Pharmacol Ther* 2001;69:5 [Abstract PI-15].
40. Krishna G, Kosoglou T, Ezzet F, et al. Effect of cimetidine on the pharmacokinetics of ezetimibe [abstract]. *AAPS Pharm Sci* 2001;3(Suppl 3):[serial online].
41. Courtney RD, Kosoglou T, Statkevich P, et al. Effect of antacid on the pharmacokinetics of ezetimibe. *Clin Pharmacol Ther* 2002;71:80.
42. Lipid Research Clinics Program. The Lipid Research Clinics Coronary Primary Prevention Trial results. II. The relationship of reduction in incidence of coronary heart disease to cholesterol lowering. *JAMA* 1984;251:365–374.
43. Brown G, Albers JJ, Fisher LD, et al. Regression of coronary artery disease as a result of intensive lipid-lowering therapy in men with high levels of apolipoprotein B. *N Engl J Med* 1990;323: 1289–1298.
44. Davidson MH, Dillon MA, Gordon B, et al. Colesevelam hydrochloride (cholestagel): a new, potent bile acid sequestrant associated with a low incidence of gastrointestinal side effects. *Arch Intern Med* 1999;159:1893–1900.

45. Jin FY, Kamanna VS, Kashyap ML. Niacin decreases removal of high-density lipoprotein apolipoprotein A-I but not cholesterol ester by Hep G2 cells. Implication for reverse cholesterol transport. *Arterioscler Thromb Vasc Biol* 1997;17:2020–2028.
46. Kamanna VS, Kashyap ML. Mechanism of action of niacin on lipoprotein metabolism. *Curr Atheroscler Rep* 2000;2:36–46.
47. Capuzzi DM, Buyton JR, Morgan JM, et al. Efficacy and safety of an extended-release niacin (Niaspan): a long term study. *Am J Cardiol* 1998;82:74U–81U.
48. Canner PL, Berge KG, Wenger NK, et al. Fifteen-year mortality in Coronary Drug Project patients: long-term benefit with niacin. *J Am Coll Cardiol* 1986;8:1245–1255.
49. Coronary Drug Project Research Group. Clofibrate and niacin in coronary heart disease. *JAMA* 1975;231:360–381.
50. Berge KG, Canner PL. Coronary drug project: experience with niacin. Coronary Drug Project Research Group. *Eur J Clin Pharm* 1991;40(Suppl):S49–S51.
51. Carlson LA, Rosenhamer G. Reduction of mortality in the Stockholm Ischaemic Heart Disease Secondary Prevention Study by combined treatment with clofibrate and nicotinic acid. *Acta Med Scan* 1988;223:405–418.
52. Brown BG, Zhao X-Q, Chait A, et al. Simvastatin and niacin, antioxidant vitamins, or the combination for the prevention of coronary disease. *N Engl J Med* 2001;345:1583–1592.
53. Gray DR, Morgan T, Chretien SD, et al. Efficacy and safety of controlled-release niacin in dyslipoproteinemic veterans. *Ann Intern Med* 1994;121:252–258.
54. Guyton JR, Goldberg AC, Kreisberg RA, et al. Effectiveness of once-nightly dosing of extended-release niacin alone and in combination for hypercholesterolemia. *Am J Cardiol* 1998;82:737–743.
55. Elam MB, Hunninghake DB, Davis KB, et al. Effect of niacin on lipid and lipoprotein levels and glycemic control in patients with diabetes and peripheral arterial disease: the ADMIT study: a randomized trial. Arterial Disease Multiple Intervention Trial. *JAMA* 2000;284:1263–1270.
56. Grundy SM, Vega GL, McGovern ME, et al. Efficacy, safety, and tolerability of once-daily niacin for the treatment of dyslipidemia associated with type 2 diabetes: results of the assessment of diabetes control and evaluation of the efficacy of niaspan trial. *Arch Intern Med* 2002;162:1568–1576.
57. Prueksaritanont T, Zhao JJ, Ma B, et al. Mechanistic studies on metabolic interactions between gemfibrozil and statins. *J Pharmacol Exp Ther* 2002;301:1042–1051.
58. Rubins HB, Robins SJ, Collins D, et al. Diabetes, plasma insulin and cardiovascular disease. Subgroup analysis from the Department of Veterans Affairs High-density lipoprotein Intervention Trial (VA-HIT). *Arch Intern Med* 2002;162:2597–2604.
59. Tenenbaum A, Motro M, Fisman EZ, Tanne D, Boyko V, Behar S. Bezafibrate for the secondary prevention of myocardial infarction in patients with metabolic syndrome. *Arch Intern Med* 2005;165:1154–1160.
60. Tenkanen L, Manttari M, Manninen V. Some coronary risk factors related to the insulin resistance syndrome and treatment with gemfibrozil. Experience from the Helsinki Heart Study. *Circulation* 1995;92:1779–1785.
61. Keech A, Simes RJ, Barter P, et al. for FIELD Study Investigators. Effects of long-term fenofibrate therapy on cardiovascular events in 9795 people with type 2 diabetes mellitus (the FIELD study): randomized controlled trial. *Lancet* 2005;366:1849–1861.
62. Nestel PJ. Fish oil and cardiovascular disease: lipids and arterial function. *Am J Clin Nutr* 2000;71:228S–231S.
63. Davidson MH. Mechanisms for the hypotriglyceridemic effect of marine omega-3 fatty acids. *Am J Cardiol* 2006;98:27i–33i.
64. Harris WS. N-3 fatty acids and serum lipoproteins: human studies. *Am J Clin Nutr* 1997;65(5 Suppl):1645S–1654S.
65. Maki KC, Van Elswyk ME, McCarthy D, et al. Lipid responses to a dietary docosahexaenoic acid supplement in men and women with below average levels of high density lipoprotein cholesterol. *J Am Coll Nutr* 2005;24:189–199.
66. Davidson MH, Macariola-Coad JR, McDonald AM, Maki KC, Hall HA. Separate and joint effects of marine oil and simvastatin in patients with combined hyperlipidemia. *Am J Cardiol* 1997;80:797–798.
67. Barter PJ, Brewer HB JR, Chapman JM, et al. Cholesteryl ester transfer protein: a novel target for raising HDL and inhibiting atherosclerosis. *Arterioscler Thromb Vasc Biol* 2003;23:160–167.
68. Bays H, McKenney JM, Davidson M. Torcetrapib/atorvastatin combination therapy. *Expert Rev Cardiovasc Ther* 2005;3:789–820.

69. Clark RW, Sutfin TA, Ruggeri RB, et al. Raising high-density lipoprotein in humans through inhibition of cholesteryl ester transfer protein: an initial multidose study of torcetrapib. *Atheroscler Thromb Vasc Biol* 2004;24:490–497.
70. Brousseau ME, Schaefer EJ, Wolfe ML, et al. Effects of an inhibitor of cholesteryl ester transfer protein on HDL cholesterol. *N Engl J Med* 2004;250:1505–1515.
71. Davidson MH, Robinson JG. Management of elevated LDL.
72. Backman JT, Luurila H, Neuvonen M, Neuvonen PJ. Rifampin markedly decreases and gemfibrozil increases the plasma concentrations of atorvastatin and its metabolites. *Clin Pharmacol Ther* 2005;78(2):154–167.
73. TriCor [package insert]. Abbott Laboratories; 2004.
74. Kyrklund C, Backman JT, Kivisto KT, Neuvonen M, Laitila J, Neuvonen PJ. Plasma concentrations of active lovastatin acid are markedly increased by gemfibrozil but not by bezafibrate. *Clin Pharmacol Ther* 2001;69:340–345.
75. Pan WJ, Gustavson LE, Achari R, et al. Lack of a clinically significant pharmacokinetic interaction between fenofibrate and pravastatin in healthy volunteers. *J Clin Pharmacol* 2000;40:316–323.
76. Backman JT, Kyrklund C, Kivisto KT, Wang JS, Neuvonen PJ. Plasma concentrations of active simvastatin acid are increased by gemfibrozil. *Clin Pharmacol Ther* 2000;68:122–129.
77. Backman JT, Kyrklund C, Neuvonen M, Neuvonen PJ. Gemfibrozil greatly increases plasma concentrations of cerivastatin. *Clin Pharmacol Ther* 2002;72:685–691.
78. Abbott Laboratories. Data on file; 2005.
79. Davidson MH. Combination therapy for dyslipidemia: safety and regulatory considerations. *Am J Cardiol* 2002;90(Suppl):50K–60K.
80. Prueksaritanont T, Tang C, Qiu Y, Mu L, Subramanian R, Lin JH. Effects of fibrates on metabolism of statins in human hepatocytes. *Drug Metab Dispos* 2002;30:1280–1287.
81. Martin PD, Dane AL, Schneck DW, Warwick MJ. An open-label, randomized, three-way crossover trial of the effects of coadministration of rosuvastatin and fenofibrate on the pharmacokinetic properties of rosuvastatin and fenofibric acid in healthy male volunteers. *Clin Ther* 2003;25:459–471.
82. Bergman AJ, Murphy G, Burke J. Simvastatin does not have a clinically significant pharmacokinetic interaction with fenofibrate in humans. *J Clin Pharmacol* 2004;44:1054–1062.
83. Davidson MH, Robinson JG. Lipid-lowering effects of statins: a comparative review. *Expert Opin Pharmacother* 2006;7:1701–1714.
84. Takemoto M, Liao JK. Pleiotropic effects of 3-hydroxy-3-methylglutaryl coenzyme a reductase inhibitors. *Arterioscler Thromb Vasc Biol* 2001;21:1712–1719.
85. Davidson MH, Toth PP. Comparative effects of lipid-lowering therapies. *Prog Cardiovasc Dis* 2004;47:73–104.
86. Pegorier JP, Le May C, Girard J. Control of gene expression by fatty acids. *J Nutr* 2004;134:2444S–2449S.
87. Spence JD, Munoz CE, Hendricks L, et al. Pharmacokinetics of the combination of fluvastatin and gemfibrozil. *Am J Cardiol* 1995;76:80A–83A.
88. Fruchart JC. Peroxisome proliferator-activated receptor-alpha activation and high-density lipoprotein metabolism. American Journal of Cardiology. 88(12A):24N–29N, 2001.

8 Effects of Thiazolidinediones on Serum Lipoproteins

Anjli Maroo, MD, RVT,
and W. H. Wilson Tang, MD, FACC

CONTENTS

> INTRODUCTION
> MECHANISMS OF TZD ACTIONS ON SERUM
>   LIPOPROTEINS
> CLINICAL EVIDENCE OF LIPID-LOWERING EFFECTS
>   OF TZDs
> META-ANALYSES OF PIOGLITAZONE
>   AND ROSIGLITAZONE TRIALS
> PROSPECTIVE COMPARATIVE CLINICAL STUDIES
> CONCOMITANT TZD AND STATIN THERAPY
> OUTCOME STUDIES
> FUTURE DIRECTIONS
> REFERENCES

INTRODUCTION

Thiazolidinediones (TZDs) are insulin-sensitizing drugs that are widely used for the lowering of blood glucose in patients with diabetes mellitus. A growing body of evidence suggests that TZDs have benefits beyond glucose control and improved insulin resistance. Historically, the TZD drug class was discovered in the midst of an intensive search for ligands with lipid-lowering effects. Emerging evidence has supported the benefits of TZDs in improving lipid profile in patients with or without diabetes mellitus. This chapter provides a critical review of the effects of TZDs on serum lipoproteins.

MECHANISMS OF TZD ACTIONS ON SERUM LIPOPROTEINS

Thiazolidinediones bind to and activate the peroxisome proliferator-activated receptor-γ (PPAR-γ), a member of the nuclear receptor superfamily of ligand-activated transcription factors. There are three distinct PPAR subtypes—α, β/δ, and γ—each of which is encoded on separate genes and has different tissue-specific expression

From: *Contemporary Cardiology: Therapeutic Lipidology*
Edited by: M. H. Davidson, P. P. Toth, and K. C. Maki © Humana Press Inc., Totowa, NJ

patterns. PPAR-γ has been shown to play an important role in glucose homeostasis, lipid metabolism, and adipogenesis. The PPAR-γ1 isoform is expressed predominantly in adipose tissue; the PPAR-γ2 isoform is expressed in the liver, heart, kidney, endothelium, and vascular smooth muscle cells (among others). PPAR-γ forms a heterodimer with the retinoid X receptor (RXR). When activated by TZD binding, this heterodimer undergoes a conformational change and recruits coactivator proteins, promoting binding to PPAR-γ response elements (PPREs) in target genes. PPREs are found in multiple genes that affect lipid metabolism and glucose homeostasis—altered transcription of these genes is thought to mediate the effects of TZDs on serum lipid levels *(1)*.

Recent studies have demonstrated that individual PPAR-γ ligands may induce unique conformational changes in the PPAR–RXR heterodimer complex, resulting in differential gene expression and distinct downstream biologic effects *(2)*. Agents capable of selectively modulating PPAR transcriptional activity are known as selective PPAR modulators. Selective PPAR modulators may also cause differential coactivator protein recruitment. Although they share a common binding domain on PPAR-γ, individual SPPARMs may have unique effects on glucose and lipid metabolism as well as distinct side effect profiles.

The effects of currently available TZDs on serum lipoproteins are not uniform. Although the exact mechanisms underlying these differences are not fully understood, specific unique properties of individual TZD agents have been recognized. TZD agents vary in their PPAR-γ-binding potencies: rosiglitazone > pioglitazone > troglitazone. Although both pioglitazone and rosiglitazone demonstrate weak PPAR-α activation, pioglitazone is associated with greater PPAR-α activation than rosiglitazone *(3)*. PPAR-α activation stimulates fatty acid oxidation in the liver, leads to increased expression of lipoprotein lipase, and increases hepatic apolipoprotein A-I (ApoA-I) and ApoA-II expression. Ultimately, PPAR-α activation leads to a reduction in plasma triglyceride (TG) levels and an elevation of plasma high-density lipoprotein cholesterol (HDL-C) levels *(4)*. Nonuniform activation of PPAR-α and PPAR-γ by TZD agents may contribute to their differential effects on serum lipoprotein profiles.

In addition to their effects on insulin sensitization and lipid metabolism, TZDs increase adipogenesis and lipid accumulation in tissues. For example, in mouse models, complete metabolomic data revealed that rosiglitazone (i) induced hypolipidemia (by dysregulating liver-plasma lipid exchange), (ii) induced de novo fatty acid synthesis, (iii) decreased the biosynthesis of lipids within the peroxisome, (iv) substantially altered free fatty acid and cardiolipin metabolism in heart, and (v) elicited an accumulation of polyunsaturated fatty acids within adipose tissue *(5)*. Similarly, pioglitazone has been shown in vivo to upregulate transcription of genes involved in lipid storage in adipocytes *(6)*. Newer SPPARMs that dissociate the insulin-sensitizing effects of TZDs from the lipogenic effects are under development *(7)*. However, at present, it remains unclear whether TZD-mediated changes in tissue lipid accumulation are a necessary component of the beneficial metabolic modulation of this class of drugs.

CLINICAL EVIDENCE OF LIPID-LOWERING EFFECTS OF TZDS

Clinical trial experience with the two currently available TZDs, pioglitazone and rosiglitazone, has been steadily accruing. More than 3000 diabetic subjects have been enrolled in 10 small randomized controlled trials evaluating pioglitazone, and more than 5000 diabetic subjects have been enrolled in small studies evaluating rosiglitazone *(8)*.

Rosiglitazone has been found to affect blood lipids in both diabetic and nondiabetic subjects. In diabetic individuals, rosiglitazone monotherapy raises serum HDL-C, with increases in the HDL_2 and HDL_3 cholesterol subfractions. It has a neutral effect on serum TGs. Serum low-density lipoprotein cholesterol (LDL-C) is raised by rosiglitazone therapy, although there is a parallel increase in the predominance of large buoyant LDL particles. Compared with metformin monotherapy, combination therapy with rosiglitazone plus metformin results in increased total cholesterol (TC), HDL-C, and LDL-C [9]. Similar results have been found for combination therapy with rosiglitazone and sulfonylureas [10]. The effect of combining rosiglitazone and fenofibrate on serum HDL-C has been inconsistent [11,12].

In nondiabetic individuals with low HDL-C and features of the metabolic syndrome, rosiglitazone did not appreciably alter HDL-C levels, although it did modestly raise TC and LDL-C. Interestingly, rosiglitazone had a favorable effect on adipokines and inflammatory markers: there was a significant increase in levels of adiponectin, a modulator of insulin sensitivity, as well as a significant reduction in the inflammatory markers, resistin, C-reactive protein (CRP), and interleukin-6 [13]. Rosiglitazone has also been used to treat nondiabetic patients with HIV lipodystrophy. Relative insulin resistance, hypertriglyceridemia, and loss of subcutaneous fat characterize this condition; the use of highly active antiretroviral therapies may contribute lipodystrophy and dyslipidemia in HIV-infected patients. Compared with metformin and placebo, respectively, rosiglitazone significantly increased subcutaneous fat, especially peripheral fat, improved insulin sensitivity, and improved metabolic indices. Rosiglitazone did not have profound effects on the dyslipidemia of HIV lipodystrophy—serum HDL-C levels were not significantly increased, and TC and serum LDL-C levels rose slightly [14,15].

Pioglitazone has been shown to have favorable effects on serum lipid profiles in multiple studies. Monotherapy with pioglitazone lowers serum TGs, raises serum HDL-C, and modestly raises serum LDL-C and TC. When added to metformin or sulfonylurea, pioglitazone therapy results in greater increases in serum HDL-C and greater decreases in serum TGs than sulfonylurea plus metformin combination therapy [16]. Although LDL-C and TC levels increased with pioglitazone, TC/HDL-C ratios decreased with both pioglitazone and metformin [17]. Importantly, in studies of pioglitazone versus sulfonylureas in patients with type 2 diabetes, pioglitazone's beneficial effects on serum lipid profile were independent of overall improvement in metabolic control, measured by HbA1c values [18,19]. These results were maintained in long-term follow-up (up to 1 year) [20]. Combination treatment with pioglitazone and sulfonylureas or pioglitazone and metformin produces an increase in both LDL-C and HDL-C particle size; the combination of pioglitazone and metformin also produced an increase in levels of ApoA1 and a decrease in levels of ApoB [21]. In nondiabetic individuals with features of the metabolic syndrome, compared with placebo, pioglitazone therapy increased serum HDL-C and increased overall HDL-C particle size. Pioglitazone did not affect serum TG or LDL-C levels, but there was an upward shift in mean LDL-C particle size. Markers of inflammation were significantly reduced: high-sensitivity CRP (hsCRP) and plasma resistin levels were reduced and plasma adiponectin levels were significantly increased [22].

META-ANALYSES OF PIOGLITAZONE AND ROSIGLITAZONE TRIALS

Recently, two carefully performed meta-analyses summarized the effects of rosiglitazone and pioglitazone therapy on serum lipid profiles in patients with type 2 diabetes mellitus *(8,23)* (Fig. 1). Chiquette et al. examined 10 randomized controlled trials of over 3000 patients treated with pioglitazone and 13 randomized controlled trials of over 5000 patients treated with rosiglitazone. Overall, patients enrolled in the pioglitazone trials had slightly higher HbA1c levels, TG levels, and LDL-C levels and lower HDL-C levels at baseline compared with patients enrolled in rosiglitazone trials. Both TZDs demonstrated similar reductions in HbA1c and similar increases in body weight. Pioglitazone significantly lowered TG level (−40 mg/dL), increased HDL-C level (+4.6 mg/dL), and showed neutral effect on LDL-C and TC levels. Rosiglitazone significantly increased HDL-C level (+2.7 mg/dL), increased LDL-C level (+15 mg/dL), increased TC level (+21 mg/dL), and demonstrated a neutral effect on TG level. Compared with rosiglitazone, pioglitazone produced a more favorable lipid profile, with greater reduction of serum TG and less elevation of TC and serum LDL-C. Plasma HDL-C concentration was raised to a similar extent by both agents.

Van Wijk et al. evaluated 19 studies, involving over 5000 patients treated with either rosiglitazone or pioglitazone. Their results mirrored those found by Chiquette et al.: after adjustment for baseline differences in dyslipidemia, pioglitazone therapy resulted in greater improvement in serum TGs, LDL-C, and TC than rosiglitazone therapy. However, they point out that there have been important differences in the baseline characteristics in the study populations that may have an impact on the findings. They suggest that the greater degree of dyslipidemia at baseline in patients treated with pioglitazone may have influenced the magnitude of the treatment effect on TG and HDL-C. They also noted that rosiglitazone combination trials have shown greater lipid effects than monotherapy trials and that in these studies, rosiglitazone monotherapy was more prevalent than pioglitazone monotherapy.

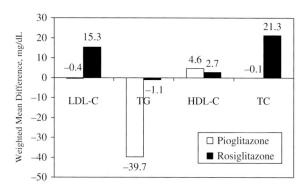

Fig. 1. Meta-analysis comparing the lipid effects of pioglitazone and rosiglitazone *(8)*. Results from a meta-analysis of 23 randomized controlled trials involving a total of 8000 patients are shown. Individually, pioglitazone significantly lowered triglyceride (TG), increased high-density lipoprotein cholesterol (HDL-C), and showed neutral effects on low-density lipoprotein cholesterol (LDL-C), and total cholesterol (TC) levels; rosiglitazone significantly increased HDL-C, LDL-C, and TC and had a neutral effect on TG. Comparatively, the treatment effects of pioglitazone on LDL-C, TG, and TC were statistically different than those of rosiglitazone (all $P < 0.01$).

The findings of these two meta-analyses have been confirmed in clinical practice settings through retrospective studies performed outside the context of controlled clinical trials *(24,25)*. Both TZDs demonstrated equivalent reduction in HbA1c. Pioglitazone demonstrated more favorable effects on lipid profiles than rosiglitazone. It was unclear from these studies whether the disparate effects on lipid profile between pioglitazone and rosiglitazone have an impact on cardiovascular outcomes.

PROSPECTIVE COMPARATIVE CLINICAL STUDIES

To directly compare the lipid effects of pioglitazone and rosiglitazone in patients with type 2 diabetes with dyslipidemia, we performed a randomized controlled head-to-head comparison study *(26)*. This study, called GLAI, enrolled subjects with type 2 diabetes aged ≥ 35 years who had fasting TG levels ≥ 150 to < 600 mg/dL and fasting LDL-C levels ≤ 130 mg/dL. Subjects received either rosiglitazone (up to 4 mg twice daily) or pioglitazone (up to 45 mg daily) for 6 months after a 4-week placebo washout period. None of the subjects enrolled took lipid-lowering agents before or during the study. All previous anti-hyperglycemic therapy consisted of either diet alone or oral monotherapy (which was discontinued before the start of the study).

In GLAI, pioglitazone was found to have greater influence on the lipid levels of patients with type 2 diabetes than rosiglitazone (Fig. 2). Pioglitazone raised serum HDL-C to a greater extent than rosiglitazone throughout the study. After 24 weeks of treatment, the 400 patients who took pioglitazone showed a 14.9% increase in their HDL-C levels or almost double the 7.8% increase seen in the 402 patients who took rosiglitazone ($P < 0.001$). The pioglitazone patients also experienced a 12% decrease in TG levels compared with a 14% increased in TG levels after rosiglitazone therapy. TC and LDL-C were both increased to a greater extent in the rosiglitazone group. Non-HDL levels rose with rosiglitazone, although they remained relatively constant with pioglitazone. LDL particle size increased with both agents, indicating reduced LDL atherogenicity, but the increase was greater with pioglitazone. LDL particle

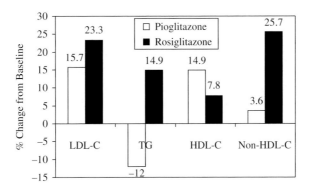

Fig. 2. Head-to-head comparison of the lipid effects of pioglitazone and rosiglitazone: GLAI trial *(26)*. Results from a randomized controlled trial involving 802 patients are shown. Pioglitazone significantly lowered triglyceride (TG), increased high-density lipoprotein cholesterol (HDL-C), and modestly increased low-density lipoprotein cholesterol (LDL-C) and non-HDL-C levels; rosiglitazone significantly increased HDL-C, LDL-C, TG, and non-HDL-C levels. Comparatively, the treatment effects of pioglitazone on all lipid parameters shown were statistically different than those of rosiglitazone (all $P < 0.0001$).

concentration and levels of ApoB and ApoC-III all increased with rosiglitazone, whereas with pioglitazone, these values decreased (LDL particle concentration and ApoC-III) or did not change (ApoB). These results were observed despite similar effects on glycemic control. Whether the difference in lipid measures seen with pioglitazone versus rosiglitazone translates into differences in major cardiovascular events has yet to be determined in a comparative outcome trial.

CONCOMITANT TZD AND STATIN THERAPY

Although proponents of pioglitazone have argued that the favorable lipid profile of pioglitazone may have added advantage in patients with diabetes mellitus (particularly in those with dyslipidemic profiles), many have questioned the clinical significance of TZD lipid effects, given the widespread use of statins in patients with diabetes mellitus. Both TZDs appeared to work synergistically with other lipid-lowering agents. Simvastatin (40 mg daily), added to rosiglitazone or pioglitazone in patients with diabetes, significantly improved LDL-C, irrespective of which TZD was used (27). Similar findings were observed when the ezetimibe/simvastatin combination was added to TZDs (28).

To address the concern that patients treated with statins were excluded from GLAI, we performed the COMPLEMENT study. This trial examined the effects of pioglitazone following conversion from rosiglitazone while maintaining stable statin therapy. COMPLEMENT was a 17-week multicenter open-label, single-arm study of 305 subjects with type 2 diabetes already taking stable doses of rosiglitazone and statins $\geq$ 90 days. Statin usage consisted of atorvastatin (51%), simvastatin (30%), pravastatin (8%), fluvastatin (5%), lovastatin (4%), or rosuvastatin (3%). In addition, 13% of patients were on an additional lipid-lowering agent, including nicotinic acid, fibrates (excluding gemfibrozil), and bile acid resin binders. The enrollment criteria stipulated TG levels $\geq$ 200 mg/dL, but < 1000 mg/dL, and HbA1c < 10.5%. Patients underwent conversion to pioglitazone, which resulted in mean changes of 15.2% lower TG, 1.8% higher HDL-C, 2.6 mg/dL reduction in ApoB, and 9% lower TC (29,30). Furthermore, there was a mean increase from baseline in LDL-C particle size of +0.2 nm and a mean decrease in LDL-C particle concentration of −189.3 nmol/mL independent of glycemic control. Thus, following conversion from rosiglitazone to pioglitazone in patients receiving stable statin therapy, there were improvements in lipid profile, independent of effects on glycemic control, measured by HbA1c.

The Rosiglitazone Study 108 was performed to assess the effects of combination therapy with rosiglitazone and atorvastatin in patients with type 2 diabetes (31) (Fig. 3). After a 4-week washout period of any prior lipid-lowering therapy, patients enrolled in this study began treatment with rosiglitazone, which was uptitrated over an 8-week run-in period. They were then randomized to receive atorvastatin 10 mg daily, atorvastatin 20 mg daily, or placebo. After the 8-week rosiglitazone run-in period, there was a mean increase in LDL-C of 9% with an increase in LDL-C relative flotation, indicative of an increase in the proportion of large buoyant LDL-C. There was an increase in HDL-C of 5.8%, with a 12.6% increase in the HDL_2 cholesterol subfraction and a 4.6% increase in the HDL_3 cholesterol subfraction. TGs were reduced 2%. Addition of atorvastatin to rosiglitazone resulted in a reduction of LDL-C (33 and 40% with atorvastatin 10 and 20 mg, respectively). There was no further change in LDL-C relative flotation. HDL-C

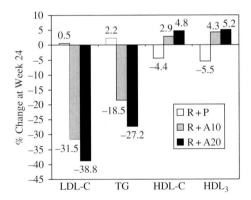

Fig. 3. Rosiglitazone study 108: effects of rosiglitazone alone and in combination with atorvastatin on the metabolic abnormalities in type 2 diabetes (31). Results from a randomized controlled trial involving 332 patients are shown. During the 8-week run-in period (not shown), rosiglitazone increased high-density lipoprotein cholesterol (HDL-C) 5.8%, increased low-density lipoprotein cholesterol (LDL-C) 9%, and lowered triglyceride (TG) 2%. In the subsequent 24 weeks of follow-up, addition of atorvastatin produced a dose-dependent significant reduction in LDL-C and TG. HDL-C rose further in the combination of rosiglitazone plus atorvastatin groups; most of the increase was due to a rise in the HDL_3 cholesterol subfraction. R + P, rosiglitazone plus placebo; R + A10, rosiglitazone plus atorvastatin 10 mg daily; R + A20, rosiglitazone plus atorvastatin 20 mg daily.

increased with rosiglitazone plus atorvastatin therapy (2.9 and 4.8% with atorvastatin 10 and 20 mg, respectively), mainly because of an increase in the HDL_3 cholesterol subfraction. Finally, TGs were significantly reduced with the addition of atorvastatin to rosiglitazone (−18.5 and −27.2% with atorvastatin 10 and 20 mg, respectively). Thus, the combination of rosiglitazone plus atorvastatin had an overall beneficial effect on plasma lipid profiles.

OUTCOME STUDIES

Pioglitazone therapy translated into clinical benefits in the large, multicenter, PROactive (prospective pioglitazone clinical trial in macrovascular events) study (32). Over half of the patients were taking concomitant lipid-lowering therapy: 43% of subjects had background statin therapy and 10% had background fibrate therapy. The 2605 subjects treated with pioglitazone had demonstrated a 10.1% relative improvement in HDL-C (+19 mg/dL absolute change), a 1.8% relative improvement in TG (−11.4 mg/dL absolute change), and a 4.9% increase in LDL-C (+7.2 mg/dL absolute change) compared with those in the placebo group. These changes in lipid parameters occurred in parallel with a 16% relative risk reduction in 3-year combined endpoint of myocardial infarction, stroke, or death with the use of pioglitazone in the PROactive trial.

Long-term effects of TZDs on cardiovascular morbidity and mortality currently are being evaluated in several ongoing large randomized controlled trials: A Diabetes Outcome Progression Trial (ADPOT), Rosiglitazone Evaluated for Cardiac Outcomes and Regulation of Glycemia in Diabetes (RECORD), Insulin Resistance after Stroke (IRIS), Diabetes REduction Assessment with ramipril and rosiglitazone Medication (DREAM), and Bypass Angioplasty Revascularization Investigation 2D (BARI 2D).

FUTURE DIRECTIONS

Several new compounds have emerged as a new class of dual PPAR-α, γ agonists known as "glitazars." It is theorized that combined treatment with PPAR-α, γ agonists may achieve improvement in insulin sensitization and lipid profiles while avoiding atherogenic dyslipidemia and weight gain. Data from the initial 24 weeks of a double-blind phase 3 clinical trial of 985 patients showed that muraglitazar therapy (0.5–20 mg) was linked to significantly lower TGs, non-HDL-C, and ApoB as well as higher HDL-C compared with those receiving pioglitazone (15 mg/day) (33). However, a post-hoc analysis of available phase 2 and phase 3 studies of muraglitazar involving 3725 patients reported an increased risk of the combined endpoint of death, nonfatal myocardial infarction, and nonfatal stroke in patients treated with muraglitazar compared with pioglitazone or placebo (relative risk 2.23) (34). The authors recommended that safety of the drug be more rigorously tested in dedicated cardiovascular events trials. Thus, when evaluating the lipid effects of existing and emerging PPAR modulators, it is important to remember that favorable changes in lipid parameters do not automatically translate into short-term and long-term cardiovascular benefits; the effects of these drugs on cardiovascular outcomes must be formally tested in well-designed clinical outcomes trials.

Tesaglitazar, another dual PPAR-α, γ agonist, showed promise in two phase 2 clinical trials reported at the ADA Scientific Sessions. The Glucose and Lipid Assessment in Diabetes (GLAD) trial was a 12-week, randomized, double-blind, dose-ranging study of 485 patients with type 2 diabetes that supported moving the 0.5-mg and 1-mg doses of tesaglitazar into phase 3 development (35). GLAD showed that tesaglitazar appeared effective in reducing fasting glucose while decreasing TG levels and increasing HDL-C. In addition, the Study in Insulin Resistance (SIR) trial was a 12-week, randomized, phase 2 study of 390 nondiabetic subjects with insulin resistance (36). The double-blind, placebo-controlled study examined the effect of tesaglitazar at 0.1–1 mg compared with placebo. Analysis revealed a dose-dependent reduction in TGs, with a reduction of 37% in fasting TGs and 41% in postprandial TGs at the 1-mg dose. Researchers also noted significant improvements in HDL-C and fatty acids. Cardiovascular outcome studies with tesaglitazar have not yet been performed, but the drug has been pulled from further development due to renal concerns.

In addition to PPAR-α and PPAR-γ, PPAR-δ is thought to have specific effects on metabolism, especially in peripheral skeletal muscle and adipose tissues (37). Although the exact actions of PPAR-δ have not been fully elucidated, this nuclear receptor is believed to influence fatty acid consumption in peripheral tissues. Future work on the role(s) of PPAR-δ, as well as potential development of pan-PPAR agents, will likely expand our understanding of both lipid metabolism and energy expenditure. Furthermore, the concept of SPPARMs can be applied to each type of PPAR: development of pharmacologic agents capable of modulating specific functions in specific tissues may allow tailored manipulation of metabolic pathways for a variety of conditions.

REFERENCES

1. Hauner H. The mode of action of thiazolidinediones. *Diabetes Metab Res Rev* 2002;18(Suppl 2): S10–S15.
2. Schupp M, Clemenz M, Gineste R, et al. Molecular characterization of new selective peroxisome proliferator-activated receptor {gamma} modulators with angiotensin receptor blocking activity. *Diabetes* 2005;54(12):3442–3452.

3. Sakamoto J, Kimura H, Moriyama S, et al. Activation of human peroxisome proliferator-activated receptor (PPAR) subtypes by pioglitazone. *Biochem Biophys Res Commun* 2000;278(3):704–711.
4. Staels B, Fruchart JC. Therapeutic roles of peroxisome proliferator-activated receptor agonists. *Diabetes* 2005;54(8):2460–2470.
5. Watkins SM, Reifsnyder PR, Pan HJ, et al. Lipid metabolome-wide effects of the PPARgamma agonist rosiglitazone. *J Lipid Res* 2002;43(11):1809–1817.
6. Bogacka I, Xie H, Bray GA, et al. The effect of pioglitazone on peroxisome proliferator-activated receptor-gamma target genes related to lipid storage in vivo. *Diabetes Care* 2004;27(7):1660–1667.
7. Berger JP, Petro AE, Macnaul KL, et al. Distinct properties and advantages of a novel peroxisome proliferator-activated protein [gamma] selective modulator. *Mol Endocrinol* 2003;17(4):662–676.
8. Chiquette E, Ramirez G, Defronzo R. A meta-analysis comparing the effect of thiazolidinediones on cardiovascular risk factors. *Arch Intern Med* 25 2004;164(19):2097–2104.
9. Wellington K. Rosiglitazone/metformin. *Drugs* 2005;65(11):1581–1592; discussion 1593–1594.
10. Derosa G, Cicero AF, Gaddi A, et al. Metabolic effects of pioglitazone and rosiglitazone in patients with diabetes and metabolic syndrome treated with glimepiride: a twelve-month, multicenter, double-blind, randomized, controlled, parallel-group trial. *Clin Ther* 2004;26(5):744–754.
11. Normen L, Frohlich J, Montaner J, et al. Combination therapy with fenofibrate and rosiglitazone paradoxically lowers serum HDL cholesterol. *Diabetes Care* 2004;27(9):2241–2242.
12. Seber S, Ucak S, Basat O, et al. The effect of dual PPAR alpha/gamma stimulation with combination of rosiglitazone and fenofibrate on metabolic parameters in type 2 diabetic patients. *Diabetes Res Clin Pract* 2006;71(1):52–58.
13. Samaha FF, Szapary PO, Iqbal N, et al. Effects of rosiglitazone on lipids, adipokines, and inflammatory markers in nondiabetic patients with low high-density lipoprotein cholesterol and metabolic syndrome. *Arterioscler Thromb Vasc Biol* 2006;26(3):624–630.
14. van Wijk JP, de Koning EJ, Cabezas MC, et al. Comparison of rosiglitazone and metformin for treating HIV lipodystrophy: a randomized trial. *Ann Intern Med* 2005;143(5):337–346.
15. Hadigan C, Yawetz S, Thomas A, et al. Metabolic effects of rosiglitazone in HIV lipodystrophy: a randomized, controlled trial. *Ann Intern Med* 2004;140(10):786–794.
16. Charbonnel B, Schernthaner G, Brunetti P, et al. Long-term efficacy and tolerability of add-on pioglitazone therapy to failing monotherapy compared with addition of gliclazide or metformin in patients with type 2 diabetes. *Diabetologia* 2005;48(6):1093–1104.
17. Schernthaner G, Matthews DR, Charbonnel B, et al. Efficacy and safety of pioglitazone versus metformin in patients with type 2 diabetes mellitus: a double-blind, randomized trial. *J Clin Endocrinol Metab* 2004;89(12):6068–6076.
18. Pfutzner A, Marx N, Lubben G, et al. Improvement of cardiovascular risk markers by pioglitazone is independent from glycemic control: results from the pioneer study. *J Am Coll Cardiol* 2005;45(12):1925–1931.
19. Satoh N, Ogawa Y, Usui T, et al. Antiatherogenic effect of pioglitazone in type 2 diabetic patients irrespective of the responsiveness to its antidiabetic effect. *Diabetes Care* 2003;26(9):2493–2499.
20. Charbonnel BH, Matthews DR, Schernthaner G, et al. A long-term comparison of pioglitazone and gliclazide in patients with type 2 diabetes mellitus: a randomized, double-blind, parallel-group comparison trial. *Diabet Med* 2005;22(4):399–405.
21. Perez A, Khan M, Johnson T, et al. Pioglitazone plus a sulphonylurea or metformin is associated with increased lipoprotein particle size in patients with type 2 diabetes. *Diab Vasc Dis Res* 2004;1(1): 44–50.
22. Szapary PO, Bloedon LT, Samaha FF, et al. Effects of pioglitazone on lipoproteins, inflammatory markers, and adipokines in nondiabetic patients with metabolic syndrome. *Arterioscler Thromb Vasc Biol* 2006;26(1):182–188.
23. van Wijk JP, de Koning EJ, Martens EP, et al. Thiazolidinediones and blood lipids in type 2 diabetes. *Arterioscler Thromb Vasc Biol* 2003;23(10):1744–1749.
24. Olansky L, Marchetti A, Lau H. Multicenter retrospective assessment of thiazolidinedione monotherapy and combination therapy in patients with type 2 diabetes: comparative subgroup analyses of glycemic control and blood lipid levels. *Clin Ther* 2003;25(Suppl B):B64–B80.
25. Peters Harmel AL, Kendall DM, Buse JB, et al. Impact of adjunctive thiazolidinedione therapy on blood lipid levels and glycemic control in patients with type 2 diabetes. *Curr Med Res Opin* 2004;20(2):215–223.
26. Goldberg RB, Kendall DM, Deeg MA, et al. A comparison of lipid and glycemic effects of pioglitazone and rosiglitazone in patients with type 2 diabetes and dyslipidemia. *Diabetes Care* 2005;28(7):1547–1554.

27. Lewin AJ, Kipnes MS, Meneghini LF, et al. Effects of simvastatin on the lipid profile and attainment of low-density lipoprotein cholesterol goals when added to thiazolidinedione therapy in patients with type 2 diabetes mellitus: a multicenter, randomized, double-blind, placebo-controlled trial. *Clin Ther* 2004;26(3):379–389.
28. Gaudiani LM, Lewin A, Meneghini L, et al. Efficacy and safety of ezetimibe co-administered with simvastatin in thiazolidinedione-treated type 2 diabetic patients. *Diabetes Obes Metab* 2005;7(1): 88–97.
29. Khan M, Perez A, Demissie S, et al. Effects of pioglitazone in combination with stable statin therapy on lipid changes in subjects with type 2 diabetes and dyslipidemia after treatment conversion from rosiglitazone: nonstatin lipid lowering results from an open-label study [Abstract 2142-PO]. *Diabetes* 2005;54(Suppl 1):A516.
30. Kupfer S, Perez A, Berhanu P, et al. Effects of pioglitazone in combination with stable statin therapy on apolipoprotein B changes in subjects with type 2 diabetes and dyslipidemia after treatment conversion from rosiglitazone: results from an open-label study [Abstract 2143-PO]. *Diabetes* 2005;54(Suppl 1):A516.
31. Freed MI, Ratner R, Marcovina SM, et al. Effects of rosiglitazone alone and in combination with atorvastatin on the metabolic abnormalities in type 2 diabetes mellitus. *Am J Cardiol* 2002;90(9): 947–952.
32. Dormandy JA, Charbonnel B, Eckland DJ, et al. Secondary prevention of macrovascular events in patients with type 2 diabetes in the PROactive Study (PROspective pioglitAzone Clinical Trial In macroVascular Events): a randomised controlled trial. *Lancet* 8 2005;366(9493):1279–1289.
33. Frederich R, Viraswamiappanna K, Rubin CJ. Effects of long-term therapy (2-year) with muraglitazar, a novel dual PPAR A/Γ agonist, on diabetic dyslipidemia in patients with type 2 diabetes: a double-blind, randomized, parallel-group study. *Diabetes* 2005;54(Suppl 1):A237.
34. Nissen SE, Wolski K, Topol EJ. Effect of muraglitazar on death and major adverse cardiovascular events in patients with type 2 diabetes mellitus. *JAMA* 23 2005;294(20):2581–2586.
35. Goldstein BJ, Rosenstock J, Anzalone D, et al. Tesaglitazar improves glucose and lipid abnormalities in patients with type 2 diabetes. *Diabetes* 2005;54(Suppl 1):A21.
36. Fagerberg B, Edwards S, Halmos T, et al. Tesaglitazar, a novel dual peroxisome proliferator-activated receptor alpha/gamma agonist, dose-dependently improves the metabolic abnormalities associated with insulin resistance in a non-diabetic population. *Diabetologia* 2005;48(9):1716–1725.
37. Fredenrich A, Grimaldi PA. PPAR delta: an uncompletely known nuclear receptor. *Diabetes Metab* 2005;31(1):23–27.

9 High-Density Lipoproteins

Peter P. Toth, MD, PhD

CONTENTS

> INTRODUCTION
> ANTIATHEROGENIC EFFECTS OF HDL
> EPIDEMIOLOGIC STUDIES EVALUATING
>   THE RELATIONSHIP BETWEEN HDL AND RISK
>   FOR CARDIOVASCULAR DISEASE
> THE DEFINITION AND PREVALENCE OF LOW HDL
> EFFECTS OF LIFESTYLE MODIFICATION ON HDL
> PHARMACOLOGIC THERAPY FOR LOW SERUM HDL-C
> HDL THERAPIES IN DEVELOPMENT
> ACKNOWLEDGMENT
> REFERENCES

INTRODUCTION

Lipoproteins are complex molecular assemblies with highly evolved, diverse functions. Lipoproteins are synthesized by the jejunum, liver, and periphery and serve to shuttle lipids (phospholipids and triglycerides), cholesterol, and cholesterol esters between a large number of cell types. Although cholesterol is pathogenic when available in excess, this crucial sterol functions to modulate cell membrane fluidity, is a precursor to steroid hormones and bile salts, and can modulate the activity of a variety of membrane-bound enzymes. Therapeutic modulation of lipid metabolism has become one of the cornerstones for preventing the development and progression of atherosclerotic disease in both the primary and the secondary prevention settings. Reducing the burden of such atherogenic lipoproteins in serum as very low-density and low-density lipoproteins (VLDLs and LDLs, respectively) is a major focus of cardiovascular medicine. Precisely defined, evidence-based targets for atherogenic lipoproteins have been promulgated by the National Cholesterol Education Program *(1,2)*.

Atherogenic lipoproteins promote the net uptake and deposition of cholesterol within macrophages localizing to the subendothelial space of arteries. A large number of clinical trials have shown that LDL reduction is associated with significant attenuation of risk for cardiovascular morbidity and mortality *(3–5)*. However, even with very aggressive reductions in serum LDL, the majority of acute cardiovascular events are still not prevented *(6,7)*. Consequently, a major focus of current investigation is to

From: *Contemporary Cardiology: Therapeutic Lipidology*
Edited by: M. H. Davidson, P. P. Toth, and K. C. Maki © Humana Press Inc., Totowa, NJ

further elucidate the role of emerging risk factors in atherogenesis and how they interact with or amplify the deleterious effects of established risk factors, such as hyperlipidemia, hypertension, diabetes mellitus, obesity, and cigarette smoking.

One of the most important risk factors for myocardial infarction (MI), stroke, sudden death, premature coronary artery disease (CAD), and peripheral arterial disease is a low serum level of high-density lipoprotein (HDL) *(8,9)*. Unlike atherogenic lipoproteins, the HDLs exert a large number of beneficial, vasculoprotective effects. This chapter will review HDL metabolism, the epidemiology linking low levels of this lipoprotein to increased risk for acute cardiovascular and cerebrovascular events, as well as established and investigational means by which to raise the serum levels of this important lipoprotein.

ANTIATHEROGENIC EFFECTS OF HDL

Reverse Cholesterol Transport

HDL particles are comprised of a charged, hydrophilic phospholipid surface. Triglycerides and cholesteryl esters (CEs) are concentrated within the hydrophobic core of these particles. HDLs carry a variety of apoproteins, the most important being apoprotein A-I (apoA-I). Other apoproteins bound to HDLs include apoproteins A-II, C-I, C-II, C-III, E, and J (clusterin). HDLs are also able to bind a variety of enzymes that perform diverse catalytic functions, such as lipid transfer and exchange, peroxide reduction, and cholesterol esterification. In older nomenclature, HDLs were described as α-lipoproteins, because of their mobility patterns in electrophoretic fields.

All somatic cells have the capacity to synthesize cholesterol. Cholesterol plays many important roles in metabolism. Cholesterol is soluble in cell membranes and modulates the fluidity of the hydrocarbon phase of phospholipid bilayers and the activity of membrane-bound enzymes. Cholesterol is a precursor to steroid hormone biosynthesis in steroidogenic organs and bile salts in hepatocytes. Cholesterol derived from the gut, liver, and systemic tissues is exchanged among lipoproteins which shuttle lipids and sterols among the various somatic lipid pools.

Unlike hepatocytes, somatic cells are unable to catabolize cholesterol. During atherogenesis, LDL migrates into the intima and subendothelial space of blood vessel walls in response to a variety of inflammatory, histologic, and thermodynamic driving forces. When exposed to modified LDL particles, macrophages upregulate scavenger receptors (SRs) (CD36, SR A) and take up LDL, promoting foam cell formation. Unless these cells can externalize intracellular cholesterol, net uptake will continue until the macrophage dies.

The HDLs regulate cholesterol balance in systemic tissues. These lipoproteins are able to interact with macrophages, promote cellular exporting of cholesterol, and deliver the cholesterol back to the liver for elimination or for repackaging into apoB-100-containing lipoproteins via a pathway defined as reverse cholesterol transport (RCT). The HDLs also deliver cholesterol to steroidogenic organs such as the adrenal, ovaries, testes, and placenta. If serum HDL levels are low or if any of the multiple steps of RCT are impaired, the capacity for systemic cholesterol clearance is reduced and risk for atherosclerotic disease is frequently increased.

ApoA-I is produced by both the jejunum and the liver and is secreted into blood in either its free or nonlipidated state or as a surface coat component of chylomicrons and VLDL. Nonlipidated apoA-I binds phospholipids and forms nascent discoidal HDL (ndHDL). Serum apoA-I levels are highly correlated with risk for acute CAD-related events in both men and women: high levels are atheroprotective, while low serum levels increase risk significantly *(10–14)*. A variety of mutations in apoA-I are associated with hypoalphalipoproteinemia and increased risk for CAD *(15)*.

One critical function of apoA-I is to interact with macrophages and stimulate the externalization of cholesterol and phospholipid. ApoA-I binds to the adenosine triphosphate-binding membrane cassette transport protein A1 (ABCA1) (Fig. 1). ABCA1 is defective in patients with Tangier's disease *(16,17)* and a variety of hypoalphalipoproteinemias *(18,19)*. In these patients, apoA-I is not lipidated normally, the HDL cannot speciate, and the excess apoA-I is eliminated via renal mechanisms after binding to cubulin or megalin. Patients with Tangier's disease have impairments in the reticuloendothelial system resulting in characteristic orange, hyperplastic tonsils and hepatosplenomegaly. A large number of mutations in ABCA1 have been identified. Gain of function mutations are associated with increased serum levels of HDL and reduced risk for CAD *(20–23)*.

Once in the extracellular space, unesterified cholesterol is converted into cholesterol esters by lecithin cholesteryl acyl transferase (LCAT), an enzyme which uses a fatty acid from the sn-2 position of phosphatidylcholine as an acyl chain donor. ApoA-I is an activator of LCAT. ApoA-I$_{Mallorca}$ has decreased capacity for activating LCAT and is associated with low serum levels of HDL *(24)*. Patients with LCAT deficiency states can develop fish-eye disease and have reduced capacity for CE formation and HDL speciation. The esterified cholesterol then follows a concentration gradient into the hydrophobic core of the HDL particle. As ndHDL incorporates more CEs, HDL particles become progressively more spherical and larger, resulting in the formation of HDL$_3$ and then HDL$_2$ particles. HDL$_3$ and HDL$_2$ particles can also promote cholesterol exporting from macrophages by interacting with ABCG1 and ABCG4 *((25)–(28))*. A wide range of HDL species can thus interact with macrophages and promote cholesterol depletion and systemic tissue homeostasis. The various HDL subfractions (encompassing the spectrum from ndHDL to large, buoyant CE-enriched HDL) can be separated and quantified using nuclear magnetic resonance imaging or gel electrophoresis (Fig. 2).

Spherical HDL particles can undergo lipid exchange, lipid transfer, and conversion reactions. Phospholipid transfer protein (PLTP) catalyzes the transfer of phospholipids from apoB-100-containing lipoproteins (LDL and VLDL) to HDL. PLTP can also mediate "HDL conversion," a fusion reaction among HDL$_3$ particles that can lead to the formation of HDL$_2$ and regeneration of ndHDL (Fig. 1) *(29)*. The in vivo importance of this conversion reaction is as yet undetermined.

The CEs in HDL can be exchanged for triglycerides from VLDL and LDL via the activity of cholesteryl ester transfer protein (CETP). Transferring CE into apoB-100-containing lipoproteins can result in indirect RCT. The CE transferred into LDL and VLDL can either be taken up by peripheral tissues or routed to hepatocytes via the LDL receptor and LDL receptor-related proteins. Cholesterol taken up by hepatocytes can then be (1) transported directly via ABCG5/G8 along bile canaliculi into bile for elimination; (2) converted into bile salts by 7α-hydroxylase, with bile salts transported into bile via ABCB11; or (3) reincorporated into VLDL and secreted by the liver into the systemic circulation. As HDL becomes progressively more enriched with triglyceride, it becomes a better target for lipolysis by hepatic lipase (HL) that converts larger, more buoyant particles into smaller, denser ones, thereby inhibiting HDL maturation and promoting HDL catabolism. The overexpression of HL is believed to cause a

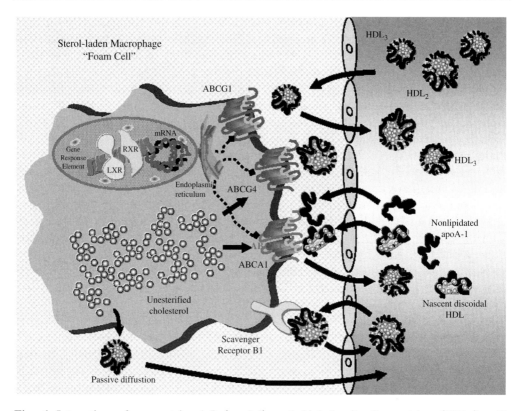

Fig. 1. Interaction of apoprotein A-I (apoA-I) and high-density lipoproteins (HDLs) with macrophages. Macrophages develop into foam cells as the amount of intracellular cholesterol increases. In order to maintain cholesterol homeostasis, the macrophage contains a number of cell membrane-bound sterol transporters that facilitate the externalization of excess intracellular cholesterol. In addition to these transporters, it is believed that some cholesterol can passively diffuse down along a concentration gradient across the cell membrane and enter the extracellular space (far left of figure). Nonlipidated apoA-I, nascent discoidal HDL (ndHDL), and spherical HDL_3 and HDL_2 particles are able to traverse the endothelial surface and access the subendothelial space. ApoA-I interacts with adenosine triphosphate-binding membrane cassette transport protein A1 (ABCA1) via its α-helical segments. Cholesterol and phospholipids are pumped out of the cell and lipidate free apoA-I and ndHDL. Nonesterified cholesterol is esterified by lecithin cholesteryl acyl transferase on the surface of the various HDL particles. The esterified cholesterol is partitioned into the hydrophobic core of HDL, making it more spherical and progressively enlarging it into HDL_3 and HDL_2. The larger HDL species can also interact with ABCG1 and ABCG4, which translocate intracellular cholesterol onto the surface of these lipoproteins. SR-BI allows for the transfer of cholesterol down along a concentration gradient onto the surface of spherical HDL species. The expression of ABC transporters is regulated by intracellular oxysterol concentrations. As the level of intracellular oxysterols rise, the cell activates transporter expression through the binding of liver X receptor-retinoid X receptor (LXR-RXR) heterodimers to the promoter regions of nuclear genes controlling the expression of the various ABC transporters. Lipidated HDL species can then reenter the plasma and deliver cholesterol to either the liver or the steroidogenic organs. Although the total amount of cholesterol mobilized from macrophages and foam cells is relatively small, mobilizing cholesterol from this histologic compartment is crucial to the prevention and/or progression of atherogenesis.

high prevalence of low serum HDL levels in both Turkish men (53%) and women (26%) *(30)*. Some evidence suggests that as HDL is more saturated with triglyceride, its capacity to drive RCT becomes progressively more impaired.

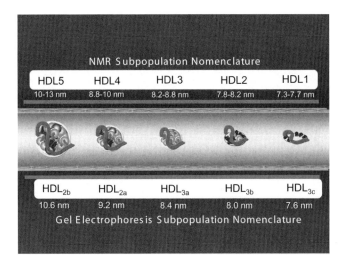

Fig. 2. Nomenclature for HDL subpopulations separated by nuclear magnetic resonance (NMR) spectroscopy and gel electrophoresis. HDL, high-density lipoprotein.

Polymorphisms that decrease CETP activity are associated with hyperalphalipoproteinemia and, in some studies, reduced risk for atherosclerotic disease. The relationship between CETP activity and risk for CAD is complex *(31)*. CETP mutations have been best studied in the Japanese, with the two most common mutations being Int14G → A and D442 → G in exon 15 *(32,33)*. Subjects with complete loss of CETP activity may be at increased risk for CAD despite having HDL levels that exceed 100 mg/dL. The HDLs of homozygotes appear to have less capacity for inducing the externalization of intracellular cholesterol, making them less efficient at RCT *(34)*. Moreover, with complete loss of CETP activity, the potential for indirect RCT is lost. A recent analysis of data from the Honolulu Heart Study suggested that subjects heterozygous for CETP deficiency have higher HDLs and lower risk for CAD compared to controls *(35)*. Among American men enrolled in the Veterans Affairs HDL Intervention Trial (VA-HIT), those with the TaqI B2B2 genotype had decreased CETP activity, increased mean serum HDL concentrations, and a 48% lower risk for cardiovascular morbidity and mortality relative to subjects with the Taq B1B1 genotype *(36)*. A CETP polymorphism that is associated with low serum HDL and increased risk for CAD is EcoN1 G/G *(37)*. Therapeutic reductions in CETP activity has clinical implications for patients in both the primary and the secondary prevention settings.

The final step in the RCT pathway is the delivery of cholesterol ester to the hepatocyte for biliary elimination or catabolic conversion into bile salts. Cholesterol uptake is mediated by a number of receptors on the hepatocyte surface (Fig. 3). The most important of these is SR class B type 1 (SR-B1) *(38–40)*. SR-B1 is a high-affinity receptor that selectively depletes HDL of CEs. Once delipidated, the HDL dissociates from SR-BI, scavenger receptor class B type 1 and can embark on another round of RCT. ApoA-I facilitates the docking of HDL to SR-BI *(41,42)*. In mice, SR-BI deficiency reduces biliary cholesterol concentrations, is associated with high CE content in HDL, increases serum levels of HDL, and increases risk for atherosclerotic disease *(43)*. The overexpression of SR-BI has the opposite effects: it is associated with reduced CE content of HDL, increased biliary cholesterol concentrations, decreased serum levels of HDL, and decreased atherosclerotic plaque burden *(44)*. These experiments in

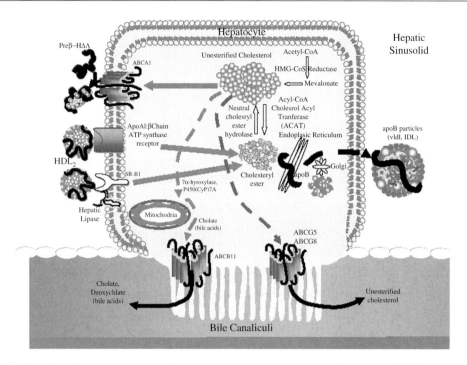

Fig. 3. Hepatic cholesterol handling. Through a series of high-density lipoprotein (HDL) maturation reactions, large spherical HDLs deliver cholesterol back to the liver in a process defined as reverse cholesterol transport (RCT). HDL can interact with SR-BI, which drives selective cholesteryl ester uptake. After the HDL particle is delipidated, it is released back into the circulation to initiate another round of RCT. HDL can also dock with the β-chain of the F1-ATP synthetase which engages in holoparticle uptake of the HDL particle. After the HDL is taken into the hepatocyte cytosol, it is catabolized. Internalized cholesteryl ester can then undergo a number of metabolic fates. It can be combined with apoB-100 and triglyceride and then exported as very low-density lipoprotein (VLDL). Alternatively, the cholesteryl ester can be hydrolyzed to free cholesterol via the activity of neutral cholesteryl ester hydrolase. This pool of unesterified cholesterol can be mobilized and exported via the activity of adenosine triphosphate-binding membrane cassette transport protein A1 (ABCA1), thereby reforming nascent discoidal HDL in the extracellular space. The liver is recognized as an important source of HDL in serum. Unesterified cholesterol can also be reesterified to a cholesteryl ester by acyl-CoA cholesterol acyl transferase.

mice support the emerging concept that it is not necessarily the absolute level of HDL in serum that confers protection against atherogenesis; rather, it is the functionality of available HDL that determines the risk of developing a net accumulation of cholesterol and lipid in vessel walls. The β-chain of the F1-ATP synthetase (usually expressed along the inner membrane of mitochondria) functions as a high-affinity HDL receptor on the surface of hepatocytes *(45)*. Unlike SR-BI, this receptor mediates holoparticle endocytosis (i.e., the HDL particle is taken up into the hepatocyte and catabolized). The glycosylphosphatidylinositol-anchored HDL-binding protein-1 is a high-affinity HDL receptor that regulates selective CE uptake in mice *(46)*. The hepatic delipidation and/or uptake of HDL is complex. It remains to be determined how these receptors are regulated and whether or not there is functional cross-talk among them.

From a purely conceptual standpoint, it can be argued that RCT is probably the most important antiatherogenic function of HDL. There is kinetic data in both humans *(47)* and animals *(48)* confirming the net flow of cholesterol via HDL-dependent pathways to the liver and gastrointestinal tract. Considerable resources are being invested in the discovery of novel means by which to raise serum levels of HDL and to augment its functionality in order to increase the throughput of RCT and decrease risk for the development and progression of atherosclerotic disease.

Prevention of Oxidative Phenomena

The oxidation of lipoproteins and molecular components of vessel walls are proatherogenic phenomena *(49)*. A variety of enzymes (myeloperoxidase, lipoprotein-associated phospholipase A2, 5'-lipoxygenase, and NAD(P)H oxidase) and transition metal cations potentiate oxidative damage *(50)*. The formation of such reactive oxidative species as peroxide and superoxide anions oxidize the fatty acids of phospholipids and triglycerides within LDL. These oxidation reactions produce reactive conjugated diene free radicals, free fatty acid radicals, and diene peroxy free radicals. Oxidized LDL adversely impacts the functioning of multiple cell lines in the vasculature *(51,52)*. For instance, macrophages exposed to oxidatively modified LDL particles increase their surface expression of scavenging receptors, promoting the formation of foam cells, fatty streaks, and atherosclerotic plaques. The exposure of endothelial cells to oxidized LDL and other lipid-free radical species induces cellular dysfunction with impaired vasoreactivity, reduced barrier function to the subendothelial space, and a proinflammatory/prothrombotic phenotype.

The HDLs exert potent antioxidative effects. In vitro assays show that HDL inhibits the genesis of LDL lipoperoxide species by up to 90%. The magnitude of inhibition varies among patients and is dependent on the concentration of HDL. Among men < 40 years of age with low serum HDL, there is a higher incidence of endothelial dysfunction and higher serum concentrations of oxidized LDL compared to age-matched controls with HDL > 40 mg/dL *(53)*. HDL harbors a number of antioxidative defenses. CETP can transfer lipid hydroperoxides from LDL into HDL where the oxidized lipid is reduced and rendered more inert. Two of these antioxidative defenses are enzymes, paraoxonase (PON) and platelet activating acetylhydrolase (PAFA). PON is an aryldialkylphosphatase that decreases LDL lipoperoxide synthesis and converts CE hydroperoxides to their less-reactive hydroxide analogs *(54)*. PAFA has phospholipase A2-like activity and hydrolyzes fatty acyl peroxides at the sn-2 position of oxidized phospholipids *(55)*. ApoA-I [Met(112) and Met(148)] and apoA-II [Met(26)] contain redox-active methionine residues that reduce cholesteryl hydroperoxides to less-reactive hydroxides *(56,57)*.

Effects on Endothelial Cell Function

The structural and functional integrity of the endothelial cell layer lining the luminal surface of arteries is crucial to the prevention of atherogenesis. Normal endothelial cells provide an antithrombotic surface and do not potentiate inflammatory and oxidative processes within blood vessel walls. As endothelial cells become exposed to proatherogenic stimuli such as dyslipidemia, hypertension, diabetes mellitus, or cigarette smoking, they become dysfunctional. Dysfunctional endothelial cells undergo a variety of biochemical changes which promote the elaboration of oxygen-free radicals (superoxide anion, peroxynitrite ion, and hydrogen peroxide), reduce nitric oxide production and capacity for vasodilatation, increase adhesion molecule expression and the binding of inflammatory white blood cells with subsequent transendothelial migration, and adversely affect the balance between tissue plasminogen activator and plasminogen activator inhibitor-1, among other changes.

HDLs exert many beneficial effects along endothelium and aid in maintaining the functional and structural integrity of these cells. The following phenomena illustrate how HDL can function as a biosensor of sorts and impact the transmission of a number of intracellular signaling pathways. Nuclear factor κB regulates the expression of a wide range of inflammatory molecules, including vascular cell adhesion molecule-1 and intercellular adhesion molecule-1 *(58)*. Sphingosine kinase catalyzes the formation of sphingosine 1-phosphate, an activator of Nuclear Factor (NF-κB). HDL inhibits sphingosine kinase, thereby reducing the surface expression of endothelial cell adhesion molecules *(59)*. C-reactive protein (CRP) can also induce the expression of endothelial cell adhesion molecules. HDL neutralizes this capacity of CRP via one of its phospholipid moieties, oxidized 1-palmitoyl-2-linoleoyl-sn-glycero-3-phosphocholine *(60)*. HDL stabilizes the mRNA transcripts for endothelial nitric synthase, facilitating the sustained production of nitric oxide. Consistent with this finding, as serum HDL levels increase, the degree of arterial vasodilation *(61)* and myocardial perfusion *(62)* increase. HDL stimulates endothelial progenitor cell recruitment and engraftment *(63,64)* as well as endothelial cell proliferation and migration *(65,66)* along injured regions of the arterial lumen. HDL inhibits endothelial cell apoptosis by producing sphingosylphosphorylcholine and lysosulfatide, inhibitors of caspases 3 and 9 *(67)*.

Effects on Thrombotic Phenomena

Platelets can influence the evolution and stability of atheromatous plaques. Platelets secrete inflammatory mediators (thrombospondin and platelet factor 4) and growth factors (e.g., platelet derived growth factor, CD40 ligand, and transforming growth factor-β) *(68,69)*. When exposed to the connective tissue elements of ruptured, ulcerated, or fissured atheromatous plaques, platelets are activated and aggregate forming clots that can occlude the arterial lumen, thereby inducing acute coronary syndromes such as unstable angina and MI *(70)*. HDL appears to potentiate urokinase-dependent fibrinolysis *(71)* and the ability of proteins C and S to inactivate coagulation factor Va *(72)*. HDL reduces thrombin-mediated platelet aggregation and fibrinogen binding *(73)*, platelet reactivity in response to collagen exposure *(74)*, and platelet thromboxane A2 synthesis *(75)*. HDL has also been shown to increase the production of prostacyclin, a prostaglandin known to inhibit platelet aggregation and stimulate vasodilatation *(76)*. HDL induces the latter effect by upregulating the expression of cyclooxygenase-2 and serving as a direct arachidonic acid donor, the substrate for prostaglandin biosynthesis. In women taking oral contraceptives, a low serum HDL is associated with increased platelet aggregability *(77)*.

Net Capacity to Affect Atherogenesis

Given the broad variety of antiatherogenic effects attributable to HDL, it is clearly of interest to establish whether or not apoA-I or HDL can antagonize atherogenesis in vivo. The intravenous infusion of apoA-I in rabbits that fed atherogenic diets significantly reduces the burden of atheromatous plaquing compared to controls *(78)*. Transfected mice overexpressing apoA-I experience regression of established atheromatous plaque (Fig. 4) *(79)*. Intravenous injection of HDL into rabbits can prevent atherogenesis and induce plaque regression in animals with established disease *(80,81)*. Such studies support the hypothesis that apoA-I and HDL are atheroprotective and, if available at high enough concentration with adequate functionality, can even induce plaque regression. The antiatherogenic effects of HDL are summarized in Table 1.

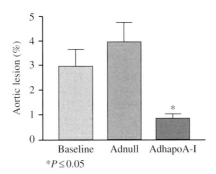

Fig. 4. Effect of increased hepatic expression of apoprotein A-I (apoA-I) on aortic atheromatous plaque. Mice that were low-density lipoprotein (LDL)-receptor deficient (LDLR$^{-/-}$) development atheromatous plaques following the administration of an atherogenic diet. Mice were then injected intrahepatically with either a control adenovirus or an adenoviral vector that contained the human gene for apoA-I. Mice in the latter group experienced a substantial elevation in serum apoA-I and high-density lipoprotein cholesterol compared to the mice in the control group. Of considerable importance in these experiments was the finding that mice treated with the apoA-I adenoviral vector experienced significant aortic atheromatous plaque regression, while the control mice experienced significant plaque progression. $^*P \leq 0.05$. Reproduced with permission from Tangirala et al. (79).

Pro-oxidative Activity by HDL

Although the preponderance of evidence favors the concept that HDL is antiatherogenic, one exception has emerged (82,83). The acute phase response is characterized as a pro-inflammatory and pro-oxidative state with a large rise in the elaboration of acute phase reactants (interleukins 1 and 6, tumor necrosis factor-α, and serum amyloid A). During the acute phase response, serum levels of HDL can drop due to increased catabolism of this lipoprotein by endothelial lipase and secretory phospholipase A2. The HDL particles also undergo significant alterations in their content of apoproteins and enzymatic constituents. ApoA-I, PON, PAFA, LCAT, CETP, and PLTP can dissociate from the HDL complex, thereby impairing its ability to drive RCT and antagonize lipid oxidation (Fig. 5). Such a loss of critical HDL functions is expected to be proatherogenic. As the HDL particle loses these key apoproteins and enzymes, it can associate with secretory phospholipase A2, apoJ, and serum amyloid A, further altering functions (Fig. 4). Work on this important area is ongoing. Assays evaluating the functionality of HDL are in development.

EPIDEMIOLOGIC STUDIES EVALUATING THE RELATIONSHIP BETWEEN HDL AND RISK FOR CARDIOVASCULAR DISEASE

The relationship between serum levels of HDL and risk for CVD was first inferred in 1951 by Barr and coworkers (84) who showed that patients with CAD had lower HDL compared to control patients who had no clinical manifestations of coronary disease. A similar relationship was subsequently noted among Israelis (85–88). Epidemiologic studies conducted around the world have affirmed that low serum levels of HDL are related independent of other risk factors to risk for developing all forms of atherosclerotic disease, while high levels appear to be atheroprotective. The Cooperative Lipoprotein Phenotyping Study included subjects from five US communities (Albany, New York; Evans County, Georgia; Framingham, Massachusetts; San Francisco, California; and Honolulu, Hawaii) and showed that there is an inverse

Table 1
Antiatherogenic Functions of High-Density Lipoproteins

Reverse cholesterol transport (net transfer of cholesterol mass from periphery back to the liver for disposal as cholesterol and bile salts)

Apoprotein donor to other lipoproteins to facilitate normal lipid and lipoprotein metabolism

Endothelial effects
1. Increase nitric oxide production and potentiate vasodilatation and myocardial perfusion
2. Suppress VCAM-1 and ICAM-1 expression by:
   (a) inhibiting sphingosine kinase and activation of NF-κB
   (b) neutralizing proinflammatory effects of CRP via oxidized 1-palmitoyl-2-linoleoyl-sn-glycero-3-phosphocholine
3. Promote endothelial progenitor cell recruitment and engraftment
4. Stimulate endothelial cell proliferation and migration
5. Inhibit apoptosis

Antioxidative effect
Reduces oxidized lipids in LDL via the activity of two antioxidative enzymes (paraoxonase and platelet activating factor acetylhydrolase) and redox active sulfide centers within the methionine residues of apoA-I and apoA-II

Antithrombotic effects
Stimulates
1. Urokinase-dependent fibrinolysis
2. The ability of proteins C and S to inactivate coagulation factor Va
3. Prostacyclin production by activating cyclooxygenase-2

Inhibits
1. Thrombin-mediated platelet aggregation via inhibition of 1,2-diacylglycerol and inositol 1,4,5-tris-phosphate formation
2. Platelet activation secondary to collagen exposure
3. Platelet thromboxane A2 production

apoA-I, apoproteinA-I; CRP, C-reactive protein; LDL, low-density lipoprotein.

relationship between HDL cholesterol (HDL-C) and CAD that is independent of LDL cholesterol (LDL-C) and triglycerides *(89)*. The Physicians' Health Study showed that a low serum HDL-C is associated with an increased risk for CAD even when total cholesterol is low *(90)*. In the Tromso Heart Study, low serum levels of HDL-C increased risk for CAD threefold more than LDL-C among 6595 men aged 20–49 years *(91)*. The Prospective Cardiovascular Münster (PROCAM) Study demonstrated that (1) subjects with HDL-C levels < 35 mg/dL had a threefold greater risk for developing CAD compared with subjects with HDL-C ≥ 35 mg and (2) this relationship was independent of LDL-C, triglycerides, and other known cardiovascular disease risk factors *(92)*. In the more recent Prospective Epidemiological Study of Myocardial Infarction, serum levels of both HDL-C and apoA-I were significantly lower among patients with established CAD compared to control subjects *(93)*. Moreover, as levels of HDL-C and apoA-I increase, risk for MI and coronary death decreased significantly.

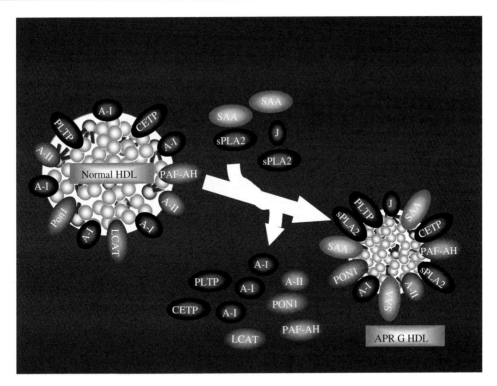

Fig. 5. Compositional alterations in high-density lipoprotein (HDL) during acute phase reactions. During an acute phase reaction, the expression of a large number of inflammatory mediators increases, including C-reactive protein, fibrinogen, serum amyloid A, and secretory phospholipase A2. The normal enzymatic and apoprotein constituents of HDL can dissociate and be replaced by apoJ, serum amyloid A, and phospholipase A2. This attenuates the ability of HDL to engage in reverse cholesterol transport and decrease oxidation and inflammation. A-I, apoprotein A-I; A-II, apoprotein A-II; J, apoprotein J; CETP, cholesteryl ester transfer protein; LCAT, lecithin:cholesteryl acyltransferase; PLTP, phospholipid transfer protein; PON1, paraoxonase 1; SAA, serum amyloid A; sPLA2, secretory phospholipase A2. Reproduced with permission from Ansell et al. *(82).*

The Framingham Study showed that at any level of LDL-C, as HDL-C decreases, there is a significant increase in risk for acute CAD-related events. (Fig. 6) In addition, for every 20 mg/dL rise in HDL, risk for CAD decreases by 50% *(94).* A combined analysis of four large American studies (Multiple Risk Factor Intervention Trial, the Framingham Heart Study, the Lipid Research Clinics Prevalence Mortality Follow-up Study, and the Lipid Research Clinics Primary Prevention Trial) showed that each 1 mg/dL increase in HDL-C confers a 2% reduction in CHD, coronary heart disease risk in men and a 3% reduction in women *(95).* Evaluating Japanese residents of Osaka reveal interesting relationships: when comparing patients in the lowest to the highest quartile of serum HDL-C, the relative risk for MI and CAD increase by 3.39 and 4.17, respectively; in addition, for every 1 mg/dL rise in HDL-C the risk for MI and CAD decreases by 6.4 and 5.7%, respectively *(96).* When evaluated in women enrolled in the Nurses' Health Study, a 17 mg/dL elevation in HDL-C is associated with a 40% reduction in CHD risk *(97).* Among men enrolled in the Framingham Offspring Study, each mg/dL rise in HDL decreased risk for CHD by 2%. Of considerable interest in this study is the observation that for each mg/dL rise in large α_1-HDL,

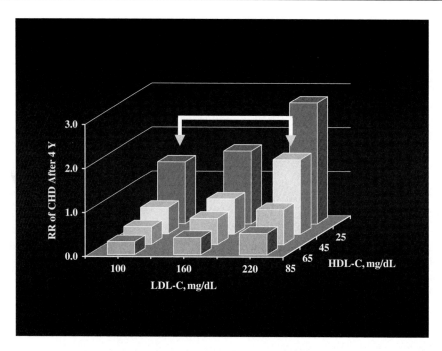

Fig. 6. Relationship between low-density lipoprotein cholesterol (LDL-C), high-density lipoprotein cholesterol (HDL-C), and relative risk of a CHD-related event over 4 years of follow-up in the Framingham Study. Note that even when LDL-C is approximately 100 mg/dL, when the HDL is 25 mg/dL, patient risk for cardiovascular disease is on par with that of a patient whose LDL-C is 220 mg/dL and HDL-C is 45 mg/dL. At any level of LDL-C, as HDL-C increases, risk increases continuously. RR, relative risk. Reproduced with permission from Castelli *(230)*.

risk for CHD decreased by 26% *(98)*. Low serum levels of HDL increase risk for MI, stroke, in-stent restenosis after angioplasty, increased carotid intima-media thickness (CIMT), sudden death, severe atherosclerotic disease in the proximal left main coronary artery, and sudden death *(99–102)*. Low HDL-C is associated with increased rates of atheromatous plaque progression *(103)*, whereas multivariate analyses show that statin-induced elevations in this lipoprotein are associated with plaque regression *(104,105)*.

A number of studies confirm the importance of HDL in elderly patients. The Cardiovascular Health Study evaluated the relationship between serum lipid levels and risk for stroke, MI, and mortality in patients ≥65 years of age over 7.5 years of follow-up *(106)*. There was an inverse relationship between HDL-C and risk for MI. High HDL-C was protective against ischemic stroke in men but not in women. Among 3904 subjects >71 years of age, a low serum level of HDL-C was a better predictor of CHD-related events than total cholesterol *(107)*. In another study evaluating 3904 American men and women older than 71 years, low HDL-C was a better predictor for incident CHD morbidity and mortality than total cholesterol. Risk for mortality increased 4.1-fold when comparing patients with baseline HDL-C <35 and >60 mg/dL, respectively. Among subjects >75 years of age enrolled in the Northern Manhattan Stroke Study, HDL-C reduced risk for stroke according to a dose–response relationship *(108)*. This benefit was apparent independent of racial or ethnic identity and protected patients of all racial and ethnic groups studied older than 75 years from stroke according to a dose–response relationship. Risk for fatal CHD and stroke is independent of LDL-C but highly correlated with low HDL-C in patients ≥85 years of age *(109)*.

THE DEFINITION AND PREVALENCE OF LOW HDL

In its third Adult Treatment Panel, the National Cholesterol Education Program (NCEP) define a low HDL-C as <40 mg/dL for all patients *(1)*. The American Heart Association has defined a normal HDL-C level in women as ≥50 mg/dL *(110,111)*. Women tend to have, on average, an HDL-C level that is 10 mg/dL higher than age-matched men. In women receiving oral contraceptive pills or postmenopausal hormone supplementation, HDL levels can be up to 20–25 mg/dL higher than age-matched men *(112)*. Although NCEP do not define a therapeutic target for HDL-C, it is recommended that patients with HDL-C <40 mg/dL be treated with therapeutic lifestyle changes (aerobic exercise, weight loss, step I diet, and smoking cessation) and medication tailored to the specific features of the lipid profile. The NCEP did not define a target for HDL-C because it was concluded that currently available medications are inadequately robust to reliably raise HDL in all patients. Moreover, the NCEP also concluded that there is as yet inadequate clinical trial data to define a target for this lipoprotein.

When calculating 10-year Framingham risk scores, an HDL-C <40 mg/dL adds 2 points to the total score; in contrast, an HDL-C ≥60 mg/dL allows for the substraction of 1 point. In general, HDL-C levels >60 mg/dL are considered atheroprotective. A question that often arises for women is whether or not elevations in LDL-C or non-HDL-C should be treated when HDL is substantially greater than 60 or even 90 mg/dL. In these clinical circumstances, the NCEP dictate that targets for atherogenic lipoproteins be based on global cardiovascular risk burden and calculation of 10-year Framingham risk scores. It should not simply be assumed that if a woman's HDL-C is high that she is protected from atherosclerotic disease per se. In contrast to NCEP, the European Consensus Panel on HDL-C *(113)* and the Expert Group on HDL Cholesterol *(114)* recommend that HDL be raised to ≥40 mg/dL in patients with CAD and those at high risk for CAD (metabolic syndrome, diabetes mellitus, and 10-year Framingham risk >20%). The American Diabetes Association has defined HDL-C targets of ≥40 mg/dL in men and ≥50 mg/dL in women *(115)* (see summary of recommendations in Table 2).

Low HDL-C is widely prevalent. In the United States with 15% of women and 39% of men aged ≥20 years have low HDL-C <40 mg/dL as defined by NCEP (<40 mg/dL) *(116)*. Among 8545 European patients with dyslipidemia, the prevalence of low HDL is 40% and 33% in women (<50 mg/dL) and in men (<40 mg/dL), respectively. Low HDL is common in patients with type 2 diabetes mellitus who have significant insulin resistance and is an important defining manifestation of the metabolic syndrome. It is anticipated that the prevalence of low HDL-C will continue to rise as the incidence of obesity and metabolic syndrome continue to increase worldwide. Approximately 1 in 4 Americans currently meet criteria for the metabolic syndrome *(117)*. Low HDL-C is frequently observed in men with CHD. The prevalence of HDL-C levels <35 mg/dL is three times higher in patients with premature CHD than in healthy controls (57 versus 19%, respectively) *(118)*. In the Quebec Cardiovascular Study, an HDL-C <35 mg/dL was an independent risk factor for coronary disease and was more prevalent in men with CAD compared to those without (50 versus 30%) *(119)*. In a sample of 8500 men with CHD, 64% had HDL-C ≤40 mg/dL *(120)*.

Table 2
Guideline Definitions for low HDL-C and Targets for Therapy

Guideline sponsor	Definition for low HDL-C
National Cholesterol Education Program	<40 mg/dL (all patients)[a]
American Heart Association	<50 mg/dL (women)
Guideline sponsor	HDL-C target for therapy
National Cholesterol Education Program	None
American Diabetes Association	≥40 mg/dL in men
	≥50 mg/dL in women
European Consensus Panel on HDL-C	≥40 mg/dL[b]
Expert Group on HDL-C	≥40 mg/dL[b]

CAD, coronary artery disease; HDL-C, high-density lipoprotein; NCEP, National cholesterol Education Program.

[a] However, when diagnosing metabolic syndrome, NCEP defines low HDL-C as <40 mg/dL in men and <50 mg/dL in women.

[b] For patients with CAD and those at high risk for CAD (metabolic syndrome, diabetes mellitus, and 10-year Framingham risk >20%).

EFFECTS OF LIFESTYLE MODIFICATION ON HDL

According to NCEP ATPIII, therapeutic lifestyle change is an important means by which to help raise low serum levels of HDL-C *(1)*. HDL levels are responsive to a number of lifestyle interventions. In patients with insulin resistance, HDL metabolism is perturbed secondary to a number of effects. First, there is an insulin response element in the promoter of the gene for apoA-I *(121)*. It is possible that as the hepatic parenchyma becomes less sensitive to the effects of insulin, nuclear expression of apoA-I decreases. Second, lipoprotein lipase is relatively inhibited during insulin resistance because the expression of its inhibitor, apoC-III, increases. Lipoprotein lipase hydrolyzes the triglycerides within large lipoproteins, such as chylomicrons and VLDL. During lipolysis, these large lipoproteins release surface coat constituents that can be used to assimilate HDL in serum. Third, as serum triglyceride levels in patients with insulin resistance rise, more triglyceride mass is transferred into HDL during an exchange reaction driven by CETP. As the HDL becomes progressively more enriched with triglyceride, it becomes a higher affinity substrate for lipolysis by HL. The HL converts large, buoyant HDL into smaller, denser HDL particles that can be lipolyzed further and completely catabolized. The apoA-I released during lipolysis from HDL can be bound by cubulin and megalin receptors in the kidney and eliminated in urine. The severity of insulin resistance can be significantly reduced and even resolved with weight loss. Insulin resistance is generally induced by increased visceral (omental, perimesenteric, perinephric, and peritoneal) adipose tissue mass.

Weight and body mass index (BMI) are highly correlated with serum levels of HDL-C. In the Framingham Offspring Study, HDL-C has a strong inverse, near-linear relationship with rising BMI in both men and women *(122)*. For every 1 kg/m^2 increase in fat mass, HDL decreases by 1 mg/dL. Weight loss is associated with increased serum levels of HDL in both men and women *(123)*, particularly if a significant degree of insulin re-sensitization is achieved among obese individuals with metabolic syndrome. During the acute phase of weight loss, HDL-C can, however, decrease. Once the patient's weight has stabilized and their metabolism has re-equilibrated, HDL usually increases. Increased saturated fat intake is associated with increased HDL-C *(124,125)*, while a low fat diet *(126,127)* and increased intake of carbohydrates *(126)* and *trans* fatty acids *(128)* decreases HDL-C. A Mediterranean type diet enriched with olive oil, legumes, fruits, and vegetables is associated with reduced insulin resistance and increased serum levels of HDL-C *(129,130)*.

There is an inverse dose–response relationship between the number of cigarettes smoked daily and serum levels of HDL-C in both men and women *(131)*. Smoking likely decreases HDL-C by promoting insulin resistance *(132)* and inhibiting LCAT activity *(133)* and reducing HDL maturation, at least in some patients. Smoking cessation is associated with increases in serum HDL-C of up to 20% in as little as 2 months *(134,135)*. Smoking cessation can, consequently, raise HDL-C as well as most of the pharmacologic interventions we currently have at our disposal to treat low HDL-C and would constitute an important first step in comprehensive lifestyle modification for these patients.

Alcohol intake is also correlated with serum levels of HDL-C. Alcohol can directly stimulate hepatocytes to increase expression of apoA-I *(136)*. Alcohol is also a weak inhibitor of CETP. The changes wrought by alcohol on serum HDL may at least partly explain the finding that moderate alcohol consumption is associated with reduced risk for acute cardiovascular events *(137–139)*. Wine is an important component of diet in many regions of the world. While increased alcohol consumption is generally not advocated by physicians due to concerns involving excess alcohol ingestion, the majority of patients can responsibly and safely consume 1–2 servings of wine with the evening meal. There is no consensus concerning the question as to whether or not red wine is better for promoting wellness than is white wine.

Exercise is an important part of any lifestyle modification regimen. As men and women age, it is generally recognized that there is "creeping" weight gain and reduced inclination for any form of exercise. There appears to be a dose–response relationship between amount of exercise and the magnitude of HDL elevation for both men and women *(140–142)*. During a weight loss program, aerobic exercise has been shown to prevent the drop in HDL-C during the acute phase of a low-fat weight reduction diet *(143)*. Exercise helps to relieve insulin resistance *(144)*, and this would be expected to decrease HDL catabolism and promote HDL secretion and maturation. As shown in the Health, Risk Factors, Exercise Training and Genetics Family Study, patients with high baseline triglycerides experience more substantial elevations in HDL-C in response to aerobic exercise *(145)*. At any BMI, women who exercise have higher mean HDL-C levels than those who do not *(146)*. Exercise is associated with risk reduction for CAD-related events *(144)*.

PHARMACOLOGIC THERAPY FOR LOW SERUM HDL-C

A number of drugs are currently available for increasing serum HDL-C levels. The capacity of these medications to raise HDL-C is summarized in Table 3. The clinical trials evaluating the effects of some of these drugs on serum lipoprotein levels as well as cardiovascular morbidity and mortality are summarized in Table 4. All of these medications impact multiple lipoprotein fractions. Consequently, it has been challenging to validate or negate the HDL hypothesis with currently available agents in clinical trials. A flow chart summarizing a suggested approach to treating low HDL-C in a variety of clinical circumstances is shown in Fig. 7.

Table 3
Expected Increase in Serum HDL in
Response to Various Pharmacologic
Therapies

Drug	% Increase in HDL
Statins	5–10
Fibrates	10–15
Niacin	10–35
Thiazolidinediones	5–20
Estrogen	10–25
α-Blockers	10–20

Niacin

Of any currently available pharmacologic agent, niacin has the greatest capacity to increase serum levels of HDL-C. Niacin induces a dose-dependent increase in HDL-C (10–35%). It has been long assumed that niacin increases HDL-C by decreasing hepatic holoparticle uptake of HDL-C without compromising RCT *(164)*. While this may still be correct, niacin appears to exert a number of its effects secondary to binding to a specific receptor on the surface of adipocytes and macrophages. In humans and mice, the niacin receptors are referred to as HM74 and PUMA-G (protein-upregulated in macrophages by interferon-γ), respectively *(165,166)*. In patients with insulin resistance, adipocyte hormone-sensitive lipase hydrolyzes stored triglyceride in a dysregulated manner. When niacin binds to the adipocyte niacin receptor, hormone-sensitive lipase is inhibited secondary to a G-protein-dependent inhibition of adenylate cyclase. This ultimately reduces the amount of fatty acid channeled into the portal circulation. With less free fatty acid available for assimilation into triglycerides, less triglyceride is available for VLDL synthesis. In the hepatocyte, niacin reduces hepatic triglyceride biosynthesis by inhibiting diacylglycerol acyltransferase *(167)*. Serum VLDL and triglycerides decrease. HDL-C likely increases because it becomes less enriched with triglyceride and is less apt to be catabolized by HL. In the macrophage, the binding of niacin to the niacin receptor increases the surface expression of ABCA1. Increased ABCA1 activity likely augments both the lipidation of apoA-I and RCT *(168)*.

Niacin has been evaluated in a number of clinical trials, as both monotherapy and in combination with a statin. The Coronary Drug Project studied the effect of crystalline niacin (3.0 g daily) versus placebo on cardiovascular event rates in men with established CAD over 5 years *(169)*. HDL levels were not measured in this study. Niacin decreased risk for stroke, MI, and revascularization by 24, 26, and 67% compared to placebo, but did not reduce mortality in a statistically significant manner during the course of the trial. A significant percentage of patients discontinued niacin therapy due to cutaneous flushing, a reaction mediated by prostaglandin D2. In the HDL Atherosclerosis Treatment Study (HATS), 160 patients with CAD (mean baseline lipids: HDL-C 31 mg/dL; LDL-C 125 mg/dL; and triglycerides 213 mg/dL) were randomized to placebo or one of three treatment arms: (1) an "antioxidant cocktail" comprised of vitamins E and C, selenium, and β-carotene; (2) simvastatin and niacin; or (3) simvastatin, niacin, and the antioxidant cocktail *(170)*. The antioxidants did not reduce risk for acute cardiovascular events compared to placebo. Relative

Table 4
Summary of Lipid Changes and Cardiovascular Outcomes in Key Fibrate, Statin, and Niacin Clinical Trials (>400 Participants)

Trial name	Study design	Patient characteristics	Active treatments	Time of lipid assessment	Lipid changes from baseline					Key cardiovascular outcome
					Total cholesterol	LDL-C	HDL-C	TGs	
Air Force/Texas Coronary Atherosclerosis Prevention Study (AFCAPS/TexCAPS) (147)	r, db, pc	Average total cholesterol and LDL-C, and below average HDL-C (45–73 years, n = 6605)	Lovastatin 20–40 mg daily	1 year	↓18%	↓25%	↑6%	↓15%	Reduced incidence of first acute major coronary event (RR 0.63; 95% CI 0.50–0.79, P < 0.001) after 5.2 years
Aggressive Lipid-Lowering Initiation Abates New Cardiac Events (ALLIANCE) (148)	r, ol	Prior acute MI, percutaneous transluminal coronary angioplasty, coronary artery bypass graft, or unstable angina plus moderate to high LDL-C (>18 years, n = 2442)	Atorvastatin titrated from 10–80 mg daily (until LDL-C ≤80 mg/dL) or	4.3 years	↓24.1%	↓34.3%	↑6.1%	↓7.4%	Reduction in risk of major CV event of 17.1% in atorvastatin treatment group compared with usual care group (hazard ratio = 0.829; P = 0.020)
			Usual care		↓15.5%	↓23.3%	↑8.5%	↓2.4%	

(Continued)

Table 4
(Continued)

Trial name	Study design	Patient characteristics	Active treatments	Time of lipid assessment	Total cholesterol	LDL-C	HDL-C	TGs	Key cardiovascular outcome
Anglo-Scandinavian Cardiac Outcomes Trial-Lipid Lowering Arm (ASCOT-LLA) (148)	r, db, pc	Hypertension (40–79 years, n = 10,305)	Atorvastatin 10 mg daily	1 year	↓24%[a]	↓35%[a]	No change	↓17%[a]	Reduction in nonfatal MI and fatal CHD (HR 0.64; 95% CI 0.50–0.83, P = 0.0005) after 3.3 years
				3.3 years	↓19%[a]	↓29%[a]	No change	↓14%[a]	
Arterial Disease Multiple Intervention Trial (ADMIT) (150)	r, db, pc	Peripheral arterial disease (≥30 years, n = 468)	Niacin up to 1500 mg BID or maximum tolerated dose. Pravastatin could be initiated from week 18 if LDL-C > 130 mg/dL	18 weeks	↓7%	↓6%	↑17%	↓16%	Cardiovascular endpoints not assessed
				1 year	11% (percent due to pravastatin not stated)	22% (10% due to pravastatin)	↑30% (all due to niacin)	28% (all due to niacin)	
Bezafibrate Infarction Prevention (BIP) (151)	r, db, pc	Prior MI or stable angina, low HDL-C, and moderately elevated total cholesterol	Bezafibrate 400 mg daily	6 years	↓5%	↓7%	↑18%	↓21%	Nonsignificant reduction in the probability of nonfatal and fatal

Study	Design	Population	Intervention	Duration				Results
Cholesterol And Recurrent Events (CARE) (4)	r, db, pc	Prior MI and average cholesterol levels (21–75 years, n = 4159)	Pravastatin 40 mg daily	5 years	↓20%[a]	↓28%[a]	↑5%[a]	↓14%[a]	MI and sudden death (7.3% reduction, P = 0.24); significant reduction in the cumulative probability of this endpoint in subgroup of patients with high baseline triglycerides (≥200 mg/dL) (39.5%, P = 0.02) Reduction in risk of fatal CHD or confirmed MI (RR 0.76; 95% CI 0.64–0.91, P = 0.003) at 5 years
Collaborative Atorvastatin Diabetes Study (CARDS) (152)	r, db, pc	Type 2 diabetes without high LDL-C (40–75 years, n = 2838)	Atorvastatin 10 mg daily	4 years	↓26%[a]	↓40%[a]	↑1%[a]	↓19%[a]	Reduction in the incidence of major cardiovascular events (HR 0.63; 95% CI 0.48–0.83, P = 0.001)

(Continued)

Table 4
(Continued)

Trial name	Study design	Patient characteristics	Active treatments	Time of lipid assessment	Total cholesterol	LDL-C	HDL-C	TGs	Key cardiovascular outcome
Coronary Drug Project (CDP) (153)	r, db, pc	Men with previous MI (30–64 years, $n = 8341$)	Clofibrate	1 year	↓5.1%	Not reported	Not reported	↓23.1%	After a mean long-term follow-up of 15 years, total mortality in the niacin treatment group was 11% lower than in the placebo group (52.0 versus 58.2%; $P = 0.0004$)
			Niacin		↓10.1%	Not reported	Not reported	↓26.9%	
Diabetes Atherosclerosis Intervention Study (DAIS) (154)	r, db, pc	Type 2 diabetes (40–65 years, $n = 731$)	Fenofibrate 200 mg daily	3 years	↓10%	↓7%	↑8%	↓29%	Significant reduction in angiographic progression of CHD (clinical endpoints not assessed)
Fenofibrate Intervention and Event Lowering in Diabetes Study (FIELD) (155)	r, db, pc	Type 2 diabetes (50–75 years, $n = 9795$)	Fenofibrate 200 mg daily	5 years	↓7%	↓6%	↑1.2%	↓22%	Nonsignificant reduction in coronary events (HR 0.89; 95% CI 0.75-1.05; $P = 0.16$);

Study	Design	Population	Intervention	Duration				Results
GREek Atorvastatin and Coronary-heart-disease Evaluation Study (GREACE) (156)		Previous MI or >70% stenosis of 1 coronary artery plus moderate to high LDL-C (<75 years, n = 1600)	Atorvastatin titrated from 10–80 mg daily (until LDL-C ≤ 100 mg/dL) or	3 years	↓36%	↓46%	↑7%	↓31%	significant reduction in non-fatal MI (HR 0.76; 95% CI 0.62–0.94; $P = 0.01$) Significant reductions in total mortality (−43%, P=0.0021), coronary mortality (−47%, $P = 0.0017$), non-fatal MI (−59%, $P = 0.0001$), unstable angina (−52%, P=0.0032), PTCA/CABG (−51% P=0.0011), CHF (−50%, P=0.021), and stroke (−47%, P=0.034)
			Usual care		↓4%	↓5%	↑2%	↓3%	
Heart Protection Study (157)	r, pc	Coronary disease, other occlusive arterial disease, or diabetes (40–80 years, n = 20,536)	Simvastatin 40 mg daily	Average after 5 years	↓∼20%[a]	↓∼30%[a]	↑∼3%[a]	↓∼15%[a]	Reduction in coronary death rate (18% proportional reduction, $P = 0.0005$); 24% reduction in first occurrence of major vascular event ($P < 0.0001$)

(Continued)

Table 4
(Continued)

Trial name	Study design	Patient characteristics	Active treatments	Time of lipid assessment	Total cholesterol	LDL-C	HDL-C	TGs	Key cardiovascular outcome
Helsinki Heart Study (HHS) (158,159)	r, db, pc	Asymptomatic primary dyslipidemia men only (40–55 years, $n = 4081$)	Gemfibrozil 600 mg BID	5 years	↓10%[a]	↓11%[a]	↑11%[a]	↓35%[a]	Reduction in the incidence of CHD (34% reduction; 95% CI, 8.2–52.6, $P < 0.02$)
Lipoprotein and Coronary Atherosclerosis Study (LCAS) (160)	r, db, pc	CHD and mild-to-moderately elevated LDL-C (35–75 years, $n = 429$)	Fluvastatin 20 mg BID ±cholestyramine	2.5 years	↓14%[b]	↓23%[b]	↑9%[b]	No change[b]	Significantly less atherosclerotic lesion progression (clinical endpoints not assessed)
Long-term Intervention with Pravastatin in Ischemic Disease (LIPID) (161)	r, db, pc	Prior acute MI or unstable angina (31–75 years, $n = 9014$)	Pravastatin 40 mg daily	5 years	↓18%[a]	↓25%[a]	↑5%[a]	↓11%[a]	Reduction in the risk of death from CHD (RR 0.76; 95% CI 0.65–0.88, $P = 0.001$)
Pravastatin or Atorvastatin Evaluation and Infection Trial (PROVE IT) (6)	r, db, dd	Acute coronary syndrome (≥18 years, $n = 4162$)	Pravastatin 40 mg daily / Atorvastatin 80 mg daily	2 years (both treatments)	Not reported / Not reported	↓10.3% / ↓42%	↑8% / ↑7%	Not reported / Not reported	High-dose atorvastatin afforded a significant risk reduction relative to pravastatin with respect to death from any cause or major cardiovascular events (HR 0.84; 95% CI 0.74–0.95, $P = 0.005$)

Study	Design	Population	Treatment	Duration	LDL-C	TG	HDL-C		Outcome
Scandinavian Simvastatin Survival Study (4S) (3)	r, db, pc	History of angina pectoris or acute MI (35–70 years, n = 4444)	Simvastatin 20–40 mg daily	6 weeks	↓28%	↓38%	↑8%	↓15%	Reduced risk of death (RR 0.70; 95% CI 0.58–0.85, P = 0.0003)
Veterans Affairs High-density lipoprotein cholesterol Intervention Trial (VA-HIT) (163)	r, db, pc	CHD and low HDL-C and LDL-C men only (<74 years, n = 2531)	Gemfibrozil 1200 mg daily	5.4 years 1 year	↓25% ↓4%[a]	↓35% No change*	↑8% ↑6%*	↓10% ↓31%[a]	Reduction in the risk of death from CHD or nonfatal MI (RR 0.78; 95% CI 0.65–0.93, P = 0.006) after 5.1 years
West Of Scotland Coronary Prevention Study (WOSCOPS) (5)	r, db, pc	Hypercholesterolemia and no history of MI men only (45–64 years, n = 6595)	Pravastatin 40 mg daily	5 years	↓20%	↓26%	↑5%	↓12%	Reduction in the risk of nonfatal MI or death from CHD (RR 0.69; 95% CI 0.57–0.83, P < 0.001)

bid, twice daily; CHD, coronary heart disease; CI, confidence interval; db, double-blind; dd, double-dummy; HDL-C, high-density lipoprotein cholesterol; HR, hazard ratio; LDL-C, low-density lipoprotein cholesterol; MI, myocardial infarction; n/a, not available; ol, open label; pc, placebo-controlled; PROBE, prospective randomized open blinded endpoint; r, randomized; RR, relative risk; TGs, triglycerides.

[a] Relative to placebo.
[b] Fluvastatin-only subgroup.
[c] All subjects completed an open-label, 8-week, run-in period of atorvastatin 10 mg/day treatment during which LDL-C decreased by 35%. Values provided show the overall change from initial screening values at initial screening.

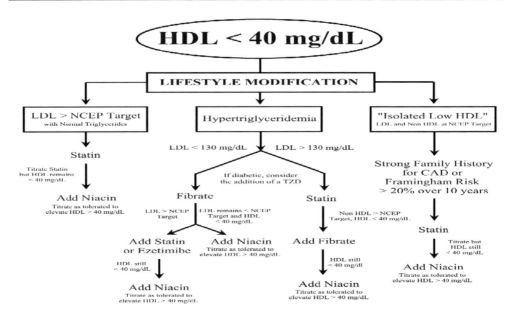

Fig. 7. Algorithm for the management of low serum levels of HDL. CAD, coronary artery disease; HDL, high-density lipoprotein; LDL, low-density lipoprotein; NCEP, National Cholesterol Education Program; TZD, thiazolidinedione. Reproduced with permission from Toth (9).

to placebo, the combination of simvastatin and niacin reduced risk for the primary composite endpoint (coronary mortality, MI, stroke, and revascularization) by 89% ($P = 0.03$) and stabilized atheromatous plaque. The other treatment groups experienced net atheromatous plaque progression. Interestingly, the addition of antioxidants to the combination of simvastatin and niacin compromised the ability of the latter two drugs to increase HDL and reduce event rates (171). It is not recommended that patients with CAD and low HDL supplement their prescription medication regimen with antioxidant vitamins. In the Arterial Biology for the Investigation of the Treatment Effects of Reducing Cholesterol 2 (ARBITER2) trial, the addition of extended-release niacin to statin therapy was shown to stabilize CIMT over 1 year of therapy in patients with CAD (Fig. 8) (172). Among patients continued on statin monotherapy in this trial, CIMT increased significantly (mean increase 0.044 ± 0.10 nm, $P < 0.001$).

Fibrates

The fibrates are synthetic peroxisomal proliferator-activated receptor-α (PPAR-α) agonists that impact HDL metabolism significantly. Subsequent to activation by a fibrate, PPAR-α can form a heterodimer with retinoid X receptor (RXR). The PPAR-α/RXR dimer binds to promoter sites (peroxisomal-proliferator response elements) on nuclear DNA to upregulate gene transcription. The fibrates increase hepatocyte expression of apoA-II and apoA-II, thereby increasing HDL biosynthesis. The fibrates also increase expression of macrophage ABCA1 and hepatocyte SR-BI, possibly increasing RCT (173–175). The fibrates increase mitochondrial fatty acid β-oxidation, decrease inflammatory tone by inhibiting NF-κB and activator protein-1 pathways, and improve the clearance of triglyceride-rich chylomicrons and VLDL by increasing the expression of lipoprotein lipase and decreasing apoC-III. As lipoprotein lipase lipolyzes VLDL and chylomicrons, apoA-I and phospholipids derived from the surface

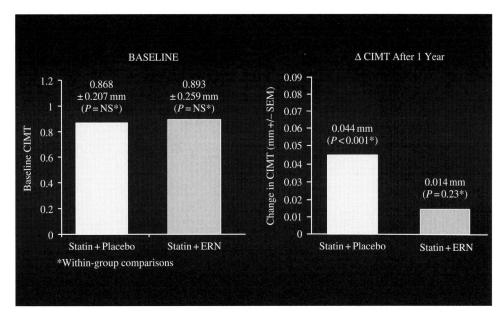

Fig. 8. Effect of supplementation with extended-release niacin on carotid intima-media thickness (CIMT) measurements in Arterial Biology for the Investigation of the Treatment Effects of Reducing Cholesterol 2 (ARBITER2). In ARBITER2, patients with coronary artery disease on statin therapy were randomized to supplementation with either placebo or niaspan 1000 mg daily. After 1 year of therapy, CIMT increased significantly in the statin monotherapy group and did not change significantly in the group receiving statin-niacin combination therapy. *Within-group comparisons. Reproduced with permission from Taylor et al. *(172)*.

coat of these lipoproteins can be transferred into HDL. With less triglyceride available in serum, HDL will become less enriched with triglyceride via CETP and will be less prone to catabolism. Treating hypertriglyceridemia is thus an important therapeutic means by which to increase and maintain appropriate serum HDL levels. However, a potential result of fibrate therapy, particularly in patients with triglycerides > 400 mg/dL, is the potential rise in LDL-C. This results from increased conversion of VLDL to LDL as lipoprotein lipase is activated. In these situations, fibrate therapy may have to be supplemented with a statin or ezetimibe in order to achieve NCEP targets for LDL-C and non-HDL-C. If a fibrate is combined with a statin, gemfibrozil should generally be avoided as this fibrate can block the glucuronidation of statins and increase risk for such adverse events as rhabdomyolysis *(176)*. Fenofibrate is a safer choice when statin-fibrate combination therapy is deemed appropriate *(177)*.

The fibrates appear to particularly benefit patients with hypertriglyceridemia. The Helsinki Heart Study (HHS) was a primary prevention trial comparing gemfibrozil (600 mg po bid) to placebo in 4081 men *(158,159,178)*. Gemfibrozil therapy was associated with a 34% reduction in the incidence of MI and coronary mortality. However, in subgroup analyses, subjects with baseline triglycerides > 200 mg/dL and HDL < 42 mg/dL experienced a 66% reduction in the primary composite endpoint. In the Bezafibrate Infarction Prevention (BIP), 3090 patients with established CAD were randomized to either placebo or 400 mg of bezafibrate (a fibrate not yet available in the United States) *(151)*. For the cohort as a whole, bezafibrate therapy for 6.2

years reduced the combined primary endpoint of fatal or nonfatal MI and sudden death relative to placebo by only 7.3% ($P = 0.24$). However, in a post hoc analysis, a significant reduction of 39.5% ($P = 0.02$) was shown in the subgroup that had an elevated triglyceride (> 200 mg/dL) and low HDL-C (< 35 mg/dL) (Fig. 9). In BIP, coronary mortality decreased significantly with increasing tertiles of on-treatment changes in HDL-C (9.5, 6.6, and 6.3% of patients sustained coronary mortality in tertiles 1, 2, and 3, respectively, for trend $P = 0.02$) *(179)*. In a multivariate analysis, on-treatment elevations in HDL-C were independently associated with a reduced risk for coronary mortality. Coronary mortality decreased by 27% for every 5 mg/dL increase in HDL-C on bezafibrate therapy ($P < 0.001$). In both HHS and BIP, particular benefit accrued to patients with features of the metabolic syndrome.

The VA-HIT study highlights the role of fibrate therapy in patients with metabolic syndrome and diabetes. VA-HIT randomized 2500 men with CAD to treatment with either gemfibrozil 600 mg po bid or placebo for a median follow-up period of 5.1 years *(180,181)*. The mean entry lipid values in this trial were LDL-C 111 mg/dL, HDL-C 31 mg/dL, and triglycerides 161 mg/dL. After the first year of therapy, there was no change in LDL-C, HDL-C increased by 6%, and triglycerides decreased by 31%. Despite there being no change in serum LDL-C, the primary composite endpoint of nonfatal MI and coronary death decreased by 22% and the composite of nonfatal MI, stroke, and coronary death decreased by 24%. A multivariate analysis demonstrated that for every 5 mg/dL increase in HDL-C, risk for acute CAD-related events decreased by 11% ($P = 0.02$). In this trial, only the elevation in HDL had predictive value for risk reduction. In addition, in subgroup analyses, only patients with the metabolic syndrome or diabetes mellitus derived statistically significant benefit from gemfibrozil therapy *(182)*. Among diabetic patients, risk for stroke decreased by 40%, coronary mortality by 41%, and the primary composite endpoint 32%. Interestingly, the fibrate

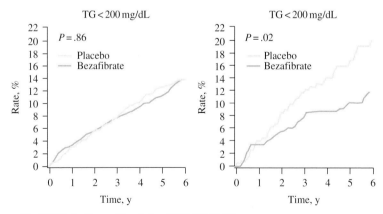

The BIP Study Group. *Circulation*. 2000;102:21-27.

Fig. 9. Effect of baseline triglycerides on risk reduction in the Bezafibrate Intervention Project (BIP). Patients in BIP were randomized to treatment with either placebo or bezafibrate 400 mg daily. When patients were stratified by baseline triglycerides, only the group with triglycerides ≥ 200 mg/dL achieved a statistically significant improvement in the primary endpoint (fatal and nonfatal myocardial infarction or sudden death). Reproduced with permission from Bezafibrate Infarction Prevention (BIP) study *(151)*.

trials, including HHS, BIP, and VA-HIT, have all failed to demonstrate a significant reduction in mortality. This has yet to be explained. Fibrate therapy has been shown to reduce rates of atheromatous plaque progression in both native coronary arteries and saphenous vein bypass grafts *(183–187)*. Fenofibrate therapy can be associated with increased serum homocysteine levels *(188)*. Increased serum homocysteine is associated with reduced expression of apoA-I in men with CAD *(189,190)*. It remains to be determined whether fenofibrate therapy induces less HDL elevation in patients with baseline hyperhomocysteinemia or if increases in homocysteine secondary to fenofibrate therapy attenuates clinical benefit with this drug.

Statins

The statins (hydroxy-methylglutaryl coenzyme A inhibitors) are lipid-lowering agents with demonstrated clinical efficacy to treat dyslipidemia and reduce cardiovascular morbidity and mortality in a broad range of patients in both the primary and secondary prevention settings (Table 4). Although these agents are primarily prescribed because of their capacity to lower serum LDL-C levels, they also reduce serum VLDL and triglycerides and increase HDL-C (typically 3–8%). There are modest differences in the HDL-C raising effects of the various statins, though rosuvastatin appears to increase it the most *(191)* (Fig. 10). In one recent trial of CAD patients treated for 2 years with rosuvastatin at 40 mg daily, HDL-C increased on average by 14.7%, which is the highest sustained elevation reported for a statin *(105)*. Statins increase HDL-C through two predominant mechanisms. First, statins inhibit factor rho and promote PPAR-α activation, with subsequent increased hepatic transcription of apoA-I and apoA-II *(192)*. Second, with less serum triglyceride available for CETP to transfer into HDL from VLDL, HDL is less enriched with triglyceride making it a less favorable target for catabolism by HL *(193)*.

Statins appear to particularly benefit patients with low baseline HDL-C. In the Air Force/Texas Coronary Atherosclerosis Prevention Study (AFCAPS/TexCAPS),

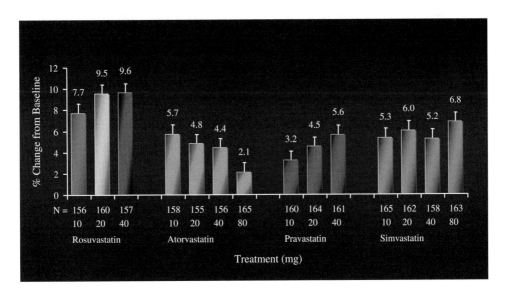

Fig. 10. Effects of various statins on serum high-density lipoprotein cholesterol levels as a function of dose. Reproduced with permission from Jones et al. *(191)*.

6605 men and women with no history of CAD were treated with either lovastatin or placebo *(194,147)*. Among subjects with an HDL ≥40 mg/dL, risk for CAD-related events decreased with lovastatin therapy by 15%. In subjects whose HDL-C was <40 mg/dL, statin therapy was associated with a 45% reduction in risk *(10)*. Secondary prevention trials also demonstrate greater benefit from statin therapy in patients with low HDL-C. In the Lipoprotein and Coronary Atherosclerosis Study (LCAS), among with patients with HDL-C <35 mg/dL (mean 31.7 mg/dL) treated with fluvastatin, rates of atheromatous plaque progression decreased significantly relative to patients given placebo *(195)*. However, among patients with baseline HDL-C ≥35 mg/dL (mean 47.4 mg/dL), rates of plaque progression decreased but not significantly relative to placebo. In the Heart Protection Study, there was a trend toward increased benefit with decreasing baseline HDL-C *(157)* (Fig. 11). The Prospective Study of Pravastatin in the Elderly at Risk (PROSPER) study evaluated the efficacy of pravastatin 40 mg daily versus placebo in 5804 high-risk patients. In this study, only the patients with baseline HDL-C <43 mg/dL derived significant risk reduction for the primary endpoint *(196,197)* (Fig. 12).

Statins should be the preferred initial therapy for patients with isolated low HDL-C or a combination of elevated LDL-C and low HDL-C. Multiple statin trials have shown a trend toward disproportionately benefiting patients with low baseline HDL-C. Statins have been shown to exert many of the same antioxidative (decrease reactive oxygen species production and LDL oxidation), anti-inflammatory (decrease adhesion molecule expression and CRP levels), and antithrombotic effects (decrease platelet reactivity and aggregability, reverse endothelial dysfunction) attributed to HDL-C *(198)*. This may at least partly explain why statins disproportionately benefit patients with low HDL-C.

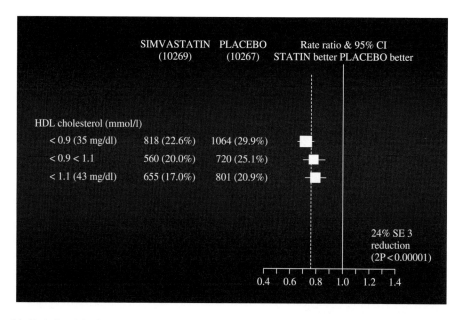

Fig. 11. Relationship between baseline serum high-density lipoprotein cholesterol (HDL-C) and risk for cardiovascular events with simvastatin therapy in the Heart Protection Study. [Based on data presented at Heart Protection Study Website, accessed 17 Aug 2006 (http://www.ctsu.ox.ac.uk/~hps/slidesfinal/HPSwebslides_files)].

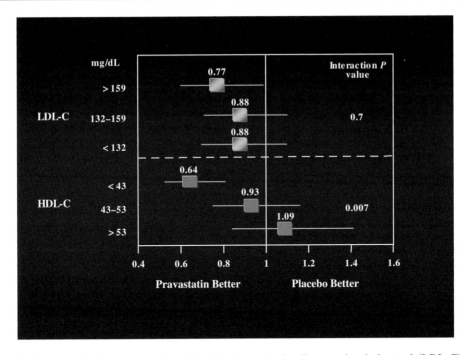

Fig. 12. Relationship between tertile of baseline low-density lipoprotein cholesterol (LDL-C) and high-density lipoprotein cholesterol (HDL-C) and risk for cardiovascular events in patients treated with pravastatin in the Prospective Study of Pravastatin in the Elderly at Risk Study. There was a statistically significant trend for benefit with statin therapy as a function of baseline HDL-C levels, but not LDL-C. Reproduced with permission from Shepherd et al. *(197)*.

Thiazolidinediones

The thiazolidinediones (TZDs) are PPAR-γ agonists that sensitize peripheral tissues to the effects of insulin. The TZDs significantly affect lipoprotein metabolism and can promote reductions in serum triglycerides; promote the conversion of small, dense LDL to its larger and more buoyant form; and raise HDL-C. Insulin resistance is a common clinical entity as suggested by the high incidence of the metabolic syndrome and type 2 diabetes mellitus in Europe and the United States. Low HDL-C is a key manifestation of insulin resistance. Both pioglitazone and rosiglitazone relieve insulin resistance and promote elevations in HDL-C. A meta-analysis of the most important TZD trials showed that pioglitazone raised HDL-C significantly more than rosiglitazone *(199)*. This conclusion was tested in a trial that directly compared the effects of these medications in patients with diabetes mellitus *(200)*. Pioglitazone at 45 mg daily and rosiglitazone at 4 mg twice daily increased HDL-C 5.2 ± 0.5 ($14.9 \pm 1.2\%$) versus 2.4 ± 0.5 ($7.8 \pm 1.2\%$) mg/dL, respectively (difference between therapies, $P < 0.001$). The addition of a TZD to a lipid-lowering regimen can be an efficacious means by which to raise HDL-C in diabetic patients. The TZDs are not yet indicated in nondiabetics, including patients with metabolic syndrome.

Estrogen

Estrogen increases serum HDL-C by promoting hepatic apoA-I expression and decreasing HL activity *(201,202)*. A number of early case control studies suggested that estrogen supplementation decreased risk for CAD-related events in postmenopausal women. However, in subsequent prospective, randomized, placebo-controlled trials, estrogen was shown to increase risk for thromboembolic phenomena and acute cardiovascular events *(203–206)*. Estrogen supplementation is not recommended as means of raising HDL-C. However, if a women is experiencing menopausal symptoms and has a favorable family history for breast cancer and no history of a uterine or other estrogen-dependent malignancy, beginning estrogen therapy can certainly relieve her menopausal symptoms and increase her HDL-C.

HDL THERAPIES IN DEVELOPMENT

Apoprotein A-I$_{Milano}$

Apo A-I$_{Milano}$ is a rare, fascinating polymorphism of apoA-I characterized by a substitution mutation of cysteine for arginine at amino acid position 173 *(207)*. Subjects with apoA-I$_{Milano}$ were identified in a small, northern Italian village named Limon sul Garda and are heterozygotes for the gene. Consequently, they produce a combination of apoA-I and apoA-I$_{Milano}$. The phenotype for this mutation is characterized by low serum levels of HDL (usually in the range of 15–25 mg/dL) and longevity with strong resistance to the development of atherosclerotic disease *(208–211)*. ApoA-I$_{Milano}$ has a long serum half-life because it is able to form homodimers and heterodimers with native apoA-I *(212)*. ApoA-I$_{Milano}$ has augmented capacity to drive RCT, block lipid oxidation, and inhibit platelet aggregation compared to wildtype apoA-I *(213)*. Animal studies showed that the infusion of apoA-I$_{Milano}$ can promote rapid atheromatous plaque regression *(214–216)*. Nissen and coworkers *(217)* took a recombinant form of apoA-I$_{Milano}$ incorporated into phospholipid vesicles and treated patients with established CAD with five once-weekly infusions. Atherosclerotic disease was evaluated at baseline and at time of completion of therapy with intravascular ultrasonography. Target lesions regressed on average by 4.3% relative to patients treated with saline. This study showed that atherosclerotic lesions can regress much more rapidly than previously thought, and plaque can be regressed with an HDL analog independent of changes in serum LDL-C. A larger confirmatory trial is underway.

*Liver-X-Receptor-*α *Agonists*

Liver-X-Receptor-α (LXR-α) is a nuclear transcription regulatory factor. After binding to intracellular oxysterols such as 27-hydroxycholesterol, LXR-α undergoes a 3D conformational change that allows it to dimerize with RXR *(218)*. The LXR-RXR heterodimer then binds to response elements and promotes the expression of a number of genes. Because LXR-α regulates the expression of ABCA1, ABCG1, ABCG5, ABCG8, CETP, and 7α-hydroxylase (among other genes), it can directly affect the rate of RCT and cholesterol elimination from the liver into the gastrointestinal tract *(219)*. In animal models, LXR-α agonists have been shown to reduce rates of atherosclerotic progression, promote RCT, and even induce plaque regression *(220–222)*. These agents have not yet entered into human clinical trials.

CETP Inhibitors

There is accumulating evidence that patients heterozygous for CETP loss of function mutations experience reduced risk for CAD secondary to increased serum levels of HDL. There are currently two agents in clinical trials that offer therapeutic inhibition of CETP. Torcetrapib increases HDL-C by up to 106% in humans *(223)*. The capacity of torcetrapib to impact carotid and coronary atherosclerotic disease is being evaluated in a number of clinical trials. JTT-705 raises HDL by approximately 37% in humans and is also being evaluated in larger trials *(224)*. The ability of these drugs to affect rates of disease progression and impact cardiovascular morbidity and mortality is yet to be established, but, if successful, will greatly expand the pharmacologic armamentarium of lipidologists and cardiologists treating atherosclerotic disease.

D4F

D4F is an apoA-I mimetic peptide that increases the functionality of HDL *(225)*. The administration of D4F to mice that fed an atherogenic diet reduces monocyte chemotaxis, LDL oxidation, and atherogenesis *(226–229)*. D4F does not increase serum levels of HDL-C. D4F is edible because it is comprised of the D-isomer of amino acids that render this polypeptide resistant to hydrolytic degradation by gastric digestive enzymes. D4F offers an entirely novel therapeutic approach to HDL-C management.

ACKNOWLEDGMENT

I would like to express my most sincere appreciation to Dr. Thomas Dayspring for enriching this manuscript with Figs. 1, 2, and 3.

REFERENCES

1. Executive Summary of The Third Report of The National Cholesterol Education Program (NCEP) Expert Panel on Detection, Evaluation, And Treatment of High Blood Cholesterol In Adults (Adult Treatment Panel III). JAMA 2001;285:2486–97.
2. Grundy SM, Cleeman JI, Merz CN, et al. Implications of recent clinical trials for the National Cholesterol Education Program Adult Treatment Panel III guidelines. Circulation 2004;110:227–39.
3. Randomised trial of cholesterol lowering in 4444 patients with coronary heart disease: the Scandinavian Simvastatin Survival Study (4S). Lancet 1994;344:1383–9.
4. Sacks FM, Pfeffer MA, Moye LA, et al. The effect of pravastatin on coronary events after myocardial infarction in patients with average cholesterol levels. Cholesterol and Recurrent Events Trial investigators. N Engl J Med 1996;335:1001–9.
5. Shepherd J, Cobbe SM, Ford I, et al. Prevention of coronary heart disease with pravastatin in men with hypercholesterolemia. West of Scotland Coronary Prevention Study Group. N Engl J Med 1995;333:1301–7.
6. Cannon CP, Braunwald E, McCabe CH, et al. Intensive versus moderate lipid lowering with statins after acute coronary syndromes. N Engl J Med 2004;350:1495–504.
7. Cannon CP, Murphy SA, Braunwald E. Intensive lipid lowering with atorvastatin in coronary disease. N Engl J Med 2005;353:93–6; author reply 93–6.
8. Gotto AM, Jr., Brinton EA. Assessing low levels of high-density lipoprotein cholesterol as a risk factor in coronary heart disease: a working group report and update. J Am Coll Cardiol 2004;43:717–24.
9. Toth PP. High-density lipoprotein and cardiovascular risk. Circulation 2004;109:1809–12.
10. Gotto AM, Jr., Whitney E, Stein EA, et al. Relation between baseline and on-treatment lipid parameters and first acute major coronary events in the Air Force/Texas Coronary Atherosclerosis Prevention Study (AFCAPS/TexCAPS). Circulation 2000;101:477–84.
11. Assmann G. Pro and con: high-density lipoprotein, triglycerides, and other lipid subfractions are the future of lipid management. Am J Cardiol 2001;87:2B–7B.
12. Khuseyinova N, Koenig W. Apolipoprotein a-I and risk for cardiovascular diseases. Curr Atheroscler Rep 2006;8:365–73.
13. Meisinger C, Loewel H, Mraz W, et al. Prognostic value of apolipoprotein B and A-I in the prediction of myocardial infarction in middle-aged men and women: results from the MONICA/KORA Augsburg cohort study. Eur Heart J 2005;26:271–8.

14. Walldius G, Jungner I, Holme I, et al. High apolipoprotein B, low apolipoprotein A-I, and improvement in the prediction of fatal myocardial infarction (AMORIS study): a prospective study. Lancet 2001;358:2026–33.
15. Sorci-Thomas MG, Thomas MJ. The effects of altered apolipoprotein A-I structure on plasma HDL concentration. Trends Cardiovasc Med 2002;12:121–8.
16. Brooks-Wilson A, Marcil M, Clee SM, et al. Mutations in ABC1 in Tangier disease and familial high-density lipoprotein deficiency. Nat Genet 1999;22:336–45.
17. Hayden MR, Clee SM, Brooks-Wilson A, et al. Cholesterol efflux regulatory protein, Tangier disease and familial high-density lipoprotein deficiency. Curr Opin Lipidol 2000;11:117–22.
18. Hong SH, Rhyne J, Zeller K, et al. ABCA1(Alabama): a novel variant associated with HDL deficiency and premature coronary artery disease. Atherosclerosis 2002;164:245–50.
19. Clee SM, Zwinderman AH, Engert JC, et al. Common genetic variation in ABCA1 is associated with altered lipoprotein levels and a modified risk for coronary artery disease. Circulation 2001;103:1198–205.
20. Vaisman BL, Lambert G, Amar M, et al. ABCA1 overexpression leads to hyperalphalipoproteinemia and increased biliary cholesterol excretion in transgenic mice. J Clin Invest 2001;108:303–9.
21. Singaraja RR, Fievet C, Castro G, et al. Increased ABCA1 activity protects against atherosclerosis. J Clin Invest 2002;110:35–42.
22. Singaraja RR, Visscher H, James ER, et al. Specific mutations in ABCA1 have discrete effects on ABCA1 function and lipid phenotypes both in vivo and in vitro. Circ Res 2006;99:389–97.
23. Van Eck M, Singaraja RR, Ye D, et al. Macrophage ATP-binding cassette transporter A1 overexpression inhibits atherosclerotic lesion progression in low-density lipoprotein receptor knockout mice. Arterioscler Thromb Vasc Biol 2006;26:929–34.
24. Martin-Campos JM, Julve J, Escola JC, et al. ApoA-I(MALLORCA) impairs LCAT activation and induces dominant familial hypoalphalipoproteinemia. J Lipid Res 2002;43:115–23.
25. Brewer HB, Jr., Santamarina-Fojo S. New insights into the role of the adenosine triphosphate-binding cassette transporters in high-density lipoprotein metabolism and reverse cholesterol transport. Am J Cardiol 2003;91:3E–11E.
26. Kennedy MA, Barrera GC, Nakamura K, et al. ABCG1 has a critical role in mediating cholesterol efflux to HDL and preventing cellular lipid accumulation. Cell Metab 2005;1:121–31.
27. Vaughan AM, Oram JF. ABCG1 redistributes cell cholesterol to domains removable by high density lipoprotein but not by lipid-depleted apolipoproteins. J Biol Chem 2005;280:30150–7.
28. Wang N, Lan D, Chen W, et al. ATP-binding cassette transporters G1 and G4 mediate cellular cholesterol efflux to high-density lipoproteins. Proc Natl Acad Sci USA 2004;101:9774–9.
29. Bruce C, Chouinard RA, Jr., Tall AR. Plasma lipid transfer proteins, high-density lipoproteins, and reverse cholesterol transport. Annu Rev Nutr 1998;18:297–330.
30. Bersot TP, Vega GL, Grundy SM, et al. Elevated hepatic lipase activity and low levels of high density lipoprotein in a normotriglyceridemic, nonobese Turkish population. J Lipid Res 1999;40:432–8.
31. Rader DJ. Inhibition of cholesteryl ester transfer protein activity: a new therapeutic approach to raising high-density lipoprotein. Curr Atheroscler Rep 2004;6:398–405.
32. Yamashita S, Maruyama T, Hirano K, et al. Molecular mechanisms, lipoprotein abnormalities and atherogenicity of hyperalphalipoproteinemia. Atherosclerosis 2000;152:271–85.
33. Tall AR, Jiang X, Luo Y, Silver D. 1999 George Lyman Duff memorial lecture: lipid transfer proteins, HDL metabolism, and atherogenesis. Arterioscler Thromb Vasc Biol 2000;20:1185–8.
34. Hirano K, Yamashita S, Nakajima N, et al. Genetic cholesteryl ester transfer protein deficiency is extremely frequent in the Omagari area of Japan. Marked hyperalphalipoproteinemia caused by CETP gene mutation is not associated with longevity. Arterioscler Thromb Vasc Biol 1997;17:1053–9.
35. Curb JD, Abbott RD, Rodriguez BL, et al. A prospective study of HDL-C and cholesteryl ester transfer protein gene mutations and the risk of coronary heart disease in the elderly. J Lipid Res 2004;45:948–53.
36. Brousseau ME, O'Connor JJ, Jr., Ordovas JM, et al. Cholesteryl ester transfer protein TaqI B2B2 genotype is associated with higher HDL cholesterol levels and lower risk of coronary heart disease end points in men with HDL deficiency: Veterans Affairs HDL Cholesterol Intervention Trial. Arterioscler Thromb Vasc Biol 2002;22:1148–54.
37. Wu JH, Lee YT, Hsu HC, et al. Influence of CETP gene variation on plasma lipid levels and coronary heart disease: a survey in Taiwan. Atherosclerosis 2001;159:451–8.

38. Krieger M. Charting the fate of the "good cholesterol": identification and characterization of the high-density lipoprotein receptor SR-BI. Annu Rev Biochem 1999;68:523–58.
39. Trigatti B, Rigotti A, Krieger M. The role of the high-density lipoprotein receptor SR-BI in cholesterol metabolism. Curr Opin Lipidol 2000;11:123–31.
40. Trigatti BL. Hepatic high-density lipoprotein receptors: roles in lipoprotein metabolism and potential for therapeutic modulation. Curr Atheroscler Rep 2005;7:344–50.
41. Williams DL, de La Llera-Moya M, Thuahnai ST, et al. Binding and cross-linking studies show that scavenger receptor BI interacts with multiple sites in apolipoprotein A-I and identify the class A amphipathic alpha-helix as a recognition motif. J Biol Chem 2000;275:18897–904.
42. Liu T, Krieger M, Kan HY, et al. The effects of mutations in helices 4 and 6 of ApoA-I on scavenger receptor class B type I (SR-BI)-mediated cholesterol efflux suggest that formation of a productive complex between reconstituted high density lipoprotein and SR-BI is required for efficient lipid transport. J Biol Chem 2002;277:21576–84.
43. Van Eck M, Twisk J, Hoekstra M, et al. Differential effects of scavenger receptor BI deficiency on lipid metabolism in cells of the arterial wall and in the liver. J Biol Chem 2003;278:23699–705.
44. Kozarsky KF, Donahee MH, Glick JM, et al. Gene transfer and hepatic overexpression of the HDL receptor SR-BI reduces atherosclerosis in the cholesterol-fed LDL receptor-deficient mouse. Arterioscler Thromb Vasc Biol 2000;20:721–7.
45. Martinez LO, Jacquet S, Esteve JP, et al. Ectopic beta-chain of ATP synthase is an apolipoprotein A-I receptor in hepatic HDL endocytosis. Nature 2003;421:75–9.
46. Ioka RX, Kang MJ, Kamiyama S, et al. Expression cloning and characterization of a novel glycosylphosphatidylinositol-anchored high density lipoprotein-binding protein, GPI-HBP1. J Biol Chem 2003;278:7344–9.
47. Eriksson M, Carlson LA, Miettinen TA, et al. Stimulation of fecal steroid excretion after infusion of recombinant proapolipoprotein A-I. Potential reverse cholesterol transport in humans. Circulation 1999;100:594–8.
48. Naik SU, Wang X, Da Silva JS, et al. Pharmacological activation of liver X receptors promotes reverse cholesterol transport in vivo. Circulation 2006;113:90–7.
49. Libby P, Ridker PM, Maseri A. Inflammation and atherosclerosis. Circulation 2002;105:1135–43.
50. Carr AC, McCall MR, Frei B. Oxidation of LDL by myeloperoxidase and reactive nitrogen species: reaction pathways and antioxidant protection. Arterioscler Thromb Vasc Biol 2000;20:1716–23.
51. Parthasarathy S, Barnett J, Fong LG. High-density lipoprotein inhibits the oxidative modification of low-density lipoprotein. Biochim Biophys Acta 1990;1044:275–83.
52. Hansson GK. Inflammation, atherosclerosis, and coronary artery disease. N Engl J Med 2005;352:1685–95.
53. Toikka JO, Ahotupa M, Viikari JS, et al. Constantly low HDL-cholesterol concentration relates to endothelial dysfunction and increased in vivo LDL-oxidation in healthy young men. Atherosclerosis 1999;147:133–8.
54. Aviram M, Hardak E, Vaya J, et al. Human serum paraoxonases (PON1) Q and R selectively decrease lipid peroxides in human coronary and carotid atherosclerotic lesions: PON1 esterase and peroxidase-like activities. Circulation 2000;101:2510–7.
55. Stremler KE, Stafforini DM, Prescott SM, et al. Human plasma platelet-activating factor acetylhydrolase. Oxidatively fragmented phospholipids as substrates. J Biol Chem 1991;266:11095–103.
56. Garner B, Waldeck AR, Witting PK, et al. Oxidation of high density lipoproteins. II. Evidence for direct reduction of lipid hydroperoxides by methionine residues of apolipoproteins AI and AII. J Biol Chem 1998;273:6088–95.
57. Garner B, Witting PK, Waldeck AR, et al. Oxidation of high density lipoproteins. I. Formation of methionine sulfoxide in apolipoproteins AI and AII is an early event that accompanies lipid peroxidation and can be enhanced by alpha-tocopherol. J Biol Chem 1998;273:6080–7.
58. de Winther MP, Kanters E, Kraal G, et al. Nuclear factor kappaB signaling in atherogenesis. Arterioscler Thromb Vasc Biol 2005;25:904–14.
59. Xia P, Vadas MA, Rye KA, et al. High density lipoproteins (HDL) interrupt the sphingosine kinase signaling pathway. A possible mechanism for protection against atherosclerosis by HDL. J Biol Chem 1999;274:33143–7.
60. Wadham C, Albanese N, Roberts J, et al. High-density lipoproteins neutralize C-reactive protein proinflammatory activity. Circulation 2004;109:2116–22.
61. Li XP, Zhao SP, Zhang XY, et al. Protective effect of high density lipoprotein on endothelium-dependent vasodilatation. Int J Cardiol 2000;73:231–6.

62. Levkau B, Hermann S, Theilmeier G, et al. High-density lipoprotein stimulates myocardial perfusion in vivo. Circulation 2004;110:3355–9.
63. Tso C, Martinic G, Fan WH, et al. High-density lipoproteins enhance progenitor-mediated endothelium repair in mice. Arterioscler Thromb Vasc Biol 2006;26:1144–9.
64. Lesnik P, Chapman MJ. A new dimension in the vasculoprotective function of HDL: progenitor-mediated endothelium repair. Arterioscler Thromb Vasc Biol 2006;26:965–7.
65. Kimura T, Sato K, Malchinkhuu E, et al. High-density lipoprotein stimulates endothelial cell migration and survival through sphingosine 1-phosphate and its receptors. Arterioscler Thromb Vasc Biol 2003;23:1283–8.
66. Tamagaki T, Sawada S, Imamura H, et al. Effects of high-density lipoproteins on intracellular pH and proliferation of human vascular endothelial cells. Atherosclerosis 1996;123:73–82.
67. Nofer JR, Levkau B, Wolinska I, et al. Suppression of endothelial cell apoptosis by high density lipoproteins (HDL) and HDL-associated lysosphingolipids. J Biol Chem 2001;276:34480–5.
68. Libby P. Current concepts of the pathogenesis of the acute coronary syndromes. Circulation 2001;104:365–72.
69. Libby P. Atherosclerosis: the new view. Sci Am 2002;286:46–55.
70. Fuster V. Elucidation of the role of plaque instability and rupture in acute coronary events. Am J Cardiol 1995;76:24C–33C.
71. Saku K, Ahmad M, Glas-Greenwalt P, et al. Activation of fibrinolysis by apolipoproteins of high density lipoproteins in man. Thromb Res 1985;39:1–8.
72. Griffin JH, Kojima K, Banka CL, et al. High-density lipoprotein enhancement of anticoagulant activities of plasma protein S and activated protein C. J Clin Invest 1999;103:219–27.
73. Nofer JR, Walter M, Kehrel B, et al. HDL3-mediated inhibition of thrombin-induced platelet aggregation and fibrinogen binding occurs via decreased production of phosphoinositide-derived second messengers 1,2-diacylglycerol and inositol 1,4,5-tris-phosphate. Arterioscler Thromb Vasc Biol 1998;18:861–9.
74. Aviram M, Brook JG. Characterization of the effect of plasma lipoproteins on platelet function in vitro. Haemostasis 1983;13:344–50.
75. Beitz J, Mest HJ. Thromboxane A2 (TXA2) formation by washed human platelets under the influence of low and high density lipoproteins from healthy donors. Prostaglandins Leukot Med 1986;23:303–9.
76. Cockerill GW, Saklatvala J, Ridley SH, et al. High-density lipoproteins differentially modulate cytokine-induced expression of E-selectin and cyclooxygenase-2. Arterioscler Thromb Vasc Biol 1999;19:910–7.
77. Bierenbaum ML, Fleischman AI, Stier A, et al. Increased platelet aggregation and decreased high-density lipoprotein cholesterol in women on oral contraceptives. Am J Obstet Gynecol 1979;134:638–41.
78. Miyazaki A, Sakuma S, Morikawa W, et al. Intravenous injection of rabbit apolipoprotein A-I inhibits the progression of atherosclerosis in cholesterol-fed rabbits. Arterioscler Thromb Vasc Biol 1995;15:1882–8.
79. Tangirala RK, Tsukamoto K, Chun SH, et al. Regression of atherosclerosis induced by liver-directed gene transfer of apolipoprotein A-I in mice. Circulation 1999;100:1816–22.
80. Badimon JJ, Badimon L, Fuster V. Regression of atherosclerotic lesions by high density lipoprotein plasma fraction in the cholesterol-fed rabbit. J Clin Invest 1990;85:1234–41.
81. Badimon JJ, Badimon L, Galvez A, et al. High density lipoprotein plasma fractions inhibit aortic fatty streaks in cholesterol-fed rabbits. Lab Invest 1989;60:455–61.
82. Ansell BJ, Fonarow GC, Fogelman AM. High-density lipoprotein: is it always atheroprotective? Curr Atheroscler Rep 2006;8:405–11.
83. Ansell BJ, Watson KE, Fogelman AM, et al. High-density lipoprotein function recent advances. J Am Coll Cardiol 2005;46:1792–8.
84. Barr DP, Russ EM, Eder HA. Protein-lipid relationships in human plasma. II. In atherosclerosis and related conditions. Am J Med 1951;11:480–93.
85. Brunner D, Altman S, Loebl K, et al. Alpha-cholesterol percentages in coronary patients with and without increased total serum cholesterol levels and in healthy controls. J Atheroscler Res 1962;2:424–37.
86. Goldbourt U, Cohen L, Neufeld HN. High density lipoprotein cholesterol: prognosis after myocardial infarction. The Israeli Ischemic Heart Disease Study. Int J Epidemiol 1986;15:51–5.

87. Goldbourt U, Medalie JH. High density lipoprotein cholesterol and incidence of coronary heart disease–the Israeli Ischemic Heart Disease Study. Am J Epidemiol 1979;109:296–308.
88. Goldbourt U, Yaari S, Medalie JH. Isolated low HDL cholesterol as a risk factor for coronary heart disease mortality. A 21-year follow-up of 8000 men. Arterioscler Thromb Vasc Biol 1997;17:107–13.
89. Castelli WP, Doyle JT, Gordon T, et al. HDL cholesterol and other lipids in coronary heart disease. The cooperative lipoprotein phenotyping study. Circulation 1977;55:767–72.
90. Stampfer MJ, Sacks FM, Salvini S, et al. A prospective study of cholesterol, apolipoproteins, and the risk of myocardial infarction. N Engl J Med 1991;325:373–81.
91. Miller NE, Thelle DS, Forde OH, et al. The Tromso heart-study. High-density lipoprotein and coronary heart-disease: a prospective case-control study. Lancet 1977;1:965–8.
92. Assmann G, Cullen P, Schulte H. The Munster Heart Study (PROCAM). Results of follow-up at 8 years. Eur Heart J 1998;19 Suppl A:A2–11.
93. Luc G, Bard JM, Ferrieres J, et al. Value of HDL cholesterol, apolipoprotein A-I, lipoprotein A-I, and lipoprotein A-I/A-II in prediction of coronary heart disease: the PRIME Study. Prospective Epidemiological Study of Myocardial Infarction. Arterioscler Thromb Vasc Biol 2002;22:1155–61.
94. Kannel WB. High-density lipoproteins: epidemiologic profile and risks of coronary artery disease. Am J Cardiol 1983;52:9B–12B.
95. Gordon DJ, Probstfield JL, Garrison RJ, et al. High-density lipoprotein cholesterol and cardiovascular disease. Four prospective American studies. Circulation 1989;79:8–15.
96. Kitamura A, Iso H, Naito Y, et al. High-density lipoprotein cholesterol and premature coronary heart disease in urban Japanese men. Circulation 1994;89:2533–9.
97. Shai I, Rimm EB, Hankinson SE, et al. Multivariate assessment of lipid parameters as predictors of coronary heart disease among postmenopausal women: potential implications for clinical guidelines. Circulation 2004;110:2824–30.
98. Asztalos BF, Cupples LA, Demissie S, et al. High-density lipoprotein subpopulation profile and coronary heart disease prevalence in male participants of the Framingham Offspring Study. Arterioscler Thromb Vasc Biol 2004;24:2181–7.
99. Shah PK, Amin J. Low high density lipoprotein level is associated with increased restenosis rate after coronary angioplasty. Circulation 1992;85:1279–85.
100. Taylor AJ, Burke AP, Farb A, et al. Arterial remodeling in the left coronary system: the role of high-density lipoprotein cholesterol. J Am Coll Cardiol 1999;34:760–7.
101. Pearson TA, Bulkley BH, Achuff SC, et al. The association of low levels of HDL cholesterol and arteriographically defined coronary artery disease. Am J Epidemiol 1979;109:285–95.
102. Burke AP, Farb A, Malcom GT, et al. Plaque rupture and sudden death related to exertion in men with coronary artery disease. JAMA 1999;281:921–6.
103. Johnsen SH, Mathiesen EB, Fosse E, et al. Elevated high-density lipoprotein cholesterol levels are protective against plaque progression: a follow-up study of 1952 persons with carotid atherosclerosis the Tromso study. Circulation 2005;112:498–504.
104. Ishikawa K, Tani S, Watanabe I, et al. Effect of pravastatin on coronary plaque volume. Am J Cardiol 2003;92:975–7.
105. Nissen SE, Nicholls SJ, Sipahi I, et al. Effect of very high-intensity statin therapy on regression of coronary atherosclerosis: the ASTEROID trial. JAMA 2006;295:1556–65.
106. Psaty BM, Anderson M, Kronmal RA, et al. The association between lipid levels and the risks of incident myocardial infarction, stroke, and total mortality: The Cardiovascular Health Study. J Am Geriatr Soc 2004;52:1639–47.
107. Corti MC, Guralnik JM, Salive ME, et al. HDL cholesterol predicts coronary heart disease mortality in older persons. JAMA 1995;274:539–44.
108. Sacco RL, Benson RT, Kargman DE, et al. High-density lipoprotein cholesterol and ischemic stroke in the elderly: the Northern Manhattan Stroke Study. JAMA 2001;285:2729–35.
109. Weverling-Rijnsburger AW, Jonkers IJ, van Exel E, et al. High-density vs low-density lipoprotein cholesterol as the risk factor for coronary artery disease and stroke in old age. Arch Intern Med 2003;163:1549–54.
110. Mosca L, Appel LJ, Benjamin EJ, et al. Summary of the American Heart Association's evidence-based guidelines for cardiovascular disease prevention in women. Arterioscler Thromb Vasc Biol 2004;24:394–6.
111. Mosca L, Appel LJ, Benjamin EJ, et al. Evidence-based guidelines for cardiovascular disease prevention in women. American Heart Association scientific statement. Arterioscler Thromb Vasc Biol 2004;24:e29–50.

112. Gardner CD, Tribble DL, Young DR, et al. Population frequency distributions of HDL, HDL(2), and HDL(3) cholesterol and apolipoproteins A-I and B in healthy men and women and associations with age, gender, hormonal status, and sex hormone use: the Stanford Five City Project. Prev Med 2000;31:335–45.
113. Chapman MJ, Assmann G, Fruchart JC, et al. Raising high-density lipoprotein cholesterol with reduction of cardiovascular risk: the role of nicotinic acid–a position paper developed by the European Consensus Panel on HDL-C. Curr Med Res Opin 2004;20:1253–68.
114. Sacks FM. The role of high-density lipoprotein (HDL) cholesterol in the prevention and treatment of coronary heart disease: expert group recommendations. Am J Cardiol 2002;90:139–43.
115. Haffner SM. Dyslipidemia management in adults with diabetes. Diabetes Care 2004;27 Suppl 1:S68–71.
116. Sempos CT, Cleeman JI, Carroll MD, et al. Prevalence of high blood cholesterol among US adults. An update based on guidelines from the second report of the National Cholesterol Education Program Adult Treatment Panel. JAMA 1993;269:3009–14.
117. Ford ES, Giles WH, Dietz WH. Prevalence of the metabolic syndrome among US adults: findings from the third National Health and Nutrition Examination Survey. JAMA 2002;287:356–9.
118. Genest J, Jr., McNamara JR, Ordovas JM, et al. Lipoprotein cholesterol, apolipoprotein A-I and B and lipoprotein (a) abnormalities in men with premature coronary artery disease. J Am Coll Cardiol 1992;19:792–802.
119. Lamarche B, Despres JP, Moorjani S, et al. Triglycerides and HDL-cholesterol as risk factors for ischemic heart disease. Results from the Quebec cardiovascular study. Atherosclerosis 1996;119:235–45.
120. Rubins HB, Robins SJ, Collins D, et al. Distribution of lipids in 8,500 men with coronary artery disease. Department of Veterans Affairs HDL Intervention Trial Study Group. Am J Cardiol 1995;75:1196–201.
121. Lam JK, Matsubara S, Mihara K, et al. Insulin induction of apolipoprotein AI, role of Sp1. Biochemistry 2003;42:2680–90.
122. Lamon-Fava S, Wilson PW, Schaefer EJ. Impact of body mass index on coronary heart disease risk factors in men and women. The Framingham Offspring Study. Arterioscler Thromb Vasc Biol 1996;16:1509–15.
123. Dattilo AM, Kris-Etherton PM. Effects of weight reduction on blood lipids and lipoproteins: a meta-analysis. Am J Clin Nutr 1992;56:320–8.
124. Hodson L, Skeaff CM, Chisholm WA. The effect of replacing dietary saturated fat with polyunsaturated or monounsaturated fat on plasma lipids in free-living young adults. Eur J Clin Nutr 2001;55:908–15.
125. Berglund L, Oliver EH, Fontanez N, et al. HDL-subpopulation patterns in response to reductions in dietary total and saturated fat intakes in healthy subjects. Am J Clin Nutr 1999;70:992–1000.
126. Appel LJ, Sacks FM, Carey VJ, et al. Effects of protein, monounsaturated fat, and carbohydrate intake on blood pressure and serum lipids: results of the OmniHeart randomized trial. JAMA 2005;294:2455–64.
127. Dansinger ML, Gleason JA, Griffith JL, et al. Comparison of the Atkins, Ornish, Weight Watchers, and Zone diets for weight loss and heart disease risk reduction: a randomized trial. JAMA 2005;293:43–53.
128. Matthan NR, Welty FK, Barrett PH, et al. Dietary hydrogenated fat increases high-density lipoprotein apoA-I catabolism and decreases low-density lipoprotein apoB-100 catabolism in hypercholesterolemic women. Arterioscler Thromb Vasc Biol 2004;24:1092–7.
129. Esposito K, Pontillo A, Di Palo C, et al. Effect of weight loss and lifestyle changes on vascular inflammatory markers in obese women: a randomized trial. JAMA 2003;289:1799–804.
130. Renaud S, de Lorgeril M, Delaye J, et al. Cretan Mediterranean diet for prevention of coronary heart disease. Am J Clin Nutr 1995;61:1360–7S.
131. Garrison RJ, Kannel WB, Feinleib M, et al. Cigarette smoking and HDL cholesterol: the Framingham offspring study. Atherosclerosis 1978;30:17–25.
132. Cullen P, Schulte H, Assmann G. Smoking, lipoproteins and coronary heart disease risk. Data from the Munster Heart Study (PROCAM). Eur Heart J 1998;19:1632–41.
133. Imamura H, Teshima K, Miyamoto N, et al. Cigarette smoking, high-density lipoprotein cholesterol subfractions, and lecithin: cholesterol acyltransferase in young women. Metabolism 2002;51:1313–6.

134. Moffatt RJ. Normalization of high density lipoprotein cholesterol following cessation from cigarette smoking. Adv Exp Med Biol 1990;273:267–72.
135. Moffatt RJ. Effects of cessation of smoking on serum lipids and high density lipoprotein-cholesterol. Atherosclerosis 1988;74:85–9.
136. Amarasuriya RN, Gupta AK, Civen M, et al. Ethanol stimulates apolipoprotein A-I secretion by human hepatocytes: implications for a mechanism for atherosclerosis protection. Metabolism 1992;41:827–32.
137. Sillanaukee P, Koivula T, Jokela H, et al. Alcohol consumption and its relation to lipid-based cardiovascular risk factors among middle-aged women: the role of HDL(3) cholesterol. Atherosclerosis 2000;152:503–10.
138. Hegsted DM, Ausman LM. Diet, alcohol and coronary heart disease in men. J Nutr 1988;118:1184–9.
139. Renaud SC, Gueguen R, Siest G, et al. Wine, beer, and mortality in middle-aged men from eastern France. Arch Intern Med 1999;159:1865–70.
140. Kokkinos PF, Fernhall B. Physical activity and high density lipoprotein cholesterol levels: what is the relationship? Sports Med 1999;28:307–14.
141. Kokkinos PF, Holland JC, Narayan P, et al. Miles run per week and high-density lipoprotein cholesterol levels in healthy, middle-aged men. A dose-response relationship. Arch Intern Med 1995;155:415–20.
142. Williams PT. High-density lipoprotein cholesterol and other risk factors for coronary heart disease in female runners. N Engl J Med 1996;334:1298–303.
143. Welty FK, Stuart E, O'Meara M, et al. Effect of addition of exercise to therapeutic lifestyle changes diet in enabling women and men with coronary heart disease to reach Adult Treatment Panel III low-density lipoprotein cholesterol goal without lowering high-density lipoprotein cholesterol. Am J Cardiol 2002;89:1201–4.
144. Thompson PD, Buchner D, Pina IL, et al. Exercise and physical activity in the prevention and treatment of atherosclerotic cardiovascular disease: a statement from the Council on Clinical Cardiology (Subcommittee on Exercise, Rehabilitation, and Prevention) and the Council on Nutrition, Physical Activity, and Metabolism (Subcommittee on Physical Activity). Circulation 2003;107:3109–16.
145. Couillard C, Despres JP, Lamarche B, et al. Effects of endurance exercise training on plasma HDL cholesterol levels depend on levels of triglycerides: evidence from men of the Health, Risk Factors, Exercise Training and Genetics (HERITAGE) Family Study. Arterioscler Thromb Vasc Biol 2001;21:1226–32.
146. Mora S, Lee IM, Buring JE, et al. Association of physical activity and body mass index with novel and traditional cardiovascular biomarkers in women. JAMA 2006;295:1412–9.
147. Downs JR, Clearfield M, Weis S, et al. Primary prevention of acute coronary events with lovastatin in men and women with average cholesterol levels: results of AFCAPS/TexCAPS. Air Force/Texas Coronary Atherosclerosis Prevention Study. JAMA 1998;279:1615–22.
148. Koren MJ, Hunninghake DB. Clinical outcomes in managed-care patients with coronary heart disease treated aggressively in lipid-lowering disease management clinics: the alliance study. J Am Coll Cardiol 2004;44(9):1772–9.
149. Sever PS, Dahlof B, Poulter NR, et al. Prevention of coronary and stroke events with atorvastatin in hypertensive patients who have average or lower-than-average cholesterol concentrations, in the Anglo-Scandinavian Cardiac Outcomes Trial–Lipid Lowering Arm (ASCOT-LLA): a multicentre randomised controlled trial. Lancet 2003;361(9364):1149–58.
150. Garg R, Elam MB, Crouse JR, III, et al. Effective and safe modification of multiple atherosclerotic risk factors in patients with peripheral arterial disease. Am Heart J 2000;140(5):792–803.
151. Bezafibrate Infarction Prevention (BIP) study. Secondary prevention by raising HDL cholesterol and reducing triglycerides in patients with coronary artery disease. Circulation 2000;102:21–7.
152. Colhoun HM, Betteridge DJ, Durrington PN, et al. Primary prevention of cardiovascular disease with atorvastatin in type 2 diabetes in the Collaborative Atorvastatin Diabetes Study (CARDS): multicentre randomised placebo-controlled trial. Lancet 2004;364(9435):685–96.
153. Canner PL, Berge KG, Wenger NK, et al. Fifteen year mortality in Coronary Drug Project patients: long-term benefit with niacin. J Am Coll Cardiol 1986;8(6):1245–55.
154. Stulc T, Ceska R. Cholesterol lowering and the vessel wall: new insights and future perspectives. Physiol Res 2001;50:461–71.

155. Keech A, Simes RJ, Barter P, et al. Effects of long-term fenofibrate therapy on cardiovascular events in 9795 people with type 2 diabetes mellitus (the FIELD study): randomised controlled trial. Lancet 2005;366(9500):1849–61.
156. Athyros VG, Papageorgiou AA, Mercouris BR, et al. Treatment with atorvastatin to the National Cholesterol Educational Program goal versus 'usual' care in secondary coronary heart disease prevention. The GREek Atorvastatin and Coronary-heart-disease Evaluation (GREACE) study. Curr Med Res Opin 2002;18(4):220–8.
157. Heart Protection Study Collaborative Group. MRC/BHF Heart Protection Study of cholesterol lowering with simvastatin in 20,536 high-risk individuals: a randomised placebo-controlled trial. Lancet 2002;360:7–22.
158. Frick MH, Elo O, Haapa K, et al. Helsinki Heart Study: primary-prevention trial with gemfibrozil in middle-aged men with dyslipidemia. Safety of treatment, changes in risk factors, and incidence of coronary heart disease. N Engl J Med 1987;317:1237–45.
159. Manninen V, Elo MO, Frick MH, et al. Lipid alterations and decline in the incidence of coronary heart disease in the Helsinki Heart Study. JAMA 1988;260:641–51.
160. Herd JA, Ballantyne CM, Farmer JA, Ferguson JJ, III, Jones PH, West MS, et al. Effects of fluvastatin on coronary atherosclerosis in patients with mild to moderate cholesterol elevations (Lipoprotein and Coronary Atherosclerosis Study [LCAS]). Am J Cardiol 1997;80(3):278–86.
161. Long-Term Intervention with Pravastatin in Ischaemic Disease (LIPID) Study Group. Prevention of cardiovascular events and death with pravastatin in patients with coronary heart disease and a broad range of initial cholesterol levels. N Engl J Med. 1998;339(19):1349–57.
162. LaRosa JC, Grundy SM, Waters DD, et al. Intensive lipid lowering with atorvastatin in patients with stable coronary disease. N Engl J Med 2005;352(14):1425–35.
163. Rubins HB, Robins SJ, Collins D, et al. Gemfibrozil for the secondary prevention of coronary heart disease in men with low levels of high-density lipoprotein cholesterol. Veterans Affairs High-Density Lipoprotein Cholesterol Intervention Trial Study Group. N Engl J Med 1999;341(6):410–8.
164. Sakai T, Kamanna VS, Kashyap ML. Niacin, but not gemfibrozil, selectively increases LP-AI, a cardioprotective subfraction of HDL, in patients with low HDL cholesterol. Arterioscler Thromb Vasc Biol 2001;21:1783–9.
165. Tunaru S, Kero J, Schaub A, et al. PUMA-G and HM74 are receptors for nicotinic acid and mediate its anti-lipolytic effect. Nat Med 2003;9:352–5.
166. Wise A, Foord SM, Fraser NJ, et al. Molecular identification of high and low affinity receptors for nicotinic acid. J Biol Chem 2003;278:9869–74.
167. Ganji SH, Kamanna VS, Kashyap ML. Niacin and cholesterol: role in cardiovascular disease (review). J Nutr Biochem 2003;14:298–305.
168. Rubic T, Trottmann M, Lorenz RL. Stimulation of CD36 and the key effector of reverse cholesterol transport ATP-binding cassette A1 in monocytoid cells by niacin. Biochem Pharmacol 2004;67:411–9.
169. Clofibrate and niacin in coronary heart disease. JAMA 1975;231:360–81.
170. Brown BG, Zhao XQ, Chait A, et al. Simvastatin and niacin, antioxidant vitamins, or the combination for the prevention of coronary disease. N Engl J Med 2001;345:1583–92.
171. Cheung MC, Zhao XQ, Chait A, et al. Antioxidant supplements block the response of HDL to simvastatin-niacin therapy in patients with coronary artery disease and low HDL. Arterioscler Thromb Vasc Biol 2001;21:1320–6.
172. Taylor AJ, Sullenberger LE, Lee HJ, et al. Arterial Biology for the Investigation of the Treatment Effects of Reducing Cholesterol (ARBITER) 2: a double-blind, placebo-controlled study of extended-release niacin on atherosclerosis progression in secondary prevention patients treated with statins. Circulation 2004;110:3512–7.
173. Duez H, Fruchart JC, Staels B. PPARS in inflammation, atherosclerosis and thrombosis. J Cardiovasc Risk 2001;8:187–94.
174. Fruchart JC. Peroxisome proliferator-activated receptor-alpha activation and high-density lipoprotein metabolism. Am J Cardiol 2001;88:24–9N.
175. Fruchart JC, Duriez P. HMG CoA reductase inhibitors and PPAR-alpha activators: are their effects on high-density lipoprotein cholesterol and their pleiotropic effects clinically relevant in prevention trials? Curr Atheroscler Rep 2002;4:403–4.
176. Prueksaritanont T, Zhao JJ, Ma B, et al. Mechanistic studies on metabolic interactions between gemfibrozil and statins. J Pharmacol Exp Ther 2002;301:1042–51.

177. Jones PH, Davidson MH. Reporting rate of rhabdomyolysis with fenofibrate + statin versus gemfibrozil + any statin. Am J Cardiol 2005;95:120–2.
178. Manttari M, Elo O, Frick MH, et al. The Helsinki Heart Study: basic design and randomization procedure. Eur Heart J 1987;8 Suppl I:1–29.
179. Goldenberg I, Goldbourt U, Boyko V, et al. Relation between on-treatment increments in serum high-density lipoprotein cholesterol levels and cardiac mortality in patients with coronary heart disease (from the Bezafibrate Infarction Prevention trial). Am J Cardiol 2006;97:466–71.
180. Robins SJ. Targeting low high-density lipoprotein cholesterol for therapy: lessons from the Veterans Affairs High-density Lipoprotein Intervention Trial. Am J Cardiol 2001;88:19N–23N.
181. Robins SJ, Collins D, Wittes JT, et al. Relation of gemfibrozil treatment and lipid levels with major coronary events: VA-HIT: a randomized controlled trial. JAMA 2001;285:1585–91.
182. Rubins HB, Robins SJ, Collins D, et al. Diabetes, plasma insulin, and cardiovascular disease: subgroup analysis from the Department of Veterans Affairs high-density lipoprotein intervention trial (VA-HIT). Arch Intern Med 2002;162:2597–604.
183. Ericsson CG, Hamsten A, Nilsson J, et al. Angiographic assessment of effects of bezafibrate on progression of coronary artery disease in young male postinfarction patients. Lancet 1996;347:849–53.
184. Frick MH, Syvanne M, Nieminen MS, et al. Prevention of the angiographic progression of coronary and vein-graft atherosclerosis by gemfibrozil after coronary bypass surgery in men with low levels of HDL cholesterol. Lopid Coronary Angiography Trial (LOCAT) Study Group. Circulation 1997;96:2137–43.
185. Despres JP. Increasing high-density lipoprotein cholesterol: an update on fenofibrate. Am J Cardiol 2001;88:30–6N.
186. Karpe F, Taskinen MR, Nieminen MS, et al. Remnant-like lipoprotein particle cholesterol concentration and progression of coronary and vein-graft atherosclerosis in response to gemfibrozil treatment. Atherosclerosis 2001;157:181–7.
187. Effect of fenofibrate on progression of coronary-artery disease in type 2 diabetes: the Diabetes Atherosclerosis Intervention Study, a randomised study. Lancet 2001;357:905–10.
188. Bissonnette R, Treacy E, Rozen R, et al. Fenofibrate raises plasma homocysteine levels in the fasted and fed states. Atherosclerosis 2001;155:455–62.
189. Mikael LG, Genest J, Jr., Rozen R. Elevated homocysteine reduces apolipoprotein A-I expression in hyperhomocysteinemic mice and in males with coronary artery disease. Circ Res 2006;98:564–71.
190. Liao D, Tan H, Hui R, et al. Hyperhomocysteinemia decreases circulating high-density lipoprotein by inhibiting apolipoprotein A-I protein synthesis and enhancing hdl cholesterol clearance. Circ Res 2006;99(6):598–606.
191. Jones PH, Davidson MH, Stein EA, et al. Comparison of the efficacy and safety of rosuvastatin versus atorvastatin, simvastatin, and pravastatin across doses (STELLAR* Trial). Am J Cardiol 2003;92:152–60.
192. Martin G, Duez H, Blanquart C, et al. Statin-induced inhibition of the Rho-signaling pathway activates PPARalpha and induces HDL apoA-I. J Clin Invest 2001;107:1423–32.
193. Schaefer EJ, Asztalos BF. The effects of statins on high-density lipoproteins. Curr Atheroscler Rep 2006;8:41–9.
194. Clearfield M, Whitney EJ, Weis S, et al. Air Force/Texas Coronary Atherosclerosis Prevention Study (AFCAPS/TexCAPS): baseline characteristics and comparison with USA population. J Cardiovasc Risk 2000;7:125–33.
195. Ballantyne CM, Herd JA, Ferlic LL, et al. Influence of low HDL on progression of coronary artery disease and response to fluvastatin therapy. Circulation 1999;99:736–43.
196. Packard CJ, Ford I, Robertson M, et al. Plasma lipoproteins and apolipoproteins as predictors of cardiovascular risk and treatment benefit in the PROspective Study of Pravastatin in the Elderly at Risk (PROSPER). Circulation 2005;112:3058–65.
197. Shepherd J, Blauw GJ, Murphy MB, et al. Pravastatin in elderly individuals at risk of vascular disease (PROSPER): a randomised controlled trial. Lancet 2002;360:1623–30.
198. Liao JK, Laufs U. Pleiotropic effects of statins. Annu Rev Pharmacol Toxicol 2005;45:89–118.
199. van Wijk JP, de Koning EJ, Martens EP, et al. Thiazolidinediones and blood lipids in type 2 diabetes. Arterioscler Thromb Vasc Biol 2003;23:1744–9.
200. Goldberg RB, Kendall DM, Deeg MA, et al. A comparison of lipid and glycemic effects of pioglitazone and rosiglitazone in patients with type 2 diabetes and dyslipidemia. Diabetes Care 2005;28:1547–54.

201. Tikkanen MJ, Nikkila EA, Kuusi T, et al. High density lipoprotein-2 and hepatic lipase: reciprocal changes produced by estrogen and norgestrel. J Clin Endocrinol Metab 1982;54:1113–7.
202. Jin FY, Kamanna VS, Kashyap ML. Estradiol stimulates apolipoprotein A-I- but not A-II-containing particle synthesis and secretion by stimulating mRNA transcription rate in Hep G2 cells. Arterioscler Thromb Vasc Biol 1998;18:999–1006.
203. Herrington DM, Reboussin DM, Brosnihan KB, et al. Effects of estrogen replacement on the progression of coronary-artery atherosclerosis. N Engl J Med 2000;343:522–9.
204. Herrington DM, Reboussin DM, Klein KP, et al. The Estrogen Replacement and Atherosclerosis (ERA) Study: study design and baseline characteristics of the cohort. Control Clin Trials 2000;21:257–85.
205. Hulley S, Grady D, Bush T, et al. Randomized trial of estrogen plus progestin for secondary prevention of coronary heart disease in postmenopausal women. Heart and Estrogen/progestin Replacement Study (HERS) Research Group. JAMA 1998;280:605–13.
206. Effects of estrogen or estrogen/progestin regimens on heart disease risk factors in postmenopausal women. The Postmenopausal Estrogen/Progestin Interventions (PEPI) Trial. The Writing Group for the PEPI Trial. JAMA 1995;273:199–208.
207. Bielicki JK, Oda MN. Apolipoprotein A-I(Milano) and apolipoprotein A-I(Paris) exhibit an antioxidant activity distinct from that of wild-type apolipoprotein A-I. Biochemistry 2002;41:2089–96.
208. Franceschini G, Frosi TG, Manzoni C, et al. High density lipoprotein-3 heterogeneity in subjects with the apo-AIMilano variant. J Biol Chem 1982;257:9926–30.
209. Franceschini G, Sirtori CR, Capurso A, II, et al. A-IMilano apoprotein. Decreased high density lipoprotein cholesterol levels with significant lipoprotein modifications and without clinical atherosclerosis in an Italian family. J Clin Invest 1980;66:892–900.
210. Marchesi M, Sirtori CR. Therapeutic use of the high-density lipoprotein protein and peptides. Expert Opin Investig Drugs 2006;15:227–41.
211. Sirtori CR, Calabresi L, Franceschini G, et al. Cardiovascular status of carriers of the apolipoprotein A-I(Milano) mutant: the Limone sul Garda study. Circulation 2001;103:1949–54.
212. Franceschini G, Calabresi L, Tosi C, et al. Apolipoprotein AIMilano. Disulfide-linked dimers increase high density lipoprotein stability and hinder particle interconversion in carrier plasma. J Biol Chem 1990;265:12224–31.
213. Franceschini G, Calabresi L, Chiesa G, et al. Increased cholesterol efflux potential of sera from ApoA-IMilano carriers and transgenic mice. Arterioscler Thromb Vasc Biol 1999;19:1257–62.
214. Chiesa G, Monteggia E, Marchesi M, et al. Recombinant apolipoprotein A-I(Milano) infusion into rabbit carotid artery rapidly removes lipid from fatty streaks. Circ Res 2002;90:974–80.
215. Shah PK, Nilsson J, Kaul S, et al. Effects of recombinant apolipoprotein A-I(Milano) on aortic atherosclerosis in apolipoprotein E-deficient mice. Circulation 1998;97:780–5.
216. Shah PK, Yano J, Reyes O, et al. High-dose recombinant apolipoprotein A-I(milano) mobilizes tissue cholesterol and rapidly reduces plaque lipid and macrophage content in apolipoprotein e-deficient mice. Potential implications for acute plaque stabilization. Circulation 2001;103:3047–50.
217. Nissen SE, Tsunoda T, Tuzcu EM, et al. Effect of recombinant ApoA-I Milano on coronary atherosclerosis in patients with acute coronary syndromes: a randomized controlled trial. JAMA 2003;290:2292–300.
218. Venkateswaran A, Laffitte BA, Joseph SB, et al. Control of cellular cholesterol efflux by the nuclear oxysterol receptor LXR alpha. Proc Natl Acad Sci USA 2000;97:12097–102.
219. Zelcer N, Tontonoz P. Liver X receptors as integrators of metabolic and inflammatory signaling. J Clin Invest 2006;116:607–14.
220. Joseph SB, McKilligin E, Pei L, et al. Synthetic LXR ligand inhibits the development of atherosclerosis in mice. Proc Natl Acad Sci USA 2002;99:7604–9.
221. Tangirala RK, Bischoff ED, Joseph SB, et al. Identification of macrophage liver X receptors as inhibitors of atherosclerosis. Proc Natl Acad Sci USA 2002;99:11896–901.
222. Levin N, Bischoff ED, Daige CL, et al. Macrophage liver X receptor is required for antiatherogenic activity of LXR agonists. Arterioscler Thromb Vasc Biol 2005;25:135–42.
223. Brousseau ME, Schaefer EJ, Wolfe ML, et al. Effects of an inhibitor of cholesteryl ester transfer protein on HDL cholesterol. N Engl J Med 2004;350:1505–15.
224. de Grooth GJ, Kuivenhoven JA, Stalenhoef AF, et al. Efficacy and safety of a novel cholesteryl ester transfer protein inhibitor, JTT-705, in humans: a randomized phase II dose-response study. Circulation 2002;105:2159–65.

225. Navab M, Anantharamaiah GM, Reddy ST, et al. Apolipoprotein A-I mimetic peptides. Arterioscler Thromb Vasc Biol 2005;25:1325–31.
226. Garber DW, Datta G, Chaddha M, et al. A new synthetic class A amphipathic peptide analogue protects mice from diet-induced atherosclerosis. J Lipid Res 2001;42:545–52.
227. Navab M, Anantharamaiah GM, Hama S, et al. Oral administration of an Apo A-I mimetic peptide synthesized from D-amino acids dramatically reduces atherosclerosis in mice independent of plasma cholesterol. Circulation 2002;105:290–2.
228. Navab M, Anantharamaiah GM, Hama S, et al. D-4F and statins synergize to render HDL antiinflammatory in mice and monkeys and cause lesion regression in old apolipoprotein E-null mice. Arterioscler Thromb Vasc Biol 2005;25:1426–32.
229. Navab M, Anantharamaiah GM, Reddy ST, et al. Oral small peptides render HDL antiinflammatory in mice and monkeys and reduce atherosclerosis in ApoE null mice. Circ Res 2005;97:524–32.
230. Castelli WP. Cholesterol and lipids in the risk of coronary artery disease–the Framingham Heart Study. Can J Cardiol 1988;4 Suppl A:5A–10A.

10 Management of Hypertriglyceridemia

Reginald Labossiere, MD
and Ira J. Goldberg, MD

CONTENTS

> INTRODUCTION
> PHYSIOLOGY
> CAUSES OF HUMAN HYPERTRIGLYCERIDEMIA
> SECONDARY CAUSES OF HYPERTRIGLYCERIDEMIA
> TREATMENTS/RECOMMENDATIONS
> CASE 1: DYSBETALIPOPROTEINEMIA/TYPE 3 HYPERLIPIDEMIA
> CASE 2: LIPOPROTEIN LIPASE DEFICIENCY
> CASE 3: HYPERTRIGLYCERIDEMIA AND RETINOIC ACID (ACCUTANE®)
> CASE 4: METABOLIC SYNDROME
> REFERENCES

INTRODUCTION

While the rationale for the treatment of elevated cholesterol, especially elevated low-density lipoprotein (LDL), has been established, the relationship between hypertriglyceridemia and cardiac risk remains obscure. Suggestive data are available, however, and have prompted national recommendations to reduce triglycerides (TGs) in patients with elevated risk for atherosclerotic vascular disease. The National Cholesterol Education Program (NCEP) has defined normotriglyceridemia as < 150 mg/dl, borderline hypertriglyceridemia as 150 to 200 mg/dl, and hypertriglyceridemia as > 200 mg/dl *(1)*. Values that markedly exceed these limits are a risk for pancreatitis and unquestionably require medical intervention.

PHYSIOLOGY

TG, a molecule created by the addition of three fatty acids to a 3-carbon glycerol backbone, is the major form of circulating fat. TG is poorly soluble in plasma and, therefore, circulates in the blood in the center of lipoproteins, primarily chylomicrons and very low-density lipoprotein (VLDL). The body either uses TG as an immediate

From: *Contemporary Cardiology: Therapeutic Lipidology*
Edited by: M. H. Davidson, P. P. Toth, and K. C. Maki © Humana Press Inc., Totowa, NJ

source of energy or stores it for later use. Adipose tissue incorporates excess TG that the body cannot promptly convert to usable material. The use of TG, versus glucose, as a source of energy, differs between tissues with physiologic and pathologic conditions. For instance, chronically exercising muscles, such as the heart or the skeletal muscle of marathon runners, use TG oxidation as a major source of energy *(2,3)*. Lipoprotein TG must be converted into fatty acids to allow their uptake into tissues. The enzyme lipoprotein lipase (LPL) mediates this process.

Dietary lipids are conveyed in chylomicrons and chylomicron remnants (Fig. 1A). Individuals in Western societies ordinarily consume 50–100 g of fat and 0.5 g of cholesterol during 3 or 4 meals. Thus, transport of dietary fats occurs continuously. Normolipidemic individuals dispose of most dietary fat in the bloodstream within 8 h of the last meal. In the intestinal mucosa, dietary TG and cholesterol are incorporated into the core of nascent chylomicrons. A chylomicron is a fat droplet containing 80–95% TGs, with a surface coat composed of phospholipid, free cholesterol, apolipoprotein B48 (apoB48), apoAI, apoCII, apoAIV, and apoE. Apolipoprotein is a term that denotes these proteins exclusive of their accompanying lipid. The small intestine

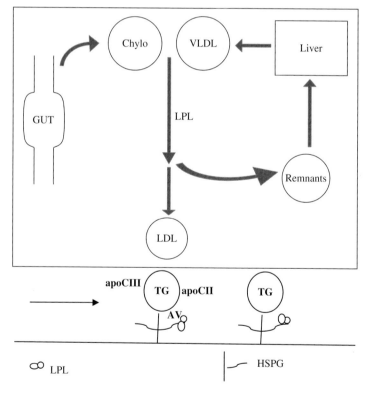

Fig. 1. Catabolism of triglyceride (TG)-rich lipoproteins. TG is a major component of gut-derived and liver-derived lipoproteins—chylomicrons (denoted chylo) and very low-density lipoproteins (VLDL), respectively. Both particles circulate in the bloodstream, and their catabolism involves an initial interaction with lipoprotein lipase (LPL). (B) Triglyceride lipolysis. LPL actions require lipoprotein margination to the vascular wall. Apolipoprotein CII (ApoCII), LPL's co-activator, and apoAV increase LPL actions. ApoAV may increase lipoprotein association with heparin sulfate proteoglycans (HSPG) on the vascular surface. ApoCIII inhibits LPL. After this step, chylomicrons are converted to remnants and many VLDLs are converted to LDL.

secretes chylomicrons into lacteals, and chylomicrons enter the circulation through the thoracic duct. In the plasma, apoCs transfer from high-density lipoproteins (HDL) to the chylomicron.

LPL hydrolyzes TGs on capillary endothelial cells in fat, heart, and muscle. LPL actions require apoCII as a cofactor, and apoCI and CIII inhibit LPL. Angiopoetin-like proteins 3 and 4, which are made by the liver and gut, also hinder LPL activity *(4)*. In addition, a newly described apolipoprotein, apoAV, increases LPL actions in animal models *(5)*. Patients with deficiencies of apoCII and apoCIII and genetically modified mouse models have confirmed the physiologic roles of these proteins *(6)*.

Although peripheral tissues take up some intact chylomicrons, most chylomicrons require an initial interaction with LPL and loss of some TG; this leads to conversion to smaller, remnant particles. The liver captures most chylomicron remnants and their accompanying TGs *(7)*. As chylomicrons contain apoB48, the shortened form of apoB that does not contain the LDL receptor-binding domain, liver uptake may occur through one of three mechanisms: (1) within the hepatic space of Disse, proteoglycans capture remnants; proteoglycans are complexes of highly negatively charged carbohydrates, e.g., heparin, attached to specific membrane or matrix proteins; (2) apoE on the lipoprotein interacts with cell-surface receptors *(8)*, such as the LDL receptor and/or the LDL receptor-related protein; and (3) other hepatic receptors recognize the remnants. These receptors could include the apoB48 receptor and scavenger receptor BI.

The liver uses the dietary cholesterol to form bile acid, incorporates the cholesterol into membranes, or re-secretes the cholesterol into the circulation as a component of VLDL.

The second source of plasma TG comes from the liver (Fig. 1A). The liver takes up fatty acids from plasma or synthesizes them from excess carbohydrate. It then assembles TG from these fatty acids. The liver also synthesizes cholesterol or receives cholesterol through the acceptance of chylomicron remnants or other lipoproteins. TG and cholesterol esters, which are in the center or core of the lipoprotein because they are poorly soluble, combine with apoB100 and phospholipids to form VLDL. TGs account for most of the mass of the VLDL (55–85% by weight), and the amount of available TG determines the size of the VLDL. Hence, the liver secretes very large, TG-rich VLDL when excess TGs have been synthesized. This occurs, for example, with caloric excess, diabetes mellitus, and alcohol consumption. Conversely, the liver releases small VLDL when fewer TGs are available. In the plasma, apoproteins CI, CII, CIII, and E attach to VLDL. The presence of both apoB100 and apoE allows hepatic and cellular uptake of VLDL and VLDL remnants by the LDL receptor.

TG lipolysis in the circulation allows for the reduction of plasma TG, the delivery of fatty acids to peripheral tissues, and the interconversion of lipoproteins (Fig. 1B). LPL converts VLDL to smaller and denser VLDL remnants called intermediate-density lipoproteins (IDLs). Like chylomicron remnants, the liver clears many VLDL particles. However, VLDL remnants also may interact with a second enzyme, hepatic lipase, to synthesize LDL. Hepatic lipase is a member of the lipase gene family. While similar to LPL in many respects, hepatic lipase does not require apoCII, has different sensitivity to salt, and prefers smaller lipoproteins than the larger chylomicrons and intact VLDL. Individuals with deficiency of apoE, hepatic lipase, or both accumulate IDL in plasma. As larger VLDL particles carry more TGs than small VLDL remnants, they are more likely to be removed from plasma without being converted to LDL.

The metabolism of TG-rich lipoproteins greatly influences HDL cholesterol levels. Lipolysis of both chylomicrons and VLDL generates surface lipids and apoproteins that are transferred to HDL. This process increases the size and lipid content of HDL. Reduced lipolysis causes a decline in HDL production, partly because HDL catabolism increases when plasma TG levels rise. Two pathways of HDL metabolism exist: the transfer of cholesteryl esters (CEs) from HDL to apoB lipoproteins and the removal of the entire HDL from plasma. The plasma VLDL and chylomicron concentrations heavily influence both processes. In the first process, cholesteryl ester transfer protein (CETP) mediates the exchange of HDL CE with TG in VLDL, chylomicrons, or both. This allows HDL to become a substrate for lipolysis by LPL, hepatic lipase, and/or endothelial lipase. The resulting smaller HDLs are removed rapidly from the bloodstream, and many of them are cleared in the kidney.

CAUSES OF HUMAN HYPERTRIGLYCERIDEMIA

The severity of hypertriglyceridemia depends on the lipoprotein particle involved. While increases in chylomicrons lead to severe cases of hypertriglyceridemia, high levels of VLDL cause milder conditions.

Chylomicronemia

Chylomicronemia is normally present in humans during the day; fasting chylomicronemia is pathologic. Almost all patients that have TG levels in excess of 1000 mg/dl have chylomicronemia. Sometimes, chylomicronemia results from a primary defect in chylomicron removal. It also occurs when LPL is saturated at plasma TG levels of approximately 500 mg/dl *(9)*.

Familial Lipoprotein Lipase Deficiency

Familial lipoprotein lipase deficiency is an autosomal recessive disorder due to the impairment or absence of LPL. Without enough LPL, chylomicrons accumulate in fasting plasma. TG levels rise above 1000 mg/dl and can exceed 10,000 mg/dl. Manifestations of LPL deficiency usually begin in infancy and include pancreatitis, eruptive xanthomas, hepatomegaly, splenomegaly, foam cell infiltration of the bone marrow, and lipemia retinalis. Although atherosclerosis once was thought to be relatively uncommon in LPL-deficient patients, cases of premature atherosclerosis in such patients have challenged this idea *(10)*. Manifestations of LPL deficiency recede dramatically when patients are placed on fat-free diets or diets primarily containing medium chain TGs (MCTs).

The diagnosis of homozygous LPL deficiency is suspected by finding a layer of cream (chylomicrons) at the top of plasma that has incubated overnight at 4°C. A physician may then confirm the diagnosis by demonstrating that LPL levels in plasma do not increase after the administration of heparin (which normally releases LPL from endothelial surfaces). While a doctor may suspect the presence of homozygous LPL deficiency as a result of a clinical situation, a more definitive diagnosis may be necessary. Such a diagnosis requires the injection of heparin and the acquisition of a postheparin blood sample and its assay at a specialized laboratory. Contraindications to this procedure include a history of coagulation disorders, gastrointestinal bleed, anemia, anti-coagulation therapy, and the use of aspirin within the previous 3 days. The patient should arrive after an overnight fast. First, the physician should obtain a preheparin serum sample. This sample can be added to a standard source of milk LPL to verify that apoCII, the LPL activator, is present. Then, 60 units/kg of heparin (usually from a standard supplier) is injected IV. Fifteen minutes later, a postheparin plasma (PHP) sample is taken in a tube without anti-coagulant. PHP should not clot if the heparinization is successful. The PHP sample must be placed on ice, the plasma separated within 30 min, and 100 µl aliquots frozen (at 70 °C if possible). Multiple (usually three) PHP aliquots and the preheparin blood should be shipped on dry ice to a lipid referral center for assay (e.g., University of Washington, Columbia University, University of Pennsylvania, University of Arkansas, University of Colorado, University of British Columbia, University of Montreal, Emory, and NIH).

Heterozygous LPL deficiency is more difficult to diagnose, as hypertriglyceridemia alone may reduce the amount of LPL associated with the capillary wall. Although genetic testing is feasible, it is not usually performed.

Over one hundred mutations of the LPL deficiency gene have been discovered (11). Presumably because the mutation occurred in a founder, this genetic disease is especially common in the French-Canadian population. Usually, the heterozygous carriers of this disorder are asymptomatic except for reduced levels of HDL. However, when patients age, develop diabetes, or become pregnant, the defective lipolysis can lead to severe hypertriglyceridemia, and even pancreatitis.

Familial Apoprotein CII Deficiency

Familial apoprotein CII deficiency is a rare autosomal recessive disorder that leads to a functional deficiency of LPL and clinical manifestations similar to those of familial LPL deficiency. ApoCII deficiency impairs hydrolysis of chylomicrons and VLDL. The diagnosis is suspected in children or adults with recurrent attacks of pancreatitis and confirmed by demonstrating absence of apoCII. A functional assay can be done to show that plasma from the patients fails to activate a source of LPL in vitro. While heterozygous carriers of apoCII deficiency have half-normal levels of apoCII and may have mild elevations of TGs, they are otherwise asymptomatic. Dietary fat restriction for apoCII-deficient patients should be lifelong.

Elevations of VLDL

Elevations of VLDL alone lead to less severe conditions of hypertriglyceridemia. Familial hypertriglyceridemia appears to be transmitted as an autosomal dominant disorder, but the underlying mutation(s) have not been identified. Both reduced catabolism of TG-rich lipoproteins and overproduction of VLDL have been reported as mechanisms of familial hypertriglyceridemia. Elevated levels of fasting plasma TGs in the range of 2.3–8.5 mmol/l (200 to 750 mg/dl) are usually associated with increased levels of VLDL TGs only. This disorder does not appear to increase risk for atherosclerosis, either because large TG-rich lipoproteins are not atherogenic or because patients generate normal amounts of LDL.

Familial Combined Hyperlipidemia

Concomitant hypercholesterolemia and hypertriglyceridemia occurs in two disorders, familial combined hyperlipidemia (FCHL) and dysbetalipoproteinemia. FCHL is inherited as an autosomal dominant disorder, and probands may have combined hyperlipidemia, isolated hypertriglyceridemia, or isolated elevated levels of LDL cholesterol.

The diagnosis of FCHL requires documentation at some time of combined hyperlipidemia in the proband. Alternatively, the diagnosis is presumed if the proband has isolated hypercholesterolemia or hypertriglyceridemia, and a first-degree relative has an increase in the other lipid. In affected individuals, the lipoprotein phenotype may change with time. The underlying defect of this disorder is unknown, but mutations or polymorphisms in the gene for LPL and the gene cluster for apoAI, apoCIII, and apoAIV may contribute to the disorder in some families. While more genetic markers have been found, they have only occurred in limited populations *(12–14)*. Many individuals with FCHL are insulin resistant; this link may result from increased free fatty acid flux driving assembly and secretion of apoB100 lipoproteins.

Most patients with FCHL have increased secretion of apoB *(15)*. The lipoprotein patterns associated with the disorder most likely are determined by polymorphisms in genes that regulate the metabolism of VLDL. For example, if the affected individual also has reduced LPL activity, hypertriglyceridemia will be present. As the hydrolysis of VLDL TGs also regulates the generation of LDL in plasma, individuals with FCHL who have inefficient catabolism of VLDL may have lower LDL. Finally, individuals with FCHL who have efficient VLDL metabolism will generate increased numbers of LDL particles and present with isolated elevations of plasma LDL cholesterol. These variations in VLDL catabolism are the basis for the different lipoprotein phenotypes in this disorder (Fig. 2).

FCHL may occur in as many as 0.5–1.0% of Americans and is the most common familial lipid disorder in survivors of myocardial infarction. Increased numbers of small, atherogenic IDL and LDL in FCHL patients may increase risk of atherosclerosis. Subjects with FCHL usually have clear plasma and do not have xanthomas or xanthelasma.

Dysbetalipoproteinemia

Dysbetalipoproteinemia, a rare disorder that affects 1 in 10,000 persons, usually results from homozygosity for apoE2, the binding-defective form of apoE. Because apoE plays a crucial role in the catabolism of chylomicrons and VLDL remnants, affected individuals possess particularly high levels of TG and cholesterol. The ratio of total cholesterol to TG within VLDL is greater than 0.30. Most patients have chylomicron remnants in fasting plasma, and occasional patients have fasting hyperchylomicronemia. Although 1% of the population is homozygous for apoE2, most have normal plasma TG and cholesterol levels. Thus, a second defect in lipid metabolism must be present in the 0.01% of individuals with dysbetalipoproteinemia. Clinical signs of dysbetalipoproteinemia include tuberous xanthomas and deposits of cholesterol in the palmar creases (striae palmaris); the latter, which appear as yellow-orange lines, are specific for dysbetalipoproteinemia. Beginning in a person's fourth or fifth decade, dysbetalipoproteinemia increases the risk for atherosclerosis. The incidence of peripheral vascular disease is higher in patients with dysbetalipoproteinemia than in those with familial hypercholesterolemia.

Hepatic Lipase Deficiency

Another rare autosomal recessive disorder, hepatic lipase deficiency, impairs the final catabolism and/or remodeling of small VLDL and IDL. Thus, subjects with hepatic lipase deficiency have elevated levels of VLDL remnants. In contrast to

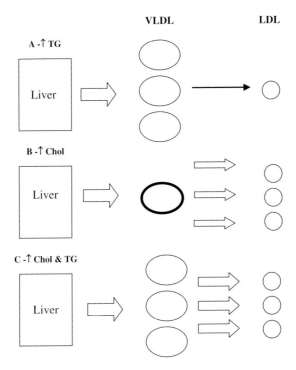

Fig. 2. Pathophysiology of familial combined hyperlipidemia (FCHL). Three lipoprotein profiles are found in FCHL, either in different family members or in the same person at different times. The condition results from liver overproduction of apolipoprotein B (apoB)-containing lipoproteins. (A) Increased very low-density lipoprotein (VLDL) production without excess conversion to LDL leads to isolated hypertriglyceridemia. (B) VLDL overproduction and increased conversion of VLDL to low-density lipoprotein (LDL) leads to isolated hypercholesterolemia. (C) In some individuals, both situations exist with an excess apoB production leading to increased blood concentrations of both VLDL and LDL, and increased plasma triglyceride (TG) and cholesterol levels.

many diseases that cause TG to increase, hepatic lipase deficiency increases HDL levels. This HDL tends to be larger as hepatic lipase removes both TG and phospholipid from HDL. Presumably due to the elevated plasma concentrations of VLDL remnants, hepatic lipase deficiency is associated with premature atherosclerosis *(16)*. A physician can diagnose hepatic lipase deficiency by assaying hepatic lipase in postheparin blood. Hepatic lipase, unlike LPL, is active in buffers containing high salt concentrations.

Hypertriglyceridemia

Hypertriglyceridemia can also result from reduced HDL. Low levels of HDL cholesterol are defined as < 35mg/dl in men and < 45mg/dl in women. Usually, hypertriglyceridemia coexists with low concentrations of HDL. However, HDL levels usually do not return to normal when fasting plasma TGs are reduced in persons with hypertriglyceridemia. Thus, many patients have an underlying hypoalphalipoproteinemia, either due to reduced production or, in most cases, increased HDL clearance. Several rare genetic disorders dramatically reduce HDL. All such diseases—lecithen Cholesterol acyl transferase (LCAT) deficiency, ATP binding Cassettee (ABC) ABCAI deficiency (Tangiers Disease), and apoAI Milano—are associated with mild-to-moderate

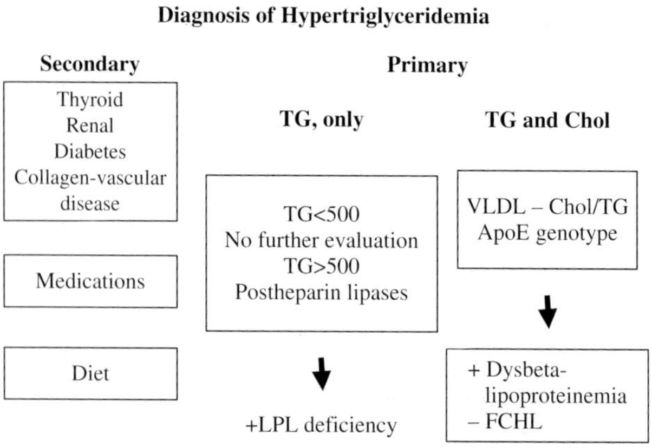

Diagnosis of Hypertriglyceridemia

```
    Secondary              Primary
  ┌──────────┐
  │ Thyroid  │
  │ Renal    │       TG, only        TG and Chol
  │ Diabetes │
  │Collagen- │    ┌──────────────┐  ┌──────────────┐
  │ vascular │    │   TG<500     │  │VLDL – Chol/TG│
  │ disease  │    │No further    │  │ApoE genotype │
  └──────────┘    │ evaluation   │  └──────┬───────┘
                  │   TG>500     │         │
  ┌──────────┐    │Postheparin   │         ▼
  │Medications│   │  lipases     │
  └──────────┘    └──────────────┘  ┌──────────────┐
                                    │+ Dysbeta-    │
  ┌──────────┐         ▼            │ lipoproteinemia│
  │   Diet   │   +LPL deficiency    │- FCHL        │
  └──────────┘                      └──────────────┘
```

Fig. 3. Diagnosis of hypertriglyceridemia. ApoE, apolipoprotein E; LDL, low-density lipoprotein; TG, triglyceride; VLDL, very low-density lipoprotein.

hypertriglyceridemia. As HDL serves as the plasma reservoir of apoCII, the very low plasma HDL in patients with these diseases may result in a relative deficiency of apoCII.

A flow diagram for the approach to patients with hypertriglyceridemia is included in Fig. 3.

SECONDARY CAUSES OF HYPERTRIGLYCERIDEMIA

Hormones

ESTROGEN

Estrogen increases TG and HDL production (Table 1). In patients with underlying hypertriglyceridemia, increased estrogen as a result of hormonal therapy or pregnancy can lead to severe exacerbation and, occasionally, pancreatitis. Exogenous estrogen increases liver production of TG and reduces hepatic lipase activity *(17)*. However, transdermal (as opposed to oral) estrogen does not have a high first pass extraction in the liver; thus, it does not lead to increased liver TG production *(18)*.

TAMOXIFEN

Tamoxifen can induce a small increase of TG up to 20–30% probably secondary to its weak estrogenic activities *(19,20)*. This increase in plasma TG is associated with a decrease in LPL activity *(21,22)*. Raloxiphene, one of the new selective estrogen receptor modulators, is not associated with increased serum TG *(23,24)*.

HYPOTHYROIDISM

Hypothyroidism is a common cause of hypercholesterolemia and must be excluded in all such patients. The increased LDL in those with hypothyroidism is likely due to a decrease in liver LDL receptors. Some hypothyroid patients develop a mixed hyperlipidemia with increased remnants (similar to that of dysbetalipoproteinemia) *(25,26)*.

Table 1
Secondary Forms of Hyperlipidemia

Increased VLDL
  Diabetes mellitus Type II
  Obesity
  Glycogen storage disease
  Lipodystrophy
  Alcohol
  Renal failure
  Stress
  Sepsis
  Cushing syndrome
  Pregnancy
  Acromegaly
  Hepatitis
  Drugs: HIV protease inhibitor, Retinoic acid, Rapamycine
  Drugs ©
    Chylomicron
    Estrogen, tamoxifen
    Thiazide, beta blocker
    Glucocorticoids,
    Bile acid-binding resins
Increased IDL
  Hypothyroidism
  Autoimmune disease, monoclonal gammopathy, multiple myeloma
Increased Chylomycron
  Autoimmune disease, Isotretinoin
Unknown
  Antipsychotic/interferon

TYPE I DIABETES

Type I diabetes often leads to marked hypertriglyceridemia. This link results from increased fatty acid return to the liver due to deficient insulin inhibition of adipose tissue hormone-sensitive lipase. In addition, plasma removal of TG may be impaired because LPL is an insulin-regulated protein. However, as insulin also stimulates HDL production, patients who control their type I diabetes may achieve HDL concentrations that exceed those of an average person.

TYPE II DIABETES

Type II diabetes is associated with a lipid pattern called diabetic dyslipidemia. In patients afflicted with diabetic dyslipidemia, hypertriglyceridemia results from increased hepatic input of substrates for TG synthesis. This heightened input subsequently leads to the production of large VLDL particles. Type II diabetes also causes decreased production of LPL, which impairs clearance of TG. Insulin therapy increases LPL activity in vivo *(27)* by increasing LPL gene transcription *(28)* and/or decreasing enzyme inactivation *(29)*. In addition, hepatic insulin resistance causes heightened production of apoCIII *(30)*. Because the proportion of VLDL converted to LDL is decreased, LDL levels are often normal. As a result of the hypertriglyceridemia, LDL is usually smaller and denser, and HDL levels are reduced.

Associated Diseases

LIVER DISORDERS

Liver disorders cause variation in lipoprotein levels. Hepatitis mildly increases TG *(31)*. Liver failure can decrease the synthesis of both cholesterol and TG *(32)*.

OBESITY

Obesity causes the metabolic syndrome, defined by the presence of three of the following factors: (1) abdominal obesity (waist circumference > 102 cm/40 inches in men and > 88 cm/35 inches in women); (2) TGs > 150; (3) low HDL cholesterol (<40 in men and <50 in women); (4) blood pressure $\geq 130/85$ mm Hg; and (5) fasting glucose ≥ 100 mg/dl *(33)*. A central abnormality in this syndrome is peripheral and hepatic insulin resistance. Reduced insulin sensitivity leads to increased activity of adipose tissue hormone-sensitive lipase, increased plasma free fatty acids, and increased hepatic production of both apoB and TG *(34)*. The underlying adipose insulin resistance might be secondary to the presence of inflammatory cells *(35)* and/or inflammatory cytokine production by adipocytes *(36)*.

LIPODYSTROPHY

Lipodystrophy, the selective loss of adipose tissue, limits the quantities of TGs that can be stored in fat depots. This condition results in excess TG accumulation in liver and skeletal muscles, which contributes to insulin resistance and hypertriglyceridemia *(37)*.

GAUCHER'S DISEASE AND GLYCOGEN STORAGE DISEASE TYPE I

Hypertriglyceridemia is a feature of both Gaucher's disease and glycogen storage disease type I *(38)*. These diseases are associated with liver dysfunction and reduced levels of hepatic lipase *(39)*.

Hypertriglyceridemia is also common in patients with renal dysfunction. Nephrotic syndrome causes hypoalbuminemia, which is associated with increased liver TG and apoB production; patients often produce excess cholesterol and TG *(40)*. The most common lipid complication associated with chronic renal failure is high levels of TG and low levels of HDL. This condition may stem partly from reduction of LPL activity due to unidentified inhibitors in the plasma *(41)*.

Beverages

ALCOHOL

Excessive consumption of ethanol commonly leads to secondary hypertriglyceridemia, especially in males. Even moderate consumption of alcohol on a regular basis results in significantly higher serum TGs than that of someone who never drinks. One postulated mechanism is that ethanol is preferentially oxidized in the liver, and free fatty acids are spared and used for TG synthesis. Withdrawal of alcohol results in a rapid decrease in TG levels. Alcohol also increases HDL cholesterol.

COFFEE

Ingestion of 1–2 l per day of boiled or percolated coffee—not the instant or filtered versions—can result in increases in serum cholesterol and TG up to 60%. Though studies have revealed that diterpenes contained in oils leached out from coffee beans mediate these effects, the mechanism is unclear *(42)*.

Medications

THIAZIDES AND B-BLOCKERS

It has been recognized for a long time that the administration of thiazide diuretics such as chlorthalidone and hydrochlorothiazide increases total cholesterol and TG. These changes probably reflect the adverse effects of these drugs on glucose tolerance.

Similarly, long-term administration of β-blockers without intrinsic sympathomimetic activity (ISA) increases serum TG and decreases HDL cholesterol. Like α-blockers, β-blockers with ISA, which possess some beta-agonist activity (e.g., pindolol and acebutolol), have less influence on serum TG. The mechanism responsible for the hypertriglyceridemic and HDL lowering effects of β-blockers may involve a decrease in LPL due to inhibition of adenyl cyclase in adipocyte. β-Blockade impairs the removal of TGs from plasma. Occasionally, markedly increased serum TGs develop in genetically predisposed individuals.

CORTICOSTEROIDS

Corticosteroids cause insulin resistance and impaired glucose tolerance, which lead to hypertriglyceridemia and a reduction of HDL cholesterol. Experimental studies suggest that a steroid-induced increase in VLDL synthesis (which also occurs in Cushing's syndrome) is one of the mechanisms involved.

RAPAMYCIN

Rapamycin or sirolimus (Rapamune), an immunosuppressive agent that inhibits the activation and proliferation of T lymphocytes, causes hypertriglyceridemia and hypercholesterolemia in 50% of patients who use the drug. The mechanism is not fully understood, but increased apoCIII and apoB100 have been proposed *(43)*.

RETINOID

Retinoid induces a marked increase in serum TGs, especially in patients with preexisting hypertriglyceridemia. This results from increased TG biosynthesis *(44)*. While the reason for this observation remains unclear, increased insulin resistance has been proposed *(45,46)*.

PROTEASE INHIBITORS

Protease inhibitors can lead to a metabolic syndrome-like disorder that results in reduced adipocyte tissue, insulin resistance, and hyperlipidemia. First generation protease inhibitors like ritonavir lead to this process with greater frequency. Protease inhibitor-induced hypertriglyceridemia and low HDL are strongly associated with apoCIII polymorphisms and excess body weight *(47–49)*.

LONG-TERM INTERFERON THERAPY

Long-term interferon therapy has also been linked to hypertriglyceridemia. Baseline dyslipidemia is not predictive. Furthermore, this side effect is not dose dependent *(50)*.

Antipsychotic Medications

Antipsychotic medications—especially clozapine, olanzapine and quetiapine—may increase TGs up to 50% from baseline; this has been observed in prospective and retrospective studies *(51)*.

TREATMENTS/RECOMMENDATIONS

Lifestyle Alterations

Physicians should instruct patients to decrease their intake of alcohol, simple sugars, and carbohydrates and to exercise regularly. Patients will more likely adhere to exercise programs that are designed to comply with their own personal goals, interests, and needs. Most patients would benefit from performing aerobic exercise that targets large muscle groups for 30 min, 4 or more times a week. Furthermore, hypertriglyceridemic patients generally should not follow diets with a fat content of less than 25% of calories; this recommendation does not apply, however, to patients with fasting hyperchylomicronemia. The response to a low fat diet (i.e., 10% of calories from fat) may be disappointing in patients with impaired glucose tolerance unless the diet is sufficiently hypocaloric. In addition, high carbohydrate diets that are isocaloric may lead to poor glycemic control and increased TG levels.

Patients with genetic LPL deficiency should receive food preparation instructions preferably from a registered nutritionist. Food oils must contain MCTs composed of fatty acid of 6–12 carbon atoms that are not incorporated in chylomicrons; in other words, the oil should not contain any long chain fatty acid (dietary fat that has 16–18 atoms of carbon). A version of this oil is made from fractionated coconut oil. Patients may order it at http://www.medicalmailorder.com/meadjohnson.html and use it as a substitute in cooking or in salad dressings.

In addition, hypertriglyceridemics should avoid drugs like estrogen, retinoid, and hydrochlorothiazide. By using transdermal estrogen instead of oral estrogen, women receiving postmenopausal hormone replacement therapy can significantly decrease their TG levels *(18)*.

A number of oils may act as treatment for hypertriglyceridemia. Omega 3 fatty acids are polyunsaturated fatty acids that are composed mainly of linolenic acid and synthesized by plant chloroplasts. Within the body of certain animals (fish and deer), linolenic acid is converted to docosahexanoic acid (DHA) and eicosapentanoic acid (EPA). These three oils increase fatty acid oxidation and may be especially potent ligands for the peroxisomal proliferator activated receptor (PPARα) transcription factor *(52)*. Flaxseed oil, which is rich in linolenic acid, and Omacor and fish oils, which contain large amounts of DHA and EPA, will decrease TG synthesis and increase fatty acid oxidation. In high doses, the ingestion of these oils may alleviate severe hypertriglyceridemia. Three to six grams a day of EPA plus DHA may decrease TG by up to 30% *(53)*. One study reported that 0.8 grams per day of linolenic acid diminished TG up to 25% *(54)*. Also, Omacor therapy in the GISSI preventive study proved to decrease sudden death in postmyocardial infarction patients *(55)*. The major drawbacks to high doses of these fatty acids include a large number of pills, eructation, and occasional diarrhea. In addition, these oils have mild anti-platelet effects that may harm patients who take warfarin and anti-platelet drugs.

Table 2
Medications

Drug	Indications	Mechanisms	Side effects
HMG-CoA reductase inhibitors (Statins)	↑ LDL	↓ Cholesterol synthesis, ↓ Hepatic LDL receptors, ↓ VLDL production	Elevated transaminases Myalgia
Nicotinic acid	Elevated LDL, Low HDL, ↑ TG	↓ VLDL hepatic Synthesis	Cutaneous flushing ↑ Glucose and uric acid
Fibric acid derivatives	Elevated TG, Elevated remnants	↑ LPL, ↓ VLDL synthesis	Elevated transaminases, myalgia
Fish oils	Elevated TG	↓ Chylomicron and VLDL production	Dyspepsia, diarrhea

HDL, high-density lipoprotein; LDL, low-density lipoprotein; TG, triglyceride; VLDL, very low-density lipoprotein.

Pharmaceuticals

When TG levels exceed 5.6 mmol/l (>500 mg/dl), the risk of developing pancreatitis increases; hence, it is necessary to decrease TGs. In less-severe hypertriglyceridemia, the recommendations for evaluating and treating hypertriglyceridemia focus on the associated LDL and HDL concentrations as guidelines for therapy (Table 2). Thus, physicians can evaluate the overall risk profile to set goals for LDL cholesterol by using a low HDL level (commonly associated with hypertriglyceridemia) as a concomitant major risk factor for atherosclerosis. The most recent NCEP guidelines include a goal for non-HDL cholesterol that is 30 mg/dl greater than that for LDL cholesterol *(1)*. Thus, if TGs are above 150, either LDL reduction should be more aggressive or the physician should employ pharmaceutical therapy directly toward reducing VLDL.

NIACIN

Niacin, or nicotinic acid, is a first-line drug for hypertriglyceridemia. It inhibits mobilization of free fatty acids from peripheral tissues, thereby reducing hepatic synthesis of TGs and secretion of VLDL. Nicotinic acid exceeds all other drugs in its ability to increase serum HDL concentrations; at the maximum dose, nicotinic acid can raise these concentrations by up to 30%. This drug causes a shift from small, dense LDL to large, buoyant LDL. Such altering of LDL probably results from the TG reduction. Nicotinic acid can decrease TG level by 45% at 2–2.5 g/day. The sustained release niacin, marketed as Niaspan, reduces TG to a similar extent as the nonprescription niacin. Many patients find the predominant side effect of nicotinic acid, flushing of the skin, intolerable. The administration of aspirin or ibuprofen 30–60 min before each dose of nicotinic acid reduces the severity of this flushing, as the side effect is thought to be prostaglandin mediated. Taking the medication with food and avoiding alcohol and spicy meals may also help. These drugs can be discontinued as tachyphylaxis develops. Other adverse effects of nicotinic acid are decreased control of blood sugar, increased amounts of uric acid, diarrhea, conjunctivitis, nasal stuffiness, acanthosis nigricans, hepatitis, and ichthyosis. The difficulties with blood sugar control can often be offset by modulation of diabetes medications *(56)*.

Gemfibrozil and Fenofibrate

Gemfibrozil and fenofibrate are the fibric acid derivatives that are available in the United States. These drugs are ligands for PPARα. They stimulate LPL production and increase hepatic fatty acid oxidation. The latter effect reduces VLDL TG entry into plasma. In addition, fibric acids reduce synthesis of apoCIII by impairing gene transcription *(57)*. Fibrate treatment is associated with 25–40% reductions in plasma TG levels. Fibric acids also reduce postprandial TG levels.

Significant increases in LDL cholesterol can accompany the TG reduction, and the HDL cholesterol increase during fibrate therapy. Such rises may require a change to another drug or the addition of a second agent.

In the short term, these drugs are well tolerated. The major adverse effect is mild gastrointestinal distress in the form of epigastric pain. Mild elevations of liver enzymes occur in 2–3% of patients but do not usually require cessation of treatment. Rarely, hepatitis can occur. Older fibrates, such as clofibrate, were associated with twofold increase in gallstone formation. In rare cases, the fibrates cause myopathy with myositis, either alone or in combination with statins. Gemfibrozil increases the risk of statin-induced myopathy by decreasing the elimination of statin acid by the glucuronidation pathway in the liver *(58–60)*.

Statin

Statin treatment decreases TG by 10–20%, possibly due to reduced secretion of VLDL by the liver. Statins are a useful adjunct in the treatment of moderate hypertriglyceridemia in patients with FCHL, but they are often insufficient.

Ezetimibe

Ezetimibe is a cholesterol absorption inhibitor that also leads to modest decreases in TGs by 9% *(61)*. Although the clinical data regarding the effects of ezetimibe are not yet available, this agent is expected to reduce the cholesterol content of chylomicrons, as it does in animals.

PPARγ Agonists

PPARγ agonists, pioglitazone and rosiglitazone, are primarily used to increase insulin sensitivity. In separate studies, pioglitazone did not change LDL cholesterol, and rosiglitazone increased LDL cholesterol by 8–16%. In both groups, HDL cholesterol increased by 10%. Pioglitazone decreased TG more than rosiglitazone did; the decreases caused by pioglitazone ranged from 14 to 26% in different studies *(62)*. Pioglitazone may act more like a partial PPARα agonist than rosiglitazone. The mechanism for the TG lowering is increased plasma TG clearance *(63)*.

Torcetrapib

Torcetrapib is a CETP inhibition that blocks transfer of VLDL TG to HDL. The effect that torcetrapib will have on plasma TG remains unclear. Brousseau et al. *(64)* reported that subjects with low HDL cholesterol treated with torcetrapib alone at 120 mg bid showed a 26% TG decrease; in combination with atorvastatin, 120 mg qd of torcetrapib decreased TG by 18%.

PLASMAPHERESIS

Plasmapheresis can be used for severe, refractory hyperchylomicronemia. Patients rarely need plasmapheresis as TG levels fall very rapidly in severely ill, non-eating patients, like those with pancreatitis.

PANCREATIC LIPASE INHIBITION

Pancreatic lipase inhibition is the pharmacologic equivalent of a low fat diet. This can be produced using tetrahydrolipstatin (Orlistat). It has been reported that the combination of Orlistat and gemfibrozil was extremely effective in reducing serum TG levels in patients with combined hyperlipidemia and predominant hypertriglyceridemia by more than 50% *(65)*. Moreover, because fat intake in patients taking Orlistat leads to bloating and diarrhea, the drug acts like an Antabuse for dietary lipid. Thus, patients should accompany Orlistat with a low fat diet.

GENE THERAPY

Gene therapy is a future approach to the treatment of several lipid disorders. For example, overexpression of LPL may be therapeutic in patients who have severe hypertriglyceridemia without any genetic perturbation. Many options have proved successful on animal models and may become useful treatment methods in the future. These include overexpression of LPL in the muscle and apoAV and VLDL receptor expression in the liver. Liver-specific genetic reduction of apoCIII, angiopoetins, and enzymes involved in TG synthesis might be other approaches.

CASE 1: DYSBETALIPOPROTEINEMIA/TYPE 3 HYPERLIPIDEMIA

A 35-year-old man has a recent history of hypertriglyceridemia. The patient, a publisher for a news magazine, has no history of coronary heart disease. He does not drink alcohol or exercise, and he presently adheres to a low fat diet. His blood test 1 month ago was TG 509 mg/dl, cholesterol 430 mg/dl, and HDL 37 mg/dl. LDL cholesterol was not calculated because TG was greater than 400 mg/dl. The patient was placed on Zetia® 10 mg po every day and referred to the lipid clinic.

The patient is taking Effexor® for depression and has no family history of severe lipidemia or coronary heart disease. He is well developed, 184 lb, and 75 inches tall. The patient has a blood pressure of 115/80, with a heart rate of 75 per minute. At the time of his visit, he had no xanthelasma or xanthomas, and his heart, lung, abdomen and pedal pulses were within normal limits.

Laboratory results on the day prior to the clinic visit showed a normal blood cell count (CBC), normal fasting glucose, and normal thyroid stimulating hormone (TSH). Lipid profile was cholesterol 280 mg/dl, LDL 169 mg/dl, HDL 46 mg/dl, and TG 324 mg/dl.

This patient presents with elevations of cholesterol and TG that can be caused by two lipoprotein abnormalities:

1. FCHL is the most common. This autosomal dominant disorder is a result of over production of apoB.
2. Dysbetalipoproteinemia is a much less common abnormality due to homozygosity for an abnormal form of apoE, apoE2. One clue to the diagnosis of dysbetalipoproteinemia (as opposed to the autosomal dominant FCHL) is that no other family member has had dyslipidemia.

The differential diagnosis between the two disorders can be made by isolating the lipoprotein particles in VLDL density range to determine the ratio of cholesterol/TG. If the ratio is >0.3, the diagnosis is dysbetalipoproteinemia. In addition, genotyping for apoE isoforms is available in some clinics.

In this patient, the VLDL cholesterol/TG ratio was >1/3. He was advised on his diet and exercise, and 160 mg/dl of fenofibrate (Tricor®) was prescribed. (A good alternative to fenofibrate would be niacin.)

Lipid profile 3 months after was cholesterol 178 mg/dl, HDL 52 mg/dl, LDL 98 mg/dl, and TG 142 mg/dl.

CASE 2: LIPOPROTEIN LIPASE DEFICIENCY

A 4-year-old boy had severe elevation of TG, noted over the last several months. He had recently been admitted for intermittent abdominal pain. At that time, his TG was 2600 mg/dl, cholesterol was 327 mg/dl, and amylase and lipase were normal. He was placed on a low fat diet. Since last admission, the patient has been doing well with TG approximately 1000 mg/dl. He takes multivitamins. His only past medical history was surgery for trigger thumb.

The patient's father and mother were in good health, and his family has no history of dyslipidemia or heart disease. He weighed 34 lb, was 43 inches tall, and his blood pressure was 100/60. The fundi were difficult to examine, and there was no eruptive xanthoma. Heart, lung, abdomen, and pedal pulses within normal limit.

Laboratory results from the day before the clinic visit showed normal fasting glucose and TSH. Lipid profile was cholesterol 125 mg/dl, TG 817 mg/dl, and HDL 23 mg/dl. LDL cholesterol was not calculated because TG was greater than 400 mg/dl.

The patient has hypertriglyceridemia, which decreased when he adopted a low fat diet. If TG exceeds 1000 mg/dl, there is risk of pancreatitis, which might have accounted for the prior episodes of abdominal pain. Dyslipidemia primarily due to increased TG at a young age is usually secondary to LPL or apoCII deficiency; all respond very well to a low fat diet. The protocol for laboratory measurement of LPL activity was described in the text. Mixing the patient's PHP with a standard source of milk LPL was used to verify that apoCII is present. In this patient, LPL was very low (10% of normal), which is consistent with type I hyperlipidemia. He was referred to a nutritionist to advise the family on diet and use of MCT oil.

CASE 3: HYPERTRIGLYCERIDEMIA AND RETINOIC ACID (ACCUTANE®)

A 16-year-old man with severe acne developed hypertriglyceridemia while on Accutane® (isotretinoin) therapy. The patient had no history of dyslipidemia, but his TG increased to more than 400 mg/dl on isotretinoin and then returned to normal after discontinuation of the Accutane®. He denies alcohol or steroids use. The patient is a high school student. He eats a normal teenage diet (rich in fast foods) and does not exercise. Six days ago, he was restarted on a low dose of Accutane® every other day, and his subsequent blood test is shown below.

The patient has no past medical history and there is no family history of coronary artery disease (CAD), diabetes, or dyslipidemia. He weighed 235 lb, was 70 inches tall, and his blood pressure was 130 / 80 with a heart rate of 88 (BMI 33). He had no xanthelasma or xanthomas. The heart, lung, abdomen, and pedal pulses were normal.

Laboratory results from the day before the clinic visit showed normal CBC, fasting glucose, and TSH. Lipid profile was cholesterol 184 mg/dl, LDL 93 mg/dl, TG 300 mg/dl, and HDL 31 mg/dl.

Retinoid is clearly the cause of hypertriglyceridemia in this patient. To allow continuation of his acne therapy, gemfibrozil and fish oil were prescribed with frequent blood monitoring. The patient's TG decreased to 150mg/dl after 3 weeks, and his Accutane® dose was increased gradually.

CASE 4: METABOLIC SYNDROME

A 44-year-old woman with past medical history of obesity went to her primary care physician for a routine follow-up. The patient has no history of CAD, had been trying to lose weight for years, and currently attempts to eat a low calorie diet. She has no time to exercise because of her work and family schedule. She denies alcohol intake. There is no history of CAD or diabetes mellitus, and her family history is positive for obesity (mother). She is G3P3 with regular menstrual period.

She weighed 208 lb, had a BMI of 36, was 63 inches tall, had a waist circumference of 38 inches, and had a blood pressure 135/88, with a heart rate of 80. She is obese with acanthosis nigricans and no xanthelasma or xanthomas. Her heart, lung, abdomen, and pedal pulses were within normal limit. Laboratory results from the day before the clinic visit showed normal CBC, fasting glucose 116 mg/dl, and normal TSH. Lipid profile was cholesterol 216 mg/dl, LDL 139 mg/dl, TG 175 mg/dl, and HDL 42 mg/dl.

This patient meets all five criteria of the metabolic syndrome for a woman. They are the following:

1. Abdominal obesity with waist circumference > 35 inches.
2. Fasting TG > 150mg/dl.
3. HDL < 50mg/dl.
4. HTN > 130/85.
5. Fasting glucose > 110 mg/dl.

In clinical practice, only three out of five criteria are necessary to diagnose the metabolic syndrome. At present, the best treatment for this patient is lifestyle changes—namely, to lose weight and exercise.

REFERENCES

1. Third Report of the National Cholesterol Education Program (NCEP) Expert Panel on Detection, Evaluation, and Treatment of High Blood Cholesterol in Adults (Adult Treatment Panel III) final report. Circulation 2002;106(25):3143–421.
2. van Loon LJ, Koopman R, Stegen JH, et al. Intramyocellular lipids form an important substrate source during moderate intensity exercise in endurance-trained males in a fasted state. J Physiol 2003;553(Pt 2):611–25.
3. Coggan AR, Raguso CA, Gastaldelli A, et al. Fat metabolism during high-intensity exercise in endurance-trained and untrained men. Metabolism 2000;49(1):122–8.
4. Backhed F, Ding H, Wang T, et al. The gut microbiota as an environmental factor that regulates fat storage. Proc Natl Acad Sci USA 2004;101(44):15718–23.
5. Voyiaziakis E, Goldberg IJ, Plump AS, et al. ApoA-I deficiency causes both hypertriglyceridemia and increased atherosclerosis in human apoB transgenic mice. J Lipid Res 1998;39(2):313–21.
6. Jong MC, Hofker MH, Havekes LM. Role of ApoCs in lipoprotein metabolism: functional differences between ApoC1, ApoC2, and ApoC3. Arterioscler Thromb Vasc Biol 1999;19(3):472–84.
7. Savonen R, Nordstoga K, Christophersen B, et al. Chylomicron metabolism in an animal model for hyperlipoproteinemia type I. J Lipid Res 1999;40(7):1336–46.

8. Datta G, Chaddha M, Garber DW, et al. The receptor binding domain of apolipoprotein E, linked to a model class A amphipathic helix, enhances internalization and degradation of LDL by fibroblasts. Biochemistry 2000;39(1):213–20.
9. van Barlingen HH, Kock LA, de Man FH, et al. In vitro lipolysis of human VLDL: effect of different VLDL compositions in normolipidemia, familial combined hyperlipidemia and familial hypertriglyceridemia. Atherosclerosis 1996;121(1):75–84.
10. Benlian P, De Gennes JL, Foubert L, et al. Premature atherosclerosis in patients with familial chylomicronemia caused by mutations in the lipoprotein lipase gene. N Engl J Med 1996;335(12):848–54.
11. Merkel M, Eckel RH, Goldberg IJ. Lipoprotein lipase: genetics, lipid uptake, and regulation. J Lipid Res 2002;43(12):1997–2006.
12. Cantor RM, de Bruin T, Kono N, et al. Quantitative trait loci for apolipoprotein B, cholesterol, and triglycerides in familial combined hyperlipidemia pedigrees. Arterioscler Thromb Vasc Biol 2004;24(10):1935–41.
13. Pajukanta P, Lilja HE, Sinsheimer JS, et al. Familial combined hyperlipidemia is associated with upstream transcription factor 1 (USF1). Nat Genet 2004;36(4):371–6.
14. Huertas-Vazquez A, del Rincon JP, Canizales-Quinteros S, et al. Contribution of chromosome 1q21-q23 to familial combined hyperlipidemia in Mexican families. Ann Hum Genet 2004;68(Pt 5):419–27.
15. de Graaf J, van der Vleuten G, Stalenhoef AF. Diagnostic criteria in relation to the pathogenesis of familial combined hyperlipidemia. Semin Vasc Med 2004;4(3):229–40.
16. Twickler T, Dallinga-Thie GM, Chapman MJ, et al. Remnant lipoproteins and atherosclerosis. Curr Atheroscler Rep 2005;7(2):140–7.
17. Schaefer EJ, Foster DM, Zech LA, et al. The effects of estrogen administration on plasma lipoprotein metabolism in premenopausal females. J Clin Endocrinol Metab 1983;57(2):262–7.
18. Sanada M, Tsuda M, Kodama I, et al. Substitution of transdermal estradiol during oral estrogen-progestin therapy in postmenopausal women: effects on hypertriglyceridemia. Menopause 2004;11(3):331–6.
19. Hozumi Y, Kawano M, Saito T, et al. Effect of tamoxifen on serum lipid metabolism. J Clin Endocrinol Metab 1998;83(5):1633–5.
20. Dullaart RP. Exogenous estrogens, antiestrogens and lipid metabolism. Neth J Med 1999;55(2):47–9.
21. Hozumi Y, Kawano M, Hakamata Y, et al. Tamoxifen inhibits lipoprotein activity: in vivo and in vitro studies. Horm Res 2000;53(1):36–9.
22. Brun LD, Gagne C, Rousseau C, et al. Severe lipemia induced by tamoxifen. Cancer 1986;57(11):2123–6.
23. Delmas PD, Bjarnason NH, Mitlak BH, et al. Effects of raloxifene on bone mineral density, serum cholesterol concentrations, and uterine endometrium in postmenopausal women. N Engl J Med 1997;337(23):1641–7.
24. Walsh BW, Kuller LH, Wild RA, et al. Effects of raloxifene on serum lipids and coagulation factors in healthy postmenopausal women. JAMA 1998;279(18):1445–51.
25. Ito M, Takamatsu J, Matsuo T, et al. Serum concentrations of remnant-like particles in hypothyroid patients before and after thyroxine replacement. Clin Endocrinol (Oxf) 2003;58(5):621–6.
26. Feussner G, Ziegler R. Expression of type III hyperlipoproteinaemia in a subject with secondary hypothyroidism bearing the apolipoprotein E2/2 phenotype. J Intern Med 1991;230(2):183–6.
27. Sadur CN, Eckel RH. Insulin stimulation of adipose tissue lipoprotein lipase. Use of the euglycemic clamp technique. J Clin Invest 1982;69(5):1119–25.
28. Ong JM, Kirchgessner TG, Schotz MC, et al. Insulin increases the synthetic rate and messenger RNA level of lipoprotein lipase in isolated rat adipocytes. J Biol Chem 1988;263(26):12933–8.
29. Simsolo RB, Ong JM, Saffari B, et al. Effect of improved diabetes control on the expression of lipoprotein lipase in human adipose tissue. J Lipid Res 1992;33(1):89–95.
30. Cohn JS, Patterson BW, Uffelman KD, et al. Rate of production of plasma and very-low-density lipoprotein (VLDL) apolipoprotein C-III is strongly related to the concentration and level of production of VLDL triglyceride in male subjects with different body weights and levels of insulin sensitivity. J Clin Endocrinol Metab 2004;89(8):3949–55.
31. Vergani C, Trovato G, Delu A, et al. Serum total lipids, lipoprotein cholesterol, and apolipoprotein A in acute viral hepatitis and chronic liver disease. J Clin Pathol 1978;31(8):772–8.
32. Koruk M, Savas MC, Yilmaz O, et al. Serum lipids, lipoproteins and apolipoproteins levels in patients with nonalcoholic steatohepatitis. J Clin Gastroenterol 2003;37(2):177–82.
33. Deen D. Metabolic syndrome: time for action. Am Fam Physician 2004;69(12):2875–82.

34. Ginsberg HN, Zhang YL, Hernandez-Ono A. Regulation of plasma triglycerides in insulin resistance and diabetes. Arch Med Res 2005;36(3):232–40.
35. Weisberg SP, McCann D, Desai M, et al. Obesity is associated with macrophage accumulation in adipose tissue. J Clin Invest 2003;112(12):1796–808.
36. Shoelson SE, Lee J, Yuan M. Inflammation and the IKK beta/I kappa B/NF-kappa B axis in obesity- and diet-induced insulin resistance. Int J Obes Relat Metab Disord 2003;27 Suppl 3:S49–52.
37. Garg A. Acquired and inherited lipodystrophies. N Engl J Med 2004;350(12):1220–34.
38. Levy E, Thibault LA, Roy CC, et al. Circulating lipids and lipoproteins in glycogen storage disease type I with nocturnal intragastric feeding. J Lipid Res 1988;29(2):215–26.
39. Muller DP, Gamlen TR. The activity of hepatic lipase and lipoprotein lipase in glycogen storage disease: evidence for a circulating inhibitor of postheparin lipolytic activity. Pediatr Res 1984;18(9):881–5.
40. Joven J, Villabona C, Vilella E, et al. Abnormalities of lipoprotein metabolism in patients with the nephrotic syndrome. N Engl J Med 1990;323(9):579–84.
41. Lee DM, Knight-Gibson C, Samuelsson O, et al. Lipoprotein particle abnormalities and the impaired lipolysis in renal insufficiency. Kidney Int 2002;61(1):209–18.
42. Mensink RP, Lebbink WJ, Lobbezoo IE, et al. Diterpene composition of oils from Arabica and Robusta coffee beans and their effects on serum lipids in man. J Intern Med 1995;237(6):543–50.
43. Tur MD, Garrigue V, Vela C, et al. Apolipoprotein CIII is upregulated by anticalcineurins and rapamycin: implications in transplantation-induced dyslipidemia. Transplant Proc 2000;32(8):2783–4.
44. Bershad S, Rubinstein A, Paterniti JR, et al. Changes in plasma lipids and lipoproteins during isotretinoin therapy for acne. N Engl J Med 1985;313(16):981–5.
45. Rodondi N, Darioli R, Ramelet AA, et al. High risk for hyperlipidemia and the metabolic syndrome after an episode of hypertriglyceridemia during 13-cis retinoic acid therapy for acne: a pharmacogenetic study. Ann Intern Med 2002;136(8):582–9.
46. Koistinen HA, Remitz A, Gylling H, et al. Dyslipidemia and a reversible decrease in insulin sensitivity induced by therapy with 13-cis-retinoic acid. Diabetes Metab Res Rev 2001;17(5):391–5.
47. Tarr PE, Taffe P, Bleiber G, et al. Modeling the influence of APOC3, APOE, and TNF polymorphisms on the risk of antiretroviral therapy-associated lipid disorders. J Infect Dis 2005;191(9):1419–26.
48. Bollens D, Guiguet M, Tangre P, et al. Major hypertriglyceridemia in HIV-infected patients on antiretroviral therapy: a role of the personal and family history. Infection 2004;32(4):217–21.
49. Fauvel J, Bonnet E, Ruidavets JB, et al. An interaction between apo C-III variants and protease inhibitors contributes to high triglyceride/low HDL levels in treated HIV patients. AIDS 2001;15(18):2397–406.
50. Wong SF, Jakowatz JG, Taheri R. Management of hypertriglyceridemia in patients receiving interferon for malignant melanoma. Ann Pharmacother 2004;38(10):1655–9.
51. Meyer JM, Koro CE. The effects of antipsychotic therapy on serum lipids: a comprehensive review. Schizophr Res 2004;70(1):1–17.
52. Sampath H, Ntambi JM. Polyunsaturated fatty acid regulation of gene expression. Nutr Rev 2004;62(9):333–9.
53. Kris-Etherton PM, Harris WS, Appel LJ. Fish consumption, fish oil, omega-3 fatty acids, and cardiovascular disease. Circulation 2002;106(21):2747–57.
54. Djousse L, Hunt SC, Arnett DK, et al. Dietary linolenic acid is inversely associated with plasma triacylglycerol: the National Heart, Lung, and Blood Institute Family Heart Study. Am J Clin Nutr 2003;78(6):1098–102.
55. Marchioli R, Barzi F, Bomba E, et al. Early protection against sudden death by n-3 polyunsaturated fatty acids after myocardial infarction: time-course analysis of the results of the Gruppo Italiano per lo Studio della Sopravvivenza nell'Infarto Miocardico (GISSI)-Prevenzione. Circulation 2002;105(16):1897–903.
56. Capuzzi DM, Guyton JR, Morgan JM, et al. Efficacy and safety of an extended-release niacin (Niaspan): a long-term study. Am J Cardiol 1998;82(12A):74U–81U; discussion 5U–6U.
57. Andersson Y, Majd Z, Lefebvre AM, et al. Developmental and pharmacological regulation of apolipoprotein C-II gene expression. Comparison with apo C-I and apo C-III gene regulation. Arterioscler Thromb Vasc Biol 1999;19(1):115–21.
58. Owczarek J, Jasinska M, Orszulak-Michalak D. Drug-induced myopathies. An overview of the possible mechanisms. Pharmacol Rep 2005;57(1):23–34.
59. Wang JS, Wen X, Backman JT, et al. Effect of albumin and cytosol on enzyme kinetics of tolbutamide hydroxylation and on inhibition of CYP2C9 by gemfibrozil in human liver microsomes. J Pharmacol Exp Ther 2002;302(1):43–9.

60. Shek A, Ferrill MJ. Statin-fibrate combination therapy. Ann Pharmacother 2001;35(7-8):908-17.
61. Jurado J, Seip R, Thompson PD. Effectiveness of ezetimibe in clinical practice. Am J Cardiol 2004;93(5):641-3.
62. Khan MA, St Peter JV, Xue JL. A prospective, randomized comparison of the metabolic effects of pioglitazone or rosiglitazone in patients with type 2 diabetes who were previously treated with troglitazone. Diabetes Care 2002;25(4):708-11.
63. Sakamoto J, Kimura H, Moriyama S, et al. Activation of human peroxisome proliferator-activated receptor (PPAR) subtypes by pioglitazone. Biochem Biophys Res Commun 2000;278(3):704-11.
64. Brousseau ME, Diffenderfer MR, Millar JS, et al. Effects of cholesteryl ester transfer protein inhibition on high-density lipoprotein subspecies, apolipoprotein A-I metabolism, and fecal sterol excretion. Arterioscler Thromb Vasc Biol 2005;25(5):1057-64.
65. Tolentino MC, Ferenczi A, Ronen L, et al. Combination of gemfibrozil and orlistat for treatment of combined hyperlipidemia with predominant hypertriglyceridemia. Endocr Pract 2002;8(3):208-12.

11 Lowering Low-Density Lipoprotein Cholesterol Levels to Reduce Atherosclerotic Coronary Heart Disease Risk

Harold E. Bays, MD, FACP

CONTENTS

> INTRODUCTION
> LDL-C PATHOPHYSIOLOGY
> BENEFITS OF REDUCING LDL-C WITH LIPID-ALTERING
>   DRUGS
> LDL-C TREATMENT GOALS
> STATINS
> EZETIMIBE
> RESINS AND POLYMERS (3)
> COMBINATION CHOLESTEROL ABSORPTION INHIBITOR
>   AND STATIN
> REFERENCES

INTRODUCTION

Dyslipidemia is an important risk factor for the development of atherosclerotic coronary artery disease (CHD). Clinical trial evidence supports that improving dyslipidemia reduces the progression of atherosclerosis, reduces CHD events, reduces CHD morbidity, and reduces CHD mortality, and several CHD outcome studies have demonstrated that therapies directed toward improvement in dyslipidemia significantly reduces overall mortality—due to reductions in CHD deaths *(1)*.

Various lipid-altering drugs are available for the treatment of dyslipidemia (Table 1). Examples of agents that predominantly lower low-density lipoprotein cholesterol (LDL-C) levels include 3-hydroxy-3-methyl-glutaryl coenzyme A reductase inhibitors ("statins"), cholesterol absorption inhibitors [(such as ezetimibe (EZE)] and bile acid binding agents/sequestrants (such as polymers and resins), as well as combination LDL-C lowering drugs such as the combination pill of EZE and statin.

From: *Contemporary Cardiology: Therapeutic Lipidology*
Edited by: M. H. Davidson, P. P. Toth, and K. C. Maki © Humana Press Inc., Totowa, NJ

Table 1
Atherosclerotic Coronary Heart Disease Event Risk Reduction: Lipid-Altering Therapies and Their General Effects Upon Lipid Blood Levels

Lipid-altering agent	Change in LDL-C (%)	Change in triglyceride (%)	Change in HDL-C (%)
Statins	↓ 18–55	↓ 7–30	↑ 5–15
Nicotinic acid (niacin)	↓ 5–25	↓ 20–50	↑ 15–35
Fibric acids (fibrates)	↓ 5–20[a]	↓ 20–50	↑ 10–20
Ezetimibe	↓ 17–22	↓ 4–11	↑ 2–5
Bile acid sequestrants	↓ 15–30	No change to increased	↑ 3–5
Fish oils	No change to increased	↓ 20–50	No change to increased
Phytosterols/phytostanols	↓ 10–15	No change to decreased	No change to increased

[a] Fibrates may increase LDL-C blood levels in some patients with hypertriglyceridaemia.
HDL-C, High density lipoprotein cholesterol; LDL-C, Low density lipoprotein cholesterol.
Adapted from Bays and Stein (2)

LDL-C PATHOPHYSIOLOGY

The cholesterol associated with LDL particles originates from predominantly two sources: synthesis (from liver and peripheral tissues) and absorption (from the intestine) (2–4) (Fig. 1). Irrespective of cholesterol origin, it is the liver that is the central regulatory organ that largely determines circulating LDL-C blood levels (Figs. 1 and 2). Decreased hepatic cholesterol synthesis (as occurs with drugs that impair hepatic cholesterol synthesis such as statins) and/or decreased intestinal cholesterol delivery to the liver (as occurs with drugs that impair cholesterol absorption and/or transport from the intestine to the liver such as EZE) (Figs. 1 and 2) all increase hepatic LDL receptor activity, which, in turn, increases clearance of LDL-C from the blood, reduces circulating LDL-C, and thus decreases the pathologic contribution of this CHD risk factor to atherogenesis.

BENEFITS OF REDUCING LDL-C WITH LIPID-ALTERING DRUGS

Most primary and secondary CHD outcome trials have demonstrated that lipid-altering drugs reduce CHD events (Table 2) (Fig. 3), even in patients with high CHD risk, such as those with diabetes mellitus (Tables 3 and 4).

LDL-C TREATMENT GOALS

Lowering the LDL-C level is currently the primary lipid treatment target toward the goal of reducing CHD risk (32,33). Dietary changes may result in modest improvement in LDL-C levels (34). But for the majority of patients with CHD, or at significant risk for CHD, drug therapy is often required. Expert panels such as the National Cholesterol Education Program (NCEP) and Adult Treatment Panel III (ATP III) have recommended the most aggressive LDL-C treatment goals for patients at highest CHD risk (Table 5).

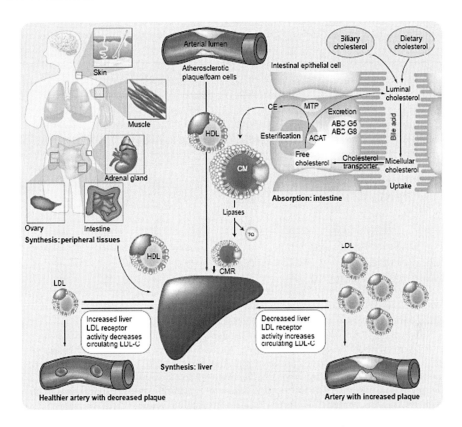

Fig. 1. Sources of cholesterol—synthesis (peripheral tissue and liver) and intestinal absorption. Cholesterol associated with atherosclerotic plaques is transported to the liver via HDL particles. ABC: Adenosine triphosphate-binding a Coenzyme A: cholesterol acyltransferase; CE: Cholesterol ester; CM: Chylomicron; CMR: Chylomicron remnant; HDL: High-density lipoprotein; MTP: Microsomal triglyceride transfer protein. Reproduced from Bays et al. *(5)*.

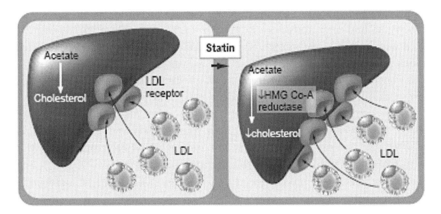

Fig. 2. Increase in low-density lipoprotein (LDL) receptor activity with statins. LDL receptor activity largely determines the concentration of circulating LDL particles, and thus largely determines LDL cholesterol (LDL-C) blood levels. A decrease in cholesterol synthesis (as might occur with statin use) or a decrease in cholesterol absorption (not shown) may upregulate LDL receptor activity, increasing LDL particle clearance, and lowering LDL blood levels. Reproduced from Bays et al. *(5)*.

Table 2
Atherosclerotic Coronary Heart Disease Event Risk Reduction in Some Illustrative Outcome Trials[a]

Trial	Drug	n	Events (n) Control group	Events (n) Statin group	Risk reduction (%)[b]	Events not avoided (%)	Ref.
4S	Simvastatin						[2]
WOSCOPS	Pravastatin						[3]
CARE	Pravastatin	30,817	2042	1490	26	74	[4]
AFCAPS	Lovastatin						[5]
LIPID	Pravastatin						[6]
HPS	Simvastatin	20,536	1212	898	26	74	[7]
PROSPER	Pravastatin	5804	356	292	19	81	[10]
ASCOT-LLA	Atorvastatin	10,305	154	100	36	64	[8]
Total		67,462	3764	2780	27	73	

[a] Nonfatal MI/CHD death; AFCAPS also included unstable angina.
[b] Weighted average. 4S, Scandinavian Simvastatin Survival Study; AFCAPS, Air Force Coronary Atherosclerosis Prevention Study; ASCOT-LLA, Anglo-Scandinavian Cardiac Outcomes Trial-Lipid Lowering Arm; CARE, Cholesterol and Recurrent Events trial; CHD, coronary heart disease; HPS, Heart Protection Study; LIPID, Long-Term Intervention with Pravastatin in Ischaemic Disease study; MI, myocardial infarction; PROSPER, PROspective Study of Pravastatin in the Elderly at Risk of vascular disease trial; WOSCOPS, West Of Scotland Coronary Prevention Study.
Reproduced from Bays (6)

STATINS

Statins (Fig. 4) inhibit 3-hydroxy-3-methyl-glutaryl coenzyme A reductase (Fig. 5), which is the rate-limiting step of cholesterol synthesis, which, in turn, increases LDL receptor activity and thus reduces circulating LDL-C levels (Figs. 1 and 2). Statins primarily lower LDL-C levels (Fig. 6), although they do have favorable effects upon triglycerides and HDL-C levels (Table 6). Statins have also been shown in both primary and secondary outcome trials to reduce CHD risk (Fig. 3) (Tables 2, 3, and 4). Statins are generally well-tolerated, but do have some potential adverse effects (such as myalgias, myopathy, and elevated liver enzymes) that require monitoring (Table 7). Due to the potential of increasing the risk of muscle toxicity (such as myopathy and rhabdomyolysis), special considerations are indicated when considering the use of statins with fibrates (Table 8). All statins are pregnancy category X drugs and should not be used in women who are pregnant or who may potentially become pregnant.

EZETIMIBE

EZE is a cholesterol absorption/transport inhibitor (Figs. 1 and 7). EZE inhibits intestinal cholesterol absorption with decreased intestinal delivery of cholesterol to the liver, increased expression of hepatic LDL receptors, and thus decreased circulating LDL-C levels (Fig. 1).

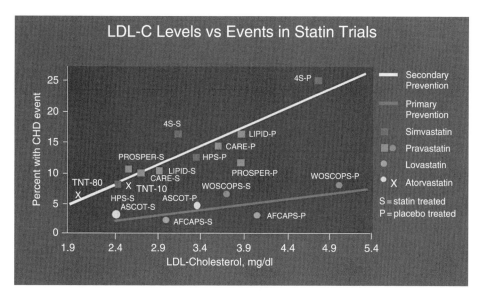

Fig. 3. Statin primary and secondary prevention trials. Reproduced from Illingworth DR. Med Clin North Am. 2000; 84: 23–42.
Note: 4S *(7)*, AFCAPS *(8)*, ASCOT *(9)*, CARE *(10)*, HPS *(11)*, LIPID *(12)*, WOSCOPS *(13)*, TNT *(14)*, PROSPER *(15)*.

Table 3
Primary Prevention Trials of Lipid–Altering Therapy Including Patients with Diabetes

Trial	Diabetic[a] (n)	Total number in study	Lipid-altering drug mg/day	CHD[a] Risk versus placebo in diabetic patients (%)
CARDS[b]	2,838	2,838	Atorvastatin 10	−37 (p = .001)
AFCAPS	155	6,605	Lovastatin 20–40[c]	−44 (NS)
HPS[d]	2,912	7,150	Simvastatin 40	−33 (p = .0003)
ASCOT	2,532	10,305	Atorvastatin 10	−16 (NS)
PROSPER	623	5,804	Pravastatin 40	+27 (NS)
HHS	135	4,081	Gemfibrozil 1200	−68 (NS)

[a] By history
[b] Prospective trial in diabetic subjects; others are subgroup analyses
[c] Mean 30 mg/day
[d] Type 1 or 2 diabetes
CARDS *(18)*,
AFCAPS *(19)*,
HPS *(20)*,
ASCOT *(21)*,
PROSPER *(22)*,
HHS *(23)*.
Adapted from Bays et al. *(16)*, copied from Bays *(17)*

Table 4
Secondary Prevention Trials of Lipid–Altering Therapy Including Patients with Diabetes

Trial	Diabetic, (n)	Total N in Study	Lipid-Altering Drug, mg/day	CHD[a] risk versus placebo in diabetic Patients, (%)
4S	202[b]	4,444	Simvastatin 20–40	−55 ($P = 0.002$)
Reanalysis	483[c]			−42 ($P = 0.001$)
CARE	586[b]	4,159	Pravastatin 40	−25 ($P = 0.05$)
LIPID	1,077[c]	9,014	Pravastatin 40	−19 (NS)
LIPS[d]	202[b]	1,677	Fluvastatin 80	−47 ($P = 0.04$)
HPS[d]	3,051[b]	13,386	Simvastatin 40	−18 ($P = 0.002$)
4D[e]	1,255[b]	1,255	Atorvastatin 20	−8 (NS)
VA-HIT	769[c]	2,351	Gemfibrozil 1,200	−32 ($P = 0.004$)
DAIS[ef]	418[b]	418	Fenofibrate 200	−23 (NS)

[a] Includes stroke in 4D and VA-HIT
[b] By history
[c] By history or glucose ≥ 126 mg/dL
[d] Type 1 or 2 diabetes
[e] Prospective trial in diabetic subjects; others are subgroup analyses
[f] Angiographic study
4S *(24)*,
4S (reanalysis) *(25)*,
CARE *(26)*,
LIPID *(27)*,
LIPS *(28)*,
HPS *(20)*,
4D *(29)*,
VA-HIT *(30)*,
DAIS *(31)*.
Adapted from Bays et al. *(16)*, copied from Bays *(17)*

EZE is absorbed through the intestine, and its active glucuronide metabolite circulates enterohepatically resulting in delivery back to the intestine, which is its site of action. This enterohepatic circulation at least partially accounts for its 22-h half-life and its limited systemic exposure (Fig. 8) that may limit its potential for drug interactions *(42)*.

Although EZE may somewhat lower triglyceride and raise high-density lipoprotein cholesterol (HDL-C) levels, EZE is mainly an LDL-C lowering drug. Studies have consistently demonstrated about an 18–22% reduction in LDL-C levels, depending on the study and depending on how the baseline lipid levels are assessed *(42)* (Fig. 9).

EZE is generally well-tolerated, and with the exception of cyclosporine, having minimal potential drug interactions (Table 9). The CHD morbidity and mortality effects of EZE monotherapy, or in combination with a statin, have not yet been established. EZE is category C and should only be used in women who are pregnant or who may become pregnant if the benefits are thought to exceed the risks.

Table 5
ATP 2004 Update: LDL-C Therapy by Risk Categories Based on Recent Clinical Trial Evidence

Risk category	LDL-C Goal	Initiate Therapeutic lifestyle changes	Consider drug therapy
High risk CHD or CHD risk equivalents (10-year risk > 20%)	< 100 mg/dl	≥ 100 mg/dl	≥ 100 mg/dl
Very high risk	Optional goal of < 70 mg/dl		
Moderately high risk: ≥ 2 risk factors (10-year risk 10–20%)	< 130 mg/dl	≥ 130 mg/dl	≥ 130 mg/dl (consider drug options if LDL-C 100–129 mg/dl)
	Optional goal < 100 mg/dl		
Moderate risk: ≥ 2 risk factors (10-year risk < 10%)	< 130 mg/dl	≥ 130 mg/dl	> 160 mg/dl
Low risk: ≤1 risk factor	< 160 mg/dl	≥ 160 mg/dl	≥ 190 mg/dl (consider drug options if LDL-C 160–189 mg/dl)

CHD, coronary heat disease; LDL-C, low-density lipoprotein cholesterol.
Adapted from Grundy et al. *(33)*. *Circulation.* 2004; 110:227–239.

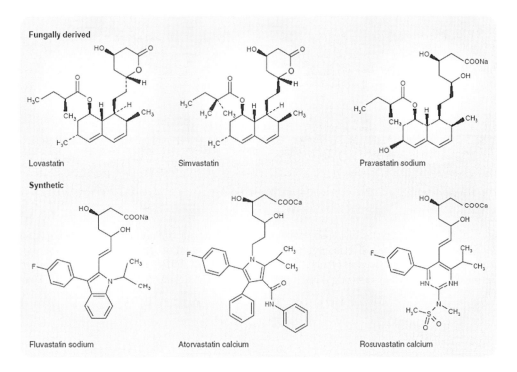

Fig. 4. The statins. Reproduced from Bays et al. *(5)*.

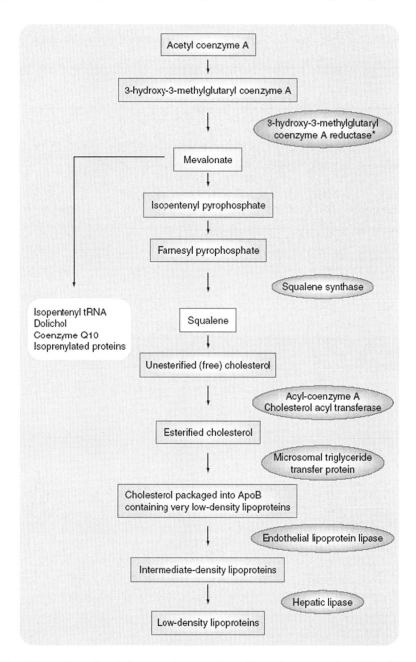

Fig. 5. Simplified schematic of cholesterol synthesis and hepatic apoprotein B (ApoB)-containing lipoprotein assembly. Animal data suggest that marked inhibition of mevalonate production might occasionally result in elevated liver enzymes, myopathy, and inhibition of renal tubular reabsorption of protein. *Rate-limiting step in cholesterol biosynthesis. Coenzyme A = Ubiquinone 10. Reproduced from Bays and Stein *(2)*.

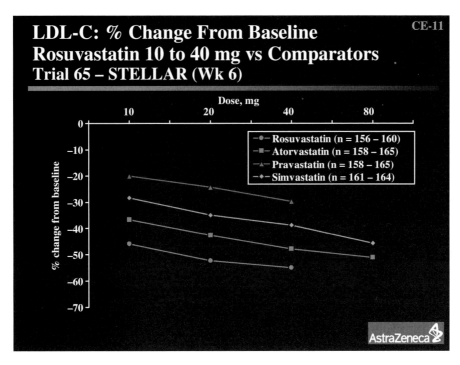

Fig. 6. Percent reduction in low-density lipoprotein cholesterol (LDL-C) with statin therapy *(35)*. $P < 0.001$ versus comparators on a mg-to-mg basis. Data are presented as means.

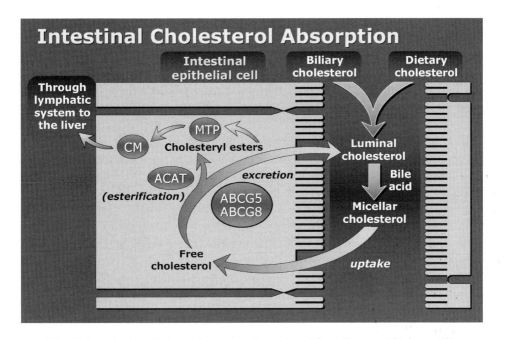

Fig. 7. Intestinal cholesterol absorption. Reproduced from Bays and Dujovne *(3)*.

Table 6
Statin Monotherapy Effects upon Lipid Parameters (36)

Lipid effects of statin

Statin	Total Cholesterol(%)	LDL-C (%)	TG (%)	HDL-C(%)
10 mg doses				
Rosuvastatin	−32.9	−45.8	−19.8	+7.7
Atorvastatin	−27.1	−36.8	−20.0	+5.7
Simvastatin	−20.3	−28.3	−11.9	+5.3
Pravastatin	−14.7	−20.1	−8.2	+3.2
20 mg doses				
Rosuvastatin	−37.6	−52.4	−23.7	+9.5
Atorvastatin	−31.8	−42.6	−22.6	+4.8
Simvastatin	−25.7	−35.0	−17.6	+6.0
Pravastatin	−17.2	−24.4	−7.7	+4.4
40 mg doses				
Rosuvastatin	−40.2	−55.0	−26.1	+9.6
Atorvastatin	−35.8	−47.8	−26.8	+4.4
Simvastatin	−38.8		−14.8	+5.2
Pravastatin	−21.5	−29.7	−13.2	+5.6
80 mg doses				
Rosuvastatin	NA	NA	NA	NA
Atorvastatin	−38.9	−51.1	−28.2	+2.1
Simvastatin	−32.9	−45.8	−18.2	+6.8
Pravastatin	NA	NA	NA	NA

HDL-C, high-density lipoprotein cholesterol; LDL-C, low-density lipoprotein cholesterol; TG, triglyceride. Reproduced from ref. (36).

RESINS AND POLYMERS (3)

Bile acid sequestrants (BASs) were the first cholesterol lowering drugs demonstrated to reduce CHD events. BASs are not significantly absorbed from the intestine into the blood and are thus truly nonsystemic. Mechanistically, the liver routinely produces and delivers bile acids to the duodenum for the purpose of digestion and absorption of cholesterol, fats and fat-soluble vitamins. Once bile acids reach the terminal ileum, 95% are normally reabsorbed and returned to the liver. If intestinal bile acids are bound by BAS, then these bile acids are lost in the feces, resulting in an upregulation of hepatic bile acid synthesis through enzymes such as cholesterol 7α-hydroxylase—the rate-limiting enzyme in bile acid biosynthesis. Because of the increased cholesterol use for bile acid production, hepatic LDL receptors are upregulated and, in turn, increase clearance of LDL-C from the blood.

BAS are predominantly LDL-C-lowering drugs, although they may raise HDL-C and triglyceride/s (TG) levels. Colesevelam HCl is a BAS with a unique polymer structure of that accounts for its improved tolerability and reduced adverse effects and less potential drug interactions, when compared to other BAS such as cholestyramine and colestipol (Table 10). BAS such as colesevelam is pregnancy category B; while

Table 7
Examples of Potential Adverse Effects and Potential Drug Interactions of Statins

Potential adverse effects

- Myalgias with/without elevations in muscle enzymes
- Elevations in (CK) blood levels with or without myalgias
- Myopathy (defined as muscle symptoms and CK elevations > 10 times the upper limits of normal)
- Rhabdomyolysis
- Elevations in liver transaminase blood levels
- Gastritis

Potential drug interactions

- Simvastatin, atorvastatin and lovastatin are metabolized by the hepatic cytochrome P450 (CYP) 3A4 enzyme system, and may have drug interactions if taken with drugs that are inhibitors (competitive or otherwise) of this same CYP 3A4 enzyme system, such as cyclosporin, macrolide antibiotics, protease inhibitors, nefazodone, and azole antifungal agents
- Pravastatin does not undergo significant metabolism through the CYP mixed oxidase system. This results in less potential for drug interactions with drugs that are inducers or inhibitors of the CYP enzyme system. However, pravastatin blood levels may increase when taken concurrently with cyclosporin
- It is sometimes recommended that some statins be administered in lower doses when used concomitantly with amiodarone, verapamil, cyclosporin, fibrates, or niacin. However, the use of statin in combination with the extended-release niacin formulation (as high as niacin 2000 mg and lovastatin 40 mg/day) has been shown to be as safe with regard to liver and muscle enzyme elevation as higher doses of statin alone (such as atorvastatin 40 mg and simvastatin 40 mg) *(40)*
- Fluvastatin blood levels decrease with rifampin. Fluvastatin blood levels increase with glyburide, phenytoin, cimetidine, ranitidine, and omeprazole. Fluvastatin is at least partially metabolized through the CYP 2C9 enzyme system and has the potential to have drug interactions with other drugs that interact with this same enzyme system
- Some statins have been suggested to interact with digoxin and increase the clotting time in patients treated with warfarin
- In general, the use of statin with fibrates may reduce clearance and increase circulating levels of all statins (except perhaps fluvastatin), increasing the risk of myopathy and possibly rhabdomyolysis. This appears to be particularly true when statins are used in combination with gemfibrozil
- Rosuvastatin does not undergo metabolism through the CYP 3A4 enzyme system to a clinically significant extent, resulting in less potential drug interactions than with statins that are significantly metabolized through this enzyme system (such as atorvastatin, simvastatin, and lovastatin). Rosuvastatin is not extensively metabolized at all, with only 10% recovered as metabolite formed through CYP 2C9. Rosuvastatin absorption may be decreased with concurrent antacids, which does not occur if antacids are taken ≥ 2 h after rosuvastatin. As with other statins, rosuvastatin blood levels may increase when taken concurrently with cyclosporin

CK, creatine kinase
Reproduced from ref. *(2)*

Table 8
General Guidelines to Maximize the Safety of Statin/Fibrate Combination Therapy

Before the combined use of statins and fibrates:

- Consider trying to achieve lipid treatment goals with a single lipid-altering agent alone or in combination with dietary compliance, lifestyle changes (such as avoidance of alcohol), regular physical exercise, and appropriate management of secondary causes of dyslipidaemia (such as diabetes mellitus, hypothyroidism, and nephrotic syndrome)
- Consider alternative combination therapies for combined hyperlipidaemia, such as fish oils and niacin (or both) for triglyceride level lowering, added to statin or ezetimibe (or both) for cholesterol level lowering
- Ensure that the potential benefits exceed potential risks
- Management of this treatment regimen should be by clinicians with a sufficient understanding of the relative risks and benefits

Avoid the combined use of statins and fibrates in patients:

- Who do not have readily available access to appropriate health care in the event that muscle symptoms occur, and who are unwilling to comply with routine clinical evaluations, as well as safety and laboratory monitoring
- Who are markedly debilitated, particularly after severe trauma, major surgery, with severe alcoholism, with overall poor or terminal prognosis, or if substantially dehydrated
- Who anticipate a marked increase in the intensity of their exercise regimen, particularly if new or additional muscle groups are to be involved
- With severe renal impairment, significant or active liver disease, or in patients with pretreatment significantly elevated baseline creatine kinase levels
- Treated with numerous concurrent drug therapies, particularly with drugs that are known to significantly interfere with the metabolism of the statin or fibrate

The combined use of statins and fibrates should be used with caution:

- In patients with mild renal insufficiency (as might occur with the elderly) or in patients with pretreatment mild liver enzyme elevation

Once the combined use of statins and fibrates is thought to be clinically indicated:

- Prescribe the lowest doses of statin and fibrates necessary to achieve adequate efficacy, with special attention to package insert recommendations for the statin and fibrate doses when these are to be used concomitantly
- Prescribe one agent at a time, followed by the addition of the other once safety has been established with the initial lipid-altering agent
- Instruct patients of the potential for myopathy and rhabodomyolysis with the combined use of statins and fibrates, as well as to report signs and symptoms of myalgias and/or muscle weakness
- Obtain creatine kinase and liver enzyme testing prior to the addition of the fibrate to the statin (or the statin to the fibrate), and then if symptoms of myopathy or liver enzyme elevation occur
- Discontinue the combined use of statin and fibrate if significant myalgias, muscle weakness, myopathy, rhabdomyolysis or if significant elevations in creatine kinase occur—particularly if otherwise unexplained
- Use fenofibrate instead of gemfibrozil if used in combination with statins

Reproduced from Bays and Stein (2).

Chapter 11 / Lowering Low-Density Lipoprotein Cholesterol Levels

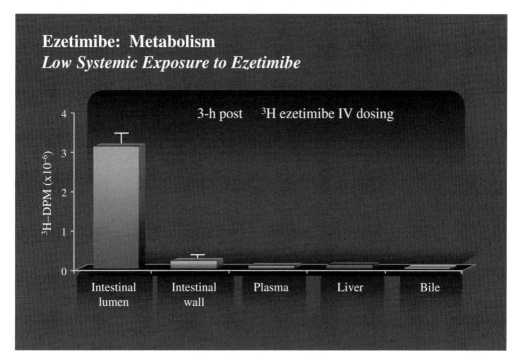

Fig. 8. Low systemic exposure to ezetimibe, even after intravenous dosing *(37)*. Reproduced from Bays *(4)*.

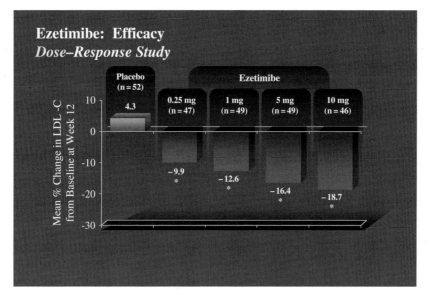

Fig. 9. Low-density lipoprotein cholesterol (LDL-C) lowering efficacy of varying doses of ezetimibe *(38)*. *$P < 0.05$ versus placebo.

Table 9
Examples of Potential Adverse Effects of Ezetimibe

Muscle

Cases of myopathy and rhabdomyolysis have been reported in patients treated with ezetimibe, with most patients having prior treatment with statins before administration of ezetimibe. It is unclear if ezetimibe was the cause of muscle toxicity in these cases.

Rhabdomyolysis has been reported very rarely with ezetimibe monotherapy or when ezetimibe has been added to fibrates. Again, it is unclear if ezetimibe was the cause of this adverse muscle experience. Patients administered ezetimibe should be instructed to promptly report muscle pain, tenderness, or weakness. Ezetimibe should be discontinued (along with any concurrent statin or fibrate) if myopathy is diagnosed or suspected.

Liver enzymes

When ezetimibe was coadministered with a statin, consecutive elevations in hepatic transaminase levels $\geq 3 \times$ upper limits of normal (ULN) were slightly higher (1.3%) than with the administration of statins alone (0.4%). Liver function tests should be performed when ezetimibe is added to statin therapy and according to statin recommendations.

Due to the unknown effects of the increased exposure to ezetimibe in patients with moderate or severe hepatic insufficiency, the ezetimibe is not recommended in these patients.

Potential drug interactions

Caution should be exercised when using ezetimibe in combination with cyclosporine because exposure to both drugs is increased, and such patients should be monitored carefully.

---

no adequate or well-controlled studies have been done in pregnant women, there is no evidence from animal studies that colesevelam harms the fetus.

COMBINATION CHOLESTEROL ABSORPTION INHIBITOR AND STATIN

The use of combination drug therapy, utilizing two lipid-altering drugs with complementary actions, has the potential to improve dyslipidemia. One example of such an approach is the EZE/simvastatin (EZE/SIMVA) combination pill.

While the combination EZE/SIMVA preparation has favorable effects upon various lipid parameters (Table 11), the main benefit of this combination agent is to reduce LDL-C blood levels better than the use of either agent alone (Fig. 10). Because it is a combination agent, its potential adverse effects are similar to the individual lipid-altering drug components (Table 12). The EZE/SIMVA tablet is pregnancy category X, because it contains a statin, and all statins are category X drugs.

Table 10
Examples of Potential Adverse Affects and Drug Interactions with Bile Acid Sequestrants

Potential adverse effects

- The bile acid resins, cholestyramine and colestipol, have a high potential for gastrointestinal effects such as dyspepsia, nausea, and constipation—especially at higher doses
- Compliance with bile acid resins may be impaired by the administration of an insoluble powder that requires mixing into a suspension with fluids or mixing with foods
- Colestipol may also be administered in tablet form, but still has a risk of gastrointestinal side effects and potential drug interactions that have not been shown to differ from the powder form
- Compliance with bile acid resins may also be impaired by the general recommendation that concurrent drugs be administered at least 1 h before, or 2–4 h afterwards
- Bile acid resins may interfere with the absorption of fat soluble vitamins such as A, D, E, and K at extremely high doses
- May cause moderate increases in triglyceride blood levels
- Clinical trials have described mild liver enzyme blood level elevations in some patients
- The bile acid polymer, colesevelam, is significantly better tolerated than the bile acid resin cholestyramine and colestipol

Potential drug interactions

- The bile acid resins cholestyramine and colestipol have been described to potentially impair the absorption of many common, concomitant drugs—particularly anionic, acidic materials. Specifically, cholestyramine has been described to potentially impair the absorption of phenylbutazone, warfarin, thiazide diuretics, propranolol, tetracycline, penicillin G, phenobarbital, phosphate supplements, hydrocortisone, oestrogen, and progesterone, thyroid and thyroxine preparations, and digitalis. Colestipol has been described to potentially impair the absorption of propranolol, thiazide diuretics, digoxin, oral phosphate supplements, and hydrocortisone
- May interfere with the pharmacokinetics of concurrent drugs that undergo enterohepatic circulation such as oestrogen and ezetimibe
- The bile acid polymer colesevelam has much less potential for drug interactions than the bile acid resins cholestyramine and colestipol. However, the potential drug interactions with all concurrent drugs has not been reported (such as with levothyroxine). Therefore, colesevelam should also be taken 1 h before, or 2–4 h after a concurrent drug if a potential drug interaction is suspected

Reproduced from Bays and stein (2)

Table 11
Least Squares Mean Percent Change in Efficacy Parameters from Baseline to End Point Presented According to Individual Treatment Groups

Efficacy parameter	SIMVA 10 mg (n = 150–155)	EZE/SIMVA 10/10 mg (n = 140–151)	SIMVA 20 mg (n = 144–147)	EZE/SIMVA 10/20 mg (n = 140–153)	SIMVA 40 mg (n = 150–154)	EZE/SIMVA 10/40 mg (n = 138–146)	SIMVA 80 mg (n = 150–156)	EZE/SIMVA 10/80 mg (n = 146–154)
LDL-C	−32.7	−44.8[ab]	−34.2	−51.9[ab]	−40.6	−55.2[ab]	−48.5	−60.2[a]
TG[b]	−17.1	−22.5	−18.1	−24.3[d]	−21.2	−22.9	−26.6	−30.7
TC	−23.1	−31.4[ab]	−24.0	−36.3[ab]	−28.9	−39.2[ab]	−34.7	−43.4[a]
HDL-C	5.4	8.0	7.4	9.8	7.5	5.5	7.1	5.6
Non-HDL-C	−30.0	−40.5[ab]	−31.7	−47.4[ab]	−37.5	−50.6[ab]	−44.5	−55.7[a]
LDL-CHDL-C	−35.5	−48.2[ab]	−37.9	−55.6[ab]	−44.0	−57.0[af]	−51.4	−61.5[a]
TCHDL-C	−26.5	−35.8[ab]	−28.3	−41.3[ab]	−33.2	−41.8[af]	−38.4	−45.5[a]
apo B	−26.2	−34.9[ab]	−27.7	−41.1[ab]	−33.2	−44.3[ab]	−39.4	−48.8[a]
apo A-1	4.6	5.5	6.0	6.3	5.7	3.6	4.5	3.9
RLP-C	−22.5	−31.8[e]	−25.6	−40.3[ae]	−31.7	−43.0[e]	−37.5	−47.4[e]
CRP[c]	−6.9	−7.9	−19.8	−34.6[ef]	−14.7	−40.0[ef]	−28.9	−34.3

apo, apolipoprotein; CRP, C-reactive protein; EZE/SIMVA = ezetimibe/simvastatin tablet; HDL-C, high-density lipoprotein cholesterol; LDL-C, low-density lipoprotein cholesterol; n = number of patients with efficacy measurements at end point; RLP-C, remnant-like particle-cholesterol; SIMVA, simvastatin monotherapy; TC, total cholesterol; TG, triglycerides.

[a] $p < 0.001$ EZE/SIMVA versus same dose of SIMVA monotherapy.
[b] $p < 0.001$ EZE/SIMVA versus next highest dose of SIMVA monotherapy.
[c] Median values.
[d] $p = 0.054$ EZE/SIMVA versus same dose of SIMVA monotherapy.
[e] $p < 0.05$ EZE/SIMVA versus next highest dose of SIMVA monotherapy.
[f] $p < 0.05$ EZE/SIMVA versus same dose of SIMVA monotherapy.

Reproduced from ref. (39)

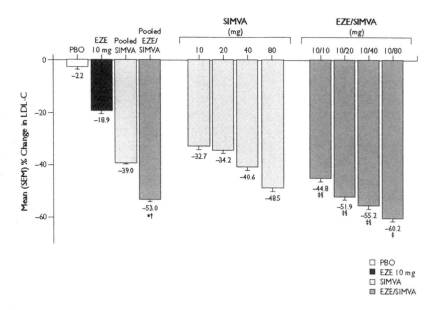

Fig. 10. Reduction in low-density lipoprotein cholesterol (LDL-C) levels with simvastatin versus the ezetimibe/simvastatin tablet. $^*P < 0.0001$ for pooled EZE/SIMVA versus pooled SIMVA, $\dagger P < 0.0001$ for pooled EZE/SIMVA versus EZE 10 mg monotherapy, $\ddagger P < 0.0001$ for EZE/SIMVA versus same dose SIMVA, $\S P < 0.0001$ for EZE/SIMVA versus same next highest dose of SIMVA. PBO, placebo; EZE, ezetimibe 10 mg; pooled SIMVA, simvastatin pooled across doses; pooled EZE/SIMVA, ezetimibe/simvastatin tablet pooled across doses. Reproduced from Bays et al. *(39)*.

Table 12
Potential Adverse Effects of the Ezetimibe/Simvastatin Combination Tablet

General potential adverse effects and possible drug interactions

- The potential adverse effects and possible drug interactions of the ezetimibe/simvastatin combination preparation are essentially what would be expected based upon the individual component drugs.

Skeletal muscle

- Myopathy and rhabdomyolysis are known adverse reactions to statins and other lipid-lowering drugs. The ezetimibe/simvastatin combination tablet contains simvastatin, which has been reported to cause myopathy, manifested as muscle pain, tenderness, or weakness with creatine kinase levels above 10 × upper limits of normal. Myopathy sometimes takes the form of rhabdomyolysis with or without acute renal failure secondary to myoglobinuria, and rare fatalities have occurred.

(Continued)

Table 12
(*Continued*)

- The risk of myopathy is dose related and is increased by higher doses of simvastatin, or higher levels of simvastatin, as might occur through drug interactions with drugs such as itraconazole, ketoconazole, erythromycin, clarithromycin, telithromycin, HIV protease inhibitors, nefazodone, or routine consumption of large quantities of grapefruit juice (> 1 quart daily).
- Similarly, the concomitant use of ezetimibe/simvastatin combination tablet and fibrates should generally be avoided.
- Although little evidence supports that niacin increases the risk of myopathy,[40,2,41] it is currently recommended that the benefit of further alterations in lipid levels by the combined use of the ezetimibe/simvastatin tablet with niacin should be carefully weighed against the potential risks of myopathy.
- The dose of the ezetimibe/simvastatin tablet should not exceed 10/10 mg daily in patients receiving cyclosporine or danazol, and 10/20 mg daily in patients taking amiodarone or verapamil, with doses higher than 10/20 mg daily in combination with amiodarone or verapamil avoided unless the clinical benefit is likely to outweigh the increased risk of myopathy.
- Patients first administered the ezetimibe/simvastatin tablet, or who have their dose increased should be advised of the risk of myopathy and told to promptly report any unexplained muscle pain, tenderness, or weakness, and discontinue therapy immediately if myopathy is diagnosed or suspected.

Liver

- The incidence of consecutive elevations ($\geq 3 \times$ upper limits of normal) in serum transaminases was 1.7% overall and appeared to be dose related, with an incidence of 2.6% for 10/80 mg. In long-term extensions (48 weeks) which included both newly treated and previously treated patients, the incidence was 1.8% overall and 3.6% for 10/80 mg. These elevations were generally asymptomatic, not associated with cholestasis, and reversible whether treatment was maintained or discontinued.
- Liver enzyme testing should be performed at treatment initiation and thereafter when clinically indicated.
- Patients titrated to 10/80 mg should receive an additional liver function test prior to titration, 3 months after titration, and periodically thereafter (e.g., semiannually) during the first year.
- If an increase in aspartate amino transferase or alanine transferase of $\geq 3 \times$ upper limits of normal persists, the drug should be discontinued.
- Due to the unknown effects of the increased exposure to ezetimibe in patients with moderate or severe hepatic insufficiency, the ezetimibe/simvatatin tablet is not recommended in these patients.

REFERENCES

1. Bays HE, McGovern ME. Time as a variable with niacin extended-release/lovastatin vs. atorvastatin and simvastatin. *Prev Cardiol* 2005;8(4):226–33.
2. Bays H, Stein EA. Pharmacotherapy for dyslipidaemia–current therapies and future agents. *Expert Opin Pharmacother* 2003 November;4(11):1901–38.
3. Bays H, Dujovne C. Colesevelam HCl: a non-systemic lipid-altering drug. *Expert Opin Pharmacother* 2003 May;4(5):779–90.
4. Bays HE. Combination lipid-altering drug therapy with statins - an update. www.lipidsonline.org, Accessed April 2005.
5. Bays H, McKenney J, Davidson M. Torcetrapib/atorvastatin combination therapy. *Expert Rev Cardiovasc Ther* 2005 September;3(5):789–820.
6. Bays HE. Extended-release niacin/lovastatin: the first combination product for dyslipidemia. *Expert Rev Cardiovasc Ther* 2004; 2(4):485–501.
7. Randomised trial of cholesterol lowering in 4444 patients with coronary heart disease: the Scandinavian Simvastatin Survival Study (4S). *Lancet* 1994 November 19;344(8934):1383–89.
8. Downs JR, Clearfield M, Weis S, et al. Primary prevention of acute coronary events with lovastatin in men and women with average cholesterol levels: results of AFCAPS/TexCAPS. Air Force/Texas Coronary Atherosclerosis Prevention Study. *JAMA* 1998 May 27;279(20):1615–22.
9. Sever PS, Dahlof B, Poulter NR, et al. Prevention of coronary and stroke events with atorvastatin in hypertensive patients who have average or lower-than-average cholesterol concentrations, in the Anglo-Scandinavian Cardiac Outcomes Trial–Lipid Lowering Arm (ASCOT-LLA): a multicentre randomised controlled trial. *Lancet* 2003 April 5;361(9364):1149–58.
10. Goldberg RB, Mellies MJ, Sacks FM, et al. Cardiovascular events and their reduction with pravastatin in diabetic and glucose-intolerant myocardial infarction survivors with average cholesterol levels: subgroup analyses in the cholesterol and recurrent events (CARE) trial. The Care Investigators. *Circulation* 1998 December 8;98(23):2513–9.
11. MRC/BHF Heart Protection Study of cholesterol lowering with simvastatin in 20,536 high-risk individuals: a randomised placebo-controlled trial. *Lancet* 2002 July 6;360(9326):7–22.
12. Prevention of cardiovascular events and death with pravastatin in patients with coronary heart disease and a broad range of initial cholesterol levels. The Long-Term Intervention with Pravastatin in Ischaemic Disease (LIPID) Study Group. *N Engl J Med* 1998 November 5;339(19):1349–57.
13. Shepherd J, Cobbe SM, Ford I, et al. Prevention of coronary heart disease with pravastatin in men with hypercholesterolemia. West of Scotland Coronary Prevention Study Group. *N Engl J Med* 1995 November 16;333(20):1301–7.
14. LaRosa JC, Grundy SM, Waters DD, et al. Intensive lipid lowering with atorvastatin in patients with stable coronary disease. *N Engl J Med* 2005 April 7;352(14):1425–35.
15. Shepherd J, Blauw GJ, Murphy MB, et al. Pravastatin in elderly individuals at risk of vascular disease (PROSPER): a randomised controlled trial. *Lancet* 2002 November 23;360(9346):1623–30.
16. Bays H, Abate N, Chandalia M. Adiposopathy: sick fat causes high blood sugar, high blood pressure, and dyslipidemia. *Future Cardiol* 2005; 1(1):39–59.
17. Bays HE. Collaborative Atorvastatin Diabetes Study (CARDS). *Lipids Online Commentaries on Research*. http://www.lipidsonline.org, Accessed 24 July 2005.
18. Colhoun HM, Betteridge DJ, Durrington PN, et al. Primary prevention of cardiovascular disease with atorvastatin in type 2 diabetes in the Collaborative Atorvastatin Diabetes Study (CARDS): multicentre randomised placebo-controlled trial. *Lancet* 2004 August 21;364(9435):685–96.
19. Downs JR, Clearfield M, Tyroler HA, et al. Air Force/Texas Coronary Atherosclerosis Prevention Study (AFCAPS/TEXCAPS): additional perspectives on tolerability of long-term treatment with lovastatin. *Am J Cardiol* 2001 May 1;87(9):1074–79.
20. Collins R, Armitage J, Parish S, et al. MRC/BHF Heart Protection Study of cholesterol-lowering with simvastatin in 5963 people with diabetes: a randomised placebo-controlled trial. *Lancet* 2003 June 14;361(9374):2005–16.
21. Sever PS, Dahlof B, Poulter NR, et al. Prevention of coronary and stroke events with atorvastatin in hypertensive patients who have average or lower-than-average cholesterol concentrations, in the Anglo-Scandinavian Cardiac Outcomes Trial–Lipid Lowering Arm (ASCOT-LLA): a multicentre randomised controlled trial. *Lancet* 2003 April 5;361(9364):1149–58.
22. Shepherd J, Blauw GJ, Murphy MB, et al. Pravastatin in elderly individuals at risk of vascular disease (PROSPER): a randomised controlled trial. *Lancet* 2002 November 23;360(9346):1623–30.

23. Koskinen P, Manttari M, Manninen V, et al. Coronary heart disease incidence in NIDDM patients in the Helsinki Heart Study. *Diabetes Care* 1992 July;15(7):820–5.
24. Pyorala K, Pedersen TR, Kjekshus J, et al. Cholesterol lowering with simvastatin improves prognosis of diabetic patients with coronary heart disease. A subgroup analysis of the Scandinavian Simvastatin Survival Study (4S). *Diabetes Care* 1997 April;20(4):614–20.
25. Haffner SM, Alexander CM, Cook TJ, et al. Reduced coronary events in simvastatin-treated patients with coronary heart disease and diabetes or impaired fasting glucose levels: subgroup analyses in the Scandinavian Simvastatin Survival Study. *Arch Intern Med* 1999 December 13;159(22):2661–7.
26. Goldberg RB, Mellies MJ, Sacks FM, et al. Cardiovascular events and their reduction with pravastatin in diabetic and glucose-intolerant myocardial infarction survivors with average cholesterol levels: subgroup analyses in the cholesterol and recurrent events (CARE) trial. The Care Investigators. *Circulation* 1998 December 8;98(23):2513–9.
27. Keech A, Colquhoun D, Best J, et al. Secondary prevention of cardiovascular events with long-term pravastatin in patients with diabetes or impaired fasting glucose: results from the LIPID trial. *Diabetes Care* 2003 October;26(10):2713–21.
28. Serruys PW, de FP, Macaya C, et al. Fluvastatin for prevention of cardiac events following successful first percutaneous coronary intervention: a randomized controlled trial. *JAMA* 2002 June 26;287(24):3215–22.
29. Wanner C, Krane V, Marz W, et al. Atorvastatin in patients with type 2 diabetes mellitus undergoing hemodialysis. *N Engl J Med* 2005 July 21;353(3):238–48.
30. Rubins HB, Robins SJ, Collins D, et al. Diabetes, plasma insulin, and cardiovascular disease: subgroup analysis from the Department of Veterans Affairs high-density lipoprotein intervention trial (VA-HIT). *Arch Intern Med* 2002 December 9;162(22):2597–604.
31. DAIS Investigators. Effect of fenofibrate on progression of coronary-artery disease in type 2 diabetes: the Diabetes Atherosclerosis Intervention Study, a randomised study. *Lancet* 2001 March 24;357(9260):905–10.
32. Third Report of the National Cholesterol Education Program (NCEP) Expert Panel on Detection, Evaluation, and Treatment of High Blood Cholesterol in Adults (Adult Treatment Panel III) final report. *Circulation* 2002 December 17;106(25):3143–421.
33. Grundy SM, Cleeman JI, Merz CN, et al. Implications of recent clinical trials for the National Cholesterol Education Program Adult Treatment Panel III guidelines. *Circulation* 2004 July 13;110(2):227–39.
34. Brunner E, Thorogood M, Rees K, et al. Dietary advice for reducing cardiovascular risk. *Cochrane Database Syst Rev* 2005 October 19;4:CD002128.
35. Jones P, Kafonek S, Laurora I, et al. Comparative dose efficacy study of atorvastatin versus simvastatin, pravastatin, lovastatin, and fluvastatin in patients with hypercholesterolemia (the CURVES study). *Am J Cardiol* 1998 March 1;81(5):582–7.
36. Jones PH, Davidson MH, Stein EA et al. Comparison of the efficacy and safety of rosuvastatin versus atorvastatin, simvastatin, and pravastatin across doses (STELLAR* Trial). *Am J Cardiol* 2003 July 15;92(2):152–60.
37. van HM, Farley C, Compton DS, et al. Comparison of the activity and disposition of the novel cholesterol absorption inhibitor, SCH58235, and its glucuronide, SCH60663. *Br J Pharmacol* 2000 April;129(8):1748–54.
38. Bays HE, Moore PB, Drehobl MA, et al. Effectiveness and tolerability of ezetimibe in patients with primary hypercholesterolemia: pooled analysis of two phase II studies. *Clin Ther* 2001 August;23(8):1209–30.
39. Bays HE, Ose L, Fraser N, et al. A multicenter, randomized, double-blind, placebo-controlled, factorial design study to evaluate the lipid-altering efficacy and safety profile of the ezetimibe/simvastatin tablet compared with ezetimibe and simvastatin monotherapy in patients with primary hypercholesterolemia. *Clin Ther* 2004 November;26(11):1758–73.
40. Bays HE, Dujovne CA, McGovern ME, et al. Comparison of once-daily, niacin extended-release/lovastatin with standard doses of atorvastatin and simvastatin (the ADvicor Versus Other Cholesterol-Modulating Agents Trial Evaluation [ADVOCATE]). *Am J Cardiol* 2003 March 15;91(6):667–72.
41. Bays HE. Commentary: Trial finds that simvastatin plus niacin is safe in people with coronary artery disease and low HDL cholesterol. *Evid Based Cardiovasc Med* 2004;8:173–6.
42. Bays H. Ezetimibe. *Expert Opin Investig Drugs* 2002 November;11(11):1587–604.

12 Lipoprotein(a) as an Emerging Risk Factor for Atherothrombosis

Principles from Bench to Bedside

Michael B. Boffa, PhD,
Santica M. Marcovina, PhD, ScD
and Marlys L. Koschinsky, PhD

CONTENTS

PERSPECTIVES
STRUCTURE OF LIPOPROTEIN(a)
   AND APOLIPOPROTEIN(a): FUNCTIONAL IMPLICATIONS
DETERMINATION OF PLASMA LP(a) LEVELS
MODULATION OF PLASMA LIPOPROTEIN(a) LEVELS
LP(a) AS AN EMERGING RISK FACTOR FOR VASCULAR
   DISEASE: CLINICAL PERSPECTIVES
THE RELATIONSHIP BETWEEN APO(a) ISOFORM SIZE
   AND CARDIOVASCULAR DISEASE RISK
PROPOSED MECHANISMS OF LP(a) PATHOGENICITY
CHALLENGES IN LP(a) MEASUREMENT
PROGRESS IN THE STANDARDIZATION OF LP(a)
   MEASUREMENT
NON-IMMUNOLOGICALLY BASED METHODS FOR LP(a)
   MEASUREMENT
THE IMPACT OF LP(a) METHOD INACCURACY
   ON THE INTERPRETATION OF LP(a) VALUES
SUGGESTED USE OF LP(a) IN CLINICAL PRACTICE
REFERENCES

PERSPECTIVES

Since its initial identification as an antigenic variant of human low-density lipoprotein (LDL), lipoprotein(a) [Lp(a)] has been extensively studied both with respect to basic research investigation into the mechanism of action of this unique lipoprotein in the process of atherothrombosis and its epidemiological association with a variety of vascular diseases. It has been estimated that approximately 37% of the US population judged to be at high risk for developing coronary heart disease (CHD), based on the National Cholesterol Education Program (NCEP) guidelines, has elevated plasma levels of Lp(a), whereas Lp(a) is elevated in only 14% of those judged to be at low risk *(1)*. Therefore, the importance of establishing a better understanding of the relative contribution of Lp(a) to the risk burden for CHD and other forms of vascular disease, as well as the underlying mechanisms, is clearly evident.

Despite extensive study in this area, the pathophysiological and physiological roles of Lp(a) continue to elude basic researchers and clinicians alike. Indeed, Lp(a) is a challenging lipoprotein to study as it has a complex structure consisting of an LDL-like moiety to which is covalently attached the unique glycoprotein apolipoprotein(a) [apo(a)]. Apo(a) contains multiply repeated kringle (K) domains which are similar to a sequence found in the fibrinolytic proenzyme plasminogen; differing numbers of K sequences in apo(a) give rise to Lp(a) isoform size heterogeneity. In addition to elevated plasma concentrations of Lp(a), apo(a) isoform size has been identified as a risk factor for CHD although studies addressing this relationship have been limited. The similarity of Lp(a) to LDL and plasminogen provides an enticing link between the processes of atherosclerosis and thrombosis although a clear demonstration of this association in vivo has not been provided. Clearly, Lp(a) is a risk factor for both atherothrombotic and purely thrombotic events; a plethora of mechanisms to explain these clinical findings has been provided by both in vitro studies and animal models for Lp(a). Challenges inherent in the measurement of Lp(a) have complicated the interpretation of epidemiological studies and have limited the determination of Lp(a) levels for risk assessment in clinical practice. Implementation of recent advances that have been made in the standardization of Lp(a) measurement should facilitate measurement of Lp(a) in the clinic. Although there are currently no methods available to lower plasma Lp(a) levels, screening of individuals at high risk for cardiovascular disease is recommended such that physicians can more aggressively treat modifiable risk factors in individuals with high Lp(a) concentrations.

STRUCTURE OF LIPOPROTEIN(a) AND APOLIPOPROTEIN(a): FUNCTIONAL IMPLICATIONS

Lp(a) is structurally similar to LDL both in protein and in lipid composition, but is distinguishable from LDL by the presence of the unique glycoprotein moiety called apo(a) (Fig. 1). It has been determined that Lp(a) particles contain apo(a) and apolipoprotein B-100 (apoB) in a 1:1 molar ratio *(2)*. In the Lp(a) particle, apo(a) is covalently linked to apoB by a single disulfide bridge *(3)*; the cysteine residues in each molecule that are involved in the disulfide linkage have been identified *(3,4)*.

A two-step model for Lp(a) particle assembly has been developed (Fig. 2) in which noncovalent interactions are required for the initial association of apo(a) and apoB *(5,6)* and precede specific disulfide bond formation. The noncovalent interactions are sensitive to the addition of lysine and lysine analogs such as ε-aminocaproic acid *(7)*. It has been demonstrated that the rate of formation of the disulfide bond is dependent on the conformational status of apo(a) (the molecule can exist in either closed or open forms), which implies the possibility for regulation of the second step of Lp(a) particle formation *(8)*. Recent studies have revealed that specific sequences in the amino-terminus of apoB are required for its noncovalent interaction with apo(a) *(7)*,

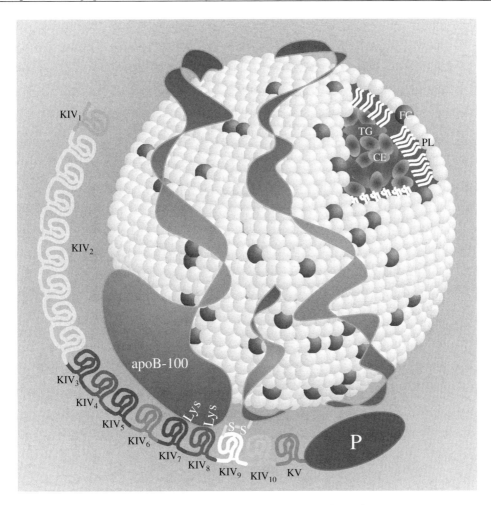

Fig. 1. Schematic representation of the structure of lipoprotein(a) [Lp(a)]. Lp(a) consists of an low-density lipoprotein (LDL)-like moiety covalently linked to apolipoprotein(a) [apo(a)] by a single disulfide bond. The LDL moiety is composed of a central core of triglycerides (TG) and cholesteryl esters (CE) and is surrounded by phospholipids (PL), free cholesterol (FC) and a single molecule of apoB-100. Apo(a) contains 10 copies of kringle (K) domains that are similar to plasminogen KIV; the KIV type 2 domain is present in varying numbers of copies that gives rise to Lp(a) isoform size heterogeneity. Also shown is the lysine-dependent noncovalent interaction between apoB-100 and apo(a) KIV types 7 and 8, which mediates the first step of Lp(a) particle formation P, protease-like domain.

while covalent bond formation involves sequences within the C-terminal of the apoB molecule *(4)*. This has been confirmed by scanning atomic force microscopy *(9)* which demonstrated that apo(a) contacts apoB at two distinct sites. The majority of studies to date suggest that the association of apo(a) and apoB to form Lp(a) particles occurs extracellularly *(3,10,11)*, perhaps on the plasma membrane of hepatocytes *(12)*.

Given that treatment of Lp(a) with reducing agents yields a lipoprotein particle that is essentially indistinguishable from LDL in its physical properties, the majority of the functionalities that have been identified for Lp(a) can be attributed to the presence of apo(a). Apo(a) is a carbohydrate-rich protein *(13)* which is synthesized primarily by the

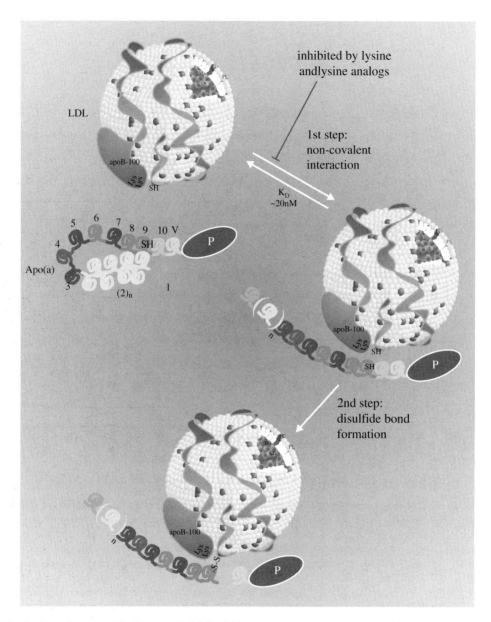

Fig. 2. Two-step model for lipoprotein(a) [Lp(a)] assembly. The first step in Lp(a) assembly involves a non-covalent interaction between apolipoprotein(a) [apo(a)] and apoB-100. This relatively high-affinity ($K_D \sim 20\,nM$) interaction has been shown to involve binding between lysine residues in the amino-terminus of apoB-100 and weak lysine-binding sites in apo(a) kringle (K)IV types 7 and 8 and also involves apo(a) KIV type 9 and carboxyl-terminal domains in apoB-100. The noncovalent step can be inhibited by lysine, lysine analogs such as ε-aminocaproic acid, and other amino acids including phenylalanine and proline. The second step involves specific disulfide bond formation between apo(a) and apoB-100. This step occurs extracellularly and is regulated, in part, by the conformational status of apo(a) and may be sensitive to apo(a) isoform size. LDL, low-density lipoprotein; Lys, lysine; SH, free sulfhydryl; P, protease-like domain.

liver *(14)*. A landmark study by McLean and colleagues *(15)* reported the characterization of a human liver apo(a) cDNA; analysis of the sequence demonstrated that apo(a) shares a high degree of homology with the fibrinolytic proenzyme plasminogen. This homology gave rise to the suggestion that Lp(a) may functionally link the processes of atherosclerosis and thrombosis, conferred by the LDL-like and plasminogen-like constituents of Lp(a) respectively.

Plasminogen contains one copy each of five structural domains called kringles (Ks), designated as KI to KV. The carboxyl-end of plasminogen contains a trypsin-like protease domain that can be cleaved by urokinase plasminogen activator (uPA) or tissue-type plasminogen activator (tPA) to generate the active enzyme plasmin. Apo(a) contains multiple-repeated copies of a sequence that closely resembles plasminogen KIV, followed by sequences which bear a high degree of sequence similarity to the KV and protease domains of plasminogen (Fig. 3) *(15)*. Interestingly, the protease-like domain of apo(a) is catalytically inactive, owing to an amino acid substitution at the uPA/tPA activation site as well as partial deletion of a critical loop which may collapse the active site *(16)*.

Apo(a) contains 10 distinct classes of plasminogen KIV-like domains [designated apo(a) KIV types 1–10] which can be distinguished from each other on the basis of amino acid sequence *(15)* (Fig. 4). It has been demonstrated that apo(a) KIV types 1 and 3–10 (as well as KV and the protease-like domain) are present in a single copy in all apo(a) molecules *(17)*, whereas the apo(a) KIV type 2 (KIV_2) sequence is present in repeated copies that vary in number from 3 to > 30 (Fig. 4); this reflects differing numbers of copies of the sequence encoding KIV_2 in the apo(a) gene *(18)*. The variable number of apo(a) KIV_2 repeats confers size heterogeneity to Lp(a) particles, with the molecular mass of apo(a) in these particles ranging from < 200 kDa to > 800 kDa in the population.

Distinct functionalities have been ascribed to some of the apo(a) K types. For example, apo(a) KIV_9 contains an unpaired cysteine residue involved in disulfide linkage to form covalent Lp(a) particles *(3)*. Apo(a) KIV_{10} contains a strong lysine-binding site (LBS) similar to the one found in plasminogen KIV *(19)*; the LBS in apo(a) KIV_{10} may mediate lysine-dependent interactions of Lp(a) with biological substrates such as fibrin. Apo(a) KIV_5–KIV_8 each contain weak LBS *(20,21)* (Fig. 5); the weak LBS in KIV_7 and KIV_8

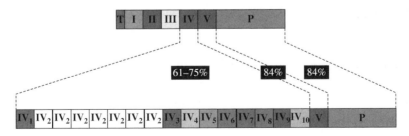

Fig. 3. Relationship between apolipoprotein(a) [apo(a)] and plasminogen. Plasminogen consists of an amino-terminal tail region (T) followed by five different kringle (K) domains (denoted by the Roman numerals I–V) and a trypsin-like protease domain (P). Apo(a) lacks the tail region and KI to KIII and consists of multiple copies of a sequence resembling plasminogen KIV, followed by domains analogous to the KV and protease domains of plasminogen. There are 10 different types of apo(a) KIV that differ in amino acid sequence. The percent amino acid identity for the respective domains is indicated.

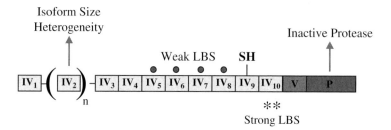

Fig. 4. Kringle (K) organization of apolipoprotein(a) [apo(a)]. Of the 10 different KIV types in apo(a), types 1 and 3–10 are each present in single copy in all apo(a) isoforms. Apo(a) KIV type 2 is present in a variable number of repeated copies, which is the molecular basis of lipoprotein(a) [Lp(a)] isoform heterogeneity. KIV type 10 has been shown to harbor a strong lysine-binding site (LBS) while weaker LBSs are present in KIV types 5–8. There is a single unpaired cysteine present in Kringle IV type 9 (SH) that has been shown to mediate disulfide bond formation with apoB to form covalent Lp(a) particles. The apo(a) protease-like domain cannot be cleaved by plasminogen activators owing to a substitution at the cryptic cleavage site and is thus inactive; the domain also contains a critical 10-amino acid deletion.

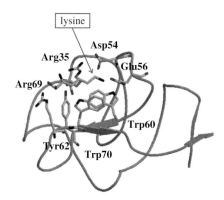

Fig. 5. The structure of a lysine-binding site. Shown is a ribbon diagram of the crystal structure of apo(a) kringle IV type 7 *(121)*, with the side chains of key lysine-binding residues shown. A molecule of L-lysine has been manually docked into the structure. The lysine-binding site consists of an anionic center (comprising Asp54 and Glu56) and a cationic center (comprising Arg35 and Arg69) that fix the side-chain amino and α-hydroxyl groups of the lysine ligand, respectively; the aliphatic side chain of lysine resides in a hydrophobic pocket defined by Trp60, Tyr62, and Trp70.

are required to mediate lysine-dependent noncovalent interactions between apo(a) and apoB which precede specific disulfide bond formation *(6)*. Interestingly, although KIV_2 lacks appreciable lysine-binding ability *(22)*, this domain has been shown to mediate interactions with β-2 glycoprotein I *(23)* and fibullin-5/DANCE *(24)*.

DETERMINATION OF PLASMA LP(A) LEVELS

Lp(a) concentrations in plasma vary over 1000-fold in the human population, ranging from undetectable to greater than 100 mg/dL, and primarily reflect differences in Lp(a) production rather than catabolism of the particle. In Caucasian populations, variability in Lp(a) levels can be largely explained at the level of the gene (approximately

90%), with size of the gene (reflecting the size of the repeated KIV type 2 domain) contributing approximately 60% of the variation *(25)*.

A general inverse correlation has been demonstrated between apo(a) isoform size and plasma Lp(a) levels *(26)*; this may result from the less-efficient secretion of larger apo(a) isoforms from hepatocytes, reflecting at least in part longer retention times in the endoplasmic reticulum, and subsequent quality control-mediated degradation associated with this compartment *(27)*. Additionally, the relationship between apo(a) isoform size and transcript stability, translational efficiency, and the efficiency of assembly into Lp(a) particles has not been studied; these processes may also contribute to the inverse correlation between apo(a) isoform size and plasma Lp(a) concentrations. It is important to note, however, that the relationship between isoform size and corresponding Lp(a) levels is not absolute. There are marked differences between ethnic populations that have been studied in this regard. For example, Black Americans have higher Lp(a) levels associated with mid-sized apo(a) isoform sizes compared to Whites *(28)*, although the molecular basis of this observation remains to be determined. Additionally, it has been reported that in individuals heterozygous for apo(a) isoform sizes, the smaller allele is not consistently dominant; this appeared to depend upon the size of the larger isoform, particularly in Caucasians *(29)*. Interesting evidence is also emerging that within individuals, each allele size affects the level associated with that allele, as well as the level associated with the other allele *(30)*. Taken together, these findings argue strongly for the measurement of allele-specific Lp(a) levels; this involves determining the ratio of intensity of apo(a) isoform sizes in a gel-based assay and the multiplication of this ratio by plasma Lp(a) levels to assign the relative contributions of each isoform to total Lp(a) levels. Elegant studies by Paultre and colleagues *(31)* have demonstrated the predictive value of computing allele-specific Lp(a) values to assess Lp(a) as a risk factor in the African Americans. These investigators demonstrated that elevated levels of Lp(a) associated with small apo(a) isoform sizes independently predict CHD risk in African American as well as in Caucasian men. This clearly demonstrates the value of determining allele-specific Lp(a) levels, as neither measurement of Lp(a) levels nor apo(a) isoform size alone predicts risk attributable to Lp(a) in the African-American population.

MODULATION OF PLASMA LIPOPROTEIN(a) LEVELS

Compared to plasma LDL, Lp(a) levels are relatively resistant to alteration by traditional pharmacologic and nonpharmacologic approaches. This follows from the large contribution of the apo(a) gene to Lp(a) levels that has been previously reported (Determination of Plasma LP(a) Levels above). However, there are reports of several interventions as well as disease conditions such as renal disease and poorly controlled diabetes mellitus which significantly increase plasma Lp(a) levels [for a review, see *(32)*]. The most effective method to decrease Lp(a) concentrations by 50% or more is by LDL apheresis or Lp(a) apheresis procedures *(33,34)*. However, these procedures are costly and are generally reserved for patients with extreme forms of familial hypercholesterolemia.

Several agents have been reported to reduce Lp(a) levels, including niacin in high doses *(35)*, L-carnitine at a dose of 2 g/day *(36)*, and ascorbic acid (3 g/day) together with L-lysine monohydrochloride (3 g/day) *(37)*.

A variety of different hormones have also been shown to modulate plasma Lp(a) concentrations. Androgens, such as danazol and tibolone, were reported to significantly reduce Lp(a) levels *(38–40)*, while estrogen treatment was reported to decrease Lp(a) levels by 50% *(41)*. Both tamoxifen and estrogen have been shown to lower Lp(a) significantly in postmenopausal women *(42)*.

Contradictory findings have been reported concerning the effect of statins on Lp(a) concentrations. A modest increase in Lp(a) concentrations was found in patients with elevated cholesterol receiving simvastatin *(43)*, and a 36% increase in plasma Lp(a) concentrations was reported in patients receiving atorvastatin *(44)*. Use of pravastatin did not influence Lp(a) levels *(45)*, while fluvastatin was found to significantly reduce Lp(a) levels *(46)*. In a study using a large patient cohort, both atorvastatin and simvastatin therapy for 6 weeks resulted in a modest but significant reduction in Lp(a) levels *(47)*. Clearly, the effect of statins on Lp(a) concentrations requires further analysis using large cohorts in order to define the roles of baseline Lp(a) concentration as well as apo(a) isoform size on the magnitude of the statin effect. Additionally, mechanistic studies need to be performed in order to identify the molecular basis for the potential modulation of Lp(a) levels by statin therapy.

Aspirin therapy (81 mg/day) has been reported to lower serum Lp(a) concentrations in a recent study of 70 patients with atherosclerotic disease *(48)*. Interestingly, the magnitude of the decrease in Lp(a) levels was larger in patients with high Lp(a) levels, irrespective of apo(a) isoform size. This has been speculated to result from a greater reduction by aspirin of apo(a) gene transcription in patients with high baseline transcriptional activity of the gene. Clearly, these intriguing results need to be confirmed with a larger patient population, and the mechanism underlying the effect of aspirin on apo(a) gene transcription requires further analysis.

In contrast to plasma LDL concentrations, Lp(a) levels are thought to be relatively resistant to diet and exercise. However, in a study of obese women with high baseline Lp(a) levels, a low calorie diet with concomitant weight loss resulted in a significant reduction in Lp(a) levels *(49)*. In terms of specific dietary effects, a recent randomized cross-over study has reported that almonds, rich in monounsaturated fat, significantly reduced Lp(a) levels *(50)*. Ginsberg and coworkers *(51)* reported a significant increase in Lp(a) levels in individuals who reduced their intake in saturated fat. Therefore, the results of these two studies support the notion that fat intake may lower Lp(a) levels.

Systematic lowering of plasma Lp(a) concentrations can provide us with a study design in which the consequences of Lp(a) reduction can be evaluated prospectively. This will allow a direct assessment of the contribution of Lp(a) concentration to atherosclerotic risk.

LP(a) AS AN EMERGING RISK FACTOR FOR VASCULAR DISEASE: CLINICAL PERSPECTIVES

Based on the Adult Treatment Panel (ATP) III guidelines, Lp(a) is currently classified as an "emerging" lipid risk factor for cardiovascular disease *(52)*. Studies have identified elevated plasma Lp(a) concentration (> 30 mg/dL in many studies) as a risk factor for a variety of atherosclerotic disorders including peripheral vascular disease *(53)*, cerebrovascular disease *(54)*, and premature CHD, the last of which corresponds to the majority of the studies performed to date. The clinical significance of Lp(a) in CHD has been consistently confirmed in numerous retrospective case-control studies performed during the past three decades *(55)*. In essentially all studies of this design, plasma Lp(a) concentrations are elevated in subjects with existing CHD compared to matched controls. These types of studies have been criticized in that patients are selected with existing CHD, which favors the inclusion of subjects with other CHD risk factors. Additionally, retrospective case-control studies cannot distinguish a causative role for Lp(a) from the possibility that Lp(a) might merely be a marker for CHD, or that plasma Lp(a) concentration might increase as a consequence of CHD. On the other hand, the results of the retrospective studies could be interpreted to suggest a strong contribution by Lp(a) to the risk for cardiovascular events.

A smaller number of prospective studies have been performed to more directly assess the contribution of plasma Lp(a) to the future development of CHD. The results from prospective studies have yielded conflicting results, however, ranging from a strong positive association between Lp(a) and CHD, to no association whatsoever *(55–57)*. It is possible that discrepant results may be attributable to variations in study design, collection and storage of samples, methods used for statistical analysis, and population differences reflecting the known ethnic variability in the distribution of Lp(a) levels and isoform sizes. Additionally, the structural heterogeneity of Lp(a) can greatly affect the accuracy of Lp(a) measurement (see section below on Challenges in Lp(a) measurement) *(57)*. Clearly, however, the majority of prospective studies performed over the past decade have demonstrated that elevated plasma Lp(a) concentration is a predictor of CHD. Meta-analysis of 12 prospective studies performed between 1991 and 1997 indicated that plasma Lp(a) concentration is an independent risk factor in both men and women; in the majority of these studies, there was a dose–response relationship between plasma Lp(a) concentration and CHD risk *(58)*. Most recently, meta-analysis of 27 prospective studies provided information on 5436 CHD cases observed during a mean follow-up time of 10 years *(59)*. Despite differences in methods to measure Lp(a), this analysis indicates that individuals in the general population with Lp(a) concentrations in the top third of baseline measurements are at approximately 70% increased risk of CHD compared to individuals in the bottom third.

The predictive strength of Lp(a) for CHD was evaluated in the Atherosclerosis Risk in Communities (ARIC) Study, with 10-year follow-up and 725 CHD events *(60)*. Lp(a) was found to be associated with modest risk ratios. In terms of population differences, the ARIC results suggest that Lp(a) confers less risk in African Americans than in Caucasians. A recent prospective study *(61)* examined a cohort of 1216 patients with a mean follow-up time of 6.7 years with total mortality and mortality due to Cardiovascular disease used as outcome variables. In this study, Lp(a) levels in excess of 30 mg/dL were present in 30% of the study population and were found to be an independent predictor of death. On this basis, the authors suggested that Lp(a) levels in excess of 30 mg/dL is associated with a poor prognosis and may serve to identify patients who would benefit from aggressive secondary prevention programs.

Although there is little or no correlation between plasma levels of Lp(a) and other vascular risk factors, evidence has been provided from a number of studies to suggest that the risk attributable to Lp(a) is dependent upon the concomitant presence of other such risk factors. In the Familial Atherosclerosis Treatment Study (FATS), Lp(a) was a strong predictor of events at baseline, but lost its predictive value when LDL cholesterol (LDL-C) was reduced to < 100 mg/dL in the treatment group *(62)*. More recently, in the Prospective Epidemiological Study of Myocardial Infarction (PRIME) *(63)*, Lp(a) was investigated as a CHD risk factor using a prospective cohort of 9133 French and Northern Irish men age 50–59, without a history of CHD. Elevated Lp(a) increased the risk for myocardial infarction (MI) and angina pectoris, and the effect was most pronounced in men with a high LDL-C level. The results of the Quebec Cardiovascular Study also suggest that Lp(a) is not an independent risk factor for ischemic heart disease in men, but increases the risk associated with elevated apoB and total cholesterol, and appears to attenuate the beneficial effects of elevated high-density lipoprotein (HDL) *(64)*. The same interactions of elevated Lp(a) with other risk factors were found in the Prospective Cardiovascular Munster (PROCAM) study. As a consequence, a high level of Lp(a) further increased the risk of MI in men with high or moderately elevated estimated global risk (i.e., risk of a coronary event > 10% in 10 years) but not in men with a low estimated global risk *(65)*.

In addition to the potential synergy of Lp(a) risk with other markers of dyslipidemia, the interaction of Lp(a) with other thrombotic risk factors has been reported. Indeed, several clinical studies suggest a role for elevated plasma Lp(a) levels in purely thrombotic disorders (i.e., those occurring in the presumed absence of underlying atherosclerosis) *(66)*. For example, Lp(a) has been identified as a risk factor for venous thromboembolism in adults *(67)* and for spontaneous *(68)* and recurrent *(69)* ischemic stroke in children; in all of these cases, the thrombotic episode would be unrelated to the presence of underlying atherosclerotic disease. Moreover, there appear to be interactions between elevated Lp(a) concentrations and established thrombotic risk factors, such as Factor V Leiden (G1691A), and protein C or antithrombin deficiencies for ischemic stroke in children *(68,69)*. There is evidence from one study of adult patients with ischemic stroke to suggest that Lp(a) may be more involved in the thrombotic phase of cardiogenic embolic stroke than in the underlying coronary atherosclerosis *(70)*.

THE RELATIONSHIP BETWEEN APO(a) ISOFORM SIZE AND CARDIOVASCULAR DISEASE RISK

Small apo(a) isoforms, as well as elevated Lp(a) levels, have been associated with vascular disease, but the inverse relationship between Lp(a) levels and apo(a) isoform size has confounded interpretation of such results. A number of studies over the last decade have thus sought to define the role of apo(a) isoform size, independent of plasma Lp(a) concentrations, in CHD risk. For example, in the Bruneck study *(71)*, it was shown that small apo(a) sizes are an independent risk factor for advanced carotid atherosclerosis, although risk is further increased in conjunction with elevated Lp(a) levels. On the other hand, these investigators reported that plasma Lp(a) levels, but not small apo(a) isoform sizes, were predictive of risk for early atherosclerosis, and this association was only present when LDL-C levels were also elevated. Although a relationship between cardiovascular disease and Lp(a) levels associated with small apo(a) isoform size in both White as well as Black men has been documented *(31)*, this association has not been consistently observed in women *(31,72,73)*. Interestingly, there are also conflicting data as to whether elevated Lp(a) levels in women are *(74,75)* or are not *(76,77)* associated with increased risk for cerebrovascular or cardiovascular diseases. Further studies on the role of both Lp(a) concentrations and apo(a) isoform size as risk factors for vascular disease in women are clearly necessary.

Results of a recent study by Wu and coworkers *(78)* suggest that small apo(a) isoform size (< 22 KIV repeats) is associated with lower endothelium-dependent, flow-mediated dilation of the brachial artery irrespective of plasma Lp(a) levels. Additionally, Emanuele and coworkers *(79)* recently reported that among patients with acute coronary syndromes, the percentage of subjects with at least one small apo(a) isoform was significantly higher in those patients who presented with acute MI versus those with unstable angina; small apo(a) isoform, but not elevated Lp(a) levels, was an independent predictor of acute MI versus unstable angina pectoris in a multivariate logistic regression model. Evidence for a role of apo(a) isoform size and risk for the development of angina was also provided by Rifai and colleagues *(80)*. In this study, they demonstrated that while both Lp(a) concentrations and small apo(a) isoforms were associated with risk for angina, only the association between apo(a) size and risk remained significant in a multivariate model.

PROPOSED MECHANISMS OF LP(a) PATHOGENICITY

Despite its recognition as a risk factor for vascular disease, the role of Lp(a) in atherogenesis remains poorly understood. It has been postulated that owing to its duality of structure, Lp(a) may provide a functional link between the processes of atherosclerosis and thrombosis. In this model, Lp(a) likely possesses both atherosclerotic (owing to its similarity to LDL) and prothrombotic properties (based on the homology between apo(a) and plasminogen) (Fig. 6); clearly, apo(a) possesses unique properties that contribute to the process of atherogenesis that are independent of its similarity to plasminogen [reviewed in *(81)*].

It is well documented that Lp(a) is present in the arterial wall at the sites of atherosclerotic lesions and that it accumulates at these sites to an extent that is proportional to plasma Lp(a) levels. Lp(a) is preferentially retained in this milieu, likely by virtue of its ability to bind to a number of arterial wall components including fibrinogen/fibrin, fibronectin, and glycosaminoglycans *(81)*. The localization of Lp(a) within the arterial wall suggests a direct causative role for Lp(a) in the initiation and/or progression of atherosclerosis.

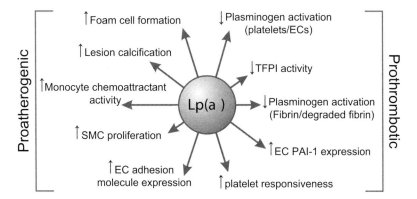

Fig. 6. Potential mechanisms of lipoprotein(a) [Lp(a)] pathogenicity. Lp(a) likely possesses both proatherogenic properties (owing to its similarity to low-density lipoprotein) and prothrombotic properties [owing to the similarity of apolipoprotein(a) (apo(a)) to plasminogen]. This diagram depicts the effects of Lp(a) that have been demonstrated by in vitro studies or in animal models of apo(a)/Lp(a). Proatherogenic mechanisms are shown to the left and prothrombotic mechanisms are shown to the right. It must be stressed that none of these mechanisms have been directly demonstrated to be mediated by Lp(a) in human disease. EC, endothelial cell; PAI-1, plasminogen activator inhibitor-1; SMC, smooth muscle cell; TFPI, tissue factor pathway inhibitor.

Insights into possible roles for Lp(a) in vascular disease have been gained through both in vitro studies and animal models. Many in vitro studies have probed structure-function relationships between different K modules of apo(a) and possible proatherosclerotic and/or prothrombotic mechanisms. These studies, which have been greatly facilitated by the use of recombinant variants of apo(a), have helped to delineate unique contributions of specific K domains and suggest that many functions of apo(a) reside within the C-terminal half of the molecule (Fig. 7) *(81)*. Analysis of apo(a)/Lp(a) function in vitro has revealed the potential for Lp(a) to inhibit fibrinolysis *(66)*, likely through the ability of apo(a) to interfere with the efficient activation of plasminogen to plasmin *(82)*; a similar conclusion was reported using a rabbit jugular vein thrombolysis model *(83)*. It had been generally accepted that apo(a) inhibits plasminogen activation on fibrin through a simple competition mechanism whereby apo(a) and plasminogen compete for the same binding sites on fibrin. However, a recent study by

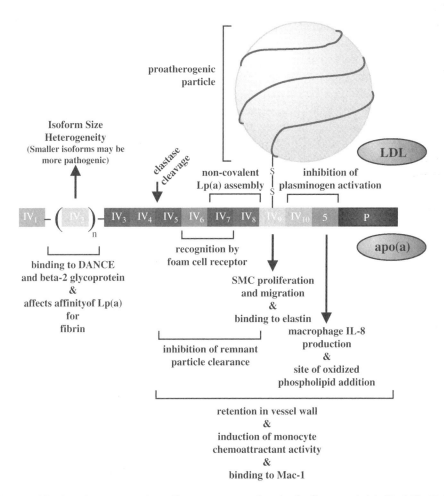

Fig. 7. Specific functions mapped to discrete structural units in lipoprotein(a) [Lp(a)]. Using a combination of the expression of recombinant forms of apolipoprotein(a) [apo(a)] and elastase cleavage of apo(a)/Lp(a), functional domains in apolipoprotein(a) have been identified. Indicated are sequences in apo(a)/Lp(a) that may mediate the promotion of atherosclerosis and inflammation, the inhibition of angiogenesis and fibrinolysis, and Lp(a) assembly. IL-8, interleukin-8.

Hancock and colleagues *(84)* has re-visited the question of the mechanism of inhibition of tPA-mediated plasminogen activation by apo(a)/Lp(a) in the presence of native fibrin and degraded fibrin (FDPs). In this study, a fluorescently labeled, enzymatically inactive form of plasminogen was used which allows plasminogen cleavage to be directly followed within a clot, in the absence of the positive feedback mechanisms normally catalyzed by plasmin. The kinetic data obtained by varying the concentrations of substrate (plasminogen), cofactor (fibrin or FDPs), and apo(a) best fit an equilibrium template model in which apo(a) can interact with all three fibrinolytic components (Fig. 8). Rather than the formation of the ternary complex of tPA, plasminogen, and fibrin as is observed in the absence of apo(a), and which leads to efficient formation of plasmin, a quaternary complex is formed in the presence of apo(a)/Lp(a) that exhibits a reduced turnover number thus resulting in inhibition of plasminogen activation. The equilibrium template model (Fig. 8) takes into account the binding interactions between apo(a)/Lp(a) and tPA, fibrin or FDPs, and plasminogen, all of which have been previously reported *(84)*, and is not consistent with a model in which apo(a) decreases the amount of plasminogen bound to fibrin.

Lp(a) may also possess the ability to inhibit plasminogen activation on the surface of vascular cells, which could contribute to intravascular thrombosis, endothelial dysfunction, and the function of arterial smooth muscle cells. The last possibility is the most well-studied phenomenon, with a role for Lp(a) in stimulating vascular smooth muscle cell proliferation demonstrated both in vitro and in vivo *(85–88)*. It has been shown that these effects arise from decreased plasmin-mediated transforming growth factor-β (TGF-β) activation that occurs as the result of decreased pericellular plasminogen activation in the presence of Lp(a). In a recent study by O'Neil and colleagues *(88)*, the domain responsible for inhibition of vascular smooth muscle cell proliferation and migration and for inhibition of TGF-β activation on these cells has been identified as the apo(a) KIV type 9 domain.

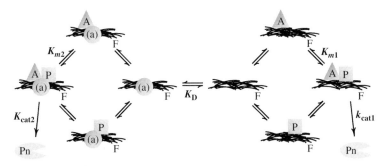

Fig. 8. Equilibrium template model for the inhibition of plasminogen activation by apolipoprotein(a) [apo(a)]. In this model *(66,84)*, apo(a) [(a)], plasminogen (P) and tissue-type plasminogen activator (tPA) (A) bind to independent sites on fibrin. Efficient catalysis to form plasmin (Pn) requires the formation of the ternary complex shown on the far right. Apo(a) itself binds to fibrin and interacts with the catalytic complex to form the quaternary complex on the far left; this complex displays a reduced turnover number (i.e., $k_{cat2} < k_{cat1}$) accounting for the observed inhibition of plasminogen activation. Although K_{m2} was found to be smaller than K_{m1}, this cannot overcome the decrease in k_{cat2} relative to k_{cat1} at physiological concentrations of plasminogen and fibrin. Also depicted is the dissociation constant for apo(a) binding to fibrin (K_D); the model predicts that apo(a)/lipoprotein(a) [Lp(a)] species with higher affinity for fibrin [such as smaller apo(a) isoforms] would be more effective inhibitors of plasminogen activation.

A recent study using human vascular endothelial cells has shown that both apo(a) and Lp(a) can elicit cytoskeletal rearrangements *(89)*. This, in turn, renders the endothelial cells more permeable, which may contribute to a dysfunctional endothelium in vivo. Given that endothelial dysfunction is an early event in the atherosclerotic process, it is interesting to speculate that consistent with the Bruneck study *(71)*, Lp(a) levels, in conjunction with elevated LDL levels, may potentiate this process. In this regard, one could envision that the increased permeability of the endothelial cells, mediated by Lp(a), could facilitate the deposition of LDL in the arterial wall. This process would be enhanced by elevated plasma LDL concentrations, which may explain the dependence of Lp(a) risk in early atherosclerosis on concomitant elevations of plasma LDL and Lp(a), but not apo(a) isoform size, as reported in the Bruneck study. Data suggesting that Lp(a) may contribute to a dysfunctional endothelium in vitro [e.g., by stimulating endothelial cell adhesion molecule expression *(90)* and increasing endothelial cell permeability *(89)*] are supported by a number of studies which demonstrate that elevated plasma Lp(a) concentrations may contribute to endothelial dysfunction in vivo *(78,91)*.

It has recently been reported that the apo(a) component of Lp(a) binds to the β2-integrin Mac-1; this in turn promotes the adhesion of monocytes to endothelial cells and enhances their Mac-1-dependent transendothelial migration *(92)*. Furthermore, the interaction between apo(a) and Mac-1 induces activation of the inflammatory transcription factor Nuclear Factor κB. Taken together, these studies define a novel role for apo(a)/Lp(a) in promoting inflammatory cell recruitment, which may represent a novel mechanism for Lp(a) atherogenicity in vivo. Interestingly, the strength of the Mac-1-dependent monocyte adhesiveness was inversely related to the number of KIV_2 sequences in apo(a). This may contribute to the increased pathogenicity that has been associated with smaller apo(a) isoform sizes in vivo. Taken together, there is evidence to suggest that Lp(a) may contribute to a dysfunctional endothelium that is both prothrombotic and proinflammatory (Fig. 9).

Recent studies have suggested a link between Lp(a) and oxidized phospholipids. Specifically, it has been shown that in human plasma, oxidized phospholipids are preferentially associated with Lp(a) compared to free LDL *(93)*. More recently, it has been demonstrated that percutaneous coronary intervention results in acute elevations of both oxidized LDL and Lp(a) and that oxidized phospholipid is rapidly transferred to and becomes predominantly associated with Lp(a) compared to other apoB-containing lipoproteins *(94)*. This is in keeping with a recent report by Edelstein and colleagues *(95)* which demonstrated that the KV domain in apo(a) itself contains oxidized phospholipids covalently bound to lysine residues present in this K. Interestingly, the KV domain was also implicated in stimulation of the production and secretion of interleukin-8 from cultured human THP-1 macrophages *(96)*; a link with the presence of oxidized phospholipids in apo(a) KV in mediating this effect remains to be determined.

What might be the mechanism(s) underlying a pathogenic role for apo(a) isoform size independent of plasma Lp(a) levels, such as has been documented in several studies *(31,71,78–80)*? Surprisingly, very few mechanistic studies have considered the role of apo(a) size. The exception to this is the effect of apo(a) isoform size on fibrin binding and plasminogen activation to plasmin. In this regard, recent studies indicate that smaller Lp(a) isoforms bind more avidly to fibrin *(97)* and inhibit plasmin formation to greater extents *(98)*. Contradictory evidence has recently been provided, however, suggesting that larger isoforms of Lp(a) as well as larger isoforms of free

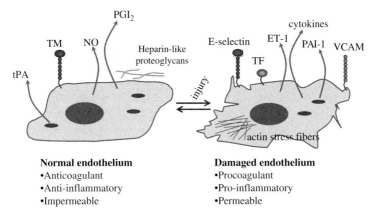

Fig. 9. Endothelial dysfunction, atherogenesis, and lipoprotein(a) [Lp(a)]. Normal endothelium is atheroprotective by virtue of its anti-coagulant and anti-inflammatory properties as well as its function as a permeability barrier between the flowing blood and the extravascular space. In response to an injurious stimulus such as oxidized lipoproteins or turbulent blood flow, however, the endothelium adopts a phenotype that, while necessary to deal with the stimulus, can lead to the promotion of atherosclerosis. Depicted are some of the changes that occur in endothelial damage or dysfunction. Of these, Lp(a) has been shown to promote stress fiber assembly, expression of adhesion molecules such as E-selectin and VCAM, stimulation of PAI-1 biosynthesis, and inhibition of NO release. ET-1, endothelin-1; NO, nitric oxide; PAI-1, plasminogen activator inhibitor-1; PGI$_2$, prostaglandin I$_2$; TF, tissue factor; TM, thrombomodulin; tPA, tissue-type plasminogen activator; VCAM, vascular cellular adhesion molecule.

apo(a) were more effective in reducing plasmin formation on fibrin *(99)*. Clearly, more studies are required in order to understand the mechanism(s) by which small apo(a) isoform size can confer risk independently of plasma Lp(a) concentrations.

Several animal models have been utilized to probe the function of apo(a)/Lp(a) in a more physiological setting *(100)*. However, these studies have been generally complicated by the unusual species distribution of this lipoprotein. Indeed, Lp(a) is only present in humans, Old World Monkeys and the European hedgehog. This has called into question the applicability of using animals such as mice and rabbits as models for Lp(a) pathogenicity given that the apo(a) gene is not present in these species. However, the balance of opinion is that single and double transgenic animals [i.e., overexpressing both human apo(a) and human LDL] can be useful tools to understand the role of Lp(a) in atherosclerosis. Indeed, transgenic apo(a) mouse and rabbit models have been used to study processes such as Lp(a) assembly, structure-function relationships in Lp(a), the regulation of expression of the apo(a) gene, and mechanisms of Lp(a) involvement in the atherosclerotic process *(100)*.

Studies using apo(a) transgenic [non-Watanabe heritable hyperlipidemic (WHHL)] rabbits are in agreement with studies in transgenic mice in that apo(a) deposition was coincident with the presence of accumulated intimal smooth muscle cells and decreased active TGF-β *(87,101)*. There was enhanced staining for markers of activated or immature smooth muscle cells in the transgenic rabbit study, suggesting that Lp(a) might modulate smooth muscle cell phenotype by promoting dedifferentiation *(101)*. A somewhat different effect of Lp(a) on smooth muscle cell phenotype was noted in WHHL transgenic apo(a) rabbits *(102)*. The advanced, complex lesions observed in the transgenic animals showed notable calcification, unlike the less-advanced lesions in the non-transgenic WHHL rabbits; examination of advanced human lesions also showed association of Lp(a) deposition with areas of calcification *(102)*. Interestingly, it was shown that Lp(a) promotes calcification of cultured smooth muscle cells, as evidenced by a stimulation of calcium uptake, promotion of an osteogenic pattern of protein expression, and promotion of an osteoblast-like phenotype (i.e., upregulated osteoblast-specific factor-2 and alkaline phosphatase activity) *(102)*. The role of apo(a) in aortic calicification remains unclear, however, particularly in light of several recent reports that suggest no relationship between coronary calcium and either Lp(a) concentrations or apo(a) isoform sizes *(103)*.

Two very recent studies using transgenic apo(a) mice have been published which may shed light on the role of Lp(a) in vivo. In the first study, mice expressing high and low levels of apo(a) (approximately 35 mg/dL and 700 mg/dL, respectively) in a transgenic human apoB background were used *(104)*. It was found that high levels of oxidized phospholipids were present in Lp(a) from the high apo(a)-expressing mice, but not in LDL from mice with human apoB alone. This likely results from preferential transfer of oxidized phospholipids to Lp(a) as has been previously suggested to occur in human plasma (see section above on proposed mechanisms of Lp(a) Pathogenicity). Although the significance of this finding in the context of how Lp(a) contributes to atherosclerosis is unclear, one can speculate that the deposition of Lp(a) containing these oxidized phospholipids in the developing lesion may contribute to both proatherosclerotic and proinflammatory processes, an effect which would be underscored by the preferential retention of Lp(a) in this milieu. This study is the first to report the use of transgenic mice expressing high levels of Lp(a); apo(a) and Lp(a) levels in previous transgenic models have been over an order of magnitude lower.

In the second study, Devlin and colleagues *(105)* reported the overexpression of a fragment of apo(a) (containing KIV types 5–8). Compared to control animals, these mice had greatly enhanced atherosclerosis and markedly elevated non-HDL cholesterol. Accordingly, these investigators found, using a perfused mouse liver model, that this four-K apo(a) species, as well as full-length apo(a) containing 17 Ks, inhibited the clearance of cholesterol-rich remnant particles. The molecular basis of this observation is not clear, but may involve binding of apo(a) to either the remnant particles themselves, their cognate receptors in the liver, or the heparin sulfate proteoglycans and/or hepatic lipase that sequester the remnants prior to uptake *(106)*.

CHALLENGES IN LP(a) MEASUREMENT

For elevated Lp(a) to be considered a "major" risk factor (i.e., in the same category as cigarette smoking, elevated LDL-C, and so on) it has to meet the following criteria: (1) robust predictive power for cardiovascular disease; (2) high prevalence of Lp(a) levels above an arbitrary risk threshold in the general population; (3) ready availability of clinical samples and a widely available, standardized, inexpensive means to measure Lp(a); and (4) evidence for a benefit of lowering Lp(a) levels *(57)*. For these reasons, it is essential to define stringent criteria for Lp(a) assay standardization as the predictive power of Lp(a) can only be fully explored if measurements are made with methods that have been evaluated for their ability to produce accurate Lp(a) values.

Measurement of Lp(a) in plasma samples has almost exclusively relied on immunological methods including ELISA, nephelometry, and immunoturbidimetry. The high degree of size heterogeneity of apo(a), derived from the variable number of KIV_2, the association of apo(a) with apoB in the Lp(a) particle, and the high degree of sequence similarity between apo(a) and plasminogen all constitute challenges to the measurement of Lp(a) using these methods. Moreover, the interpretation of results from clinical studies addressing the role of Lp(a) as a risk factor for CHD have been complicated by the lack of standardization of Lp(a) measurement. As such, approaches to the standardization of Lp(a) measurement have represented a major focus of recent study.

The major challenge in Lp(a) measurement is principally the presence of a variable number of identical epitopes (specifically, the variable size of the identically repeated KIV_2 region underlying apo(a) isoform size heterogeneity) which may interact with the antibodies used in immunoassays to measure the concentration of Lp(a) in serum or plasma. If this is the case, then the antibodies in the assay will not possess the same immunoreactivity per particle if the apo(a) isoform size of the sample and calibrator are not the same. Indeed, the choice of apo(a) size in the calibrator is arbitrary and, irrespective of the choice, the size would not be representative of the size of apo(a) in most of the samples. As such, the immunoassays using antibodies sensitive to apo(a) isoform size will tend to underestimate Lp(a) levels in samples with apo(a) sizes smaller than the size of the apo(a) calibrator, while overestimating Lp(a) levels in samples with larger apo(a) isoforms.

Another major factor that constitutes a challenge to the accurate measurement of Lp(a) is differences in the properties of Lp(a) assays from different manufacturers. Assay kits may vary in terms of antibody properties, assay precision, and sensitivity of the assays to sample handling and storage conditions. In order to allow different assays to be appropriately optimized with respect to these factors, an international standard reference material for assay calibration is required. Significant progress in the development of such as standard has recently been made *(107,108)*. Nonetheless, it is clear that no reference material would be able to eliminate differences in Lp(a) values obtained by methods that are affected by apo(a) size heterogeneity and/or are not properly optimized.

PROGRESS IN THE STANDARDIZATION OF LP(a) MEASUREMENT

In a study designed to document the impact of apo(a) size on the accuracy of Lp(a) measurements, Marcovina and colleagues *(109)* generated and characterized a variety of monoclonal antibodies (MAbs) directed to different apo(a) epitopes. An MAb directed to an epitope present in KIV_2 [whose immunoreactivity per particle would vary depending on apo(a) size] and another MAb specific for a unique epitope located in apo(a) KIV_9 were used to develop two enzyme immunoassay (ELISA) methods to measure Lp(a) in a large number of samples *(109)*. Both assays were calibrated with the same serum containing an apo(a) species with a total of 21 KIV-like motifs. The Lp(a) protein value was assigned to the assay calibrator using a primary preparation, and the values were expressed in nmol/L to reflect the number of Lp(a) particles in plasma. As expected, highly comparable values were obtained by the two ELISA methods in the samples with the same apo(a) size as that of the assay calibrator, while Lp(a) values measured using the KIV_2-specific antibody were either underestimated or overestimated in samples with apo(a) containing < 21 or > 21 KIV repeats, respectively, compared to values obtained by the KIV_9-specific MAb. The KIV_9 MAb-based ELISA has been carefully validated in a large number of individuals and has been used as a reference method for the standardization of Lp(a) measurement.

Extensive work has been conducted on the development of a reference material for the purpose of achieving comparability of values amongst different Lp(a) assays.

Based on analyses performed by the International Federation of Clinical Chemistry and Laboratory Medicine (IFCC) Working Group on Lp(a), a lyophilized serum pool preparation was selected as secondary reference material for Lp(a) *(108)*. As a collaboration between the recipients of an NIH/National Heart, Lung, and Blood Institute contract and the IFCC Working Group on Lp(a), a consensus reference method using the MAb a40-based ELISA described above, and calibrated with two purified Lp(a) preparations, was used to assign a target value to the IFCC secondary reference material *(107)*. The IFCC preparation has been shown to have excellent stability and commutability properties and has been accepted by the WHO Expert Committee on Biological Standardization as the "First WHO/IFCC International Reference Reagent for Lipoprotein(a) for Immunoassay" *(110)*. This represents the achievement of a worldwide consensus on the calibration of Lp(a) test systems and will allow the determination of cutoff points for increased risk of CHD in different populations to be determined based on a molar concentration of Lp(a), as recommended *(57)*. The issue of isoform-size bias in different assay methodologies still needs to be addressed; this may necessitate the development of a range of Lp(a) isoform sizes for use as assay calibrators to correct for the inherent bias in a particular assay.

To evaluate the performance of commercially available methods for Lp(a) immunoassay, the IFCC reference preparation was used to transfer an accuracy-based value to the various immunoassay calibrators *(107)*. Among the 22 Lp(a) test systems evaluated, 10 were turbidimetric, 6 were nephelometric, 2 were fluorescence-based, 1 was an electroimmunodiffusion assay, and 1 was ELISA-based. Most methods used polyclonal antibodies against apo(a) to measure Lp(a). After uniformity of calibration was demonstrated in the 22 evaluated systems, Lp(a) was measured in 30 fresh-frozen samples covering a wide range of Lp(a) levels and apo(a) sizes with values assigned by the reference method. A significant apo(a) size-dependent bias was observed for all the evaluated systems. Three systems showed a minimal extent of bias, and of these, only one displayed a high level of concordance with the reference method over the full range of apo(a) sizes *(107)*. As such, despite the use of a common reference preparation, no harmonization in Lp(a) values among the different methods was achieved. This indicates that the impact of apo(a) isoform size on Lp(a) levels varies among the different methods as a function of the apo(a) size of the assay calibrators.

Studies have been conducted to determine how to overcome the problem of isoform-size dependence in Lp(a) measurement methods. Instead of using serial dilutions of a single calibrator, five fresh-frozen samples with different apo(a) sizes from small to large and suitable Lp(a) levels were used to calibrate a turbidimetric assay affected by apo(a) size variation *(57)*. Analyses were performed in parallel using the original assay calibrator and the five-sample calibrator on a large number of samples. A consistent apo(a) size-dependent bias was observed with the original assay calibrator. In contrast, good comparability was demonstrated between the observed values and the values obtained by the reference method (the ELISA employing the MAb directed against KIV_9) when the five independent samples were used to calibrate the assay *(57)*. While this approach appears promising for reducing assay inaccuracy, it may not be equally effective in all the methods or in all the samples. Therefore, its potential use for assay standardization will require further evaluation using multiple methods and a large number of samples with a good representation of single and double apo(a) isoforms.

Progress has also been made in establishing a primary reference standard for Lp(a). In particular, Edelstein and Scanu *(111,112)* have examined the effects of freezing and lyophilization of Lp(a) on its immunological properties and have described methods for preserving purified Lp(a) in a form suitable for a primary reference material. When isolated Lp(a) was lyophilized in the presence of suitable cryopreservatives and then reconstituted, it was found to be indistinguishable from the starting material with respect to chemical and biological properties. Moreover, the reconstituted products exhibited unchanged immunochemical properties and thus appeared to have all of the requisites to be used as a primary reference material *(112)*.

NON-IMMUNOLOGICALLY BASED METHODS FOR LP(a) MEASUREMENT

A potential approach to bypass the problems of immunochemical determination of Lp(a) concentrations involves quantification of Lp(a) by measuring its cholesterol content. Early methods used continuous flow analysis of lipoprotein classes separated by ultracentrifugation *(113)* or by lectin affinity chromatography to separate Lp(a) from other lipoproteins *(114)*; Lp(a) cholesterol was then measured by enzymatic assay. A more recent method involves plasma electrophoresis followed by detection of Lp(a) cholesterol by in situ enzymatic assay and densitometry *(115)*. Importantly, it was found that an ultracentrifugation step intended to eliminate β-VLDL that might overlap with Lp(a) on the gels which did not materially affect the outcome of the assays *(115)*. Further studies are required to compare the clinical significance of Lp(a) cholesterol with that of Lp(a) mass.

THE IMPACT OF LP(a) METHOD INACCURACY ON THE INTERPRETATION OF LP(A) VALUES

In the majority of clinical studies, Lp(a) levels have been determined by methods affected by apo(a) size heterogeneity. Therefore, for the conclusions of these studies to be valid, the assumption has to be made that the distribution of apo(a) isoforms was similar between cases and controls, thus minimizing the potential that method-dependent overestimation or underestimation of Lp(a) values contributed to the observed difference or lack thereof between cases and controls. Despite the importance of this topic, few studies have been performed to evaluate the impact of method inaccuracy on the interpretation of clinical data (57).

Marcovina and colleagues *(57)* determined Lp(a) levels (using the reference method described above) as well as apo(a) isoform sizes in 2940 samples from the participants in the Framingham study collected during the fifth cycle. During the same cycle, Lp(a) levels were also determined in other laboratories in 2556 of the samples by a turbidimetric method and in 2662 of the samples by a commercially available ELISA. Depending on the assay used, 5–15% of the subjects were misclassified as being at higher risk (false positive) while approximately 1% were misclassified as being not at risk (false negative). The vast majority of the misclassifications observed by the turbidimetric method were explained by the overestimation or underestimation of Lp(a) values based on the apo(a) size in the samples; this in turn was attributable to the small apo(a) size in the assay calibrator and by the high frequency of samples in the general population with apo(a) sizes larger than that in the calibrator. On the other hand, the large number of false positive values generated by the ELISA method was not explained by the apo(a) sizes in these samples. In fact, a high degree of variability in Lp(a) values was observed with this ELISA method. In addition, the values obtained from frozen samples were generally higher than those from fresh samples, but the magnitude of the increase was sample-dependent. These findings clearly indicate that assay standardization can only be achieved if each assay is properly optimized in addition to being evaluated for its sensitivity to apo(a) size polymorphism *(57)*.

In a separate study by Marcovina and colleagues *(57,80)*, they directly compared the ability of the ELISA reference method and a commercially available latex-based nephelometric method to predict future angina pectoris in men participating in the Physicians' Health Study. Apo(a) isoform size was determined and plasma Lp(a) concentration was simultaneously measured by the ELISA reference method and in a different laboratory by the nephelometric method. Analyses were performed in samples from 195 study participants who subsequently developed angina and from paired controls, matched for age and smoking, who remained free of reported vascular disease. Baseline median level of Lp(a) in cases, as determined by the reference method, was approximately 35% higher ($P = 0.02$) than that in controls. Additionally, Lp(a) was associated with increased relative risk for angina, and this association was strengthened after controlling for lipid risk factors. However, the median Lp(a) level determined by the commercial method was not statistically different between cases and controls, and the association with angina was also not significant. Very high Lp(a) values obtained by the nephelometric assay (> 95th percentile of controls) were a significant predictor of angina even though the relative risk was not as strong than that ascertained using the reference method. These results clearly indicate the importance of using suitable and standardized methods for risk assessment and for the interpretation of clinical outcomes.

SUGGESTED USE OF LP(A) IN CLINICAL PRACTICE

In assessing appropriate cardiovascular preventive measures, it is necessary to consider whether intervention to lower Lp(a) is clinically warranted. At present, Lp(a) is not an established cardiovascular risk factor and there are no guidelines recommending intervention *(52,57)*. Our current level of understanding would suggest that Lp(a) lowering might be beneficial in some subgroups of patients with high Lp(a) levels, but we still lack enough details on how to define such subgroups with regard to Lp(a) levels, apo(a) size, and presence of other risk factors. Moreover, there does not currently exist a well-characterized, well-tolerated and sufficiently safe therapeutic agent capable of adequately lowering plasma Lp(a) concentrations *(57,116)*. In any event, the lack of knowledge of Lp(a) metabolism both regarding its formation and catabolism raises considerable challenges in devising strategies to lower Lp(a) levels *(117)*. As apo(a) synthesis is of major importance in regulating Lp(a) levels, inhibition of Lp(a) particle formation at the step of particle assembly from apo(a) and apoB-100 may offer intervention possibilities *(3,118,119)*. Furthermore, model studies in transgenic animals and cell cultures have suggested a hepatic elimination pathway that might be a fruitful target *(120)*. However, the current lack of a well-defined metabolic pathway for Lp(a) has retarded progress in the development of agents that interfere with either formation or catabolism.

Although screening for Lp(a) elevation in the general population is not suggested at this time, measurement of Lp(a) is recommended in individuals with an increased risk of CVD, particularly in those with borderline LDL-C or high apoB. In these individuals, allele-specific Lp(a) measurement should be considered. Individuals with high (> 30 mg/dL or 75 nmol/L, i.e., above the 80th percentile) Lp(a) levels should be more aggressively managed with respect to modifiable risk factors.

REFERENCES

1. Superko HR, Hecht HS. Metabolic disorders contribute to subclinical coronary atherosclerosis in patients with coronary calcification. *Am J Cardiol* 2001;88:260–4.
2. Albers JJ, Kennedy H, Marcovina SM. Evidence that Lp(a) contains one molecule of apo(a) and one molecule of apo B: evaluation of amino acid analysis data. *J Lipid Res* 1996;37:192–6.
3. Koschinsky ML, Côté GP, Gabel BR, van der Hoek YY. Identification of the cysteine residue in apolipoprotein(a) that mediates extracellular coupling with apolipoprotein B-100. *J Biol Chem* 1993;268:19819–25.
4. Callow MJ, Rubin EM. Site-specific mutagenesis demonstrates that cysteine 4326 of apolipoprotein B is required for covalent linkage with apolipoprotein(a) *in vivo*. *J Biol Chem* 1995;270:23914–7.
5. Trieu VN, McConathy WJ. A two-step model for lipoprotein(a) formation. *J Biol Chem* 1995;270:15471–4.
6. Becker L, Cook PM, Wright TG, Koschinsky ML. Quantitative evaluation of the contribution of weak lysine-binding sites present within apolipoprotein(a) kringle IV types 6–8 to lipoprotein(a) assembly. *J Biol Chem* 2004;279:2679–88.
7. Becker L, McLeod RS, Marcovina SM, Yao Z, Koschinsky ML. Identification of a critical lysine residue in apolipoprotein B-100 that mediates noncovalent interaction with apolipoprotein(a). *J Biol Chem* 2001;276:36155–62.
8. Becker L, Webb BA, Chitayat S, Nesheim ME, Koschinsky ML. A ligand-induced conformational change in apolipoprotein(a) enhances covalent Lp(a) formation. *J Biol Chem* 2003;278:14074–81.
9. Xu S. Apolipoprotein(a) binds to low density lipoprotein at two distant sites in lipoprotein(a). *Biochemistry* 1998;37:9284–94.
10. White AL, Rainwater DL, Lanford RE. Intracellular maturation of apolipoprotein(a) and assembly of lipoprotein(a) in primary baboon hepatocytes. *J Lipid Res* 1993;34:509–17.
11. Frank S, Krasznai K, Durovic S, Lobentanz EM, Dieplinger H, Wagner E, Zatloukal K, Cotten M, Utermann G, Kostner GM, et al. High-level expression of various apolipoprotein(a) isoforms by "transferrinfection": the role of kringle IV sequences in the extracellular association with low-density lipoprotein. *Biochemistry* 1994;33:12329–39.
12. White, AL, Lanford, RE. Cell surface assembly of lipoprotein(a) in primary cultures of baboon hepatocytes. *J Biol Chem* 1994;269:28716–23.
13. Fless GM, ZumMallen ME, Scanu AM. Physicochemical properties of apolipoprotein(a) and lipoprotein(a–) derived from the dissociation of human plasma lipoprotein(a). *J Biol Chem* 1986;261:8712–8.
14. Kraft HG, Menzel HJ, Hoppichler F, Vogel W, Utermann G. Changes of genetic apolipoprotein phenotypes caused by liver transplantation. Implications for apolipoprotein synthesis. *J Clin Invest* 1989;83:137–40.
15. McLean JW, Tomlinson JE, Kuang W-J, Eaton DL, Chen EY, Fless GM, Scanu AM, Lawn RM. cDNA sequence of human apolipoprotein(a) is homologous to plasminogen. *Nature* 1987;330:132–7.
16. Gabel BR, Koschinsky ML. Analysis of the proteolytic activity of a recombinant form of apolipoprotein(a). *Biochemistry* 1995;34:15777–84.
17. van der Hoek YY, Wittekoek ME, Beisiegel U, Kastelein JJ, Koschinsky ML. The apolipoprotein(a) kringle IV repeats which differ from the major repeat kringle are present in variably sized isoforms. *Hum Mol Genet* 1993;2:361–6.
18. Lackner C, Cohen JC, Hobbs HH. Molecular definition of the extreme size polymorphism in apolipoprotein(a). *Hum Mol Genet* 1993;2:933–40.
19. Sangrar W, Marcovina SM, Koschinsky, ML. Expression and characterization of apolipoprotein(a) kringle IV types 1, 2 and 10 in mammalian cells. *Protein Eng* 1993;7:723–31.
20. Gabel BR, May LF, Marcovina SM, Koschinsky ML. Lipoprotein(a) assembly. quantitative assessment of the role of apo(a) kringle IV types 2–10 in particle formation. *Arterioscler Thromb Vasc Biol* 1996;16:1559–67.
21. Ernst A, Helmhold M, Brunner C, Petho-Schramm A, Armstrong VW, Muller HJ. Identification of two functionally distinct lysine binding sites in kringle 37 and in kringles 32–36 of human apolipoprotein(a). *J Biol Chem* 1994;270:6227–34.

22. Rahman MN, Becker L, Petrounevitch V, Hill BC, Jia Z, Koschinsky ML. Comparative analyses of the lysine binding site properties of apolipoprotein(a) kringle IV types 7 and 10. *Biochemistry* 2002;41:1149–55.
23. Kochl S, Fresser F, Lobentanz E, Baier G, Utermann G. Novel interaction of apolipoprotein(a) with beta-2 glycoprotein I mediated by the kringle IV domain. *Blood* 1997;90:1482–9.
24. Kapetanopoulos A, Fresser F, Millonig G, Shaul Y, Baier G, Utermann G. Direct interaction of the extracellular matrix protein DANCE with apolipoprotein(a) mediated by the kringle IV-type 2 domain. *Mol Genet Genomics* 2002;267:440–6.
25. Boerwinkle E, Leffert CC, Lin J, Lackner C, Chiesa G, Hobbs HH. Apolipoprotein(a) gene accounts for greater than 90% of the variation in plasma lipoprotein(a) concentrations. *J Clin Invest* 1990;90:52–60.
26. Utermann G, Menzel HJ, Kraft HG, Duba HC, Kemmler HG, Seitz C. Lp(a) glycoprotein phenotypes. Inheritance and relation to Lp(a)-lipoprotein concentrations in plasma. *J Clin Invest* 1987;80:458–65.
27. Brunner C, Lobentanz EM, Petho-Schramm A, Ernst A, Kang C, Dieplinger H, Muller HJ, Utermann G. The number of identical kringle IV repeats in apolipoprotein(a) affects its processing and secretion by HepG2 cells. *J Biol Chem* 1996;271:32403–10.
28. Marcovina SM, Albers JJ, Wijsman E, Zhang Z, Chapman NH, Kennedy H. Differences in Lp(a) concentrations and apo(a) polymorphs between Black and White Americans. *J Lipid Res* 1996;37:2569–85.
29. Rubin J, Paultre F, Tuck CH, Holleran S, Reed RG, Pearson TA, Thomas CM, Ramakrishnan R, Berglund L. Apolipoprotein(a) genotype influences isoform dominance pattern differently in African Americans and Caucasians. *J Lipid Res* 2002;43:234–44.
30. Bergund L, Ramakrishnan R. Lipoprotein(a). An elusive cardiovascular risk factor. *Arterioscler Thromb Vasc Biol* 2004;24:2219–26.
31. Paultre F, Pearson TA, Weil HF, Tuck CH, Myerson M, Rubin J, Francis CK, Marx HF, Philbin EF, Reed RG, et al. High levels of Lp(a) with a small apo(a) isoform are associated with coronary artery disease in African American and White men. *Arterioscler Thromb Vasc Biol* 2000;20:2619–24.
32. Marcovina SM, Koschinsky ML. Lipoprotein(a): structure, measurement and clinical significance. In: Rifai N, Warnick GR, Dominiczak MH, eds. *Handbook of Lipoprotein Testing*. 2nd ed. Washington, DC: AACC Press, 2000:345–85.
33. Armstrong VW, Schleef J, Thiery J, Muche R, Schuff-Werner P, Eisenhauer T, Seidel D. Effect of HELP-LDL-apheresis on serum concentrations of human lipoprotein(a): kinetic analysis of the post-treatment return to baseline levels. *Eur J Clin Invest* 1989;19:235–40.
34. Pokrovsky SN, Adamova I, Afanasieva OY, Benevolenskaya GF. Immunosorbent for selective removal of lipoprotein(a) from human plasma: in vitro study. *Artif Organs* 1991;15:136–40.
35. Crouse JR, III. New developments in the use of niacin for treatment of hyperlipidemia: new considerations in the use of an old drug. *Coron Artery Dis* 1996;7:321–6.
36. Sirtori CR, Calabresi L, Ferrara S, Pazzucconi F, Bondioli A, Baldassarre D, Birreci A, Koverech A. L-carnitine reduces plasma lipoprotein(a) levels in patients with hyper Lp(a). *Nutr Metab Cardiovasc Dis* 2000;10:247–51.
37. Dalessandri KM. Reduction of lipoprotein(a) in postmenopausal women. *Arch Intern Med* 2001;161:772–3.
38. Crook D, Sidhu M, Seed M, O'Donnell M, Stevenson JC. Lipoprotein Lp(a) levels are reduced by danazol, an anabolic steroid. *Atherosclerosis* 1992;92:41–7.
39. Haenggi W, Riesen W, Birkhaeuser MH. Postmenopausal hormone replacement therapy with Tibolone decreases serum lipoprotein(a). *Eur J Clin Chem Clin Biochem* 199;31:645–50.
40. Rymer J, Crook D, Sidhu M, Chapman M, Stevenson JC. Effects of tibolone on serum concentrations of lipoprotein(a) in postmenopausal women. *Acta Endocrinol (Copenh)* 1993;128:259–62.
41. Henriksson P, Angelin B, Berglund L. Hormonal regulation of serum Lp(a) levels. Opposite effects after estrogen treatment and orchidectomy in males with prostatic carcinoma. *J Clin Invest* 1992;89:1166–71.
42. Shewmon DA, Stock JL, Rosen CJ, Heiniluoma KM, Hogue MM, Morrison A, Doyle EM, Ukena T, Weale V, Baker S. Tamoxifen and estrogen lower circulating lipoprotein(a) concentrations in healthy postmenopausal women. *Arterioscler Thromb* 1994;14:1586–93.
43. Plenge JK, Hernandez TL, Weil KM, Poirier P, Grunwald GK, Marcovina SM, Eckel RH. Simvastatin lowers C-reactive protein within 14 days: an effect independent of low-density lipoprotein cholesterol reduction. *Circulation* 2002;106:1447–52.

44. Dujovne CA, Harris WS, Altman R, Overhiser RW, Black DM. Effect of atorvastatin on hemorheologic-hemostatic parameters and serum fibrinogen levels in hyperlipidemic patients. *Am J Cardiol* 2000;85:350–3.
45. Cobbaert C, Jukema JW, Zwinderman AH, Withagen AJ, Lindemans J, Bruschke AV. Modulation of lipoprotein(a) atherogenicity by high density lipoprotein cholesterol levels in middle-aged men with symptomatic coronary artery disease and normal to moderately elevated serum cholesterol. Regression Growth Evaluation Statin Study (REGRESS) Study Group. *J Am Coll Cardiol* 1997; 30:1491–9.
46. Duriez P, Dallongeville J, Fruchart JC. Lipoprotein(a) as a marker for coronary heart disease. *Br J Clin Pract Suppl* 1996;77A:54–61.
47. Gonbert S, Malinsky S, Sposito AC, Laouenan H, Doucet C, Chapman MJ, Thillet J. Atorvastatin lowers lipoprotein(a) but not apolipoprotein(a) fragment levels in hypercholesterolemic subjects at high cardiovascular risk. *Atherosclerosis* 2002;164:305–11.
48. Akaike M, Azuma H, Kagawa A, Matsumoto K, Hayashi I, Tamura K, Nishiuchi T, Iuchi T, Takamori N, Aihara K, et al. Effect of aspirin treatment on serum concentrations of lipoprotein(a) in patients with atherosclerotic diseases. *Clin Chem* 2002;48:1454–9.
49. Kiortsis DN, Tzotzas T, Giral P, Bruckert E, Beucler I, Valsamides S, Turpin G. Changes in lipoprotein(a) levels and hormonal correlations during a weight reduction program. *Nutr Metab Cardiovasc Dis* 2001;11:153–7.
50. Jenkins DJ, Kendall CW, Marchie A, Parker TL, Connelly PW, Qian W, Haight JS, Faulkner D, Vidgen E, Lapsley KG, et al. Dose response of almonds on coronary heart disease risk factors: blood lipids, oxidized low-density lipoproteins, lipoprotein(a), homocysteine, and pulmonary nitric oxide: a randomized, controlled, crossover trial. *Circulation* 2002;106:1327–32.
51. Ginsberg HN, Kris-Etherton P, Dennis B, Elmer PJ, Ershow A, Lefevre M, Pearson T, Roheim P, Ramakrishnan R, Reed R, et al. Effects of reducing dietary saturated fatty acids on plasma lipids and lipoproteins in healthy subjects: the DELTA Study, protocol 1. *Arterioscler Thromb Vasc Biol* 1998;18:441–9.
52. Executive Summary of the Third Report of The National Cholesterol Education Program (NCEP) Expert Panel on Detection, Evaluation, And Treatment of High Blood Cholesterol In Adults (Adult Treatment Panel III). *J Am Med Assoc* 2001;285:2486–97.
53. Cheng SW, Ting AC, Wong J. Lipoprotein(a) and its relationship to risk factors and severity of atherosclerotic peripheral vascular disease. *Eur J Vasc Endovasc Surg* 1997;14:17–23.
54. Peng DQ, Zhao SP, Wang JL. Lipoprotein(a) and apolipoprotein E epsilon 4 as independent risk factors for ischemic stroke. *J Cardiovasc Risk* 1999;6:1–6.
55. Marcovina SM, Hegele RA, Koschinsky ML. Lipoprotein(a) and coronary heart disease risk. *Curr Cardiol Rep* 1999;1:105–11.
56. Marcovina SM, Koschinsky ML. Lipoprotein(a) as a risk factor for coronary artery disease. *Am J Cardiol* 1998;82:57U-66U
57. Marcovina SM, Koschinsky ML, Albers JJ, Skarlatos S. Report of the National Heart, Lung, and Blood Institute Workshop on Lipoprotein(a) and Cardiovascular Disease: recent advances and future directions. *Clin Chem* 2003;49:1785–96.
58. Craig WY, Neveux LM, Palomaki GE, Cleveland MM, Haddow JE. Lipoprotein(a) as a risk factor for ischemic heart disease: meta-analysis of prospective studies. *Clin Chem* 1998;44:2301–6.
59. Danesh J, Collins R, Peto R. Lipoprotein(a) and coronary heart disease. Meta-analysis of prospective studies. *Circulation* 2000;102:1082–5.
60. Sharrett AR, Ballantyne CM, Coady SA, Heiss G, Sorlie PD, Catellier D, Patsch W, Atherosclerosis Risk in Communities Study Group. Coronary heart disease prediction from lipoprotein cholesterol levels, triglycerides, lipoprotein(a), apolipoproteins A-I and B, and HDL density subfractions: The Atherosclerosis Risk in Communities (ARIC) Study. *Circulation* 2001;104:1108–13.
61. Glader CA, Birgander LS, Stenlund H, Dahlen GH. Is lipoprotein(a) a predictor for survival in patients with established coronary artery disease? Results from a prospective patient cohort study in northern Sweden. *J Intern Med* 2002;252:27–35.
62. Maher VM, Brown BG, Marcovina SM, Hillger LA, Zhao XQ, Albers JJ. Effects of lowering elevated LDL cholesterol on the cardiovascular risk of lipoprotein(a). *J Am Med Assoc* 1995;274:1771–4.
63. Luc G, Bard JM, Arveiler D, Ferrieres J, Evans A, Amouyel P, Fruchart JC, Ducimetiere P, PRIME Study Group. Lipoprotein (a) as a predictor of coronary heart disease: the PRIME Study. *Atherosclerosis* 2002;163:377–84.

64. Cantin B, Gagnon F, Moorjani S, Despres JP, Lamarche B, Lupien PJ, Dagenais GR. Is lipoprotein(a) an independent risk factor for ischemic heart disease in men? The Quebec Cardiovascular Study. *J Am Coll Cardiol* 1998;31:519–25.
65. von Eckardstein A, Schulte H, Cullen P, Assmann G. Lipoprotein(a) further increases the risk of coronary events in men with high global cardiovascular risk. *J Am Coll Cardiol* 2001;37:434–9.
66. Marcovina SM, Koschinsky ML. Evaluation of lipoprotein(a) as a prothrombotic factor: progress from bench to bedside. *Curr Opin Lipidol* 2003;14:361–6.
67. Marcucci R, Liotta AA, Cellai AP, Rogolino A, Gori AM, Giusti B, Poli D, Fedi S, Abbate R, Prisco D. Increased plasma levels of lipoprotein(a) and the risk of idiopathic and recurrent venous thromboembolism. *Am J Med* 2003;115:667–8.
68. Nowak-Gottl U, Strater R, Heinecke A, Junker R, Koch HG, Schuierer G, von Eckardstein A. Lipoprotein(a) and genetic polymorphisms of clotting Factor V, prothrombin, and methylenetetrahydrofolate reductase are risk factors of spontaneous ischemic stroke in childhood. *Blood* 1999;94:3678–82.
69. Strater R, Becker S, von Eckardstein A, Heinecke A, Gutsche S, Junker R, Kurnik K, Schobess R, Nowak-Gottl U. Prospective assessment of risk factors for recurrent stroke during childhood. *Lancet* 2002;360:1540–5.
70. Dahl T, Kontny F, Slagsvold CE, Christophersen B, Abildgaard U, Odegaard OR, Morkrid L, Dale J. Lipoprotein(a), other lipoproteins and hemostatic profiles in patients with ischemic stroke: the relation to cardiogenic embolism. *Cerebrovasc Dis* 2000;10:110–7.
71. Kronenberg F, Kronenberg MF, Kiechl S, Trenkwalder E, Santer P, Oberhollenzer F, Egger G, Utermann G, Willeit J. Role of lipoprotein(a) and apolipoprotein(a) phenotype in atherogenesis: prospective results from the Bruneck study. *Circulation* 1999;100:1154–60.
72. Wild SH, Fortmann SP, Marcovina SM. A prospective case-control study of lipoprotein(a) levels and apo(a) size and risk of coronary heart disease in Stanford Five-City Project participants. *Arterioscler Thromb Vasc Biol* 1997;17:239–45.
73. Paultre F, Tuck CH, Boden-Albala B, Kargman DE, Todd E, Jones J, Paik MC, Sacco RL, Berglund L. Relation of Apo(a) size to carotid atherosclerosis in an elderly multiethnic population. *Arterioscler Thromb Vasc Biol* 2002;22:141–6.
74. Shai I, Rimm EB, Hankinson SE, Cannuscio C, Curhan G, Manson JE, Rifai N, Stampfer MJ, Ma J. Lipoprotein(a) and coronary heart disease among women: beyond a cholesterol carrier? *Eur Heart J* 2005;26:1633–9.
75. Orth-Gomer K, Mittleman MA, Schenck-Gustafsson K, Wamala SP, Eriksson M, Belkic K, Kirkeeide R, Svane B, Ryden L. Lipoprotein(a) as a determinant of coronary heart disease in young women. *Circulation* 1997;95:329–34.
76. Ariyo AA, Thach C, Tracy R. Lp(a) lipoprotein, vascular disease, and mortality in the elderly. *N Engl J Med* 2003;349:2108–15.
77. Ridker M, Hennekens CH, Buring JE, Rifai N. C-reactive protein and other markers of inflammation in the prediction of cardiovascular disease in women. *N Engl J Med* 2000;342:836–43.
78. Wu HD, Berglund L, Dimayuga C, Jones J, Sciacca RR, Di Tullio MR, Homma S. High lipoprotein(a) levels and small apolipoprotein(a) sizes are associated with endothelial dysfunction in a multiethnic cohort. *J Am Coll Cardiol* 2004;43:1828–33.
79. Emanuele E, Peros E, Minoretti P, D'Angelo A, Montagna L, Falcone C, Geroldi D. Significance of apolipoprotein(a) phenotypes in acute coronary syndromes: relation with clinical presentation. *Clin Chim Acta* 2004;350:159–65.
80. Rifai N, Ma J, Sacks FM, Ridker PM, Hernandez WJ, Stampfer MJ, Marcovina SM. Apolipoprotein(a) size and lipoprotein(a) concentrations and future risk of angina pectoris with evidence of severe coronary atherosclerosis in men: The Physician's Health Study. *Clin Chem* 2004;50:1364–71.
81. Koschinsky ML, Marcovina SM. Structure-function relationships in apolipoprotein(a): insights into lipoprotein(a) assembly and pathogenicity. *Curr Opin Lipidol* 2004;15:167–74.
82. Sangrar W, Bajzar L, Nesheim ME, Koschinsky ML. Antifibrinolytic effect of recombinant apolipoprotein(a) in vitro is primarily due to attenuation of tPA-mediated Glu-plasminogen activation. *Biochemistry* 1995; 34:5151–7.
83. Biemond BJ, Friederich PW, Koschinsky ML, Levi M, Sangrar W, Xia J, Buller HR, ten Cate JW. Apolipoprotein(a) attenuates endogenous fibrinolysis in the rabbit jugular vein thrombosis model in vivo. *Circulation* 1997;96:1612–5.

84. Hancock MA, Boffa MB, Marcovina SM, Nesheim ME, Koschinsky ML. Inhibition of plasminogen activation by lipoprotein(a): critical domains in apolipoprotein(a) and mechanism of inhibition on fibrin and degraded fibrin surfaces. *J Biol Chem* 2003;278:23260–9.
85. Kojima S, Harpel PC, Rifkin DB. Lipoprotein (a) inhibits the generation of transforming growth factor beta: an endogenous inhibitor of smooth muscle cell migration. *J Cell Biol* 1991;113:1439–45.
86. Grainger DJ, Kirschenlohr HL, Metcalfe JC, Weissberg PL, Wade DP, Lawn RM. Proliferation of human smooth muscle cells promoted by lipoprotein(a). *Science* 1993;260:1655–8.
87. Grainger DJ, Kemp PR, Liu AC, Lawn RM, Metcalfe JC. Activation of transforming growth factor-beta is inhibited in transgenic apolipoprotein(a) mice. *Nature* 1994;370:460–2.
88. O'Neil CH, Boffa MB, Hancock MA, Pickering JG, Koschinsky ML. Stimulation of vascular smooth muscle cell proliferation and migration by apolipoprotein(a) is dependent on inhibition of transforming growth factor-beta activation and on the presence of kringle IV type 9. *J Biol Chem* 2004;279:55187–95.
89. Pellegrino M, Furmaniak-Kazmierczak E, LeBlanc JC, Cho T, Cao K, Marcovina SM, Boffa MB, Côté GP, Koschinsky ML. The apolipoprotein(a) component of lipoprotein(a) stimulates actin stress fiber formation and loss of cell-cell contact in cultured endothelial cells. *J Biol Chem* 2004; 279:6526–33.
90. Allen S, Khan S, Tam S, Koschinsky M, Taylor P, Yacoub M. Expression of adhesion molecules by Lp(a): a potential novel mechanism for its atherogenicity. *FASEB J* 1998;12:1765–76.
91. Schachinger V, Halle M, Minners J, Berg A, Zeiher AM. Lipoprotein(a) selectively impairs receptor-mediated endothelial vasodilator function of the human coronary circulation. *J Am Coll Cardiol* 1997;30:927–34.
92. Sotiriou SN, Orlova VV, Al-Fakhri N, Ihanus E, Economopoulou M, Isermann B, Bdeir K, Nawroth PP, Preissner KT, Gahmberg CG, et al. Lipoprotein(a) in atherosclerotic plaques recruits inflammatory cells through interaction with Mac-1 integrin. *FASEB J* 2006;20:559–61. .
93. Tsimikas S, Bergmark C, Beyer RW, Patel R, Pattison J, Miller E, Juliano J, Witztum JL. Temporal increases in plasma markers of oxidized low-density lipoprotein strongly reflect the presence of acute coronary syndromes. *J Am Coll Cardiol* 2003;41:360–70.
94. Tsimikas S, Lau HK, Han KR, Shortal B, Miller ER, Segev A, Curtiss LK, Witztum JL, Strauss BH. Percutaneous coronary intervention results in acute increases in oxidized phospholipids and lipoprotein(a). *Circulation* 2004;109:3164–70.
95. Edelstein C, Pfaffinger D, Hinman J, Miller E, Lipkind G, Tsimikas S, Bergmark C, Getz GS, Witztum JL, Scanu AM. Lysine-physphatidylcholine adducts in kringle V impart unique immunological and protein pro-inflammatory properties to human apolipoprotein(a). *J Biol Chem* 2003; 278:52841–7.
96. Klezovitch O, Edelstein C, Scanu AM. Stimulation of interleukin-8 production in human THP-1 macrophages by apolipoprotein(a). Evidence for a critical involvement of elements in its C-terminal domain. *J Biol Chem* 2001;276:46864–9.
97. Kang C, Dominguez M, Loyau S, Miyata T, Durlach V, Anglés-Cano E. Lp(a) particles mold fibrin-binding properties of apo(a) in size-dependent manner: a study with different-length recombinant apo(a), native Lp(a), and monoclonal antibody. *Arterioscler Thromb Vasc Biol* 2002; 22:1232–8.
98. Anglés-Cano E, de la Peña Díaz A, Loyau S. Inhibition of fibrinolysis by lipoprotein(a). *Ann N Y Acad Sci* 2001;936:261–75.
99. Knapp JP, Herrmann W. In vitro inhibition of fibrinolysis by apolipoprotein(a) and lipoprotein(a) is size- and concentration-dependent. *Clin Chem Lab Med* 2004;42:1013–9.
100. Boffa MB, Marcovina SM, Koschinsky ML. Lipoprotein(a) as a risk factor for atherosclerosis and thrombosis: mechanistic insights from animal models. *Clin Biochem* 2004;37:333–43.
101. Fan J, Shimoyamada H, Sun H, Marcovina S, Honda K, Watanabe T. Transgenic rabbits expressing human apolipoprotein(a) develop more extensive atherosclerotic lesion in response to a cholesterol-rich diet. *Arterioscler Thromb Vasc Biol* 2001;21:88–94.
102. Sun H, Unoki H, Wang X, Liang J, Ichikawa T, Arai Y, Shiomi M, Marcovina SM, Watanabe T, Fan J. Lipoprotein(a) enhances advanced atherosclerosis and vascular calcification in WHHL transgenic rabbits expressing human apolipoprotein(a). *J Biol Chem* 2002;277:47486–92.
103. Guerra R, Yu Z, Marcovina S, Peshock R, Cohen JC, Hobbs HH. Lipoprotein(a) and apolipoprotein(a) isoforms: no association with coronary artery calcification in the Dallas Heart Study. *Circulation* 2005;111:1471–9.

104. Schneider M, Witztum JL, Young SG, Ludwig EH, Miller ER, Tsimikas S, Curtiss LK, Marcovina SM, Taylor JM, Lawn RM, et al. High-level lipoprotein(a) expression in transgenic mice: evidence for oxidized phospholipids in lipoprotein(a) but not in low density lipoproteins. *J Lipid Res* 2005;46:769–78.

105. Devlin CM, Lee SJ, Kuriakose G, Spencer C, Becker L, Grosskopf I, Ko C, Huang LS, Koschinsky ML, Cooper AD, et al. An apolipoprotein(a) peptide delays chylomicron remnant clearance and increases plasma remnant lipoproteins and atherosclerosis in vivo. *Arterioscler Thromb Vasc Biol* 2005;25:1704–10.

106. Yu KC, Cooper AD. Postprandial lipoproteins and atherosclerosis. *Front Biosci* 2001;6:D332–54.

107. Marcovina SM, Albers JJ, Scanu AM, Kennedy H, Giaculli F, Berg K, Couderc R, Dati F, Rifai N, Sakurabayashi I, et al. Use of a reference material proposed by the International Federation of Clinical Chemistry and Laboratory Medicine to evaluate analytical methods for the determination of plasma lipoprotein(a). *Clin Chem* 2000;46:1956–67.

108. Tate JR, Berg K, Couderc R, Dati F, Kostner GM, Marcovina SM, Rifai N, Sakurabayashi I, Steinmetz A. International Federation of Clinical Chemistry and Laboratory Medicine (IFCC) Standardization Project for the Measurement of Lipoprotein(a). Phase 2: selection and properties of a proposed secondary reference material for lipoprotein(a). *Clin Chem Lab Med* 1999;37:949–58.

109. Marcovina SM, Albers JJ, Gabel B, Koschinsky ML, Gaur VP. Effect of the number of apolipoprotein(a) kringle 4 domains on immunochemical measurements of lipoprotein(a). *Clin Chem* 1995;41:246–55.

110. Dati F, Tate JR, Marcovina SM, Steinmetz A, International Federation of Clinical Chemistry and Laboratory Medicine, IFCC Working Group for Lipoprotein(a) Assay Standardization. First WHO/IFCC International Reference Reagent for Lipoprotein(a) for Immunoassay–Lp(a) SRM 2B. *Clin Chem Lab Med* 2004;42:670–6.

111. Edelstein C, Hinman J, Marcovina S, Scanu AM. Properties of human free apolipoprotein(a) and lipoprotein(a) after either freezing or lyophilization in the presence and absence of cryopreservatives. *Anal Biochem* 2001;288:201–8.

112. Scanu AM, Hinman J, Pfaffinger D, Edelstein C. Successful utilization of lyophilized lipoprotein(a) as a biological reagent. *Lipids* 2004;39:589–93.

113. Kulkarni KR, Garber DW, Marcovina SM, Segrest JP. Quantification of cholesterol in all lipoprotein classes by the VAP-II method. *J Lipid Res* 1994;35:159–68.

114. Seman LJ, Jenner JL, MC Namara JR, Schaefer EJ. Quantification of lipoprotein(a) in plasma by assaying cholesterol in lectin – bound plasma fraction. Clin Chem 1994;40:400–3.

115. Baudhuin LM, Hartman SJ, O'Brien JF, Meissner I, Galen RS, Ward JN, Hogen SM, Branum EL, McConnell JP. Electrophoretic measurement of lipoprotein(a) cholesterol in plasma with and without ultracentrifugation: comparison with an immunoturbidimetric lipoprotein(a) method. *Clin Biochem* 2004;37:481–8.

116. Angelin B. Therapy for lowering lipoprotein (a) levels. *Curr Opin Lipidol* 1997;8:337–41.

117. Hobbs HH, White AL. Lipoprotein(a): intrigues and insights. *Curr Opin Lipidol* 1999;10:225–36.

118. Steyrer E, Durovic S, Frank S, Giessauf W, Burger A, Dieplinger H, Zechner R, Kostner GM. The role of lecithin: cholesterol acyltransferase for lipoprotein(a) assembly. Structural integrity of low density lipoproteins is a prerequisite for Lp(a) formation in human plasma. *J Clin Invest* 1994;94:2330–40.

119. White AL, Lanford RE. Biosynthesis and metabolism of lipoprotein(a). *Curr Opin Lipidol* 1995;6:75–80.

120. Hrzenjak A, Frank S, Wo X, Zhou Y, Van Berkel T, Kostner GM. Galactose-specific asialoglycoprotein receptor is involved in lipoprotein (a) catabolism. *Biochem J* 2003;376:765–71.

121. Rahman MN, Becker L, Petrounevitch V, Hill BC, Jia Z, Koschinsky ML. Comparative analyses of the lysine binding site properties of apolipoprotein(a) kringle IV types 7 and 10. *Biochemistry* 2002;41:1149–55.

13 Familial Hypercholesterolemia and Lipid Apheresis

Patrick M. Moriarty, MD,
Cheryl A. Gibson, PhD,
and Klaus Flechsenhar, MD

CONTENTS

INTRODUCTION
FAMILIAL HYPERCHOLESTEROLEMIA
TREATMENT GUIDELINES
VASCULAR EFFECTS IN ADDITION TO LDL LOWERING
  AND PLAQUE REGRESSION
ALTERNATIVE APPLICATIONS OF LDL APHERESIS
SUMMARY
REFERENCES

INTRODUCTION

Primary hypercholesterolemia is a relatively common condition of elevated serum cholesterol concentrations that has been clearly established as a major risk factor for the development of atherosclerosis and premature cardiovascular disease *(1)*. In industrialized countries, it is generally manifested as moderate hypercholesterolemia (240–350 mg/dL) with serum triglyceride concentrations within the normal reference range. The global health burden among Western society is significant, with more than 50% of all deaths linked to atherosclerotic cardiovascular diseases *(2)*. The pathogenesis of atherosclerosis results from a combination of genetic and environmental factors with inherited disturbances in the low-density lipoprotein receptor (LDLR) and similar lipid-related abnormalities accounting for the majority of these deaths *(3)*. The National Cholesterol Education Program (NCEP) Expert Panel recommends reducing plasma concentrations of LDL cholesterol (LDL C) as the primary target of lipid-lowering therapy *(1)*. Clinical trial data support the benefits of reducing LDL-C with reductions of heart attacks by 33% and deaths related to heart attacks by nearly 40% with drug treatment among high-risk individuals *(4–6)*.

From: *Contemporary Cardiology: Therapeutic Lipidology*
Edited by: M. H. Davidson, P. P. Toth, and K. C. Maki © Humana Press Inc., Totowa, NJ

FAMILIAL HYPERCHOLESTEROLEMIA

Familial hypercholesterolemia (FH) is an autosomal co-dominant disorder characterized by elevated plasma LDL-C with normal triglycerides, tendon xanthomas, and premature coronary atherosclerosis *(7)*. This disorder causes severe elevations in total cholesterol and LDL-C. FH is generally caused by mutations in the LDLR gene (over 700 variants) with a frequency of 0.2% in most populations. This estimate was based on the frequency of FH in survivors of myocardial infarction in the United States and supported by other studies in the United Kingdom *(8,9)*. Internationally, the prevalence of heterozygous FH in Europe approximates that of the United States. However, in selected populations such as Afrikaners, Christian Lebanese, and French Canadians, the disease is more frequent due to the founder effect *(10)*. As described by Austin et al. *(11)*, a founder effect occurs when a subpopulation is formed through the immigration of a small number of "founder" subjects, followed by a population expansion.

A substantial change in the number or functional status of LDLR directly influences serum cholesterol levels. With mutations in the protein coding for the LDLR, the control of LDL uptake and cholesterol homeostasis in hepatocytes is severely affected, causing uncontrolled synthesis of cholesterol within the liver. The total cholesterol and LDL-C levels of individuals with two mutated LDLR alleles (FH homozygotes) are much more affected than those with one mutant allele (FH heterozygotes), with levels typically twice the population average.

Homozygous FH is a rare disorder, occurring in approximately 1 case per 1 million persons in the United States and Europe. Patients with homozygous FH can be classified into one of two groups based on the amount of LDLR activity: (1) patients with < 2% of normal LDLR activity (receptor negative) and (2) patients with 2–25% of normal LDLR activity (receptor-defective mutations). Homozygous FH patients display extremely high levels of circulating, cholesterol-rich, apolipoprotein B (apoB)-containing lipoproteins with typical total cholesterol levels ranging between 500 mg/dL and 1000 mg/dL *(12)*. As a consequence, complications of homozygous FH are manifested in premature cardiovascular disease and mortality in childhood or early adulthood. Without treatment, receptor-negative homozygous FH patients seldom live to reach 30 years of age. Patients with LDLR defects have a better prognosis but most develop atherosclerotic disease by the third decade.

The clinical diagnosis of homozygous FH is usually made in childhood (i.e., during the first decade of life) when cutaneous and tendinous xanthomas appear (typically on hands, wrists, elbows, knees, heels, or buttocks) *(12)*. Tendon xanthomas are not present in persons with non-FH. If tendon xanthomas are present, FH or familial defective apoB-100 is the correct diagnosis. Other clinical features typically present include deposits of fatty material around the eyes (xanthelasmas) and the cornea (arcus corneae).

Heterozygous FH is caused by the inheritance of one mutant LDLR allele and occurs in approximately 1 in 500, roughly 500,000 individuals in the United States and more than 10 million persons worldwide *(12,13)*. It is one of the most common genetic disorders, which is characterized by autosomal inheritance of increased total cholesterol, elevated plasma LDL-C (200–400 mg/dL), and normal triglyceride levels, primarily attributable to mutations in the LDLR gene. Heterozygous FH patients have hypercholesterolemia from birth, although the clinical manifestations of the disease, such as xanthomas, are not usually detected in childhood despite having elevated levels of LDL-C. Instead, clinical features of the disease are most often detected in adulthood after a first cardiac event, usually between 40 and 50 years of age.

Currently, there are three groups that have developed diagnostic classification tools for FH patients: the Simon Broome Register in the United Kingdom, the Dutch Lipid Clinic Network, and the MedPed (Make Early Diagnosis to Prevent Early Deaths) Program in the United States *(11)*. The diagnostic criteria of the Simon Broome Register for FH include cholesterol levels, clinical features, molecular diagnostic technologies, and family history. Based on these criteria, individuals can be classified as a "definite" or "probable" FH (Table 1). Patients with elevated cholesterol levels and the presence of tendinous xanthomata or an identified mutation in the LDLR gene or the apoB-100 gene are classified as "definite." Patients with elevated cholesterol levels and family history of hypercholesterolemia or heart disease are classified as "probable" *(14,15)*. Similar to the Simon Broome Register, the Dutch Lipid Clinic Network criteria assign different point values for family history of heart disease or hypercholesterolemia, clinical history [i.e., patients with premature coronary artery disease (CAD) or cerebral or peripheral vascular disease], presence of tendinous xanthomata or arcus cornealis, elevated LDL-C levels, and/or an identified mutation in the LDLR gene. A diagnosis is based on the total number of points obtained. A "definite" FH diagnosis requires more than 8 points, "probable" FH diagnosis requires 6–8 points, and a "possible" FH diagnosis requires 3–5 points (Table 2). The MedPed diagnostic criteria use cutpoints for total cholesterol levels specific to an individual's age and family history *(16)* with cutpoints differing for individuals with first-degree, second-degree, and third-degree relatives with FH and for the general population (see Table 3).

Table 1
Simon Broome Register Diagnostic Criteria for Familial Hypercholesterolemia (FH)

Criteria	*Description*
A	Total cholesterol concentration >7.7 mmol/L in adults or a total cholesterol concentration >6.7 mmol/L in children <16 years of age
	Low-density lipoprotein cholesterol concentration >4.9 mmol/L in adults or >4.0 mmol/L in children
B	Tendinous xanthomata in the patient or a first-degree relative
C	DNA-based evidence of mutation in the low-density lipoprotein receptor or apolipoprotein B gene
D	Family history of myocardial infarction before age 50 years in a second-degree relative or before age 60 years in a first-degree relative
E	Family history of raised total cholesterol concentration >7.5 mmol/L in a first-degree or second-degree relative
Diagnosis	
Definite	A "definite" FH diagnosis requires either criteria A and B or criterion C
Probable	A "probable" FH diagnosis requires either criteria A and D or criteria A and C

Source: Anonymous *(14)* and Anonymous *(15)*.

Table 2
Dutch Lipid Clinic Network Diagnostic Criteria for Familial Hypercholesterolemia (FH)

Criteria	Points
Family history	
First-degree relative with known premature (men <55 years; women <60 years) coronary and vascular disease, or	
First-degree relative with known LDL-C above the 95th percentile	1
First-degree relative with tendinous xanthomata and/or arcus cornealis, or	
children aged <18 years with LDL-C > 95th percentile	2
Clinical history	
Patient with premature (men <55 years; women <60 years) coronary artery disease	2
Patient with premature (men <55 years; women <60 years) cerebral or peripheral vascular disease	1
Physical examination	
Tendinous xanthomata	6
Arcus cornealis prior to age 45 years	4
Cholesterol levels (mmol/L)	
LDL-C ≥ 8.5	8
LDL-C 6.5–8.4	5
LDL-C 5.0–6.4	3
LDL-C 4.0–4.9	1
DNA analysis	
Functional mutation in the LDLR gene	8
Diagnosis is based on the total number of points obtained	
"Definite" FH diagnosis requires more than 8 points	
"Probable" FH diagnosis requires 6–8 points	
"Possible" FH diagnosis requires 3–5 points	

World Health Organization. Familial hypercholesterolemia—report of a second WHO Consultation. Geneva, Switzerland: World Health Organization, 1999. (WHO publication no. WHO/HGN/FH/CONS/99.2). (15)

LDL-C, low-density lipoprotein cholesterol; LDLR, low-density lipoprotein cholesterol.

TREATMENT GUIDELINES

In heterozygous FH, the cumulative risk of a coronary event by age 60 without effective treatment is at least 50% in men and about 30% in women. Aggressive treatment to lower plasma levels of LDL-C for heterozygous FH patients should be undertaken. In addition, recommendations to follow the NCEP Therapeutic Lifestyle Changes (TLC) diet should be given. The TLC diet is a low saturated fat, low cholesterol diet, which has been shown to help reduce blood cholesterol levels although most patients with heterozygous FH will inevitably require lipid-lowering therapy. The prognosis of heterozygous FH patients has improved substantially with the widespread use of HMG-CoA reductase inhibitors (statins) *(17,18)*. However, even the strongest statins at their maximum doses will be inadequate for FH patients, and additional cholesterol-lowering medication is needed. In such cases, combination therapy with the addition of a bile acid sequestrant or nicotinic acid or cholesterol

Table 3
US MedPed Program Diagnostic Criteria for Familial Hypercholesterolemia (FH)

Age (years)	Total cholesterol cutpoints (mmol/L)			
	First-degree relative with FH	Second-degree relative with FH	Third-degree relative with FH	General population
< 20	5.7	5.9	6.2	7.0
20–29	6.2	6.5	6.7	7.5
30–39	7.0	7.2	7.5	8.8
≥ 40	7.5	7.8	8.0	9.3

Note: Diagnosis (FH is diagnosed if total cholesterol levels exceed the cutpoint)
Source: Williams et al. *(16).*

absorption inhibitors have a major role in LDL lowering when statin monotherapy is not sufficient.

When diet, exercise, and pharmacotherapy prove to be unsuccessful at reaching LDL-C goals, current treatment options for homozygous FH patients include portocaval shunting, liver transplantation, plasmapheresis, or LDL apheresis. Liver transplantation, although rarely performed, provides normal LDLRs and causes dramatic decreases in LDL-C level. However, considerable risks are associated with organ transplantation and long-term immunosuppression. Plasmapheresis, though effective in lowering LDL-C, is non-selective and can lower both high-density lipoprotein-cholesterol (HDL-C) and apo A-1 by 30% in addition to lowering albumin and immunoglobulins (Igs) *(19).*

LDL apheresis is the current treatment of choice for homozygous FH patients because the procedure selectively removes lipoproteins that contain apo-B from the circulation, promoting regression of xanthomas and slowing the progression of atherosclerosis. In 1997, treatment was approved by the Food and Drug Administration (FDA) in the United States for three groups of individuals who have not responded adequately to maximum drug and diet therapy: Group A functional homozygous hypercholesterolemia having LDL-C levels above 500 mg/dL; Group B patients with LDL-C ≥ 300 mg/dL; and Group C functional hypercholesterolemic heterozygotes having LDL-C ≥ 200 mg/dL with documented coronary heart disease (CHD).

Although once reserved for patients with homozygous FH only, heterozygous FH patients which corresponds to approximately 1 per 500 of the population, who cannot be adequately controlled on combination drug therapy are candidates for LDL apheresis *(20).* Initiation of LDL apheresis should be delayed until approximately 5 years of age and/or at least 27 kg in weight except when evidence of atherosclerotic vascular disease is present.

LDL Apheresis Methods

Due to their elevated cholesterol levels and resistance to diet and pharmacotherapy, FH patients are the most common group to benefit from LDL apheresis. LDL apheresis is a weekly or bimonthly process that takes approximately 1.5–2 h to acutely remove apoB-containing lipoproteins from the blood. In a 1967 case report, de Gennes and colleagues successfully applied nonspecific plasmapheresis to symptomatic homozygous FH patients for the reduction of plasma cholesterol *(21)*. The first systematic clinical trial using plasma exchange in homozygous FH patients was performed by Thompson in 1975 *(22)*. However, it was not until the early 1980s when apheresis systems were widely introduced, with systems of varying degrees of selectivity and effectiveness *(23)*. With all methods, a 60–70% reduction in LDL-C per treatment is achieved.

Presently, two methods of LDL apheresis are performed in the United States. They include the heparin-induced extracorporeal low-density lipoprotein precipitation (HELP) system (B. Braun, Melsungen, Germany) and the dextran sulfate cellulose LDL adsorption (DSA) Liposorber LA-15 system (Kaneka Corporation, Osaka, Japan). Table 4 displays the technical features of the different systems.

Currently, the most commonly available apheresis procedures throughout the world include the DSA, HELP, direct adsorption of lipoproteins, double-membrane filtration, and immunoadsorption. Each of these apheresis procedures will be described briefly below.

DEXTRAN SULFATE CELLULOSE ADSORPTION

The Liposorber LA-15 apheresis system is an extracorporeal process that involves using columns containing dextran sulfate cellulose (Liposorber, LA-15, Kaneka Corp., Osaka, Japan). It is comprised of a tubing set, a hollow fiber plasma separator (Sulflux FS-05), and two dextran sulfate LDL-adsorption columns (Liposorber LA-15), which mediate adsorption of atherogenic lipoproteins by cellulose beads covalently linked with dextran sulfate (Fig. 1).The patient's blood is withdrawn through a venous access and enters the plasma separator. As blood flows through the hollow fibers of the plasma filter, the plasma is separated and pumped into one of the two LDL adsorption columns. As the plasma passes through the column, the apoB-containing lipoproteins [LDL, lipoprotein(a) (Lp(a)), and very low-density lipoprotein (VLDL)] are selectively adsorbed by the dextran sulfate cellulose beads within the column. There is minimal effect on other plasma components such as HDL and albumin. The LDL-depleted plasma exits the column and is recombined with the blood cells exiting the separator, all of which is returned to the patient through a second venous access. When the first

Table 4
Acute Percentage Reductions in Lipids and Lipoproteins Following Treatment with The HELP and Liposorber® LA-15 Apheresis Systems

	Acute percentage reductions (%)	
Parameter	*HELP*	*Liposorber® LA-15*
Total cholesterol	49–57	61–71
LDL-C	60–64	73–83
HDL-C	13–17	3–14
Lp(a)	47–72	53–76
Triglycerides	36–47	47–86

HDL-C, high-density lipoprotein cholesterol; HELP, heparin-induced extracorporeal low-density lipoprotein precipitation; LDL-C, low-density lipoprotein cholesterol.

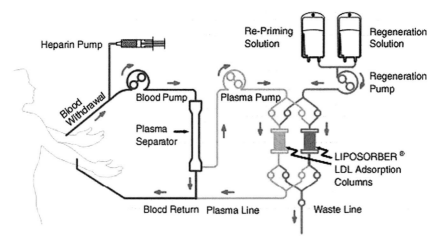

Fig. 1. The Liposorber system.

column has completed adsorbing LDL, the computer-regulated machine automatically switches the plasma flow to the second column. The plasma remaining in the first column is returned to the patient. The column is then regenerated, eluting the LDL, Lp(a), and VLDL to the waste lines. After elution, the column is reprimed completely and ready for the next cycle of adsorption, allowing continuous treatment. A typical treatment takes 2–4 h and must be repeated every 1–2 weeks.

The DSA system is contraindicated for patients taking angiotensin-converting enzyme inhibitors due to possible anaphylactic reactions, including hypotension, associated with flushing, dyspnea, and bradycardia caused by increased bradykinin levels (24). The negative charges of dextran sulfate cellulose activate the intrinsic coagulation pathway which is accompanied by bradykinin production (25). Risk of anaphalactoid-like reaction may be minimized by the temporary cessation of angiotensin converting enzyme (ACE) inhibitors for at least 24–48 h before each LDL apheresis procedure or changing the ACE inhibitor to an angiotensin receptor antagonist (26). The HELP system does not affect bradykinin levels and is safe to perform on patients using ACE inhibitors.

LDL-C will increase immediately after treatment at a nonlinear rate such that 7 days posttherapy, the LDL-C levels may have rebounded 40–70% to baseline. Due to the rapid rebound of cholesterol and clinical trial results, patients in groups 1 and 2 (i.e., patients with functional homozygous hypercholesterolemia having LDL-C levels > 500 mg/dL and patients with LDL-C ≥ 300 mg/dL) should be treated once every week, or at least every week for the first few months to lower baseline LDL-C levels, while patients in group 3 (i.e., functional hypercholesterolemic heterozygotes having LDL-C ≥ 200 mg/dL with documented CHD) can be treated every 2 weeks.

The long-term safety and clinical efficacy of the Liposorber system have been supported in a long-term study of patients with heterozygous and homozygous FH (27). In this study, 64 patients with FH (10 homozygotes and 54 heterozygotes) whose LDL-C was not adequately lowered by diet and maximum tolerated drug therapy, LDL apheresis reduced the cardiovascular event rate by 44% compared with the event rate during the 5 years before apheresis was initiated.

HEPARIN-INDUCED EXTRACORPOREAL LOW-DENSITY LIPOPROTEIN PRECIPITATION

HELP (Braun, Melsungen, Germany) is an FDA-approved procedure for specifically removing apoB-containing lipoproteins from the body *(28)*. This procedure is based on the precipitation of positively charged LDL and other β lipoproteins when heparin is added. The lipoprotein complexes are precipitated at a reaction pH of 5.12 (Fig. 2). Unlike other apheresis procedures, treatment by the HELP system removes a substantial quantity of fibrinogen (approximately 50% reduction) *(29)*. The safety and long-term applicability of the HELP system have been proven in more than 100,000 treatments, and serious complications have not been observed *(30,31)*. Problems of hypotension are rare provided that a well-trained operator carries out therapy. There is extensive experience with HELP therapy in patients with hypercholesterolemia who have been diagnosed with premature CHD, as well as in patients after coronary artery bypass surgery or post-cardiac transplantation *(32,33)*. HELP therapy can be given to patients who are on oral anticoagulation therapy as well as on medications that affect platelet aggregation, such as aspirin or ticlopidine. The elimination of fibrinogen can be precisely monitored, which results in fewer bleeding complications. In addition, HELP therapy can be tolerated by patients who have decreased cardiac ejection fractions because the flow and volume can also be precisely regulated, so that there is no more than 100 mL/min of plasma removed from the patient.

A postmarketing surveillance study of HELP therapy that included 628 patients with hypercholesterolemia and/or CAD treated multiple times over 5 years in 82 centers in Germany yielded only eight bleeding complications. Most complications were associated with difficulty in establishing venous access. In addition, plasma electrolytes, hormones, vitamins, enzymes, and Ig concentrations essentially stay the same after each LDL apheresis treatment as well as on long-term utilization of HELP therapy, and in combination with HMG-CoA reductase inhibitors *(23,28,34,35)*.

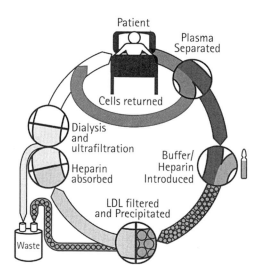

Fig. 2. Flow diagram of the heparin-induced extracorporeal low-density lipoprotein precipitation (HELP) system.

DIRECT ADSORPTION OF LIPOPROTEINS BLOOD PERFUSION SYSTEM

The direct adsorption of lipoproteins [direct adsorption of low-density lipoproteins (DALI)] system (Fresenius, St. Wendel, Germany) is the first whole blood apheresis system in regular clinical use by which atherogenic LDL and Lp(a) can be selectively removed from whole blood without plasma separation *(36)*. Blood is perfused through an adsorber, which contains polyacrylate-coated beads. LDL-C and Lp(a) are eliminated by adsorption onto the polyacrylate-coated beads. Polyacrylate consists of polyanions with negatively charged carboxylate groups. The binding mechanism is related to an electrostatic interaction between the negatively charged carboxylate groups of the polyanionic polyacrylate ligands and the positive charge of the apoB moiety of LDL *(37)*. Unlike the Liposorber and HELP systems, in which anticoagulation of the blood is achieved by heparin infusion, the DALI system performs anticoagulation of the extracorporeal circuit by continuous application of a citrate solution after an initial heparin bolus *(38)*. Similar to the DSA system, DALI is contraindicated for patients taking angiotensin-converting enzyme inhibitors due to possible anaphylactic reactions.

Since 2002, Kaneka has also offered a whole blood LDL apheresis system (Liposorber D) *(39)*. Like the DALI system, the Liposorber D has not received FDA approval for use in the United States.

DOUBLE-MEMBRANE FILTRATION

The membrane filtration technique, mainly used in Europe and Japan, involves the filtration of blood by filters of different pore sizes. It is a selective process, in which these filters are used sequentially in a process called double or cascade filtration, enabling removal of specific LDL and Lp(a) and discrimination of smaller molecules such as albumin and HDL-C *(40)*. However, the semi-selective process can significantly remove HDL-C, alpha 2-macroglobulin and IgM *(41)*.

IMMUNOADSORPTION

Immunoadsorption was the first selective LDL apheresis system used clinically *(42)*. Plasma LDL is removed by reusable columns of polyclonal sheep antibodies to human apoB-100. Anti-sheep IgG can be detected in most patients receiving treatments. One variation of immunoadsorption (i.e., Lipopak) specifically removes Lp(a) *(43)*. Similar to whole blood and the Liposorber system, little if any fibrinogen is removed by the device.

LDL Apheresis and Lipid-Lowering Therapy

The combination of LDL apheresis with high doses of HMG-CoA reductase inhibitors (i.e., statins) has been the best currently available therapy for the prevention of cardiovascular disease in patients with heterozygous FH. For example, Mabuchi and colleagues *(44)* performed one of the most significant event reduction trials related to coronary heart disease (CHD) and LDL apheresis. In their 6-year study, heterozygous FH patients with angiographically documented CHD were treated with LDL apheresis combined with cholesterol-lowering drugs ($n = 43$) or cholesterol-lowering drug therapy alone ($n = 87$). LDL apheresis significantly reduced LDL-C levels from 7.42 ± 1.73 to 3.13 ± 0.80 mmol/L (-58%) compared with the drug therapy group, from 6.03 ± 1.32 to 4.32 ± 1.53 mmol/L (-28%). Kaplan–Meier analyses revealed that the rate of total coronary events, including nonfatal myocardial infarction, percutaneous transluminal coronary angioplasty, coronary artery bypass grafting, and death from CHD, was 72% lower in the LDL apheresis group (10%) than in the drug therapy group (36%) ($P = 0.0088$).

VASCULAR EFFECTS IN ADDITION TO LDL LOWERING AND PLAQUE REGRESSION

In addition to LDL-C, apheresis modifies a host of other vascular markers and pathologic processes associated with CAD. The mechanism for the reductions of these markers is not fully understood but believed to be similar to the removal of lipoproteins. This section will address these pathophysiological properties of only FDA-approved devices (Liposorber and HELP).

Inflammation

An accumulation of clinical evidence has demonstrated that markers of inflammation correlate with cardiovascular disease. For example, elevated levels of the acute phase reactant, C-reactive protein (CRP), can predict prospectively the risk of CAD. In fact, clinical research (PROVE-IT, A to Z) *(45,46)* suggests that during the acute coronary syndrome it may be more advantageous to reduce inflammation, as measured by CRP, than LDL-C. LDL apheresis treatment acutely lowers CRP by up to 65% *(47–50)*, and chronic therapy reduces mean baseline levels by 50% *(51)*. Post-apheresis filtrate has been found to contain both free and LDL-C-bound CRPs *(49)*. Other markers related to vascular inflammation found to be improved by apheresis include lipoprotein-associated phospholipase A_2 *(52)*, fibrinogen, E-selectin, vascular cellular adhesion molecule-1, intercellular adhesion molecule-1 *(53)*, monocyte chemoattractant protein-1, lipopolysaccharide-binding protein *(54,55)*, matrix metalloproteinase-9, and tissue inhibitor of metalloproteinase-1 *(56)*. Table 5 lists various markers of inflammation that have shown improvement after LDL apheresis.

Lp(a), Small Dense LDL, Oxidized LDL, and Vascular Motion

Lp(a) is a powerful predictor of premature CVD in patients with simultaneous hypercholesterolemia *(57)*. Mechanisms of Lp(a) association with atherogenesis include induction of adhesion molecules, foam cell formation, plaque inflammation, inhibition of nitric oxide (NO), and increased thrombosis *(58)*. LDL apheresis is presently the only treatment that significantly reduces Lp(a) levels.

Epidemiological data indicate that small condensed LDL is a more atherogenic risk than large, buoyant LDL *(59)* and a strong predictor of coronary stenosis *(60)*. Schamberger and colleagues *(61)* demonstrated that 32 patients who received LDL apheresis (i.e., immunoadsorption, DSA, HELP, and cascade filtration) experienced a relative increase in large LDL subfractions and a concomitant decrease of small dense LDL. The type of apheresis device did not influence the outcome.

Table 5
Proinflammatory Markers: % Changes Following Apheresis

Marker	Acute percentage reductions (%)	
	HELP	Liposorber® LA-15
MCP-1	–15	–20
MMP-9	ND	–20
TIMP-1	ND	–30
ET-1	–25	–75
LBP	–27	ND
Lp-PLA$_2$	–22	ND
VCAM-1	–20	–10
ICAM-1	–10	–10
E-Selectin	–31	6
Fibrinogen	–65	–20
Oxidized LDL	–65	–65
CRP	–65	–65

CRP, C-reactive protein; ET-1, endothelin-1; HELP, heparin-induced extracorporeal low-density lipoprotein precipitation; LBP, lipopolysaccharide-I binding protein; LP-PLA$_2$, lipoprotein-associated phospholipase A$_2$; MCP-1, monocyte chemoattractant protein-1; MMP-9, matrix metalloproteinase-9; ND, no data available; TIMP-1, tissue inhibitor of metalloproteinase-1; VCAM-1, vascular cellular adhesion molecule-1.

Proinflammatory oxidized LDL (Ox-LDL) may be the unifying link between inflammation and lipid deposition into the vessel wall. Ox-LDL has been strongly associated with CAD and acute coronary syndrome *(62)*. LDL apheresis significantly lowers Ox-LDL *(63,64)*, which correlates with the increase of NO levels and vasodilatation *(65)*. Besides NO, other mediators of vasomotion improved by LDL apheresis include bradykinin, endothelin-1, and prostaglandin I_2 (PGI$_2$) *(66)*. Table 6 provides the percent changes following apheresis.

Table 6
Vasomotion markers: % Changes Following Apheresis

Marker	Acute percentage reductions (%)	
	HELP	Liposorber® LA-15
Nitric oxide	ND	45
Bradykinin	0	> 2000
Endothelin-1	–25	–75
PGI$_2$	ND	300

ND, no data available; HELP, heparin-induced extracorporeal low-density lipoprotein precipitation; PGI$_2$, prostaglandin I$_2$.

Thrombosis and Fibrinolysis

Blood coagulation and thrombosis result in the critical event following a rupture of the atheroma. The release of certain factors such as tissue factor and von Willebrand factor by monocytes can activate the prothrombotic pathway resulting in thrombus formation. LDL apheresis immediately lowers coagulation factors (Table 7) such as tissue factor *(54)*, von Willebrand factor *(67)*, Factor V, Factor VII, soluble CD40 ligand, homocysteine, and fibrinogen. The Liposorber system primarily influences the intrinsic coagulation pathway (IX, XI, and prekallkrein). The HELP system preferentially reduces factors from the extrinsic clotting pathway (II, VII, and X) and fibrinogen *(68,69)*. Apheresis can also lower markers that promote the fibrinolytic cascade (Table 8) including plasminogen, protein S, protein C, and antithrombin *(67,69)*. Most thrombotic/fibrinolytic markers return to baseline levels within 24 h. The return to baseline is generally dependent on the magnitude of reduction by apheresis and the plasma half-life of the particular protein. For example, fibrinogen, which is reduced by 65% after HELP apheresis and has a half-life > 100 h, does not return to baseline levels until the 3rd or 4th day. No acute-phase rebound above baseline levels has been identified with either system.

Table 7
Thrombotic Markers: % Changes Following Apheresis

Marker	Acute percentage reductions (%)	
	HELP	*Liposorber® LA-15*
Tissue Factor	−26	ND
von Wilebrand Factor	−56	−48
Thrombin	−55	ND
Factor V	−57	−74
Factor VII	−35	−35
Factor XI	−56	−82
Factor XII	−50	−73
sCD40L	−16	ND
Homosystine	−25	−15
Fibrinogen	−65	−20

ND, no data available.

Table 8
Fibrinolytic Markers: % Changes Following Apheresis

Marker	Acute percentage reductions (%)	
	HELP	*Liposorber® LA-15*
Plasminogen	−50	−23
Protein S	−35	−27
Protein C	−48	−32
Antithrombin	−25	−11

HELP, heparin-induced extracorporeal low-density lipoprotein precipitation.

Despite the reduction of plasma thrombotic and fibrinolytic mediators, 20 years of clinical use with apheresis have not demonstrated any evidence of increased complications of clotting or bleeding. The lack of adverse effect s (i.e., thrombosis/fibrinolysis) following a treatment of apheresis may be attributed to the equal removal of these mediators by the system.

Hemorheology (Blood Viscosity)

The arterial system is controlled and subjugated by the body's need for oxygen. This requirement is met by the ability of the vasculature to adjust its resistance and consequently change the flow of blood. As described by Poiseuille:

$$\text{Vascular Resistance}(R) = 8\eta L/\pi r^4,$$

where L = length, r = radius, and η = blood viscosity.

In blood vessels where change in diameter is not significant, such as capillaries or atherosclerotic arteries, viscosity dictates the resistance and blood flow during moments of increased oxygen demand *(70,71)*. The ability to acutely improve viscosity can have an immediate effect on vascular flow. Mediators of blood viscosity (besides hematocrit, shear forces, and temperature) include red blood cell (RBC) deformability, RBC aggregation, and plasma viscosity.

LDL apheresis, after one treatment, reduces blood viscosity by 20% *(72)* and remains lowered for at least 1 week *(73)*. As summarized in Table 9, the reduction of blood viscosity by LDL apheresis is achieved through its capability of improving RBC aggregation/deformability and plasma viscosity *(74)*. The HELP system can acutely reduce blood viscosity more effectively than the Liposorber. The rheologic advantage of HELP is largely due to its dramatic reduction *(75)* of fibrinogen, which can represent 20% of plasma viscosity *(76)*.

The improved coronary microcirculation and altered hemorheology, particularly blood viscosity, may help explain the rapid recovery in myocardial blood flow and minimum coronary resistance seen in patients with CAD after one treatment of LDL apheresis (Table 10) *(77)*.

Table 9
Hemorheology Markers: % Changes Following Apheresis

Markers	Acute percentage reductions (%)	
	HELP	Liposorber® LA-15
Plasma Viscosity	15	11
Blood Viscosity	15	5
RBC Aggregation	52	31
RBC Deformity	45	45
Fibrinogen	62	11

Note: HELP, heparin-induced extracorporeal low-density lipoprotein precipitation.
Source: Comparison from head to head study from Grutzmacher et al. *(114)*.

Table 10
Vasodilation Capacity After Single LDL Apheresis (HELP) (18 h posttherapy)

Flow dynamics	Pre-apheresis	Post-apheresis	Difference
MBF (ml/min per 100g)	173	226	23%[a]
CFR	1.91	2.48	23%[b]
MCR (mmHg per 100g/mL)	0.61	0.43	29%[b]

CFR, coronary flow reserve:
MBF, myocardial blood flow;
MCR, minimal coronary resistance.
[a] $p < 0.01$; [b] $p < 0.02$

ALTERNATIVE APPLICATIONS OF LDL APHERESIS

For decades plasmapheresis and now LDL apheresis have been utilized in non-FH patients to treat other complex vascular diseases. Due to its pleiotropic effects on plasma levels of inflammation, coagulation, rheology, lipids, and vasoreactivity, the use of apheresis has been considered in acute and chronic situations in which standard therapy has failed. This section reviews the clinical data related to the usefulness of apheresis in specific vascular diseases.

Sudden Idiopathic Hearing Loss

The pathogenesis of sudden hearing loss is not well understood. A suggested cause *(78)* has been attributed to local hypoperfusion from increased viscosity due to elevated plasma levels of fibrinogen or cholesterol, resulting in altered blood flow and hearing loss. Inflammation, coagulation, and autoimmune mechanisms have also been associated with the disease, which has been effectively treated with apheresis *(79)*.

Suckfull et al. *(80)* randomized 201 patients with sudden-onset sensorineural hearing loss in one ear to standard therapy (10 days of prednisolone and infusion with plasma expander) or a one-time treatment with LDL apheresis (HELP). Results demonstrated LDL apheresis was as effective as conventional therapy and could be used as an alternative therapy particularly in patients when plasma fibrinogen and LDL-C were elevated.

Renal Disease

Diabetic nephropathy is a leading cause of end-stage renal disease. Nakamura et al. *(113)* used LDL apheresis (Liposorber) 12 times over a 9-week period in 8 patients with long standing type 2 diabetes and nephrotic syndrome. In comparison to the control group, the apheresis patients revealed a significant reduction of creatinine, blood urea nitrogen, urinary protein excretion, number of urinary podocytes, and increased creatinine clearance. Muso et al. *(81)* using LDL apheresis (Liposorber) treated 17 patients with steroid-resistant nephrotic syndrome, due to focal segmental glomerulosclerosis, again 12 times over a 9-week period. Apheresis treatment significantly lowered urinary protein and increased serum albumin when compared to the steroid treated control group.

Patients with end-stage renal disease, requiring hemodialysis, have an elevated incidence of atherosclerosis CAD mortality, the cause of which still remains undefined. Nakamura et al. *(56)* studied 30 hemodialysis patients with arteriosclerosis obliterans

(ASO). Eleven patients underwent weekly LDL apheresis (Liposorber) for 10 weeks and unlike the hemodialysis only group demonstrated significant decrease in carotid intima-media thickness and improved pulse wave velocity of the posterior tibial arteries.

Hemodialysis increases Ox-LDL-C levels, impairs endothelial function *(83)*, and worsens the rheological profile by increasing plasma viscosity and fibrinogen *(84)*. The potential benefit LDL apheresis to dialysis patients may be related to the reduction of lipids, inflammation, and blood viscosity.

Cardiac Transplant

The development of CAD in cardiac transplant (CTX) patients is not well understood. Complications such as hyperlipidemia, rejection crises and hypertension accelerate atherosclerosis in comparison to the general population *(85)*. Elevated serum concentrations of Lp(a), irrespective of other factors, is an independent risk factor for the development of accelerated CAD among patients with CTX *(86)*.

Park et al. *(32)*, using weekly LDL apheresis (HELP), treated eight CTX patients with cardiac allograft vasculapathy (CAV), as documented by intravascular ultrasound. Angiographic measurements of all coronary artery segments following 1–2.5 years of therapy revealed significant increase in mean luminal diameter. In another CTX patient population ($n = 10$) with CAV and a mean LDL-C of 170mg/dL, Matschke et al. *(87)* discovered one treatment of LDL apheresis (HELP) increased intramuscular (anterior tibial) partial pressure (pO_2) by over 150% at values similar to those found in healthy subjects. Jaeger et al. *(31)* compared graft vessel disease in CTX patients ($n = 10$) on lipid-lowering therapy and weekly LDL apheresis (HELP) to a similar numbered control group on medication alone. After a 3.6-year period, coronary angiography revealed 70% of the control group had an increase of vessel disease compared with only 10% ($P = 0.006$) in the apheresis group.

Ocular Microcirculatory Disturbances

Occlusion of the retinal vein and artery system, particularly nonarteritic acute anterior ischemic optic neuropathy (NAION), is the most frequent acute disease of the optic nerve in adults over the age of 50 *(88)* and often leads to irreversible vision loss *(89)*. No present therapy (steroids, platelet antiaggregants, and hemorheologic drugs) has gained acceptance due to lack of significant improvement *(90)*. Risk factors include atherosclerosis, diabetes, and blood viscosity *(91,92)*.

Walzl et al. *(93)*, using LDL apheresis (HELP), treated five patients with visual loss due to microcirculatory disease (two branch retinal artery occlusion, one central retinal artery and vein occlusion, and one NAION). After six treatments, each patient experienced an improvement of visual field and visual acuity. Ramunni et al. *(94)* treated 11 patients suffering acute NAION with mean LDL-C levels of 144mg/dL. Patients received three treatments of LDL apheresis (HELP) over a 17-day period. After the three treatments and 3 months later, all patients experienced improvement in their visual acuity (3.7/10 versus 7.9/10, $P = 0.002$), while in three controlled group patients, on standard therapy, visual acuity declined (9.3/10 versus 3.3, $P = 0.0001$). In one case study *(95)* of a 29-year-old female with central nervous system vasculitis and progressive bilateral loss of vision, due to retinal artery occlusion, four treatments of LDL apheresis (HELP) over a 2-week period stabilized her vision and significantly improved blood flow to both the inferior and superior branches of the retinal artery

(Figs. 3 and 4). Besides lowering serum lipids, the treatments dramatically reduced both fibrinogen (82%) and blood viscosity (32%).

Age-related macular degeneration (AMD) is the leading cause of irreversible visual impairment and blindness in persons over 60 years of age *(96)*. The exact cause of AMD is unknown. Pathophysiologically relevant risk factors include LDL-C, von Willebrand factor, and fibrinogen *(97)*. Studies *(98)* have confirmed that the use of cholesterol-lowering medications may reduce the risk of developing AMD. In a multi-center, prospective, randomized, double-masked, placebo-controlled trial, 43 patients with AMD were treated with either a double-membrane filtration apheresis system (Rheopheresis), a non-FDA-approved apheresis system, or sham therapy 8 times over

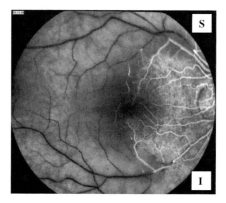

Fig. 3. Retinal angiogram, before low-density lipoprotein apheresis, reveals a decrease in perfusion (white) for both the superior (S) and inferior (I) temporal branches of the retinal artery.

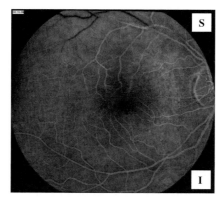

Fig. 4. Retinal angiogram taken after 2 weeks and after 4 treatments of low-density lipoprotein apheresis demonstrates the complete resolution of flow in the inferior (I) branch with an increased perfusion of the superior (S) branch.

a 10-week period *(99)*. At 12 months, apheresis demonstrated statistically significant ($P = 0.001$) and clinically relevant effects on best-corrected visual acuity when compared with the placebo controls.

Peripheral Vascular Disease

LDL apheresis has demonstrated its ability to slow the progression of atherosclerosis in peripheral atherosclerosis lesions. The LDL Apheresis Atherosclerosis Regression Study (LAARS), *(100)* a prospective randomized, single-center investigation, assessed the effect of apheresis plus medication (simvastatin 40 mg) or medication alone in 42 men with hypercholesterolemia and CHD. The study included a subset trial *(101)* in which one-half of the patients ($n = 21$) with CHD and peripheral vascular disease (PVD) were treated with biweekly apheresis (Liposorber) plus simvastatin 40mg/day or simvastatin 40mg/day, for 2 years. In the apheresis group, the number of patients with hemodynamically significant stenoses in the aortotibial tract decreased from 9 to 7; while in the simvastatin group, the number increased from 6 to 13 ($P = 0.002$). Also, mean carotid intima-media thickness decreased by a mean of 0.05 mm in the apheresis group and increased by 0.06 mm in the simvastatin-only group ($P < 0.001$).

Kobayashi et al. *(55)* administered LDL apheresis (Liposorber) 10 times over a 5-week period to 28 patients with PVD (15 patients with diabetes and 19 patients undergoing hemodialysis). They confirmed significant rates of symptomatic improvement of foot chillness or numbness at 82%, claudication at 54%, and foot ulcers at 14%, while ankle-brachial pressure index increased from 0.69 to 0.85 ($P < 0.005$) before and 3 months after LDL apheresis.

The strength of these and other clinical trials demonstrating the benefit of apheresis therapy to drug-resistant ASO patients has convinced Japanese health care providers to cover the costs of 10 treatments over a 3-month period *(102)*.

Cerebral Vascular Disease

Cerebral vascular disease (CVD), particularly stroke, is the third most common cause of death *(103)*. Ninety percent of strokes are ischemic and in most cases linked with atherosclerosis. As demonstrated in CAD, inflammation, particularly plasma CRP, may play an important role in the disease. CRP levels *(104)* measured in the first week after an ischemic stroke, correlated significantly with brain infarct volume and mortality. Prospective studies have demonstrated that statins, which can lower both lipids and CRP, reduce stroke occurrence by 20–25% *(105)*. Schwartz et al. *(106)* demonstrated initiation of early (< 72 h) and aggressive statin therapy (atorvastatin 80mg/dL) in patients experiencing the acute coronary syndrome, resulted in a 50% reduction of cerebral vascular events after 16 weeks. In addition to inflammation, hemorheologic disturbances can occur in $> 40\%$ of patients with ischemic cerebrovascular diseases *(107)*.

The influence of LDL apheresis on the cerebrovascular system has been analyzed in clinical trials. Pfefferkorn et al. *(108)* measured CO_2 reactivity, a marker of cerebral vasoreactivity, in 13 patients (7 with CTX) with CAD and hyperlipidemia (mean LDL-C 244 mg/dL) following LDL apheresis (HELP). After a single treatment, patients CO_2 reactivity increased by 14% ($P < 0.05$) and plasma viscosity was reduced by 14% ($P < 0.001$). Walzl *(109)* studied 26 patients with acute embolic stroke and 22 with multi-infarct dementia (MID). Thirty patients, 16 with stroke and 14 with MID, were part of the control group. All patients were subjected to two sessions of apheresis

Table 11
Impact on Rating Scales of HELP Treatment in Relation to Controls

Scale	First treatment before HELP/Control	After HELP/Control	After second treatment HELP/Control	Day 11 HELP/Control
MS	85.8/85.3	90.9[a]/84.5	93.5[itc]/85.9	94.3/87.3[f]
MMSE	26.0/26.1	27.6[a]/25.8[itb]	29.1[c]/26.2[e]	29.3/27.7[f]
ADL	85.5/84.9	89.8[a]/86.2[b]	92.5[d]/86.4[d]	93.2/86.9[f]

ADL, activities of daily living scale; HELP, Heparin-induced extracorporeal low-density lipoprotein precipitation; MMSE, mini-mental state examination; MS, Mathew scale.
HELP group, $n = 48$; control group, $n = 30$.
[a] $P < .05$ before and after first HELP treatment.
[b] $P < .05$ between HELP group and control group.
[c] $P < .01$ before first HELP treatment.
[d] $P < .05$ after first HELP treatment.
[e] $P < .01$, $P < .05$ between HELP group and control group after second HELP treatment.
[f] $P < .05$ between HELP group and control group at day 11.

(HELP) within 8 days. Control patients were connected to the machine and blinded to the sham treatment. The treated group demonstrated improvement relative to the control group ($P < 0.05$) in the Mathew scale, Mini-mental State Examination, and activities of daily living test scores (Table 11). Apheresis also produced an immediate and significant reduction of rheological markers (fibrinogen 34%, blood viscosity 17%, plasma viscosity 16%, and red cell transit time 17%).

SUMMARY

LDL apheresis is a safe and effective means of lowering plasma cholesterol and reducing cardiovascular events in patients with uncontrolled hypercholesterolemia. Despite evidence-based assistance and approval for reimbursement for this high-risk patient population, only a fraction (< 200) of potentially eligible candidates (> 6000) receive apheresis therapy in the United States.

Recently, large multicenter clinical studies *(45,110,111)* in patients with CAD have demonstrated three major points of interest related to the treatment of plasma cholesterol levels and the prevention of vascular disease: (1) lipid-lowering therapy, for example statins, can significantly improve clinical outcomes irrespective of baseline plasma cholesterol levels; (2) further reduction of LDL-C beyond previously recognized national guidelines levels may result in additional benefit to the patient; and (3) the level of systemic inflammation, such as CRP and Ox-LDL, may have a consequence on the outcome of both acute and chronic events related to atherosclerosis. All of these points reinforce the advantage of using LDL apheresis for qualified patients, as designated by the present guidelines, and support its potential use in acute and chronic vascular disease irrespective of baseline cholesterol levels.

REFERENCES

1. Stamler J, Wentworth D, Neaton JD. Is relationship between serum cholesterol and risk of premature death from coronary heart disease continuous and graded? Findings in 356,222 primary screenees of the Multiple Risk Factor Intervention Trial (MRFIT). JAMA 1986; 256:2823–8.

2. Murray CJ, Lopez AD. Mortality by cause for eight regions of the world: Global Burden of Disease Study. [see comment]. Lancet 1997; 349:1269–76.
3. Hazzard WR, Goldstein JL, Schrott MG, et al. Hyperlipidemia in coronary heart disease. 3. Evaluation of lipoprotein phenotypes of 156 genetically defined survivors of myocardial infarction. J Clin Invest 1973; 52:1569–77.
4. Scandinavian Simvastatin Survival Study G. Randomised trial of cholesterol lowering in 4444 patients with coronary heart disease: the Scandinavian Simvastatin Survival Study (4S). Lancet 1994; 344:1383–9.
5. Shepherd J, Cobbe SM, Ford I, et al. Prevention of coronary heart disease with pravastatin in men with hypercholesterolemia. West of Scotland Coronary Prevention Study Group. [see comment]. N Engl J Med 1996; 333:1301–7.
6. Sacks FM, Pfeffer MA, Moye LA, et al. The effect of pravastatin on coronary events after myocardial infarction in patients with average cholesterol levels. Cholesterol and Recurrent Events Trial investigators. [see comment]. N Engl J Med 1996; 335:1001–9.
7. Goldstein JL, Hobbs HH, Brown MS. Familial hypercholesterolemia. In: Scriver CR, Beauder WS, Sly WS, Valle D, eds. The Metabolic Basis of Inherited Diseases. New York: McGraw-Hill Publishing Co, 1995:1981–2030.
8. Patterson D, Slack J. Lipid abnormalities in male and female survivors of myocardial infarction and their first-degree relatives. Lancet 1972; 1:393–9.
9. Slack J. Inheritance of familial hypercholesterolemia. Atheroscler Rev 1979; 5:35–6.
10. Leitersdorf E, Tobin EJ, Davignon J, et al. Common low-density lipoprotein receptor mutations in the French Canadian population. J Clin Invest 1014; 85:1014–23.
11. Austin MA, Hutter CM, Zimmern RL, et al. Genetic causes of monogenic heterozygous familial hypercholesterolemia: a HuGE prevalence review. Am J Epidemiol 2004; 160:407–20.
12. Goldstein J, Hobbs H, Brown M. Familial hypyercholesterolemia. In: Scriver C, Sly W, Childs B, Valle, eds. The Metabolic and Molecular Bases of Inherited Diseases. New York: McGraw-Hill Companies, Inc., 2001:2863–914.
13. Schuster H, Luft FC. Clinical criteria versus DNA diagnosis in heterozygous familial hypercholesterolemia: is molecular diagnosis superior to clinical diagnosis? [comment]. Arterioscler Thromb Vasc Biol 1998; 18:331–2.
14. Anonymous. Mortality in treated heterozygous familial hypercholesterolaemia: implications for clinical management. Scientific Steering Committee on behalf of the Simon Broome Register Group. Atherosclerosis 1999; 142:105–12.
15. Anonymous. Risk of fatal coronary heart disease in familial hypercholesterolaemia. Scientific Steering Committee on behalf of the Simon Broome Register Group. [see comment]. BMJ 1991; 303:893–6.
16. Williams RR, Hunt SC, Schumacher MC, et al. Diagnosing heterozygous familial hypercholesterolemia using new practical criteria validated by molecular genetics. Am J Cardiol 1993; 72:171–6.
17. Slack J. Risks of ischaemic heart-disease in familial hyperlipoproteinaemic states. Lancet 1969; 2:1380–2.
18. Stone NJ, Levy RI, Fredrickson DS, et al. Coronary artery disease in 116 kindred with familial type II hyperlipoproteinemia. Circulation 1974; 49:476–88.
19. Barbagallo CM, Averna MR, DiMarco T, et al. Effectiveness of cascade filtration plasmapheresis in two patients affected by familial hypercholesterolemia. J Clin Apher 1995; 10:96–100.
20. Knopp RH, McKenney JM. Lipid metabolism and lipid disorders. In: McKenney JM, Hawkins D, eds. Handbook on the Management of Lipid Disorders. St. Louis, MO: National Pharmacy Cardiovascular Council, 2001:34–58.
21. de Gennes JL, Touraine R, Maunand B, et al. [Homozygous cutaneo-tendinous forms of hypercholesteremic xanthomatosis in an exemplary familial case. Trial of plasmapheresis as heroic treatment]. Bull Mem Soc Med Hop Paris 1967; 118:1377–402.
22. Thompson GR, Lowenthal R, Myant NB. Plasma exchange in the management of homozygous familial hypercholesterolaemia. Lancet 1975; 1:1208–11.
23. Thiery J, Seidel D. Safety and effectiveness of long-term LDL-apheresis in patients at high risk. Curr Opin Lipidol 1998; 9:521–6.
24. Hershcovici T, Schechner V, Orlin J, et al. Effect of different LDL-apheresis methods on parameters involved in atherosclerosis. J Clin Apher 2004; 19:90–7.
25. Kojima S, Yoshitomi Y, Saotome M, et al. Effects of losartan on low-density lipoprotein apheresis. Ther Apher 1999; 3:303–6.
26. Sinzinger H, Chehne F, Ferlitsch A, et al. Angiotensin receptor antagonists during dextran sulfate LDL-apheresis are safe. Thromb Res 2000; 100:43–6.

27. Gordon BR, Kelsey SF, Dau PC, et al. Long-term effects of low-density lipoprotein apheresis using an automated dextran sulfate cellulose adsorption system. Liposorber Study Group. Am J Cardiol 1998; 81:407–11.
28. Seidel D. H.E.L.P. apheresis therapy in the treatment of severe hypercholesterolemia: 10 years of clinical experience. Artif Organs 1996; 20:303–10.
29. Thiery J, Seidel D. [New strategies in treatment of severe hypercholesterolemia in coronary patients: HMG-CoA reductase inhibitors and H.E.L.P.-LDL apheresis]. Versicherungsmedizin 1992; 44:186–92.
30. Seidel D, Armstrong VW, Schuff-Werner P. The HELP-LDL-apheresis multicentre study, an angiographically assessed trial on the role of LDL-apheresis in the secondary prevention of coronary heart disease. I. Evaluation of safety and cholesterol-lowering effects during the first 12 months. HELP Study Group. Eur J Clin Invest 1991; 21:375–83.
31. Jaeger BR, Meiser B, Nagel D, et al. Aggressive lowering of fibrinogen and cholesterol in the prevention of graft vessel disease after heart transplantation. Circulation 1997; 96:154–8.
32. Park JW, Merz M, Braun P. Regression of transplant coronary artery disease during chronic low-density lipoprotein-apheresis. J Heart Lung Transplant 1997; 16:290–7.
33. Blessing F, Wang Y, Walli AK, et al. Heparin-mediated extracorporeal low-density lipoprotein precipitation: rationale for a specific adjuvant therapy in cardiovascular disease. Transfus Apher Sci 2004; 30:255–66.
34. Seidel D. The HELP system: an efficient and safe method of plasmatherapy in the treatment of severe hypercholesterolemia. Ther Umsch 1990; 47:514–9.
35. Thiery J. Maximaltherapie der Hypercholesterinaemie bei koronarer Herzkrankheit. Therapiewoche 1988; 38:3424–37.
36. Bosch T, Lennertz A, Schenzle D, et al. Direct adsorption of low-density lipoprotein and lipoprotein(a) from whole blood: results of the first clinical long-term multicenter study using DALI apheresis. J Clin Apher 2002; 17:161–9.
37. Bosch T. Low-density lipoprotein hemoperfusion using a modified polyacrylate adsorber: in vitro, ex vivo, and first clinical results. Artif Organs 1996; 20:344–5.
38. Krieter DH, Steinke J, Kerkhoff M, et al. Contact activation in low-density lipoprotein apheresis systems. Artif Organs 2005; 29:47–52.
39. Otto C, Kern P, Bambauer R, et al. Efficacy and safety of a new whole-blood low-density lipoprotein apheresis system (Liposorber D) in severe hypercholesterolemia. Artif Organs 2003; 27:1116–22.
40. Matsuda Y, Sueoka A, Nose Y. An effective LDL removal filter for the treatment of hyperlipidemia. Artif Organs 1995; 19:129–34.
41. Olbricht CJ. Extracorporeal treatment of hypercholesterolaemia. Nephrol Dial Transplant 1993; 8:814–20.
42. Stoffel W, Demant T. Selective removal of apolipoprotein B-containing serum lipoproteins from blood plasma. Proc Natl Acad Sci USA 1981; 78:611–5.
43. Bambauer R. Low-density lipoprotein apheresis: clinical results with different methods. Artif Organs 2002; 26:133–9.
44. Mabuchi H, Koizumi J, Shimizu M, et.al. Long-term efficacy of low-density lipoprotein apheresis on coronary heart disease in familial hypercholesterolemia. Am J Cardiol 1998; 82:1489–95.
45. Cannon CP, Braunwald E, McCabe CH, et al. Intensive versus moderate lipid lowering with statins after acute coronary syndromes. [see comment]. N Engl J Med 2004; 350:1495–504.
46. de Lemos JA, Blazing MA, Wiviott SD, et al. Early intensive vs a delayed conservative simvastatin strategy in patients with acute coronary syndromes: phase Z of the A to Z trial. [see comment]. JAMA 2004; 292:1307–16.
47. Moriarty P, Gibson C. C-reactive protein and other markers of inflammation among patients undergoing HELP LDL apheresis. Atherosclerosis 2001; 158:495–98.
48. Kobayashi J, Katsube S, Shimoda M, et al. Single LDL apheresis improves serum remnant-like particle-cholesterol, C-reactive protein, and malondialdehyde-modified-low-density lipoprotein concentrations in Japanese hypercholesterolemic subjects. Clin Chim Acta 2002; 321:107–12.
49. Kojima S, Shida M, Yokoyama H. Changes in C-reactive protein plasma levels during low-density lipoprotein apheresis. Ther Apher Dial 2003; 7:431–4.
50. Wieland E, Schettler V, Armstrong VW. Highly effective reduction of C-reactive protein in patients with coronary heart disease by extracorporeal low density lipoprotein apheresis. Atherosclerosis 2002; 162:187–91.

51. Otto C, Geiss HC, Empen K, et al. Long-term reduction of C-reactive protein concentration by regular LDL apheresis. Atherosclerosis 2004; 174:151–6.
52. Moriarty PM, Gibson CA. Effect of low-density lipoprotein apheresis on lipoprotein-associated phospholipase A2. Am J Cardiol 2005; 95:1246–7.
53. Empen K, Otto C, Brodl UC, et al. The effects of three different LDL-apheresis methods on the plasma concentrations of E-selectin, VCAM-1, and ICAM-1. J Clin Apher 2002; 17:38–43.
54. Wang Y, Blessing F, Walli AK, et al. Effects of heparin-mediated extracorporeal low-density lipoprotein precipitation beyond lowering proatherogenic lipoproteins–reduction of circulating proinflammatory and procoagulatory markers. Atherosclerosis 2004; 175:145–50.
55. Kobayashi S, Moriya H, Maesato K, et al. LDL-apheresis improves peripheral arterial occlusive disease with an implication for anti-inflammatory effects. J Clin Apher 2005. 20(4):239–43.
56. Nakamura T, Matsuda T, Suzuki Y, et al. Effects of low-density lipoprotein apheresis on plasma matrix metalloproteinase-9 and serum tissue inhibitor of metalloproteinase-1 levels in diabetic hemodialysis patients with arteriosclerosis obliterans. ASAIO J 2003; 49:430–4.
57. Stein JH, Rosenson RS. Lipoprotein Lp(a) excess and coronary heart disease. [see comment]. Arch Intern Med 1997; 157:1170–6.
58. Deb A, Caplice NM. Lipoprotein(a): new insights into mechanisms of atherogenesis and thrombosis. Clin Cardiol 2004; 27:258–64.
59. Hirano T, Ito Y, Koba S, et al. Clinical significance of small dense low-density lipoprotein cholesterol levels determined by the simple precipitation method. [Miscellaneous Article]. Arterioscler Thromb Vasc Biol March 2004; 24:558–63.
60. Williams PT, Superko HR, Haskell WL, et al. Smallest LDL particles are most strongly related to coronary disease progression in men. [Miscellaneous Article]. Arterioscler Thromb Vasc Biol February 2003; 23:314–21.
61. Schamberger B, Geiss H, Ritter M, et al. Influence of LDL apheresis on LDL subtypes in patients with coronary heart disease and severe hyperlipoproteinemia. J Lipid Res 2000; 41:727–33.
62. Tsimikas S, Brilakis ES, Miller ER, et al. Oxidized phospholipids, Lp(a) lipoprotein, and coronary artery disease. [see comment]. N Engl J Med 2005; 353:46–57.
63. Leitinger N, Pirich C, Blazek I, et al. Decreased susceptibility of low-density lipoproteins to in-vitro oxidation after dextran-sulfate LDL-apheresis treatment. Atherosclerosis 1996; 126:305–12.
64. Napoli C, Ambrosio G, Scarpato N, et al. Decreased low-density lipoprotein oxidation after repeated selective apheresis in homozygous familial hypercholesterolemia. Am Heart J 1997; 133:585–95.
65. Tamai O, Matsuoka H, Itabe H, et al. Single LDL apheresis improves endothelium-dependent vasodilatation in hypercholesterolemic humans. Circulation 1997; 95:76–82.
66. Mii S, Mori A, Sakata H, et al. LDL apheresis for arteriosclerosis obliterans with occluded bypass graft: change in prostacyclin and effect on ischemic symptoms. Angiology 1998; 49:175–80.
67. Jaeger BR. Evidence for maximal treatment of atherosclerosis: drastic reduction of cholesterol and fibrinogen restores vascular homeostatis. Ther Apher 2001; 5 (3):207–11.
68. Julius U, Metzler W, Pietzsch J, et al. Intraindividual comparison of two extracorporeal LDL apheresis methods: lipidfiltration and HELP. Int J Artif Organs 2002; 25:1180–8.
69. Knisel W, Di Nicuolo A, Pfohl M, et al. Different effects of two methods of low-density lipoprotein apheresis on the coagulation and fibrinolytic systems. [see comment]. J Intern Med 1993; 234:479–87.
70. Kaul S, Ito H. Microvasculature in acute myocardial ischemia: part II: evolving concepts in pathophysiology, diagnosis, and treatment. Circulation 2004; 109:310–5.
71. Jayaweera A, Wei K, Coggins M, et al. Role of capillaries in determining CBF reserve: new insights using myocardial contrast echocardiography. Am J Physiol 1999; 277:H2363–72.
72. Moriarty P, Gibson C, Kensey K, et al. LDL apheresis and its effect on whole blood viscosity., 3rd Annual Conference on Arteriosclerosis, Thrombosis, and Vascular Biology, Salt Lake, UT, 2002.
73. Rubba P, Iannuzzi A, Postiglione A, et al. Hemodynamic changes in the peripheral circulation after repeat low density lipoprotein apheresis in familial hypercholesterolemia. Circulation 1990; 81:610–6.
74. Iannuzzi A, Bianciardi G, Faccenda F, et al. Correction of erythrocyte shape abnormalities in familial hypercholesterolemia after LDL-apheresis: does it influence cerebral hemodynamics? Heart Vessels 1997; 12:234–40.
75. Schuff-Werner P, Schutz E, Seyde WC, et al. Improved haemorheology associated with a reduction in plasma fibrinogen and LDL in patients being treated by heparin-induced extracorporeal LDL precipitation (HELP). Eur J Clin Invest 1989; 19:30–7.

76. Harkness J. Measurement of plasma viscosity. In: Lowe G, Barbanei J, Forbes C, eds. Clinical Aspects of Blood Viscosity and Cell Deformability. Berlin: Springer-Verlag, 1981:79–87.
77. Mellwig K, Baller D, Gleichmann U, et al. Improvement of coronary vasodilatation capacity through single LDL apheresis. Atherosclerosis 1998; 139:173–8.
78. Zheng S, Jiang R, Fan M. [Hemorheological disorders in patients with sudden deafness]. Sheng Wu Yi Xue Gong Za Zhi/J Biomed Eng; 14:11–4.
79. Valbonesi M, Mora F, Mora R, et al. Rheopheresis for sudden hearing loss (SHL). Int J Artif Organs 2004; 27:806–9.
80. Suckfull M. Fibrinogen and LDL apheresis in treatment of sudden hearing loss: a randomised multicentre trial. Lancet 2002; 360:1811–7.
81. Muso E, Mune M, Fujii Y, et al. Significantly rapid relief from steroid-resistant nephrotic syndrome by LDL apheresis compared with steroid monotherapy. Nephron 2001; 89:408–15.
82. Grutzmacher P, Vallbracht C, Scheuermann E, et al. Combined LDL apheresis and hemodialysis in a patient with end-stage renal disease and accelerated coronary atherosclerosis. ASAIO Trans 1991; 37:M435–6.
83. Miyazaki H, Matsuoka H, Itabe H, et al. Hemodialysis impairs endothelial function via oxidative stress: effects of vitamin E-coated dialyzer. [see comment]. Circulation 2000; 101:1002–6.
84. Martinez M, Vaya A, Alvarino J, et al. Hemorheological alterations in patients with chronic renal failure. Effect of hemodialysis. Clin Hemorheol Microcirc 1999; 21:1–6.
85. Wenke K. Management of hyperlipidaemia associated with heart transplantation. Drugs 2004; 64:1053–68.
86. Barbir M, Kushwaha S, Hunt B, et al. Lipoprotein(a) and accelerated coronary artery disease in cardiac transplant recipients. Lancet 1992; 340:1500–2.
87. Matschke K, Mrowietz C, Sternitzky R, et al. Effect of LDL apheresis on oxygen tension in skeletal muscle in patients with cardiac allograft vasculopathy and severe lipid disorder. Clin Hemorheol Microcirc 2004; 30:263–71.
88. Vignal-Clermont C. [Acute anterior, ischemic optic neuropathy]. Rev Prat 2001; 51:2202–5.
89. Arnold AC, Hepler RS. Natural history of nonarteritic anterior ischemic optic neuropathy. J Neuroophthalmol 1994; 14:66–9.
90. Fraser S, Siriwardena D. Interventions for acute non-arteritic central retinal artery occlusion. In: Library TC, ed. Vol. CD001989: Oxford, 2002.
91. Beatty S, Eong K. Acute occlusion of the retinal arteries: current concepts and recent advances in diagnosis and management. J Accid Emerg Med 2000; 17:324–9.
92. Williamson TH, Rumley A, Lowe GD. Blood viscosity, coagulation, and activated protein C resistance in central retinal vein occlusion: a population controlled study. [see comment]. Br J Ophthalmol 1996; 80:203–8.
93. Walzl M, Lechner P, Walzl B, et al. First experiences with the heparin-induced extracorporeal low-density lipoprotein precipitation in the treatment of critical limb ischaemia: a new therapeutical approach? Haemostasis 1993; 23:237–43.
94. Ramunni A, Giancipoli G, Saracino A, et al. LDL-apheresis in acute anterior ischemic optic neuropathy. Int J Artif Organs 2004; 27:337–41.
95. Moriarty PM, Whittaker TJ. Treatment of acute occlusion of the retinal artery by LDL-apheresis. J Clin Apher 2005; 20:88–92.
96. Friedman DS, O'Colmain BJ, Munoz B, et al. Prevalence of age-related macular degeneration in the United States. [see comment]. Arch Ophthalmol 2004; 122:564–72.
97. Klingel R, Fassbender C, Fassbender T, et al. Clinical studies to implement rheopheresis for age-related macular degeneration guided by evidence-based-medicine. Transfus Apher Sci 2003; 29:71–84.
98. McGwin G, Jr., Xie A, Owsley C. The use of cholesterol-lowering medications and age-related macular degeneration. Ophthalmology 2005; 112:488–94.
99. Pulido JS, Multicenter Investigation of Rheopheresis for AMDSG. Multicenter prospective, randomized, double-masked, placebo-controlled study of rheopheresis to treat nonexudative age-related macular degeneration: interim analysis. Trans Am Ophthalmol Soc 2002; 100:85–106; discussion 106–7.
100. Kroon A, Aengevaeren W, vanderWerf T, et al. LDL Apheresis Atherosclerosis Regression Study (LAARS). Effect of aggressive versus conventional lipid lowering treatment on coronary atherosclerosis. Circulation 1996; 93:1826–33.

101. Kroon AA, van Asten WN, Stalenhoef AF. Effect of apheresis of low-density lipoprotein on peripheral vascular disease in hypercholesterolemic patients with coronary artery disease. Ann Intern Med 1996; 125:945–54.
102. Kawashima A. Low-density lipoprotein apheresis in the treatment of peripheral arterial disease. Ther Apher 2003; 7:413–8.
103. Goldstein LB, Adams R, Becker K, et al. Primary prevention of ischemic stroke: a statement for healthcare professionals from the Stroke Council of the American Heart Association. [see comment]. Stroke 2001; 32:280–99.
104. Smith CJ, Emsley HC, Gavin CM, et al. Peak plasma interleukin-6 and other peripheral markers of inflammation in the first week of ischaemic stroke correlate with brain infarct volume, stroke severity and long-term outcome. BMC Neurol 2004; 4:2.
105. Moonis M, Kane K, Schwiderski U, et al. HMG-CoA reductase inhibitors improve acute ischemic stroke outcome. Stroke 2005; 36:1298–300.
106. Schwartz GG, Olsson AG, Ezekowitz MD, et al. Effects of atorvastatin on early recurrent ischemic events in acute coronary syndromes: the MIRACL study: a randomized controlled trial. [see comment] [summary for patients in Curr Cardiol Rep. 2002 Nov;4(6):485; PMID: 12379168]. JAMA 2001; 285:1711–8.
107. Szapary L, Horvath B, Marton Z, et al. Hemorheological disturbances in patients with chronic cerebrovascular diseases. Clin Hemorheol Microcirc 2004; 31:1–9.
108. Pfefferkorn T, Knuppel H, Jaeger B, et al. Increased cerebral CO_2 reactivity after heparin-mediated extracorporeal LDL precipitation (HELP) in patients with coronary heart disease and hyperlipidemia. Stroke 1999; 30:1802–6.
109. Walzl M. Effect of heparin-induced extracorporeal low-density lipoprotein precipitation and bezafibrate on hemorheology and clinical symptoms in cerebral multiinfarct disease. Haemostasis 1993; 23:192–202.
110. Heart Protection Study Collaborative G. MRC/BHF Heart Protection Study of cholesterol lowering with simvastatin in 20,536 high-risk individuals: a randomised placebo-controlled trial. [see comment] [summary for patients in Curr Cardiol Rep. 2002 Nov;4(6):486–7; PMID: 12379169]. Lancet 2002; 360:7–22.
111. Nissen SE, Tuzcu EM, Schoenhagen P, et al. Effect of intensive compared with moderate lipid-lowering therapy on progression of coronary atherosclerosis: a randomized controlled trial. [see comment]. JAMA 2004; 291:1071–80.
112. Nakamura T, Kawagoe Y, Ogawa H, et al. Effect of low-density lipoprotein apheresis on urinary protein and podocyte excretion in patients with nephrotic syndrome due to diabetic nephropathy. Am J Kidney Dis 2005; 45:48–53.
113. Gruztmacher P, Landgraf H, Esser R, et al. In vivo rheologic effects of lipid apheresis techniques: comparison of dextran sulfate LDL absorption and heparin-induced LDL precipitation. ASAIO Trans 1990; 36:M327–30.
114. Nakamura T, Ushiyama C, Osada S, Shimada N, Koide H. Effect of low-density lipoprotein apheresis on plasma endothelin-1 levels in diabetic hemodialysis patients with arterisclerosis obliterans. J Diabetics Complications 2003; 17(6):349–54.

14 Phytosterolemia
Synthesis, Absorption, Trafficking, and Excretion of Cholesterol and Noncholesterol Sterols

Thomas Dayspring, MD, FACP

CONTENTS

    INTRODUCTION
    CHOLESTEROL
    NONCHOLESTEROL STEROLS
    STANOLS
    INTESTINAL ABSORPTION OF STEROLS
    PHARMACOLOGIC MODULATION OF STEROL ABSORPTION
    SITOSTEROLEMIA (PHYTOSTEROLEMIA)
    REFERENCES

INTRODUCTION

Sterols are waxy insoluble substances or lipids synthesized from acetyl coenzyme A (CoA). They are steroid-based alcohols having a hydrocarbon (aliphatic) side chain of 8–10 carbons at the 17β position and a hydroxyl group (-OH) at the 3β position (making it an alcohol). Because of the hydrophilicity at the -OH end and hydrophobicity at the hydrocarbon side chain, sterols can be incorporated into the lipid bilayers of the cytoplasmic membrane.

The most notable sterol in humans is cholesterol, but over 40 "noncholesterol" sterols, which are major components of plants, fungi, yeasts, and shellfish, are present in the human diet. The hallmark of atherosclerosis is the presence of cholesterol-laden macrophages or "foam cells," but in 1974 two patients were described relating tendon xanthomas to a new "lipid storage disease" characterized by elevated plasma sitosterol concentrations *(1)*. Afterwards several patients have been described with premature atherosclerosis and high "sitosterol" levels leading to the disease originally called sitosterolemia, and later termed phytosterolemia. In actuality, atherosclerosis is a disease caused by arterial wall accumulation of sterol-laden foams cells and that sterol does not have to be cholesterol. Much is now known about how sterols are

From: *Contemporary Cardiology: Therapeutic Lipidology*
Edited by: M. H. Davidson, P. P. Toth, and K. C. Maki © Humana Press Inc., Totowa, NJ

absorbed, synthesized, and excreted, and such knowledge helps us better understand atherogenesis and novel ways of preventing or treating patients.

CHOLESTEROL

Cholesterol is a 27-carbon compound that can be acquired by both dietary and endogenous sources (Fig. 1). Cholesterol homeostasis is very tightly regulated by multiple genes affecting numerous biologic pathways. It is required for cellular membrane integrity and function, as well as bile acid (BA), vitamin D, and steroid synthesis. Humans typically manufacture far more daily cholesterol (800–1200 mg) than they ingest (300–500 mg) or acquire from shedding of intestinal epithelia (300 mg daily).

The liver is responsible for about 15% of cholesterol synthesis and the remainder is extrahepatic *(2)*. Cholesterol synthesis is a four-step process starting from its precursor acetate: CH_3-COO^- *(3)*. In the first step, acetyl-CoA condenses with acetoacetyl-CoA upon the action of cytosolic hydroxymethylglutaryl (HMG) synthase and becomes HMG-CoA. The catalytic action (rate-limiting state) of HMG-CoA reductase, an integral part of the smooth endoplasmic reticulum, forms mevalonate (a six-carbon intermediate) with NADPH serving as the reductant. HMG-CoA reductase transcription is tightly regulated by a membrane-bound transcription factor designated as sterol regulatory element-binding protein-2 (SREBP-2) *(4)*.

In the second step, mevalonate is phosphorylated from ATP to isoprene units or isoprenoids, namely isopentyl pyrophosphate, which can isomerize or interconvert to dimethylallyl pyrophosphate. In the third step, isoprenoids react with each other to form geranyl pyrophosphate. Condensation with another isopentyl-PP yields farnesyl pyrophosphate. Squalene synthase catalyzes the condensation of two molecules of farnesyl-PP with reduction by NADPH to make squalene. The fourth step involves conversion of the linear squalene molecule to the four-ringed steroid nucleus. Upon

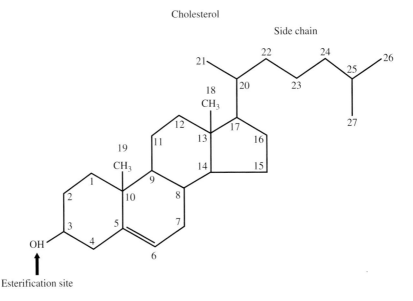

Fig. 1. Cholesterol structure.

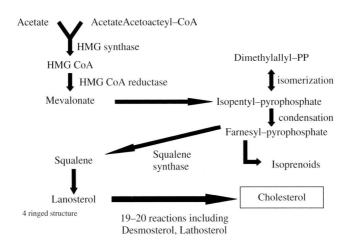

Fig. 2. Cholesterol synthesis.

catalysis by squalene epoxidase and oxidocyclase, squalene becomes the four-ringed steroid lanosterol. Conversion to cholesterol takes about 19–20 reactions through enzymes (cytochrome P_{450}) in the mitochondria and endoplasmic reticulum that include migration and removal of methyl groups *(3)*. There are several intermediary sterols such as desmosterol and cholestenol (lathosterol), and measurement of the latter is often used as a measure of cholesterol synthesis *(5)*. The sterol intermediates of cholesterol synthesis undergo catabolism or further conversion to cholesterol to prevent their pathological accumulation in peripheral tissues. Other metabolites of cholesterol, such as cholestanol, are used as markers of cholesterol absorption *(6)*. Intermediary products of the cholesterol synthesis pathway such as farnesyl-PP is a precursor for other isoprenoids used in the synthesis of other compounds such as dolichol or ubiquinone or prenylation of cellular proteins, many of which are involved with cell signaling (Fig. 2).

NONCHOLESTEROL STEROLS

Plants, fungi, and yeasts convert squalene not to cholesterol but rather to other related sterols: stigmasterol, sitosterol, campesterol, ergosterol, brassicasterol, avenosterol, etc. Shellfish manufacture fucosterol. Plant (phyto) sterols differ from cholesterol by the presence of additional methyl or ethyl groups or by additional double bonds. The most prevalent plant sterols, sitosterol, campesterol, and stigmasterol, are structurally similar to cholesterol in that all are 4-desmethylsterols (no methyl group at carbon 4). They are present in western diets in amounts equal to that of cholesterol (150–350 mg daily) and all are similarly absorbed by the intestinal epithelium *(7)*. Phytosterols are in nuts, seeds, and vegetable oils, and obviously vegetarians consume larger amounts. Henceforth, phytosterols and shellfish and yeast sterols will collectively be termed either noncholesterol sterols or simply phytosterols (Fig. 3).

Because unlike cholesterol no phytosterol serves any human physiologic function, it is not generally incorporated into intestinal chylomicrons in preparation for systemic entry. Phytosterols occur in human plasma in concentrations normally < 0.5% that

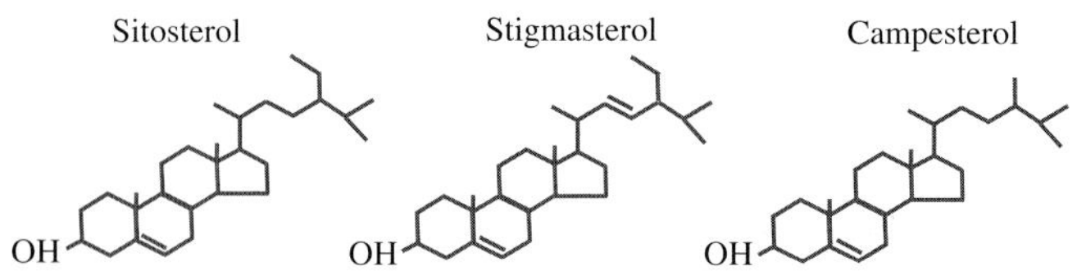

Fig. 3. Phytosterols.

of cholesterol *(7)*. Their entry into the plasma is tightly regulated by enterocytes and hepatocytes and a variety of genes regulating sterol trafficking and excretion.

As far back as the 1950s, researchers noted that oral ingestion of large amounts of plant sterols resulted in reduced cholesterol levels *(8)*. Today, numerous phytosterol products are used therapeutically to lower cholesterol and have received the recommendation of the National Cholesterol Education Program, Adult Treatment Panel III, as part of therapeutic lifestyle changes to lower cholesterol levels *(9)*.

STANOLS

Stanols are saturated sterols that also exist in the diet, mainly from plant sources with an intake of about 50 mg daily. Saturation (hydrogenation) of the Δ5 double bonds of sterols results in 5α-stanols: in other words stanols differ from sterols because they lack the Δ5 double bond in their B-ring. Saturation of sitosterol and campesterol, respectively, forms sitostanol and campestanol. Stanols are not absorbed, and like noncholesterol sterols, can interfere with the absorption of cholesterol and reduce plasma cholesterol concentrations. Esterified stanol products, which are fat-soluble, can be incorporated into a variety of substances including margarines and have been developed commercially as cholesterol-lowering agents *(10)* (Fig. 4).

INTESTINAL ABSORPTION OF STEROLS

Sterol and stanol absorption is a complicated, tightly regulated, multiple-step process involving uptake into intestinal epithelial cells, absorption or incorporation into lipoproteins, delivery to the liver, and re-excretion into the intestinal lumen. Sterol uptake and absorption is regulated by multiple genes, and most of it occurs in the duodenum or proximal jejunum. Cholesterol absorption is incomplete, variable, and not related to gender or age with a mean value of 56% in humans with individual rates being highly reproducible *(11)* (Fig. 5).

Close examination of the graph reveals that significant numbers of humans are either hypoabsorbers or hyperabsorbers of cholesterol. The latter was found to be associated

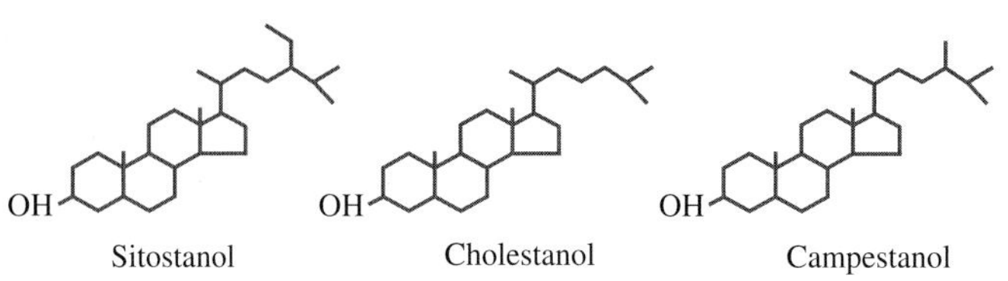

Stanols are saturated sterols

Fig. 4. Stanols.

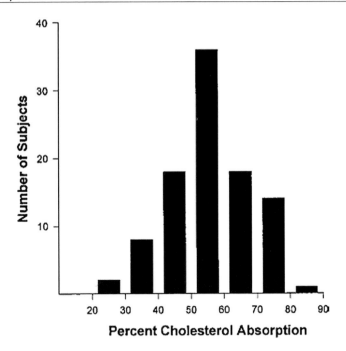

Fig. 5. Distribution of values for percent cholesterol absorption by decile *(11)*.

> **Summary #1**
>
> - Sterols are a group steroid-based alcohols including cholesterol and a number of noncholesterol members such as sitosterol, campesterol, etc.
> - Cholesterol which can be synthesized de nova or absorbed intestinally is required by humans for cell membranes, steroid and vitamin D production.
> - Noncholesterol sterols serve no physiologic function
> - All sterols are atherogenic if they accumlate in arterial wall macrophages
> - Stanols are saturated sterols. They are not absorbed and esterified versions can be used therapeutially to reduce cholesterol absorption

Fig. 6. Sterols and Stanols.

with the apolipoprotein E (apoE) polymorphisms E4/3 and E4/4 in a Finnish study *(12)* but not in other trials *(13)*. Noncholesterol sterol absorption is significantly more limited than that of cholesterol. As noted above, humans absorb less than 5% of dietary sitosterol resulting in plasma levels of sitosterol that are very low (< 1 mg/dL) in normal individuals *(14,15)*.

After a meal, from both oral and biliary sources, the intestine contains a collection of fatty acids and monoglycerols formed from hydrolyzed triglycerides (TGs), free cholesterol (FC), formed from hydrolyzed cholesteryl ester (CE), noncholesterol sterols, and phospholipids. These lipids are emulsified by biliary phospholipids and then enwrapped by amphipathic BAs forming mixed biliary micelles. The nonpolar end of the BA surrounds the lipids, and the polar end bulges outward making the micellar collection of lipids soluble in aqueous intestinal fluids. The micelles transport or "ferry" the lipids through a diffusion barrier consisting of unstirred water and mucous coat layer *(16)* to the brush border (microvilli) of the intestinal epithelium, where their delipidation will occur. Fatty acids flux into the enterocyte by both passive diffusion and fatty acid transport protein *(17)*.

Cellular entry or uptake of sterols is facilitated by several proteins formerly referred to as sterol permeases. Many facets of the molecular mechanisms responsible for sterol absorption remain to be discovered but several candidates have been identified, including the Niemann–Pick C1-Like 1 protein (NPC1L1) *(18,19)*, caveolae or lipid rafts *(20)*, and aminopeptidase N *(21)*. Other factors such as scavenger receptor-B type 1 (SR-B1) may be involved but are not obligatory *(22)*.

The NPC1L1 protein, which is expressed at the jejunal enterocyte apical surface and hepatocyte biliary surface, has a sterol-sensing domain that interfaces with biliary and dietary cholesterol and noncholesterol sterols *(23)*. NPC1L1 protein may be influenced (downregulated) by peroxisome proliferator-activated receptor (PPAR)-δ *(24)*. A membrane-bound ectoenzyme aminopeptidase N (alanyl)-aminopeptidase (APN) facilitates the endocytosis of cholesterol-rich membrane microdomains *(21)*. Cholesterol becomes part of caveolae which are sophisticated lipid rafts serving as cholesterol and sphingolipid-rich complex membrane domains regulating cell membrane transport and cellular signal transduction *(20)*. Caveolin-1 (CAV1) is a protein that binds with cholesterol and helps shape caveolae *(25,26)*. CAV1 also can complex with annexin-2 (ANX2), cyclophilin A, and cyclophilin 40 to traffic cholesterol from caveolae to the endoplasmic reticulum, where esterification will occur *(27,28)*.

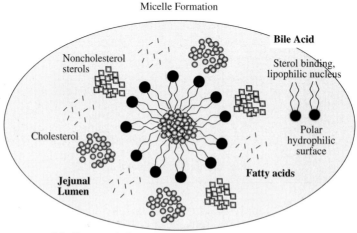

Fig. 7. Micelle formation.

Once inside the enterocyte, FC and noncholesterol sterols have several trafficking options. FC can undergo esterification at the endoplasmic reticulum. The primary enzyme responsible facilitating cholesterol esterification in the intestine and liver is acyl-CoA:cholesterol acyltransferase 2 (ACAT2) *(29)*. Esterification involves attaching a fatty acid to the alcohol group (3-OH) of cholesterol. The majority of cholesterol is transported in lipoproteins, including chylomicrons as its esterified form (CE). Formation and activity of ACAT2 is involved in the regulation of cholesterol absorption. Indeed, ACAT2 is also linked to the preferential absorption of cholesterol over noncholesterol sterols. Phytosterols are not suitable substrates for ACAT *(30)*. Deficiency of ACAT2 is associated with impaired cholesterol absorption *(31)*. Because cholesterol is the preferred sterol substrate for ACAT2, there is a de facto ACAT2-mediated sorting of sterols after absorption (Fig. 8).

Cholesterol proceeds down esterification pathways, whereas noncholesterol sterols are directed to sterol efflux transporters *(32)*. In mice, combined ACAT2 and low-density lipoprotein (LDL) receptor (LDLr) deficiency leads to the redirecting of cholesterol to intestinal ATP-binding cassette transport A1 (ABCA1), which effluxes the cholesterol to apoA-I making high-density lipoprotein (HDL) particles part of the cholesterol absorption process *(33)*. Most of the absorbed cholesterol in preparation for entry into plasma is incorporated into the formation of TG-rich lipid lipoproteins termed chylomicrons *(34)*. With the aid of microsomal TG transfer factor (MTF), CE joins with intestinally assembled TGs at the endoplasmic reticulum and joins apoB48 (which consists of the N-terminal 2,152 amino acids of hepatic produced apoB100) in the formation of chylomicrons *(35)*. This truncated apoB is the main apolipoprotein that conveys structure, stability, and surface hydrophilicity to chylomicrons. ApoB48 is synthesized

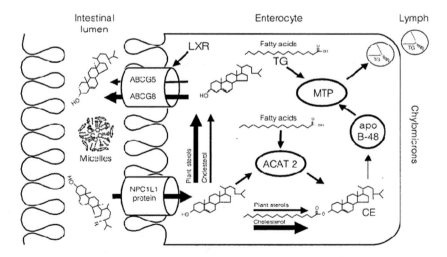

Fig. 8. Major pathways in absorption and intracellular traffic of cholesterol and phytosterols in the enterocyte [Klaus von Bergmann et al. Am J Cardiol 2005; 95(suppl) 10D–14D].

in the endoplasmic reticulum and then transported through the Golgi-apparatus acquiring phospholipids in the process *(36)*. Co-transcriptional and post-transcriptional regulation of apoB influence apoB production as does apoA-IV *(37)*. ApoA-IV expression in humans is limited to the intestine, and the physical association between apoA-IV and apoB increases the nascent chylomicron residence time within a lipoprotein expansion compartment, thereby facilitating particle expansion with TGs *(38)*.

FC can also gain systemic entry via efflux through the ATP-binding cassette transporter A1 (ABCA1) onto a cholesterol acceptor protein like apoA-I, which is also synthesized and secreted by enterocytes, or onto apoE. This HDL pathway is absent in apoA-I$^{-/-}$ mice. The intestine thus has more than one option in lipoprotein-mediated absorption of cholesterol and no doubt the sterol export mechanisms work together or compensate for one another *(33)* (Fig. 9).

The enterocyte has additional regulatory elements that deny noncholesterol sterols' systemic entry and control how much FC will undergo esterification. Because noncholesterol sterols serve no known physiologic functions and can be pathological, teleological forces have made sure they are denied entry into the plasma. ATP-binding cassette half transporters G5 and G8 (ABCG5 and ABCG8) also called sterolin 1 and sterolin 2 *(39)*, respectively, are expressed in intestinal and hepatic cells. The ABCG transporter subfamily is composed of six half transporters with "reverse" proteins that have an ATP-binding cassette at the amino terminus and a transmembrane (TM) helix at the carboxy terminus *(40)*. These membrane proteins translocate sterols across membranes. The ABCG5–ABCG8 heterodimer functions to increase intestinal excretion and biliary secretion of noncholesterol sterols *(41)*. The genes controlling these transporters are located in a head-to-head orientation on chromosome 2p and are expressed exclusively in the intestine and liver. ABCG5 and ABCG8 actively efflux noncholesterol sterols and some cholesterol from enterocyte to the intestinal lumen

Intestinal Lipoprotein Assembly and Secretion

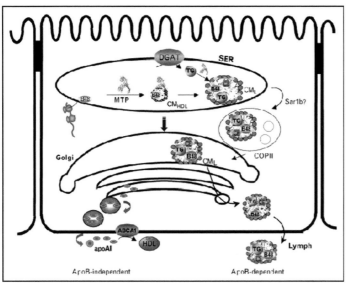

MTP = microsomal triglyceride transfer protein

DGAT = diacylglycerol acyltransferase

B48 = intestinal apoB

SER = smooth endoplasmic reticulum

CM_{HDL} and CM_L = primordial phospholipid-rich lipoproteins

Sar1b & COPII = transport vesicles

Fig. 9. Intestinal lipoprotein assembly and secretion *(34)*.

where they can be excreted in the stool. Should a noncholesterol sterol make it to the liver as part of chylomicron or HDL sterols, the hepatocyte ABCG5/ABCG8 will export it into the bile excretion. It was recently hypothesized that plant stanols and sterols upregulate ABCG5 and ABCG8 *(42)* (Fig. 10).

The homozygous mutation or absence of *ABCG5* and *ABCG8* genes was discovered to be the cause of the rare genetic condition sitosterolemia (phytosterolemia) known to be associated with premature atherosclerosis *(43)*. Both copies of either gene (ABCG5 or ABCG8) have to be defective to cause the disease *(39)*. Polymorphisms or sequence variants of *ABCG5* or *ABCG8* genes also exist, and systemic absorption of noncholesterol sterols will vary among individuals. The *ABCG5* gene is principally mutated in Asians and the *ABCG8* gene in Caucasians suggesting that the proteins form both heterodimers and homodimers to transport the wide range of dietary sterols present in the diet *(41)*. Sitosterol concentrations do not correlate well with dietary intake and vary fivefold to tenfold among individuals on similar diets *(44)*. Two such groups are persons with strong family histories of premature atherosclerosis *(45)* and postmenopausal women *(46)*. Statin therapy is also associated with hyperabsorption of cholesterol and noncholesterol sterols, such as sitosterol and campesterol levels *(47)*.

Chylomicrons transport intestinally acquired lipids (CE, FC, noncholesterol sterols, and TG) through lymphatic channels and plasma to the liver. During passage through myocyte and adipocyte beds, lipoprotein lipase (LPL)-mediated hydrolysis of TG creates smaller TG-poor chylomicron remnants: about 10–30% of the TG remains. Chylomicron remnants are internalized at the hepatocyte by fixation with hepatic lipase, heparan sulfate proteoglycans, LDLrs, and LDLr-related protein *(48)*. Lysosomal degradation of the chylomicron releases the sterols. CE is hydrolyzed by hormone-sensitive lipase (a cholesterol esterolase) into FC *(49)*. Sterols are then transported by both vesicular (endosomal) and nonvesicular pathways mediated by carrier proteins *(50)*. Depending on cholesterol balance, the increase in FC delivery to the liver may cause a downregulation of HMG-CoA reductase and a decrease in FC synthesis. Several other distinct pathways exist for their further sterol trafficking.

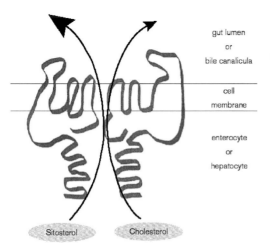

Fig. 10. Model of ABCG transporter *(39)*.

As in the intestine, FC can be esterified by hepatic ACAT2. There is a FC threshold at which ACAT2 production is upregulated *(51)*. MTF protein helps to package newly synthesized TGs, phospholipids and apoB (a single molecule) with CE, and FC into a TG-rich VLDL that is released by the liver. Lipolysis of VLDL by LPL occurs peripherally, and the resultant smaller particle (LDL) can traffic cholesterol to other tissues like the adrenals it may be needed peripherally or back to the liver for internalization by LDLrs in a process called indirect reverse cholesterol transport. The LDL can also acquire through the protein CE transfer protein (CETP) additional CE from HDL particles *(52)*.

Hepatocytes can secrete FC and/or noncholesterol sterols into the bile canaliculi through ABCG5/G8. HDL particles are major sources of cholesterol for biliary excretion *(53)*. The sterols enter the biliary system to join with bile salts and phospholipids and proceed into the gall bladder and ultimately the small intestine. ABCG5 and ABCG8 are important mechanism to excrete noncholesterol hepatic sterols thus limiting their incorporation into hepatic produced lipoproteins *(54)*. The liver ABCA1 transporters, regulated by liver X receptors (LXR) *(55,56)*, also secretes FC and phospholipids into unlipidated apoA-I, forming a pre-β-HDL. When FC is delivered to apoA1, forming a pre-β-HDL, the FC is esterified by lecithin cholesterol acyl transferase (LCAT), forming larger, lipidated HDL particles (α-HDL). The vast majority of cholesterol in HDL particles originates in the liver *(57)*. The large HDL particles are delipidated of CE by SR-B1 receptors in steroidogenic endocrine glands. Large HDL can also transfer CE to the liver either directly (SR-B1 delipidation) or indirectly by CETP-mediated transfer to apoB particles *(52)*.

Biliary excretion of cholesterol can also involve its transformation into the BAs chenodeoxycholic and cholic acids. The cholesterol used for BA synthesis originates mostly from LDL particles, not from endogenous synthesis *(58,59)*. BA synthesis occurs in multiple steps through two pathways: the major "neutral pathway" of cholesterol 7α-hydroxylase (CYP7A1) or the "alternative" path utilizing mitochondrial 27-sterol hydroxylase (CYP27A1) *(60)* (Fig. 11).

Many proteins such as steroidogenic acute regulatory protein (StAR) and sterol carrier protein-2 (SCP-2) participate in cholesterol trafficking from intracellular locations to the mitochondria, where BA synthesis occurs *(61)*. BA and phospholipids are transported across the liver cell by a P-glycoprotein called mdr2 (multidrug resistance protein) toward the canalicular membrane, where they are endocytosed into the membrane of the canalicular transporters *(54,62)*. Spgp (sister of P-glycoprotein) or ABCB11 acts as an ATP-dependent bile salt export pump and transports bile salts into the biliary tree *(63)*. Hepatobiliary secretion of BAs becomes the major stimulus for bile formation. Intestinal bile salts after micelle delipidation are extensively reabsorbed at the ileum by the ileal bile salt transporter (IBAT), which is also transcriptionally regulated by BA concentrations *(64)*.

In summary, sterols are present in the gut through dietary ingestion and/or biliary excretion where they mix with hepatic secreted BAs. The sterols are either excreted in stool or gathered for intestinal uptake in biliary micelles. Micelles are delipidated of fatty acids and sterols at enterocyte microvilli by sterol permeases including the NPC1L1. Entry into the cell is influenced by lipid rafts and other proteins such as CAV1 and annexin and aminopeptidase N. FC can be esterified by ACAT2, and unesterified noncholesterol sterols can be re-excreted into the gut lumen by ABCG5/G8 heterodimers or incorporated into chylomicrons. FC is also effluxed to unlipidated apoA-I by ABCA1. After micelle delipidation, the vast majority of bile salts are

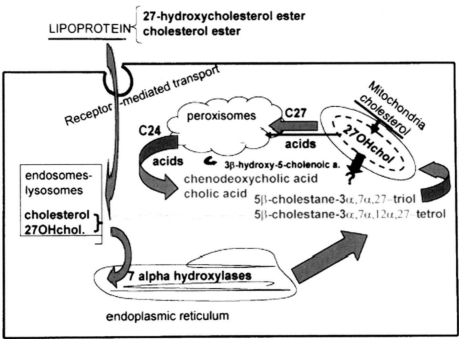

Fig. 11. Bile acid synthesis *(60)*.

reabsorbed at the ileum. Free or esterified cholesterol and noncholesterol trafficking in the hepatocyte is complex and highly regulated. Cholesterol can be esterified and secreted in apoB lipoproteins, effluxed to apoA-I by ABCA1, excreted into bile by ABCG5/G8, or transformed into Bas, which are excreted by ABCB11. Noncholesterol sterols are also effluxed to bile through ABCG5/G8 (Fig. 12).

PHARMACOLOGIC MODULATION OF STEROL ABSORPTION

Both cholesterol and noncholesterol sterol absorption and excretion at the intestine and liver can be manipulated by dietary modification and/or prescription and nonprescription medications, including plant stanols and sterols, ezetimibe, BA sequestrants (BAS), fibrates, and statins.

Sterols and Stanols

Plant sterols have been used therapeutically for five decades and stanols for two to reduce total and LDL-cholesterol (LDL-C). Numerous double-blind studies are testament to their cholesterol-lowering benefit with an LDL-C reduction of approximately 10–14%. Younger persons may have better cholesterol-lowering responses than do elderly, and the benefit minimizes at doses above 2 g daily. The use of sterols and stanols became practical when they were esterified with long chain fatty acids and added to margarines *(65–68)*. Current data reveal that single-dose regimens of plant stanols are as efficacious as two or three daily doses *(69)*. Phytostanol esters and statins provide synergistic cholesterol lowering *(70)* (Fig. 13).

> **Summary #2**
>
> - The majority of the cholesterol in the gut is unesterified and of endogenous origin delivered via the biliary system
>
> - Sterols are delivered by biliary micelles to the enterocyte microvilli where they are internalized by a variety of complex mechanisms
>
> - Human enterocytes typically absorb about 50–55% of intestinal sterols
>
> - Once in the enterocyte cholesterol is trafficked for esterification and incorporation into chylomicrons, effluxed to prebeta HDL particles or effluxed back to the gut via ABCG5/G8 transporters
>
> - Non cholesterol sterols are returned (effiuxed) to the gut via ABCG5/G8 transporters (enterocyte or hepatic)

Fig. 12. Enterocyte Handling of Sterols.

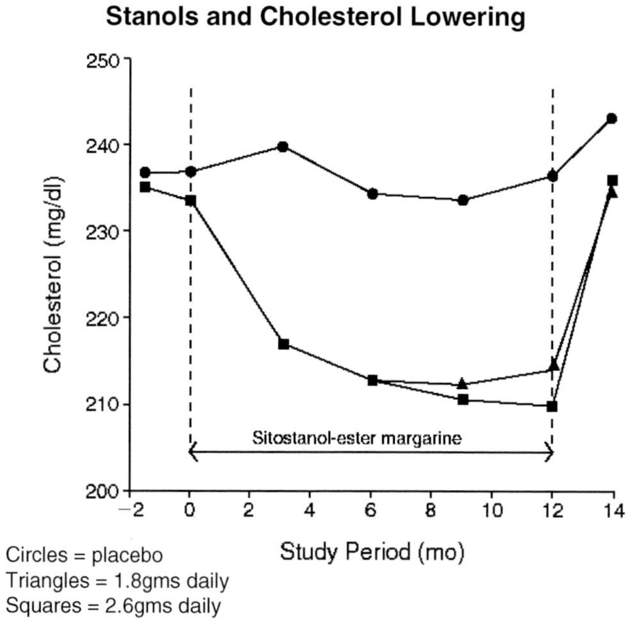

Fig. 13. Stanols and cholesterol lowering *(68)*.

As biliary micelles are being formed, both sterols and stanols compete for cholesterol for inclusion. The cholesterol (from oral or biliary sources) that does not enter micelles is excreted in the stool. Less cholesterol is delivered to the brush border of the enterocyte, thereby reducing the amount of cholesterol that can be trafficked to the endoplasmic reticulum and incorporated into chylomicrons *(65)*. The over-absorption of noncholesterol sterols may have an upregulating effect on ABCG5/G8 transporters in both the intestine and liver leading to additional sterol secretion into the gut and biliary tree. The decreased delivery of cholesterol by chylomicrons to the liver has two effects: increased hepatic cholesterol synthesis and upregulation of LDLrs. Despite the increase in cholesterol synthesis, the increased LDLr will lead to removal of apoB lipoproteins from plasma causing a net decrease in apoB and LDL-C *(42)*. Pant sterols but not stanols suppress BA synthesis, and this may also lessen their LDL-C lowering efficacy over time as less hepatic cholesterol would be required for BA synthesis *(71)* (Fig. 14).

The plant sterols that are absorbed in place of cholesterol are potentially atherogenic, but the actual atherosclerosis potential is probably low because the amount of absorbed sterols is small (5% of β-sitosterol, 15% of campesterol, and less than 1% of dietary stanols are absorbed) *(66)*. Because sterols achieve more significant plasma levels than stanols, long-term worry over their systemic effect may be a concern *(45)*. Plant sterols but not stanols are sensitive to oxidation and such oxidized sterols are potentially atherogenic *(72)*. Perhaps caution with sterol but not stanol administration may be called for in persons known to have increased plasma sterol levels: postmenopausal women, persons on statins *(73)*, and kindreds with strong family history of coronary heart disease (CHD). In some studies, sterols and stanols lower blood concentrations of β-carotene by about 25%, concentrations of α-carotene by 10%, and concentrations of vitamin E by 8%. This may be of some concern as lack of these nutrients may adversely affect LDL oxidation *(67)*. A more recent study with sitostanol ester did not effect fat-soluble vitamin concentrations *(70)*.

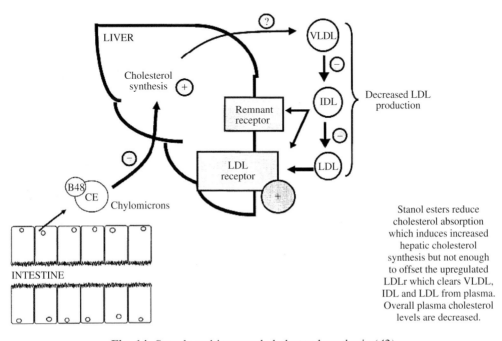

Fig. 14. Stanols and increased cholesterol synthesis *(42)*.

Ezetimibe

Ezetimibe is the first of a new group of drugs that reduce intestinal sterol absorption in humans with no effect on fat-soluble vitamins *(74)*. Ezetimibe is a synthetic 2-azetidinone whose full chemical name is 1-(4-fluorophenyl)-3(R)-[3-(4-fluorophenyl)-3(S)-hydroxypropyl]-4(S)-(4-hydroxyphenyl)-2-azetidinone *(75)* (Fig. 15).

Ezetimibe is glucuronidated by a variety of intestinal and hepatic UDP-Uridine diphosphate-glucuronosyltransferases *(76)*. Ezetimibe undergoes enterohepatic recirculation, which allows for it to be present at its site of action with negligible systemic exposure. As a result of enterohepatic recirculation, the half-life of ezetimibe is 22 h, which thereby allows for once-daily dosing *(74)*.

Several clinical trials using ezetimibe monotherapy in humans have revealed LDL-C-lowering effects in the range of 17–20% at a dose of 10 mg per day. The reductions in sitosterol and campesterol are even more significant at 48% for campesterol and 41% for sitosterol. Ezetimibe seems to reduce cholesterol absorption on mean average by 54% compared with placebo. Fecal sterol excretion increased by 72%. Interestingly, cholesterol synthesis was increased by 72% compared with placebo, as indicated by the lathosterol/cholesterol ratio (an indicator of hepatic HMG-CoA reductase activity) *(77)* (Fig. 6). Safety and tolerability of ezetimibe has been excellent in numerous trials *(78,79)*. In a dose–response study, 5- and 10-mg doses of ezetimibe significantly reduced LDL-C levels by 15.7 and 18.5%, respectively ($P < 0.01$ versus placebo) and significantly increased HDL-C levels by 2.9 and 3.5%, respectively ($P < 0.05$ versus placebo). A reduction in plasma TG levels was observed ($P = $ NS) *(79)*. There are case reports that ezetimibe lowers LDL particle number and has no effect on LDL size *(80)*. There is considerable interindividual variation in the response to ezetimibe. Pharmacogenetic results suggest that nonsynonymous NPC1L1 variation is associated with interindividual variation in response to ezetimibe treatment *(81)*.

The exact mechanism of action of ezetimibe is still under study. Ezetimibe-binding localizes at the brush border of the jejunum. Evidence is suggestive that it binds to the NPC1L1 protein *(18)* and prohibits sterol entry into the enterocyte. Ezetimibe also binds to the membrane-bound ectoenzyme APN from the lumen of the small intestine, thereby blocking endocytosis of cholesterol-rich membrane microdomains, which would decrease intestinal cholesterol absorption *(22)*. Ezetimibe by binding to CAV1 also effectively disrupts the CAV1–ANX2 heterocomplex in mice *(28)*.

Ezetimibe is most often utilized in clinical practice in combination with statins because the dual mechanisms of action (hindering cholesterol absorption and synthesis) lead to synergistic LDLr upregulation and reductions in apoB and LDL-C and

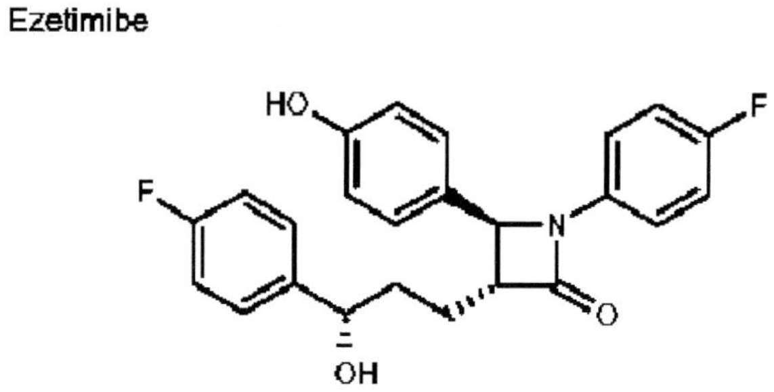

Fig. 15. Ezetimibe structure *(75)*.

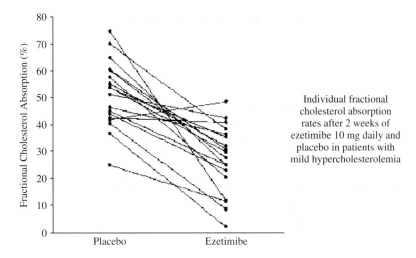

Fig. 16. Ezetimibe and cholesterol absorption rates *(77)*.

C-reactive protein *(82,83)*. Low-dose statin plus ezetimibe has the same cholesterol-lowering effect of the high-dose statins *(84)*. In a small pharmacokinetic study, ezetimibe 10 mg and rosuvastatin 10 mg reduced LDL-C by 61.4% within 2 weeks *(85)* (Figs. 17–19).

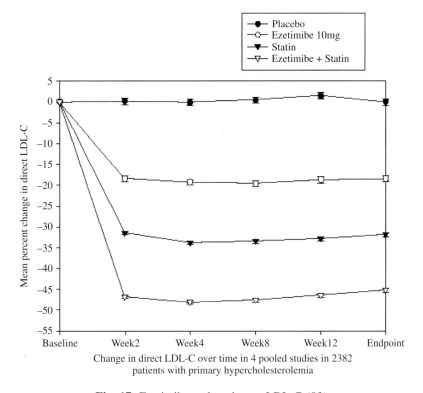

Fig. 17. Ezetimibe and statins on LDL-C *(83)*.

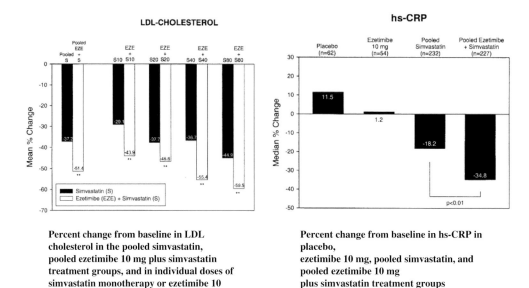

Percent change from baseline in LDL cholesterol in the pooled simvastatin, pooled ezetimibe 10 mg plus simvastatin treatment groups, and in individual doses of simvastatin monotherapy or ezetimibe 10 mg plus individual doses of simvastatin cohorts. **p <0.01 for combination versus simvastatin alone.

Percent change from baseline in hs-CRP in placebo, ezetimibe 10 mg, pooled simvastatin, and pooled ezetimibe 10 mg plus simvastatin treatment groups

Fig. 18. Percent change from baseline in LDL cholesterol in the pooled simvastatin, pooled ezetimibe 10 mg plus simvastatin treatment groups, and in individual doses of simvastatin monotherapy or ezetimibe 10 mg plus individual doses of simvastatin cohorts. **$P < 0.01$ for combination versus simvastatin alone. Percent change from baseline in hs-CRP in placebo, ezetimibe 10 mg, pooled simvastatin, and pooled ezetimibe 10 mg plus simvastatin treatment groups *(82)*.

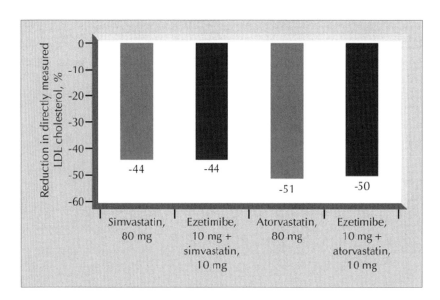

Fig. 19. Ezetimibe and statins on LDL-C *(84)*.

Ezetimibe can be particularly beneficial when administered to persons with less than predicted LDL-C-lowering responses to statins or in persons know to be hyperabsorbers of cholesterol. One study of familial hypercholesterolemic patients showed poor responders to statins have decreased rates of cholesterol synthesis that may be secondary to a genetically determined increase in cholesterol absorption, associated with increased apoE4 phenotypes *(86)*. In patients having a > 40% additional LDL-C reduction with the addition of ezetimibe to a statin, the LDL-C response to the statin was < 60% of the predicted lowering range (3–60%), i.e., statin hyporesponders have exaggerated responses to ezetimibe *(87)*. In the Scandinavian Simvastatin Survival Study (4S), the LDL-C-lowering ability of simvastatin was positively related to hypoabsorbers of cholesterol and negatively related to hyper-absorbers of cholesterol *(88)* (Fig. 20).

In such statin hyporesponders, statin responsiveness could be enhanced by reducing dietary cholesterol intake or inhibiting absorption with plant stanols or ezetimibe. No outcome trials have been completed looking at clinical event reduction with ezetimibe or ezetimibe/statin combination. However, using the Framingham risk equation, Davies et al. estimate co-administration of ezetimibe with statin therapy will provide an additional 13.7% reduction in predicted 5-year CHD risk when compared with statin monotherapy *(89)* (Fig. 21).

Statins and Cholesterol Absorption

Because cholesterol absorption, synthesis, and excretion are tightly regulated, it is not surprising that drugs that affect cholesterol homeostasis may have variable effects on cholesterol absorption and cholesterol lowering in the plasma. Short-term studies of statin use reveal an initial decrease in cholesterol and noncholesterol sterols' plasma levels *(90)*. After sometime (approximately 6 weeks), noncholesterol sterol levels rise in patients on statin therapy *(65,91)*. The reasons behind this effect are unknown. Miettinen demonstrated atorvastatin decreases cholesterol synthesis (indicated by increased lathosterol levels) but at the same time increases the absorption of cholesterol as indicated by increased campesterol levels (twofold to sixfold), and these sterols ultimately appear in lipoproteins *(92)*. A study in metabolic syndrome patients had similar findings *(93)* (Fig. 22).

The increased noncholesterol sterol absorption associated with statins may have clinical effects: in a study of carotid endarterectomy tissue in patients on statins versus those not on statins, serum plant sterols correlated with cholesterol absorption and

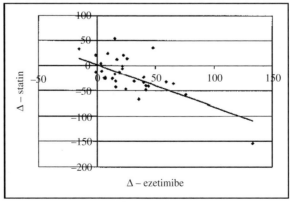

Regression plot of change in ezetimibe (x-axis) vs change in statin (y axis)

Fig. 20. Statins' LDL-C response predicts ezetimibe response *(87)*.

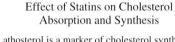

Fig. 21. Therapeutic modulation of cholesterol absorption.

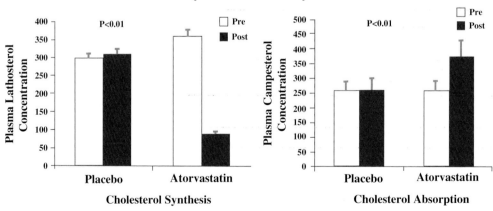

Fig. 22. Effect of statins on cholesterol absorption and synthesis *(93)*.

the plaque in the statin patients had increased campesterol levels. The sterols in the plaque were of dietary origin transported mostly in LDL. No definitive conclusions as to role of noncholesterol sterols in atherosclerosis can be taken from this study *(94)* (Fig. 23). However, there is very interesting data from the 4S study that patients

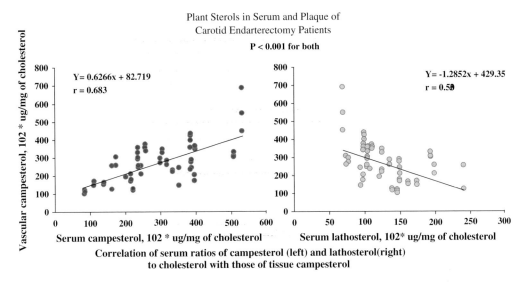

Fig. 23. Plant sterols in serum and plaque of carotid endarterectomy patients *(94)*.

with coronary atherosclerosis with hyperabsorption of sterols and low synthesis of cholesterol (as indicated by increased cholestanol:cholesterol ratio) did not respond to statin treatment. The patients with the lowest markers of absorption had the best event reduction with simvastatin, and those with the highest sterol absorption had no statistically significant event reduction on simvastatin. The incidence of coronary events was unrelated to sterol levels in the placebo group *(95)*. Thus, patients likely to be responsive to statins can be identified before treatment by measuring a marker of cholesterol absorption like cholestanol concentration. The lack of response to statins in the cholesterol hyperabsorbers is due to the decreased hepatic cholesterol synthesis in such patients. Indeed, analysis of 4S baseline sterol levels strongly indicated simvastatin suppresses the synthesis of cholesterol markedly more effectively in subjects with high rather than low baseline synthesis, but reduces respective serum cholesterol levels less markedly than synthesis *(96)* (Fig. 24).

This also suggests such patients would benefit from a combination therapy of statins with stanols or ezetimibe to lower more effectively their serum cholesterol levels and prevent an increase in the levels of plant sterols.

Fibrates

Fibrates, which are PPAR-α agonists, are lipid-modulating drugs that have long been known to effect hepatic and biliary cholesterol homeostasis, and their use has been associated with cholelithiasis *(97)*. This is thought to be secondary to both decreased BA synthesis and excretion and increased biliary excretion of cholesterol obtained through SR-B1 delipidation of HDL particles *(98,99)*. PPAR-α agonism causes downregulation of cholesterol 7α-hydroxylase and sterol 27-hydroxylase *(100)*.

PPAR-α activation is also associated with increased Mdr2 expression levels leading to an increased phospholipid output into bile. Upon activation of LXR, hepatic and intestinal ABCG5/G8 expression is increased leading to an increased biliary cholesterol output. Bile lithogenicity is likely increased by all the above mechanisms *(101,102)*.

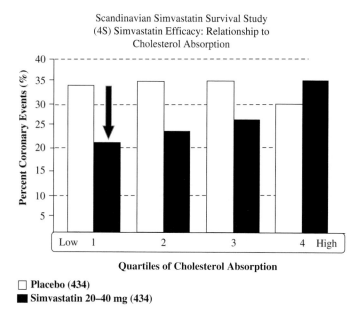

Fig. 24. Scandinavian simvastatin survival study (4S) simvastatin efficacy: relationship to cholesterol absorption *(95)*.

A study by Nigon et al. combined fibrates with plant sterol ester supplemented-spread and reported a potentiation of lipid-lowering effects. The plasma total cholesterol and LDL-C were significantly lower in the subgroup of patients currently treated with fibrates after consumption of phytosterol ester-enriched spread as compared with controls. The mechanism of action was hypothesized to be sterol-mediated upregulation of LDLrs and the reduced reabsorption of the increased fibrate-induced biliary cholesterol *(103)*.

There have also been trials of fibrates and ezetimibe *(104)* showing complimentary lipid benefits: combination therapy reduced LDL-C by 20.4%, non-HDL-C by 30.4%, and TG by 44.0%, and increased HDL-C by 19.0% and caused just over 60% of patients to achieve non HDL-C goals (Fig. 25). There was also a shift of LDL size from small to large in a majority of the patients and a decrease in high sensitivity C-reactive protein (hs-CRP). Co-administration of ezetimibe and fenofibrate has not demonstrated any clinically significant pharmacokinetic adversities *(105)*. Similar to the proposed effect seen with plant sterols and fibrates, one would expect fibrates to increase the biliary excretion of cholesterol and ezetimibe to prevent much of that cholesterol from being reabsorbed. Stool cholesterol excretion would likely be significantly increased.

SITOSTEROLEMIA (PHYTOSTEROLEMIA)

β-Sitosterolemia is a disease characterized by premature atherosclerosis and aortic stenosis where the composition of the plaque is not cholesterol but rather noncholesterol or phytosterols. The true incidence is unknown but estimated at 1 per 1,000,000 births and was first described in two sisters with extensive tendon xanthomas in 1974. Both sisters had increased absorption as well as elevated plasma levels of sitosterol, campesterol, and stigmasterol but normal levels of cholesterol. The majority of the

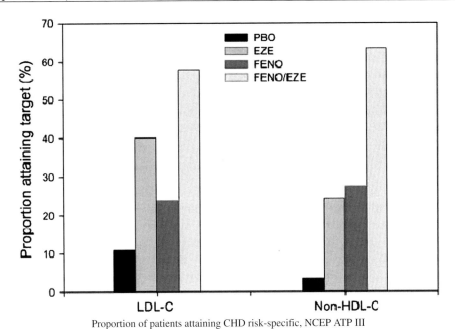

Proportion of patients attaining CHD risk-specific, NCEP ATP III LDL-C and Non HDL-C treatment targets at study endpoint

Fig. 25. Fenofibrate and ezetimibe study *(104)*.

sterols were transported in LDL and the rest in HDL: almost none were present in VLDL. The red blood cells and the xanthomas also had abnormal sterol levels *(1)*.

Subsequent cases were associated with xanthomas (Achilles and hands), premature atherosclerosis resembling familial hypercholesterolemia *(106)*, hemolytic anemia (due to erythrocyte fragility), arthritis, and arthralgias. The disease has been found in a number of differing ethnicity and geographic locations *(106–108)*.

Subsequent research has determined that the defect in these patients is pathological uptake of noncholesterol sterols (predominantly phytosterols) caused by a mutation or recessive trait (thus there is no family history of premature atherosclerosis) causing an absence of the ABCG5 and ABCG8 transporters. As much as 60% of absorbed noncholesterol sterols are absorbed compared with < 5% in normal humans *(109)*. There is impaired biliary secretion of the sterols. The mutation has been linked to chromosome 2p21, between microsatellite Markers *D2S1788* and *D2S1352* (maximum lodscore 4.49, $\Phi = 0.0$) *(110)*. The diagnosis of phytosterolemia is made by the detection of elevated phytosterols in the plasma using either capillary gas liquid or high-performance liquid chromatography *(39,111)*.

Heterozygotes are clinically normal but have somewhat elevated sterol levels (10-fold to 20-fold less than the homozygotes) *(109)*. Sitosterol concentrations range from 10 to 65 mg/dL with an average of 35 mg/dL. The reference range is 0.3–1 mg/dL. Owing to endogenous metabolism, there are increased amounts of 5α-dihydro derivative of cholesterol (cholestanol) and the 5α-dihydro plant sterol derivatives, 5α-campestanol and 5α-sitostanol in plasma and tissues *(112)*. Cholesterol levels vary but are usually normal or somewhat higher than matched controls as are apoB levels. Sitosterolemic LDLs are indistinguishable from normal LDLs under electron microscopy. The blood–brain barrier prevents the sterols from entering brain tissue. Less cholesterol and sterols are excreted into the bile. There is a downregulation of

HMG-CoA reductase synthesis leading to decreased cholesterol synthesis *(113)*. The increased hepatic phytosterols competitively inhibits cholesterol 7α-hydroxylase, the rate-determining enzyme for BA synthesis, which may eventually lead to decreased BA production and deficient pool size *(114)* (Fig. 26).

The association between sitosterolemia and the half transporters ABCG5 and ABCG8 transporters (sterolins 1 and 2) was first described by Berge et al in 2000 *(43)*. These sterol transporters limit the enterocyte uptake and increase enterocyte and hepatic excretion of noncholesterol sterols. Mutations will cause sterol accumulation. The translational start sites of *ABCG5* and *ABCG8* are separated by 374 base pairs, and the two genes are arranged in a head-to-head orientation. Mutation in either gene results in phytosterolemia. Of concern is whether more subtle polymorphisms of these transporters would explain variable sterol absorption seen among humans *(43)*. It has been reported that Asians with phytosterolemia may more commonly have *ABCG8* mutations *(115)*. There are now 12 different *ABCG5* mutations and 24 different *ABCG8* mutations *(21–28)* in sitosterolemia *(116)* (Fig. 27).

The treatment of sitosterolemia classically has involved BAS *(117)* and more recently ezetimibe. Statins have not been very effective. However, treatment is somewhat variable between different patients. As mentioned, BA enzymes sterol 27-hydroxylase and cholesterol 7α-hydroxylase are downregulated in patients with sitosterolemia. BAS use will deplete the concentration of hepatic BAs, which induces 7α-hydroxylase synthesis. The substrate cholesterol is needed for BAS synthesis and because HMG-CoA reductase is also downregulated (preventing cholesterol synthesis) LDLrs are upregulated *(118)*. Hepatic removal of LDL particles from the plasma will provide the needed cholesterol but also reduce plasma sterol levels. Sterol reductions of 50% have been reported *(116,119)*.

Ezetimibe will reduce sterol levels by preventing their intestinal absorption. Reductions of 21% have been reported *(120)* and the drug has FDA approval to lower sitosterol levels in patients with sitosterolemia. A trial is in recruitment by the NIH to test the safety and efficacy of 40 mg versus 10 mg of ezetimibe in sitosterolemic patients *(121)*. Because of the HMG-CA reductase downregulation, statins are of limited or

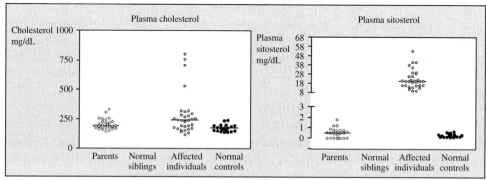

The plasma sitosterol levels were determined using capillary gas liquid or high-performance liquid chromatography. In general, most unaffected individuals had plasma sitosterol levels of < 1 mg/dL. Of the three parents and three siblings who had values > 1 mg/dL, none was > 2 mg/dL. All of the affected individuals had plasma sitosterol values > 8 mg/dL, and many of these values were obtained while the patients were on treatment.

Fig. 26. Plasma sitosterol and cholesterol levels in affected individuals, their parents, clinically unaffected siblings, and normal controls *(39)*.

5 kb
A
5′
3′
3′
ABCG5
5′
ABCG8
B
R408stop
COOH
NH3
ATP
L596R
G574R
Y658stop
R412stop
COOH
W361stop
R263Q
NH3
ATP

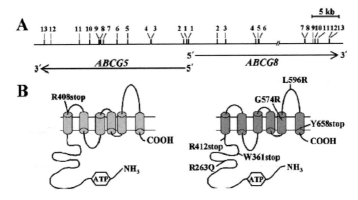

Genomic structure (**A**), putative topology (**B**), and predicted amino acid sequences ofABCG5 and ABCG8 (**C**). *ABCG5* and *ABCG8* are located on chromosome 2p21 between markers D2S177 and D2S119. (A) *ABCG5* and *ABCG8* are tandemly arrayed in a head-to-head orientation separated by 374 base pairs. *ABCG5* and *ABCG8* are both encoded by 13 exons and each spans ;28 kb.

Fig. 27. Genomic structure (**A**), putative topology (**B**), and predicted amino acid sequences of ABCG5 and ABCG8 (**C**). *ABCG5* and *ABCG8* are located on chromosome 2p21 between markers D2S177 and D2S119. (**A**) *ABCG5* and *ABCG8* are tandemly arrayed in a head-to-head orientation separated by 374 base pairs. *ABCG5* and *ABCG8* are both encoded by 13 exons and each spans 28 kb *(43)*.

no use *(87)*. Indeed, failure to respond to a statin should raise the suspicion of sitosterolemia. Some patients are treated with all three medications. Plant stanols and ileal bypass surgery has also been used successfully in the treatment of sitosterolemia *(122)*. Treatment does reduce xanthomas, aortic stenosis murmurs, angina, and arthritis *(123)* (Fig. 28).

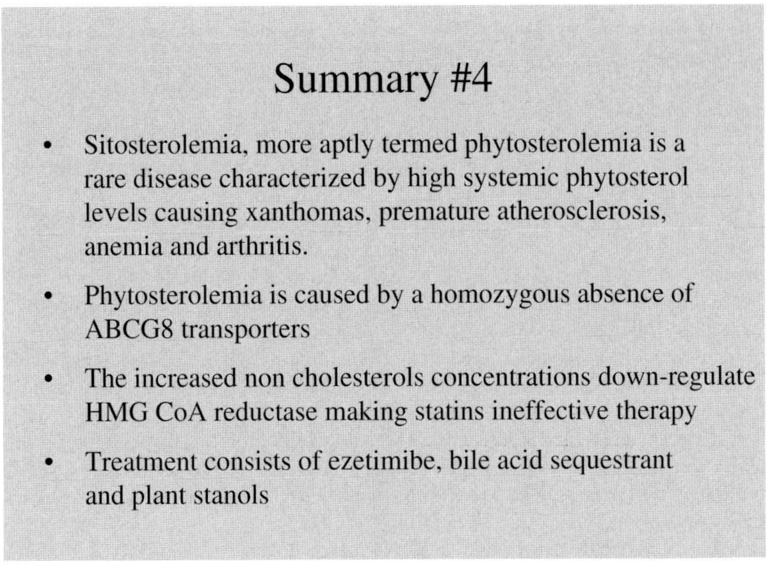

Summary #4
• Sitosterolemia, more aptly termed phytosterolemia is a rare disease characterized by high systemic phytosterol levels causing xanthomas, premature atherosclerosis, anemia and arthritis.
• Phytosterolemia is caused by a homozygous absence of ABCG8 transporters
• The increased non cholesterols concentrations down-regulate HMG CoA reductase making statins ineffective therapy
• Treatment consists of ezetimibe, bile acid sequestrant and plant stanols

Fig. 28. Phytosterolemia.

REFERENCES

1. Bhattacharyya AK and Connor WE. *J Clin Invest* 1974;53:1033–1043.
2. Lammert F and Wang DQ. New insights into the genetic regulation of cholesterol absorption. *Gastroenterology* 2005;129:718–734.
3. Nelson DL and Cox MM. *Lehninger Principles of Biochemistry*, 4th edition (Chapter 21.4.). WH Freeman and Company, New York.
4. Horton JD, Goldstein L, and Brown MS. SREBPs: activators of the complete program of cholesterol and fatty acid synthesis in the liver. *J Clin Invest* 2002;109:1125–1131.
5. Herman M, Kempen FC, Glatz A, et al. Serum lathosterol concentration is an indicator of whole-body cholesterol synthesis in humans. *J Lipid Res* 1988;29:1149–1155.
6. Miettinen TA, Tilvis RS, and Kesaniemi YA. Serum plant sterols and cholesterol precursors reflect cholesterol absorption and synthesis in volunteers of a randomly selected male population. *Am J Epidemiol* 1990;131:20–31.
7. Thompson GR and Grundy SM. History and development of plant sterol and stanol esters for cholesterol-lowering purposes. *Am J Cardiol* 2005;96(suppl):3D–9D.
8. Pollak OJ. Reduction of blood cholesterol in man. *Circulation* 1953;7:702–706.
9. National Cholesterol Education Program (NCEP) Expert Panel on Detection, Evaluation, and Treatment of High Blood Cholesterol in Adults (Adult Treatment Panel III). Third Report of the National Cholesterol Education Program (NCEP) Expert Panel on Detection, Evaluation, and Treatment of High Blood Cholesterol in Adults (Adult Treatment Panel III) (final report). *Circulation* 2002;106:3143–3421.
10. Katan MB, Grundy SM, Jones P, Law M, Miettinen T, and Paoletti R, for the Stresa Workshop Participants. Efficacy and safety of plant stanols and sterols in the management of blood cholesterol levels. *Mayo Clin Proc* 2003;78:965–978.
11. Bosner MS, Lange LG, Stenson WF, and Ostlund RE, Jr. Percent cholesterol absorption in normal women and men quantified with dual stable isotopic tracers and negative ion mass spectrometry. *J Lipid Res* 1999;40:302–308.
12. Kesaniemi YA, Ehnholm C, and Miettinen TA. Intestinal cholesterol absorption efficiency in man is related to apoprotein E phenotype. *J Clin Invest* 1987;80:578–581.
13. von Bergmann K, Lütjohann D, Lindenthal B, and Steinmetz A. Efficiency of intestinal cholesterol absorption in humans is not related to apoE phenotype. *J Lipid Res* 2003;44:193–197.
14. Salen G, Ahrens EH, Jr, and Grundy SM. Metabolism of beta-sitosterol in man. *J Clin Invest* 1970;49:952–967.
15. Lutjohann D, Bjorkhem I, Beil UF, et al. Sterol absorption and sterol balance in phytosterolemia evaluated by deuterium-labeled sterols: effect of sitostanol treatment. *J Lipid Res* 1995;36:1763–1773.
16. Dietschy JM, Sallee VL, and Wilson FA. Unstirred water layers and absorption across the intestinal mucosa. *Gastroenterology* 1971;61:932–934.
17. Guyton AC and Hall JE. *Textbook of Medical Physiology*, 10th edition (Chapter 65). WB Saunders Co., Philadelphia, PA.
18. Davis, HR, Jr, Zhu L, Hoos LM, et al. Niemann-Pick C1 like 1 (NPC1L1) is the intestinal phytosterol and cholesterol transporter and a key modulator of whole-body cholesterol homeostasis. *J Biol Chem* 2004;279:33586–33592.
19. Klett EL and Patel SB. Will the real cholesterol transporter please stand up? *Science* 2004;303:1149–1150.
20. Uittenbogaard A, Everson WV, Matveev SV, and Smart EJ. Cholesteryl ester is transported from caveolae to internal membranes as part of a caveolin-annexin II lipid-protein complex. *J Biol Chem* 2002;277:4925–4931.
21. Kramer W, Girbig F, Corsiero D, et al. Aminopeptidase N (CD13) is a molecular target of the cholesterol absorption inhibitor ezetimibe in the enterocyte brush border membrane. *J Biol Chem* 2004;280:1306–1320.
22. Rigotti A, Miettinen HE, and Krieger M. The role of the high-density lipoprotein receptor SR-BI in the lipid metabolism of endocrine and other tissues. *Endocr Rev* 24(3):357–387.
23. Altmann SW, Davis HR, Jr, Zhu L, et al. Niemann-Pick C1 like 1 protein is critical for intestinal cholesterol absorption. *Science* 2004;303:1201–1204.
24. van der Veen JN, Kruit JK, Havinga R, et al. Reduced cholesterol absorption upon PPAR-activation coincides with decreased intestinal expression of NPC1L1. *J Lipid Res* 2005;46:526–534.
25. Marx J. Caveolae: a once-elusive structure gets some respect. *Science* 2001;294:1862–1865.

26. Uittenbogaard A and Smart EJ. Palmitoylation of caveolin-1 is required for cholesterol binding chaperone complex formation, and rapid transport of cholesterol to caveolae. *J Biol Chem* 2000;275:25595–25599.
27. Uittenbogaard A, Everson WV, Matveev SV, and Smart EJ. Cholesterol is transported from caveolae to internal membranes as part of a caveolin-annexin II lipid-protein complex. *J Biol Chem* 2002;277:4925–4931.
28. Smart EJ, De Rose RA, and Farber SA. Annexin 2–caveolin 1 complex is a target of ezetimibe and regulates intestinal cholesterol transport. *Proc Natl Acad Sci USA* 2004;101(10):3450–3455.
29. Anderson RA, Joyce C, Davis M, et al. Identification of a form of acyl-CoA:cholesterol acyltransferase specific to liver and intestine in nonhuman primates. *J Biol Chem* 1998;273:26747–26754.
30. Field FJ and Mathur SN. Beta-sitosterol: esterification by intestinal acylcoenzyme A:cholesterol acyltransferase (ACAT) and its effect on cholesterol esterification. *J Lipid Res* 1983;24:409–417.
31. Buhman KK, Accad M, Novak S, et al. Resistance to diet-induced hypercholesterolemia and gallstone formation in ACAT2-deficient mice. *Nat Med* 2000;6:1341–1347.
32. Temel RE, Gebre AK, Parks JS, and Rudel LL. Compared with acyl-CoA:cholesterol *O*-acyltransferase (ACAT) 1 and lecithin:cholesterol acyltransferase, ACAT2 displays the greatest capacity to differentiate cholesterol from sitosterol. *J Biol Chem* 2003;278(48):47594–47601.
33. Iqbal J and Hussain MM. Evidence for multiple complementary pathways for efficient cholesterol absorption in mice. *J Lipid Res* 2005;46:1491–1501.
34. Hussain MM, Fatmaa S, Pana X and Iqbala J. Intestinal lipoprotein assembly. *Curr Opin Lipidol* 2005;16:281–285.
35. van Greevenbroek MMJ, Robertus-Teunissen MG, Erkelens DW, and de Bruin TWA. Participation of the microsomal triglyceride transfer protein in lipoprotein assembly in Caco-2 cells: interaction with saturated and unsaturated dietary fatty acids. *J Lipid Res* 1998;39:173–185.
36. Innerarity TL, Bore'n J, Yamanaka S, and Olofsson S-O. Biosynthesis of apolipoprotein B48-containing lipoproteins. *J Biol Chem* 1996;271:2353–2356.
37. Weinberg RB, Cook VR, DeLozier JA, and Shelness GS. Dynamic interfacial properties of human apolipoproteins A-IV and B-17 at the air/water and oil/water interface. *J Lipid Res* 2000;41:1419–1427.
38. Gallagher JW, Weinberg RB, and Shelness GS. ApoA-IV tagged with the ER retention signal KDEL perturbs the intracellular trafficking and secretion of apoB. *J Lipid Res* 2004;45:1826–1834.
39. Lee M-H, Lu K, and Patel SB. Genetic basis of sitosterolemia. *Curr Opin Lipidol* 2001;12:141–149.
40. Lorkowski S and Cullen P. ABCG subfamily of human ATP-binding cassette proteins. *Pure Appl Chem* 2002;74:2057–2081
41. Dean M, Hamon Y, and Chimini G. The human ATP-binding cassette (ABC) transporter superfamily. *J Lipid Res* 2001;42:1007–1017.
42. Plat J and Mensink RP. Plant stanol and sterol esters in the control of blood cholesterol levels: mechanism and safety aspects. *Am J Cardiol* 2005;96(suppl):15D–22D.
43. Berge KE, Tian H, Graf GA, Yu L, Grishin NV, Schultz J, Kwiterovich P, Shan B, Barnes R, and Hobbs HH. Accumulation of dietary cholesterol in sitosterolemia caused by mutations in adjacent ABC transporters. *Science* 2000;290;1771–1775.
44. Kempen HJ, de Knijff P, Boomsma DI, van der Voort HA, Leuven JAG, and Havekes L. Plasma levels of lathosterol and phytosterols in relation to age, sex, anthropometric parameters, plasma lipids, and apolipoprotein E phenotype, in 160 Dutch families. *Metab Clin Experiment* 1991;40:604–611.
45. Sudhop T, Gottwald BM, and von Bergmann K. Serum plant sterols as a potential risk factor for coronary heart disease. *Metabolism* 2002;51:1519–1521.
46. Rajaratnam RA, Gylling H, and Miettinen TA. Independent association of serum squalene and noncholesterol sterols with coronary artery disease in postmenopausal women. *J Am Coll Cardiol* 2000;35:1185–1191.
47. Miettinen TA and Gylling H. Synthesis and absorption markers of cholesterol in serum and lipoproteins during a large dose of statin treatment. *Eur J Clin Invest* 2003;33(11):976–982.
48. Mortimer B-C, Tso P, Phan CT, Beveridge DJ, Wen J, and Redgrave TG. Features of cholesterol structure that regulate the clearance of chylomicron-like lipid emulsions. *J Lipid Res* 1995;36:2038–2053.

49. Kraemer FB and Shen W-J. Hormone-sensitive lipase control of intracellular tri-(di-)acylglycerol and cholesteryl ester hydrolysis. *J Lipid Res* 2002;43:1585–1594.
50. Hao M, Lin SX, Karylowski OJ, Wustner D, McGraw TE, and Maxfield FR. Vesicular and non-vesicular sterol transport in living cells. *J Biol Chem* 2002;277:609–671.
51. Xu X-X and Tabas I. Lipoproteins activate acyl-coenzyme A:cholesterol acyltransferase in macrophages only after cellular cholesterol pools are expanded to a critical threshold level. *J Biol Chem* 1991;266:17040–17048.
52. Lewis GF and Rader DJ. New insights into the regulation of HDL metabolism and reverse cholesterol transport. *Circ Res* 2005;96:1221–1232.
53. Bothman KM and Bravo E. The role of lipoprotein cholesterol in biliary steroid secretion: studies with in vivo experimental models. *Prog Lipid Res* 1995;34:71–97.
54. Langheim S, Yu L, von Bergmann K, Lütjohann D, Xu F, Hobbs HH, and Cohen JC. ABCG5 and ABCG8 require MDR2 for secretion of cholesterol into bile. *J Lipid Res* 2005;46:1732–1738.
55. Venkateswaran A, Laffitte BA, Joseph SB, Mak PA, Wilpitz DC, Edwards PA, and Tontonoz P. Control of cellular cholesterol efflux by the nuclear oxysterol receptor LXR alpha. *Proc Natl Acad Sci USA* 2000;97:12097–12102.
56. Costet P, Luo Y, Wang N, and Tall AR. Sterol-dependent transactivation of the *ABC1* promoter by the liver x receptor/retinoid x receptor. *J Biol Chem* 2000;275:28240.
57. Basso F, Freeman L, Knapper CL, Remaley A, Stonik J, Neufeld EB, Tansey T, Amar MJA, Fruchart-Najib J, Duverger N, Santamarina-Fojo S, and Brewer HB, Jr. Role of the hepatic ABCA1 transporter in modulating intrahepatic cholesterol and plasma HDL cholesterol. *J Lipid Res* 2003;44:296–302.
58. Hillebrant CG, Nyberg G, Einarsson K, and Eriksson M. The effect of plasma low density lipoprotein apheresis on the hepatic secretion of biliary lipids in humans. *Gut* 1997;41:700–704.
59. Carey MC. Homing-in on the origin of biliary steroids. *Gut* 1997;41:721–722.
60. Javitt NB. 25R,26-Hydroxycholesterol revisited: synthesis, metabolism, and biologic roles. *J Lipid Res* 2002;43:665–670.
61. Ren S, Hylemon P, Marques D, Hall E, Redford K, Gil G, and Pandak WM. Effect of increasing the expression of cholesterol transporters (StAR, MLN64, and SCP-2) on bile acid synthesis. *J Lipid Res* 2004;45:2123–2131.
62. Suchy FJ, Sippel CJ, and Ananthanarayanan M. Bile acid transport across the hepatocyte canalicular membrane. *FASEB J* 1997;11:199–205.
63. Figge A, Lammert F, Paigen B, Henke A, Matern S, Korstanje R, Shneider BL, Chen F, Stoltenberg E, Spatz K, Hoda F, Cohen DE, and Richard M. Green. Hepatic overexpression of murine Abcb11 increases hepatobiliary lipid secretion and reduces hepatic steatosis. *J Bio Chem* 2004;279:2790–2799.
64. Chen F, Ma L, Dawson PA, Sinal CJ, Sehayek E, Gonzalez FJ, Breslow J, Ananthanarayanan M, and Shneider BL. Liver receptor homologue-1 mediates species- and cell line-specific bile acid-dependent negative feedback regulation of the apical sodium-dependent bile acid transporter. *J Biol Chem* 2003;278:19909–19916.
65. Law M. Plant sterols and stanols in health. *BMJ* 2000;320:861–864.
66. Jones PJH, MacDougall DE, Ntanios F, and Vanstone CA. Dietary phytosterols as cholesterollowering agents in humans. *Can J Physiol Pharmacol* 1997;75:217–227.
67. Hallikainen MA, Sarkkinen ES, and Uusitupa MIJ. Effects of lowfat stanol ester enriched margarine on concentrations of serum carotenoids in subjects with elevated serum cholesterol concentrations. *Eur J Clin Nutr* 1999;53:966–969.
68. Miettinen TA, Puska P, Gylling H, Vanhanen H, and Vartiainen E. Reduction of serum choletserol with sitostanol-ester margarine in a midly hypercholesterolemic populaion. *N Engl J Med* 1995;333:1308–1312.
69. Plat J, van Onselen EN, van Heugten MM, and Mensink RP. Effects on serum lipids, lipoproteins and fat soluble antioxidant concentrations of consumption frequency of margarines and shortenings enriched with plant stanol esters. *Eur J Clin Nutr* 2000;54:671–677.
70. Blair S, Capuzzi DM, Gottlieb SO, Nguyen T, Morgan JM, and Cater NB. Incremental reduction of serum total cholesterol and low-density lipoprotein cholesterol with the addition of plant stanol ester containing spread to statin treatment. *Am J Cardiol* 2000;86:46–52.
71. O'Neill FH, Sanders TAB, and Thompson GR. Comparison of efficacy of plant stanol ester and sterol ester: short-term and longer-term studies. *Am J Cardiol* 2005;96(suppl):29D–36D.
72. Brown AJ and Jessup W. Oxysterols and atherosclerosis. *Atherosclerosis* 1999;142:1–28.

73. Miettinen TA, Gylling H, Lindbohm N, Miettinen TE, Rajaratnam RA, and Relas H, for the Finnish Treat-to-Target Study Investigators. Serum noncholesterol sterols during inhibition of cholesterol synthesis by statins. *J Lab Clin Med* 2003;141:131–137.
74. van Heek M, Farley C, Compton DS, et al. Ezetimibe selectively inhibits intestinal cholesterol absorption in rodents in the presence and absence of exocrine pancreatic function. *Br J Pharmacol* 2001;134:409–417.
75. Toth PP and Davidson MH. Simvastatin plus ezetimibe: combination therapy for the management of dyslipidaemia. *Expert Opin Pharmacother* 2005;1:131–139.
76. Ghosal A, Hapangama N, Yuan Y, et al. Identification of human UDPglucuronosyltransferase enzymes responsible for the glucuronidation of ezetimibe. *Drug Metab Dispos* 2004; 32:314–320.
77. Sudhop T, Lujohann D, Kodal A et al. Inhibition of intestinal cholesterol absorption by ezetimibe in humans. *Circulation* 2002;106:1943–1948.
78. Ezzet F, Wexler D, Statkevich P, et al. The plasma concentration and LDL-C relationship in patients receiving ezetimibe. *J Clin Pharmacol* 2001;41:943–949.
79. Gagne C, Bays HE, Weiss SR, et al. Efficacy and safety of ezetimibe added to ongoing statin therapy for treatment of patients with primary hypercholesterolemia. *Am J Cardiol* 2002;90:1084–1091.
80. Al-Shaer MH. The effects of ezetimibe on the LDL-cholesterol particle number. *Cardiovasc Drugs Ther* 2004;18:327–328.
81. Hegele RA, Guy1 J, Ban1 MR, and Wang J. NPC1L1 haplotype is associated with inter-individual variation in plasma low-density lipoprotein response to ezetimibe. *Lipids Health Dis* 2005; 4:2–20.
82. Sager PT, Melani L, Lipka L, Strony J, Yang B, Suresh R, and Veltri E, for the Ezetimibe Study Group. Effect of coadministration of ezetimibe and simvastatin on high-sensitivity c-reactive protein. *Am J Cardiol* 2003;92:1414–1418.
83. Davidson MH, Ballantyne CM, Kerzner B, Melani L, Sager PT, Lipka L, Strony J, Suresh R, Veltri E, for the Ezetimibe Study Group. Efficacy and safety of ezetimibe coadministered with statins: randomised, placebo-controlled, blinded experience in 2382 patients with primary hypercholesterolemia. *Int J Clin Pract* 2004;58:746–755.
84. Ballantyne CM. Role of selective cholesterol absorption inhibition in the management of dyslipidemia. *Curr Atheroscler Rep* 2004;6:52–59.
85. Kosoglou1 T, Statkevich P, Yang B, Suresh R, Zhu Y, Boutros T, Maxwell SE, Tiessen R, and Cutler DL. Pharmacodynamic interaction between ezetimibe and rosuvastatin. *Curr Med Res Opin* 2004;20:1185–1195.
86. O'Neill FH, Patel DD, Knight BL, Neuwirth CKY, Bourbon M, Soutar AK, Taylor GW, Thompson GR, and Naoumova RP. Determinants of variable response to statin treatment in patients with refractory familial hypercholesterolemia. *Arterioscler Thromb Vasc Biol* 2001;21:832–837.
87. Ziajka PE, Reis M, Kreul S, and King H. Initial low-density lipoprotein response to statin therapy predicts subsequent low-density lipoprotein response to the addition of ezetimibe. *Am J Cardiol* 2004;93:779–780.
88. Miettinen TA, Strandberg TE, and Gylling H, for the Finnish Investigators of the Scandinavian Simvastatin Survival Study Group. Noncholesterol sterols and cholesterol lowering by long-term simvastatin treatment in coronary patients relation to basal serum cholestanol. *Arterioscler Thromb Vasc Biol* 2000;20:1340–1346.
89. Davies GM, Cook JR, Erbey J, Alemao E, and Veltri EP. Projected coronary heart disease risk benefit with ezetimibe. *Atherosclerosis* 2005;179:375–378.
90. Ntanios FY, Jones PJH, and Frohlich JJ. Effect of 3-hydroxy-3-methylglutaryl coenzyme A reductase inhibitor on sterol absorption in hypercholesterolemic subjects. *Metabolism* 1999; 48:68–73.
91. Miettinen TA, Gylling H, Lindbohm N, Miettinen TE, Rajaratnam RA, and Relas H, for the Finnish treat to target study (3T) investigators. Serum noncholesterol sterols during inhibition of cholesterol synthesis by statins. *J Lab Clin Med* 2003;141:131–137.
92. Miettinen TA and Gylling H. Synthesis and absorption markers of cholesterol in serum and lipoproteins during a large dose of statin treatment. *Eur J Clin Invest* 2003;33:976–982.
93. Watts GF, Chan DC, Barrett1 PHR, O' Neill FH, and Thompson GR. Effect of a statin on hepatic apolipoprotein B-100 secretion and plasma campesterol levels in the metabolic syndrome. *Int J Obes Relat Metab Disord* 2003;27:862–865.
94. Miettinen TA, Railo M, Lepäntalo M, and Gylling H. Plant sterols in serum and in atherosclerotic plaques of patients undergoing carotid endarterectomy. *J Am Coll Cardiol* 2005;45:1794–1801.

95. Miettinen TA, Gylling H, Strandberg T, and Sarna S, for the Finnish 4S Investigators. Baseline serum cholestanol as predictor of recurrent coronary events in subgroup of Scandinavian simvastatin survival study. *BMJ* 1998;316:1127–1130.
96. Miettinen TA and Gylling H. Ineffective decrease of serum cholesterol by simvastatin in a subgroup of hypercholesterolemic coronary patients. *Atherosclerosis* 2002;164:147–152.
97. Raedsch R, Plachky J, Wolf N, and Simonis G. Biliary lipids, lithogenic index and biliary drug concentration during etofibrate and bezafibrate treatment. *Eur J Drug Metab Pharmacokinet* 1995;20:113–118.
98. Zanlungo S, Rigotti A, and Nervi F. Hepatic cholesterol transport from plasma into bile: implications for gallstone disease. *Curr Opin Lipidol* 2004;15:279–286.
99. Thijs C, Knipschild P, and Brombacher P. Serum lipids and gallstones: a case–control study. *Gastroenterology* 1990;99:843–849.
100. Post SM, Duez H, Gervois PP, Staels B, Kuipers F, and Princen HMG. Fibrates suppress bile acid synthesis via peroxisome proliferator-activated receptor-mediated downregulation of cholesterol 7alpha-hydroxylase and sterol 27-hydroxylase expression. *Arterioscler Thromb Vasc Biol* 2001;21:1840–1845.
101. Kok T, Bloks VW, Wolters H, Havinga R, Jansen PL, Stals B, and Kuipers F. Peroxisome proliferator-activated receptor α (PPARα)-mediated regulation of multidrug resistance 2 (Mdr2) expression and function in mice. *Biochem J* 2003;369:539–547.
102. Roglans N, Vasquez-Carrera M, Alegret M, Novell F, Zambon D, Ros E, Laguna JC, and Sanchez RM. Fibrates modify the expression of key factors involved in bile-acid synthesis and biliary-lipid secretion in gallstone patients. *Eur J Clin Pharmacol* 2004;59:855–861.
103. Nigon F, Serfaty-Lacrosnière C, Beucler I, Chauvois D, Neveu C, Giral P, Chapman MJ, and Bruckert E. Plant sterol-enriched margarine lowers plasma LDL in hyperlipidemic subjects with low cholesterol intake: effect of fibrate treatment. *Clin Chem Lab Med* 2001;39:634–640.
104. Farnier M, Freeman MW, Macdonell G, Perevozskaya I, Davies MJ, Mitchel YB, and Gumbiner B, for the Ezetimibe Study Group. Efficacy and safety of the coadministration of ezetimibe with fenofibrate in patients with mixed hyperlipidaemia. *Eur Heart J* 2005;26:897–905.
105. Kosoglou T, Statkevich P, Fruchart JC, Pember LJC, Reyderman L, Cutler DL, Guillaume M, Maxwell SE, and Veltri EP. Pharmacodynamic and pharmacokinetic interaction between fenofibrate and ezetimibe. *Curr Med Res Opin* 2004;20:1197–1207.
106. Salen G, Horak I, Rothkopf M, Cohen JL, Speck J, Tint GS, Shore V, Dayal B, Chen T, and Shefer S. Lethal atherosclerosis associated with abnormal plasma and tissue sterol composition in sitosterolemia with xanthomatosis. *J Lipid Res* 1985;26:1126–1133.
107. Bjorkhem I and Boberg KM. Inborn errors in bile acid biosynthesis and storage of sterols other than cholesterol. In: *The Metabolic Basis of Inherited Disease*, 7th edition, vol. 2. Scriver CR, Beaudet AL, Sly WS, and Valle D (editors). New York: McGraw-Hill Inc.; 1995. pp. 2073–2102.
108. Steiner RD and Campbell P. Sitosterolemia. Last updated July 2005. http://www.emedicine.com/ped/ topic2110.htm Last accessed 10/25/05.
109. Salen G, Shefer S, Nguyen L, Ness GC, Tint GS, and Shore V. Sitosterolemia. *J Lipid Res* 1992;33:945–955.
110. Patel SB, Salen G, Hidaka H, Kwiterovich PO, Jr, Stalenhoef AFH, Miettinen TA, Grundy SM, Lee M-H, Rubenstein JS, Polymeropoulos MH, and Brownstein MJ. Mapping a gene involved in regulating dietary cholesterol absorption. The sitosterolemia locus is found at chromosome 2p21. *J Clin Invest* 1998;102:1041–1044.
111. Kuksis A, Myher JJ, Marai L, et al. Usefulness of gas chromatographic profiles of plasma total lipids in diagnosis of phytosterolemia. *J Chromatogr* 1986;381:1–12.
112. Salen G, Kwiterovich PO, Jr, Shefer S, Tint GS, Horak I, Shore V, Dayal B, and Horak E. Increased plasma cholestanol and 5a-saturated plant sterol derivatives in subjects with sitosterolemia and xanthomatosis. *J Lipid Res* 1985;26:203–209.
113. Nguyen L, Shefer S, Salen G, Ness G, Tint GS, Zaki FG, and Rani I. Molecular defect in cholesterol synthesis in sitosterolemia with xanthomatosis. *J Clin Invest* 1990;86:926–931.
114. Shefer S, Salen G, Nguyen L, Batta AK, Packin V, Tint GS, and Hauser S. Competitive inhibition of bile acid synthesis by endogenous cholestanol and sitosterol in sitosterolemia with xanthomatosis. *J Clin Invest* 1988;82:1833–1839.
115. Sehayek E, Yu HJ, von Bergmann K, Lutjohann D, Stoffel M, Duncan EM, Garcia-Naveda L, Salit J, Blundell ML, Friedman JM, and Breslow JL. Phytosterolemia on the island of Kosrae:

founder effect for a novel ABCG8 mutation results in high carrier rate and increased plasma plant sterol levels. *J Lipid Res* 2004;45:1608–1613.
116. Wang J, Joy T, Mymin D, Frohlich J, and Hegele RA. Phenotypic heterogeneity of sitosterolemia. *J Lipid Res* 2004;45:2361–2367.
117. Belamarich PF, Deckelbaum RJ, Starc TJ, Dobrin BE, Tint GS, and Salen G. Response to diet and cholestyramine in a patient with sitosterolemia. *Pediatrics* 1990;86:977–981.
118. Nguyen L, Shefer S, Salen G, Horak I, Tint GS, and McNamara DJ. The effect of abnormal plasma and cellular sterol content and composition on low density lipoprotein uptake and degradation by monocytes and lymphocytes in sitosterolemia with xanthomatosis. *Metabolism* 1988;37: 346–351.
119. Parsons HG, Jamal R, and Baylis B. A marked and sustained reduction in LDL sterols by diet and cholestyramine in sitosterolemia. *Clin Invest Med* 1995;18:389–400.
120. Salen G, von Bergmann K, Lutjohann D, Kwiterovich P, Kane J, Patel SB, Musliner T, Stein P, and Musser B. Ezetimibe effectively reduces plasma plant sterols in patients with sitosterolemia. *Circulation* 2004;109:966–971.
121. National Institutes of Health Clinical Center. Higher-dose ezetimibe to treat homozygous sitosterolemia. http://www.clinicaltrials.gov/ct/gui/show/NCT00099996 (Last accessed 10/25/05).
122. Nguyen LB, Cobb M, Shefer S, et al. Regulation of cholesterol biosynthesis in sitosterolemia: effects of lovastatin, cholestyramine, and dietary sterol restriction. *J Lipid Res* 1991;32:1941–1948.
123. Liitjohann D, Bjiirkhem I, Beil UF, and von Bergmann K. Sterol absorption and sterol balance in phytoterolema evalated by deuterium-labeled sterols: effect of sitostanol treatment. *J Lipid Res* 1995;36:1763–1773.
124. Bays HE, Moore PB, Drehobl MA, Rosenblatt S, Toth PD, Dujovne CA, Knopp RH, Lipka LJ, LeBeaut AP, Yang B, Mellars LE, Cuffie-Jackson C, Veltri EP, for the Ezetimibe Study Group. Effectiveness and tolerability of ezetimibe in patients with primary hypercholesterolemia: pooled analysis of two phase ii studies. *Clin Ther* 2001;23:1209–1230.

15 Utilization of Lipoprotein Subfractions

William C. Cromwell, MD, and James D. Otvos, PhD

Contents

INTRODUCTION
PHYSIOLOGIC ORIGINS OF LIPOPROTEIN
   PARTICLE HETEROGENEITY
LIPOPROTEIN ASSAY METHODS
ANALYTIC IMPLICATIONS OF LIPOPROTEIN HETEROGENEITY
UTILIZATION OF LIPOPROTEIN INFORMATION
   FOR INDIVIDUAL PATIENT MANAGEMENT
CONCLUSION
REFERENCES

INTRODUCTION

It is widely appreciated that the interaction of lipoprotein particles with the arterial wall affects coronary heart disease (CHD) risk. Low-density lipoprotein (LDL) particles [and to a lesser extent very-low-density lipoprotein (VLDL) and remnant particles] promote atherosclerosis by entering the artery wall, becoming oxidized, and subsequently being ingested by macrophages to create cholesterol-rich foam cells which develop into atherosclerotic plaques. High-density lipoprotein (HDL) particles entering the artery wall antagonize this process by, among other actions, inhibiting the oxidation of LDL particles and removing cholesterol from the foam cells for delivery back to the liver. The overall risk of cardiovascular disease (CVD) depends on the balance between these atherogenic and anti-atherogenic particles.

Historically, the clinical management of lipoprotein-related CHD risk has been framed as a cholesterol, not a lipoprotein, issue. The surrogate relationship of lipids [cholesterol and triglycerides (TGs)] to lipoproteins was described by Fredrickson et al. 35 years ago *(1)*. They noted that "all abnormalities in plasma lipid concentrations, or dyslipidemia, can be translated into dyslipoproteinemia" and "the shift of emphasis to lipoproteins offers distinct advantages in the recognition and management of such disorders." The reason that lipids, rather than lipoproteins, are the traditional focus of clinical attention was also discussed: "there is no single test that infallibly separates all those who have dyslipoproteinemia from those who do not...the majority of laboratories still employ a combination of chemical measurements of plasma lipids for this purpose." Thus, for reasons that were related primarily to the difficulty of measuring lipoprotein particles directly, circulating TGs have come to serve as a surrogate measure of VLDL levels, whereas LDL-cholesterol (LDL-C) and HDL-cholesterol (HDL-C) values serve as indicators of the concentrations of LDL and HDL particles.

Few people have regarded the surrogate relationship of lipids to lipoproteins as a clinical limitation. Over the past 40 years, an extensive body of genetic, epidemiologic, and clinical intervention trial data has amassed showing that, at a population level, abnormal lipid levels are strongly related to atherosclerosis and CHD events. However, at an individual patient level, significant variability in CHD risk is present across a wide range of cholesterol values in prospective epidemiologic trials, as well as among patients followed in placebo and active therapy groups of clinical intervention trials *(2–7)*. Furthermore, on-trial lipid values often are weak predictors of CHD risk in intervention studies *(7–10)*.

Recently, trials have begun to investigate relations of CHD with quantitative measures of lipoproteins other than those based on how much lipid a particular particle contains. What is under-appreciated is that the amount of cholesterol carried inside lipoprotein particles is highly variable among individuals with the same measured cholesterol levels. As a consequence of the magnitude and prevalence of this lipid compositional variability within lipoprotein particles, even the most accurate cholesterol measurements will, for many individuals, provide an inaccurate measure of circulating lipoproteins particles *(7,11–13)*.

Given the recent growth in availability of specialized lipoprotein tests available through commercial clinical laboratories, it is important to understand whether the CHD risk of individual patients might be assessed and managed more effectively by measuring the lipoprotein mediators of the disease process rather than the lipids carried within them. This chapter will focus on essential analytic, physiologic, and cardiovascular outcome considerations that influence strategies for the clinical utilization of lipoprotein subclass information. Key topics include:

1. Discussion of the physiologic origins of lipoprotein particle heterogeneity;
2. Review of qualitative and quantitative data supplied from commercially available lipoprotein assay technologies;
3. Examination of prevalence of lipoprotein heterogeneity among patients at increased CHD risk and the potential limiting effect this exerts on traditional lipid testing;
4. Review of univariate and multivariate cardiovascular outcome associations with qualitative and quantitative lipoprotein parameters;
5. Discussion of clinical utilization of lipoprotein information for individual patient management.

PHYSIOLOGIC ORIGINS OF LIPOPROTEIN PARTICLE HETEROGENEITY

Because cholesterol and TGs are hydrophobic lipids that are immiscible in plasma, intravascular transport cannot occur with these lipids in their free state. Consequently, cholesterol esters and TGs are packaged in globular (or sometimes discoidal) lipoprotein particles configured, so that the outer surface is hydrophilic and the inner core is hydrophobic. The surface of lipoprotein particles is composed of a phospholipid monolayer, nonesterified cholesterol, and various apolipoproteins. The core of lipoprotein particles contains variable amounts of cholesterol ester and TGs. Human plasma lipoproteins comprise a spectrum of macromolecules that exhibit a wide range of particle size (diameter 5–1200 nm), hydrated density (0.93–1.21 kg/L), molecular weight (0.15×10^6 to 500×10^6), and variability of lipid and protein composition *(14)*. Previous chapters have discussed many aspects of normal and abnormal lipoprotein metabolism. Here, we will focus attention on issues responsible for the extensive heterogeneity of particle number, size, density, and core lipid content commonly encountered in clinical practice.

Lipoprotein Heterogeneity in Exogenous Lipid Transport

Exogenous lipid transport represents the multi-step integration of dietary fat absorption, lipid re-synthesis and assembly of chylomicrons in intestinal enterocytes, and subsequent intravascular metabolism of chylomicrons following entry into the circulation. Chylomicron assembly and secretion proceeds in a sequential fashion. First, microsomal transfer protein (MTP) mediates the partial lipidation of apolipoprotein (apo) B-48 with phospholipids on the smooth endoplasmic reticulum (ER) membrane to form dense, "primordial" particles *(15)*. Subsequently, the lipid core of primordial particles expands by progressive addition of TGs. These enlarging chylomicron particles move to the Golgi apparatus, where additional TGs, phospholipids, cholesterol esters, and apo A-I and apo A-IV are incorporated to form a population of mature chylomicron particles of varying size *(16)*.

Mature chylomicron particles enter the circulation through the mesenteric lymph, where they face a multi-step process of intravascular remodeling and catabolism. First, chylomicrons and HDL particles participate in apolipoprotein transfer with apo A-I and apo A-IV moving from chylomicrons to HDL in exchange for apo C-I, apo C-II, apo C-III, and apo E *(17–19)*. Second, chylomicrons undergo remodeling following lipoprotein lipase (LPL) catalyzed hydrolysis of core TGs. Apo C-II activates LPL activity, whereas apo C-III inhibits LPL-mediated TG hydrolysis. The progressive loss of core TGs results in an approximate 20-fold decrease in chylomicron mass *(14)*. Additionally, phospholipid transfer protein (PLTP) facilitates the shedding of surface phospholipids from chylomicrons that are subsequently incorporated into the surface of maturing HDL particles. As a result of core TG hydrolysis and lose of surface phospholipid, a spectrum of smaller sized chylomicron remnant particles are produced. Third, cholesterol ester transfer protein (CETP) facilitates the pair-wise exchange of TGs and cholesterol ester between TG-rich VLDL, intermediate density lipoprotein (IDL), and remnant particles and cholesterol-rich (LDL and HDL) particles resulting in significant heterogeneity of core cholesterol and TG composition (Fig. 1). The degree to which this imposes an analytic limitation on the use of plasma lipids to assess the quantity of lipoprotein particles present will be addressed later in this chapter. Finally, hepatic receptors bind apo E and remove chylomicron remnants from the circulation. Approximately, 35% of chylomicron remnants clear through hepatic apo B/E receptors, with the remaining particles clearing by LDL-receptor-related protein receptors *(20,21)*. Apo C-I and apo C-III attenuate binding of remnants to apo E receptors, thus limiting premature clearance of chylomicrons and remnants from plasma *(22)*.

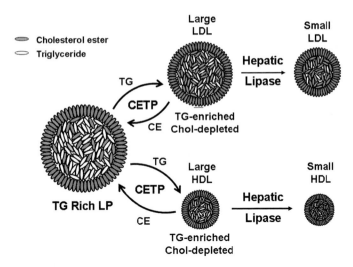

Fig. 1. Schematic representation of the metabolic origins of compositional change in low-density lipoprotein (LDL) and high-density lipoprotein (HDL) particles. TG, triglyceride; LP, lipoprotein; Chol, cholesterol; CETP, cholesterol ester transfer protein; HL, hepatic lipase.

Lipoprotein Heterogeneity in Endogenous Lipid Transport

Endogenous lipid transport constitutes the integrated processes of hepatic lipid and apolipoprotein synthesis, VLDL particle assembly and secretion, and intravascular metabolism of VLDL, IDL, and LDL particles. VLDL particle assembly begins with the synthesis of apo B-100 in the rough ER of hepatocytes. Following synthesis, apo B-100 translocates from the cytosolic to the luminal side of the ER, where it undergoes one of two metabolic fates. When sufficient lipid is available, MTP catalyzes the partial lipidation of apo B-100 leading to entry into the VLDL synthetic cascade. Otherwise, nonlipidated apo B-100 is degraded. Maturation of partially lipidated apo B-100 continues by MTP-mediated accumulation of additional TGs, phospholipids, and cholesterol esters. Subsequently, enlarging VLDL particles move to the Golgi apparatus, where additional TGs, phospholipids, cholesterol esters, and apolipoproteins (apo E, apo C-II, and apo C-III) are incorporated to form a population of mature VLDL particles of varying size *(23,24)*. Finally, mature VLDL particles are packaged into secretory vesicles and transported to the surface of hepatocytes, where they enter the circulation through into the space of Dissie.

The production and metabolism of VLDL particles is dynamic and modifiable. First, hepatic TG availability affects VLDL particle production. When hepatic TG concentration is increased, greater numbers of VLDL particles are formed driven by differential production of large TG-rich VLDL particles *(25)*. Second, through parallel processing pathways, distinct IDL and LDL particles are produced from individual VLDL precursors through a non-CETP-dependent process *(25–27)*. Specifically, very large TG-enriched VLDL particles give rise to VLDL remnants and IDL particles that clear rapidly by hepatic LDL-receptor-related protein receptors. IDL particles not cleared appear to be metabolized to small-sized LDL particles. By contrast, smaller VLDL particles containing less core TGs give rise to VLDL remnants and IDL particles that clear less efficiently. IDL particles from these VLDL precursors are associated with larger size LDL particles.

Intravascular metabolism of VLDL particles follows a similar sequence as described for chylomicrons. Initial remodeling occurs through the combined effects of LPL-mediated core TG hydrolysis and PLTP-mediated liberation of surface phospholipids. Second, surface constituents liberated from VLDL are exchanged for apo C-I, apo C-II, apo C-III, and apo E from HDL particles. As progressive loss of core TG volume and surface phospholipids occurs, VLDL particles also shed surface apo C proteins. This results in the generation of a continuum of smaller, increasingly cholesterol-enriched VLDL particles, as well as IDL particles that have little or no surface apo C proteins. Normally, the majority of IDL particles undergo further lypolysis by LPL and hepatic lipase to form LDL particles. Alternatively, IDL particles can be cleared by hepatic apo B:E receptors. Third, CETP facilitates the pair-wise exchange of TGs and cholesterol ester between cholesterol-rich (LDL and HDL) and TG-rich (VLDL, IDL, and remnants) lipoprotein particles resulting in LDL particles that are partially depleted in core cholesterol and enriched in core TGs. In a physiologic attempt to reestablish the normal ratio of cholesterol/TG in the particle core, these compositionally abnormal particles become a substrate for hepatic lipase and endothelial lipase, which partially hydrolyzes core TGs. In response to the efflux of fatty acids from the core and remodeling of the surface shell, particles are transformed into smaller, denser LDL particles.

As a result of these reactions, four different types of LDL particles are likely to be seen in individuals depending on their metabolic circumstances: (1) large LDL particles with a normal core lipid content, (2) large LDL particles with relatively cholesterol-deficient, TG-rich lipid cores, (3) small LDL particles with a normal lipid content, and (4) small LDL particles with relatively cholesterol-deficient, TG-rich lipid cores (Fig. 2) *(11)*.

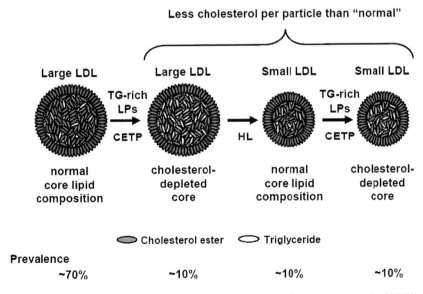

Fig. 2. Schematic representation of the metabolic origins of low-density lipoprotein (LDL) particles containing less cholesterol than normal. CETP, cholesterol ester transfer protein; HL, hepatic lipase; LPs, lipoproteins; TG, triglyceride. Reproduced with permission *(11)*.

Lipoprotein Heterogeneity in Reverse Cholesterol Transport

HDL particles serve to acquire and redistribute cholesterol esters among lipoproteins and to various cells throughout the body. HDL particles originate by secretion of a poorly lipidated apo A-I phospholipid disc (also termed "nascent" HDL or pre-β HDL) from the liver and intestine. Alternatively, surface constituents (primarily apo A-I and phospholipids) generated during intravascular remodeling of chylomicrons and VLDL particles also serve as a substrate precursor for HDL particles. Pre-β HDL generally contain two molecules of apo A-I per particle *(28)* and undergoes initial lipidation through transfer of free cholesterol from peripheral cells (principally hepatocytes and, to a lesser extent, macrophages) by ATP-binding cassette transporter (ABC) A1 *(29)*. Macrophages can also efflux cholesterol to mature HDL particles by ABCG1 and ABCG4 transport pathways.

Maturation of pre-β HDL proceeds by the lecithin-cholesterol acyltransferase (LCAT)-mediated transfer of two acyl groups from lecithin to free cholesterol on the particle surface, generating cholesterol ester and lysolecithin. Owing to its hydrophobicity, esterified cholesterol moves to the core of pre-β HDL resulting in expansion of the particle to form small, spherical HDL_3. Alternatively, HDL particles may be formed by PLTP-mediated fusion of discoid and small HDL particles. Particle maturation also requires incorporation of additional surface phospholipid, as well as apo A-I and, to a lesser extent, apo A-II via PLTP. This process continues with the generation of progressively larger HDL_2 particles that contain three to four apo A-I proteins per particle, as well as increasing amounts of core cholesterol ester.

Intravascular remodeling of HDL_2 particles occurs via the CETP-mediated exchange of TG molecules from the core of TG-rich lipoproteins (TRLs) (mainly VLDL) with cholesterol ester molecules in the core of HDL particles. In a fashion analogous to that previously described for LDL, HDL particles become partially depleted in core cholesterol and enriched in core TGs. These compositionally abnormal particles become a substrate for hepatic lipase and endothelial lipase, which partially hydrolyze core TGs. With the loss of core lipid volume and remodeling of the surface shell (including shedding of some apo A-I), particles are transformed into smaller, denser HDL_3 particles.

LIPOPROTEIN ASSAY METHODS

Our understanding that chylomicrons, VLDL, IDL, LDL, and HDL particles are linked in a continuous metabolic cascade resulting in the generation of discrete particle subclasses varying in size, density, and core lipid content represents the synthesis of several decades of innovative research. Prior to 1940, little was known regarding lipoprotein structure, composition, or physiologic function, although many investigators postulated that serum cholesterol must travel intravascularly as an emulsion or associated with serum proteins. During the 1940s large-scale human serum fractionation by Cohn et al. and Oncley et al. demonstrated that serum lipids were clustered into two major classes having α1- and β-mobility, respectively.

These experiments set the stage for landmark work by Gofman and Lindergen in the early 1950s *(30–32)*. Utilizing a combination of preparative and analytic ultracentrifugation, they successfully characterized the major classes of human lipoproteins based on their density and flotation characteristics [Svedberg sedimentation units (Sf)] in plasma (Table 1). Although a robust research tool, this approach to lipoprotein

Table 1
Characteristics of Major Lipoproteins *(14)*

Lipoprotein	Density (kg/L)	Particle diameter (nm)	Flotation rate (Sf)	Electrophoretic mobility
Chylomicrons	< 0.95	80–1200	> 400	Origin
Very low density lipoprotein	0.95–1.006	30–80	60–400	Pre-beta
Intermediate density lipoprotein	1.006–1.019	23–35	20–60	Broad beta
Low-density lipoprotein	1.019–1.063	18–25	0–20	Beta
High-density lipoprotein	1.063–1.21	5–12	0–9	Alpha

characterization was found to be too costly and labor intensive for routine clinical application. Because no unique chemical entity was found that distinguished one particle population from another, efforts shifted to applying other analytic techniques that could separate particle populations and allow these classes to be individually analyzed.

Use of electrophoresis extended the opportunity for lipoprotein separation by characterizing particles based on their mobility in various media under the influence of an electric current. A major clinical advance occurred in 1967 when Fredrickson et al., utilizing paper electrophoresis for lipoprotein separation, published the first classification of hyperlipoproteinemias based on the class(es) of lipoprotein present in relative excess: type 1 (chylomicron); type 2 [β-lipoproteins (LDL)]; type 3 [broad-β lipoproteins (VLDL remnants)]; type 4 [pre-β lipoproteins (VLDL)]; and combined lipoprotein elevations termed type 2B (excess LDL and VLDL) and type 5 (excess VLDL and chylomicrons) *(33)*. Subsequently, electrophoresis of plasma lipoproteins was performed on a number of support media including starch gel, cellulose acetate, agarose, and polyacrylamide gel.

In addition to these separation-based lipoprotein assays, a nonseparation method employing nuclear magnetic resonance (NMR) spectroscopy has been developed to quantify the number of particles present in various VLDL, LDL, and HDL subclasses. Because the spectroscopically distinct lipid methyl group signal amplitudes of individual lipoprotein subclass particles are unaffected by variations in core cholesterol ester or TG levels, this technique has expanded our understanding of the quantitative relationships of lipoprotein subclass particle numbers with CHD risk.

A comprehensive review of each of these methodologies is beyond the scope of this chapter. Instead, our goal is to provide a clinically relevant overview of the information produced by lipoprotein subclass methods commonly available through commercial clinical laboratories, including gradient gel electrophoresis (GGE), density gradient ultracentrifugation (DGU), and NMR spectroscopy. A basic understanding of these technologies, and the information each supplies, is needed before discussing clinical factors that influence the application of lipoprotein information in clinical management of individual patients.

Gradient Gel Electrophoresis

GGE entails the migration of charged particles, under the influence of an electric current, through a semi-solid gel matrix created by the polymerization of polyacrylamide across a defined concentration range. As the concentration of the polyacrylamide increases from the top to the bottom of the gel, cross-linkages forming the matrix become progressively tighter resulting in a series of increasingly smaller "pores" through which lipoproteins travel. Movement of particles through the gel continues until lipoproteins encounter pores too small to allow continued migration. Particle diameters estimated by GGE are derived by a referencing procedure that compares the distance(s) traveled in the gel by a subject's lipoprotein particles with the distances traveled by nonlipoprotein size standards (various proteins and latex beads of known diameter). The resolving power of this technology allows for the identification of up to seven unique LDL subclasses (using a 2–16% polyacrylamide gel matrix) and five unique HDL subclasses (using a 4–30% polyacrylamide gel matrix). Pore sizes created using the lowest concentration of polyacrylamide also allow for partial separation of VLDL particles (those < 40 nm in diameter) yielding restricted information regarding major VLDL subclasses present.

At the completion of the separation phase, a lipid or protein stain is used to identify the lipoproteins within the gel. Because plasma proteins can limit the unique staining of lipoproteins, lipid stains are used routinely to visualize lipoproteins. Subclass information is gleaned from optical density scanning, which reveals the presence of a series of major and minor peaks in various regions of the stained gel. Lipoprotein size information is reported by different laboratories in a variety of ways including: peak particle size (based on the largest peak present); average particle size (weighted average of all subclasses identified as a continuous variable), dichotomous characterization of the predominant particle population present (large versus small), and percentages of particles present in predefined size ranges (determined by comparing the area of a predefined region with the integrated area under the curve for all subclasses identified on optical density scanning). Owing to variability in stain uptake, GGE is not able to reliably measure the absolute concentrations (numbers) of lipoprotein particles present in a given subclass.

Density Gradient Ultracentrifugation

DGU is a process of separating lipoprotein particles of differing densities by spinning them in a defined salt solution at very high speeds. This spinning creates a progressively increasing density (gradient) of salt from the top to the bottom of the tube. In response, lipoprotein particles migrate to that position of the tube where the density of the salt solution matches the density of the lipoprotein particle. The degree of subclass separation (resolving power) that can be achieved for LDL, HDL, and VLDL particles is determined by the density range of the salt solution employed. In order for individual subclasses of VLDL, LDL, or HDL particles to be measured with precision, separate salt solutions must be used that have narrow density ranges that match the unique density of VLDL, LDL, and HDL particles, respectively. If a single solution gradient is used for ultracentrifugation separation, general partitioning of HDL, LDL, and VLDL classes is achieved. However, individual HDL, LDL, and VLDL subclasses blur together limiting the identification of individual lipoprotein particle subclasses. Thus, the greater the density range of the salt solution from top to bottom of the tube, the less the resolution of lipoprotein subfractions.

Following ultracentrifugation, lipid testing is performed to quantify the cholesterol or TGs present in the separated solution. One approach is to perform lipid testing on aliquots of the solution corresponding to VLDL, LDL, and HDL classes. Another approach is to flow the ultracentrifuged solution into a spectrophotometer to yield a continuous absorbance tracing used to determine cholesterol concentrations across the spectrum of lipoproteins present. The use of software to deconvolute data embedded in the parent tracing has also been used to gain additional information regarding IDL cholesterol and Lp(a) cholesterol values from the same ultracentrifuged sample. However, because the amount of cholesterol or TGs carried inside lipoprotein particles varies significantly between individuals, lipid testing performed following separation by ultracentrifugation is not capable of measuring the absolute concentration (number) of lipoprotein particles present.

Nuclear Magnetic Resonance Spectroscopy

Two phenomena make NMR quantification of numbers of lipoprotein subclass particles possible: (1) lipoprotein subclasses of different size in plasma emit distinctive NMR signals whose individual amplitudes can be accurately and reproducibly measured and (2) measured subclass signal amplitudes are directly proportional to the numbers of subclass particles emitting the signal, irrespective of variation in particle lipid composition. Each lipoprotein subclass signal emanates from the aggregate number of terminal methyl groups on the lipids contained within the particle, with the cholesterol esters and TGs in the particle core each contributing three methyl groups and the phospholipids and unesterified cholesterol in the surface shell each contributing two methyl groups *(34)*. Because the methyl signals from these lipids are indistinguishable from each other, they overlap to produce a bulk lipid "particle signal." The amplitude of each lipoprotein particle signal serves as a measure of the concentration of that lipoprotein. What makes it possible to exploit the methyl lipid signal for lipoprotein subclass quantification (without separating the subclasses first) is a magnetic property specific to lipoproteins that causes the lipids in larger particles to broadcast signals that are characteristically different in frequency and shape from the lipid signals emitted by smaller particles *(35)*.

It is important to note that the total number of methyl groups contained within a subclass particle is, to a close approximation, determined solely by the particle's diameter and is not affected by differences in lipid composition arising from such sources as variability in the relative amounts of cholesterol ester and TG in the particle core, varying degrees of unsaturation of the lipid fatty acyl chains, or varying phospholipid composition. For this reason, the methyl NMR signal emitted by each subclass serves as a convenient and direct measure of the number of particles in that subclass.

The lipoprotein particle information from a single NMR test includes total LDL particle concentration (LDL-P), as well as the subclass particle concentrations of VLDL (large VLDL-P, medium VLDL-P, and small VLDL-P), IDL-P, LDL (large LDL-P and small LDL-P), and HDL (large HDL-P, medium HDL-P, and small HDL-P). Although levels of total LDL, HDL, and VLDL have always been measured with high precision by NMR (variability $< 3\%$), recent modifications to the computational algorithm used in the analysis process have improved the reproducibility of the quantification of individual LDL subclasses (variability $< 10\%$) *(36)*.

ANALYTIC IMPLICATIONS OF LIPOPROTEIN HETEROGENEITY

As previously noted, patients with elevated TG levels typically have cholesterol-poor LDL and HDL particles, because they are smaller in size, have a core lipid content enriched in TGs, and depleted of cholesterol ester, or both. Termed the "atherogenic lipid phenotype" or the "lipid triad," the combination of elevated TG, low HDL-C, and normal or minimally elevated LDL-C levels is commonly encountered in patients with insulin resistance, metabolic syndrome, and type 2 diabetes mellitus. Over the past two decades, several groups have described additional lipoprotein features underlying this lipid phenotype including a predominance of small LDL particles, relative decrease in large HDL particles, and increased concentration of apo B *(37–38)*. Because the quantity of LDL (assessed by LDL-C) appeared not to be elevated, many authors concluded increased numbers of non-LDL apo B-containing lipoproteins (VLDL, IDL, and remnant particles) were responsible for the elevated apo B levels observed. Confounding this conclusion are data indicating that, with the exception of type III dyslipoproteinemia, more than 90% of apo B is bound to LDL, even in the setting of elevated TG levels *(39,40)*. This is supported by data from the Framingham Offspring Study (Fig. 3) demonstrating discordance between low LDL-C and increased LDL particle number (LDL-P) among subjects with elevated TG or reduced HDL-C levels *(11)*.

Recent trials utilizing NMR lipoprotein quantification have provided additional insight into the quantitative lipoprotein abnormalities present in these clinical settings. Garvey et al. *(41)* measured VLDL, LDL, and HDL subclass particle numbers in nondiabetic individuals with a wide range of insulin sensitivity as defined by the hyperinsulinemic–euglycemic clamp, as well as in patients with type 2 diabetes mellitus. When compared with insulin-sensitive subjects, the insulin-resistant and diabetes subgroups had: (1) a twofold to threefold increase in large VLDL particle number (VLDL-P) (without change in medium or small VLDL-P) resulting in increased TG levels; (2) a twofold to threefold increase in the number of small LDL-P and a reduction in large LDL-P, resulting in a decrease in LDL particle size; (3) a progressive increase in total LDL-P, despite no difference (IS versus IR) or a minimal difference (IS versus diabetes) in LDL-C; and (iv) a decrease in number of large HDL particles (HDL-P) combined with an increase in small HDL-P resulting in no significant difference in HDL-C. The strongest relations with insulin resistance and diabetes were found for large HDL-P (inversely), large VLDL-P, and total and small LDL-P, all independent of glucose and glycemic control. Overall, the authors concluded that these insulin resistance-induced changes in the lipoprotein particle concentrations were not

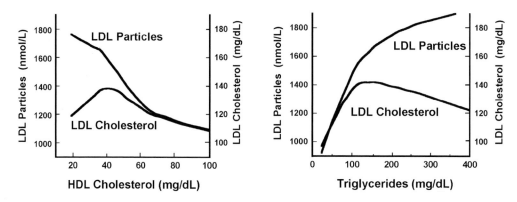

Fig. 3. Relations in the Framingham Offspring Study ($n = 3437$) of nuclear magnetic resonance (NMR)-measured low-density lipoprotein (LDL) particles (LDL-P) and Friedewald-calculated LDL-cholesterol (LDL-C) to HDL-C and triglycerides. Reproduced with permission *(12)*.

reflected in traditional LDL-C, HDL-C, and TG values. In the much larger Insulin Resistance Atherosclerosis Study (IRAS), very similar strong subclass associations were found with insulin resistance *(42)* and incident type 2 diabetes mellitus *(43)*. This trial also showed that VLDL particle size and small HDL-P predicted incident diabetes independently of lipids and insulin sensitivity measured by frequently sampled intravenous glucose tolerance testing.

Two recent NMR trials have also addressed the degree to which LDL-C is discordant with LDL-P measured in metabolic syndrome and type 2 diabetic patients. In the Framingham Offspring Study, the sex-specific relations of the metabolic syndrome with chemically-measured LDL-C versus NMR-measured LDL-P and small LDL-P were reported in 2993 participants (mean age 51 years, 53% women) without CVD (Fig. 4) *(44)*. Age-adjusted total LDL-P was markedly higher in individuals with the metabolic syndrome than in those without. Owing to a compositional change in LDL, small LDL-P increased progressively in men and women with an increasing number of components of the metabolic syndrome while large LDL-P declined. The combination of greatly increased small LDL-P (with less cholesterol per particle) and a numerically smaller decrease in large LDL-P (with more cholesterol per particle) resulted in LDL-C values that were not significantly different in men with and without the metabolic syndrome and only modestly higher in metabolic syndrome women.

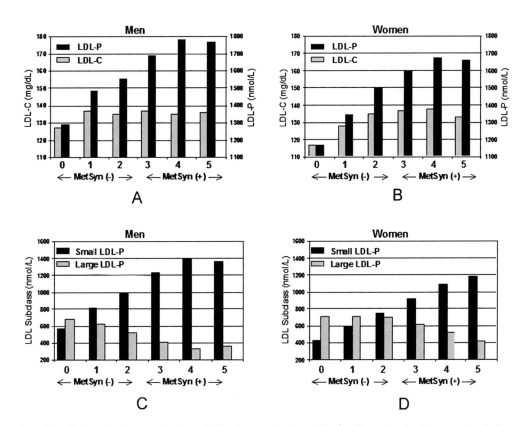

Fig. 4. Relations in the Framingham Offspring Study ($n = 2993$) of low-density lipoprotein cholesterol (LDL-C) and total LDL particle number (**A** men, **B** women), as well as small and large LDL particle number (**C** men, **D** women) to the number of components of the metabolic syndrome. LDL-C, LDL-cholesterol; LDL-P, LDL particle number. [Based on Kathiresan S, et al. (2006) *(44)*].

A substantial degree of LDL heterogeneity has also been documented in the ambulatory care setting. Split sample comparisons of Friedewald-calculated LDL-C and NMR-measured LDL-P were reported recently for 2355 patients with type 2 diabetes mellitus seen in clinical practice, confirmed to have LDL-C < 100 mg/dL (Fig. 5) *(45)*. Patients were categorized according to their LDL-P values, using cutpoints corresponding to the 20th, 50th, 80th, and 95th percentile values of a reference population consisting of more than 6900 subjects enrolled in the Multiethnic Study of Atherosclerosis (MESA). A high percentage (61%) of patients with type 2 diabetes mellitus and optimal LDL-C values < 100 mg/dL (20th percentile) had suboptimal LDL-P levels > 1000 nmol/L (20th percentile). Even among patients with very low LDL-C levels (< 70 mg/dL) (< 5th percentile), a large number (40%) had LDL-P > 1000 nmol/L (20th percentile).

Collectively these data demonstrate that patients with elevated TG and decreased HDL-C harbor substantial qualitative and quantitative lipoprotein abnormalities, the magnitude of which is not discernable from traditional lipid testing. The degree to which these lipoprotein abnormalities might be more strongly associated with CHD risk versus traditional lipid values, as well as the association of on-trial lipoprotein values with future CHD events among those on lipid-lowering therapy, helps to define the potential clinical utility of lipoprotein information.

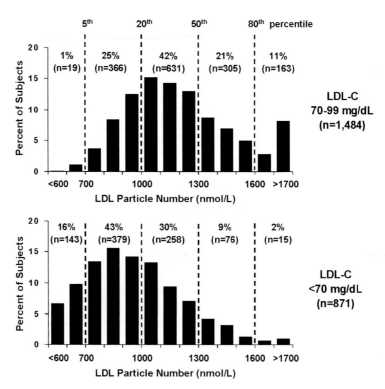

Fig. 5. Distribution of low-density lipoprotein particle number (LDL-P) among 2355 patients with type 2 diabetes mellitus and LDL cholesterol 70–99 mg/dL (top) or < 70 mg/dL (bottom). LDL-P concentrations of 700, 1000, 1300, and 1600 correspond closely to the 5th, 20th, 50th, and 80th percentile values of subjects enrolled in the Multi-Ethnic Study of Atherosclerosis (MESA) reference population. Reproduced with permission *(45)*.

Relationship of LDL Subfractions with CHD Outcomes

The associations of CVD risk with LDL particle size and LDL particle number in more than 70 cross-sectional and prospective epidemiologic and clinical intervention trials were reviewed recently *(13)*. With few exceptions, small LDL particle size (pattern B) was found to be significantly associated with CVD risk in univariate analyses. However, the origin of this risk association remains controversial. Many authors cite indirect lines of evidence that implicate atherogenic properties of small-sized LDL particles. Various data indicate that small LDL more easily enters the arterial wall, undergoes localized retention due to binding with arterial wall proteoglycans, exhibits enhanced oxidizability in several in vitro models, and directly participates in the production of subendothelial macrophage foam cells *(46)*. Collectively, these findings imply that small LDL is a potent atherogenic lipoprotein, the measurement of which may be useful to enhance CVD risk prediction and better evaluate response to lipid therapy *(26,47,48)*.

However, small-sized LDL particles are most commonly present as a component of a broader pathophysiology characterized by high TGs, low HDL-C, increased LDL particle number, obesity, insulin resistance, diabetes, and the metabolic syndrome *(43,49–51)*. As a result, it is unclear if the increased risk associated with small LDL size in univariate analyses is a reflection of an increased atherogenic potential of small LDL particles, or simply a consequence of the broader pathophysiology of which small LDL is a part. Following multivariate adjustment for these confounding risk factors, LDL size was rarely found to be a significant, independent predictor of CVD risk.

An alternative explanation for the higher CVD risk observed among pattern B individuals is the increased quantity of LDL particles carried by these individuals. Total plasma apo B has been used historically to estimate numbers of circulating LDL particles, because there is one apo B molecule per LDL, VLDL, and IDL particle, and approximately 95% of plasma apo B is bound to LDL. Many prospective epidemiologic and clinical intervention trials have documented that cardiovascular events are significantly more strongly associated with apo B than with LDL-C *(6,7)*. The use of NMR spectroscopy to quantify LDL particle subclasses has provided further insight into the quantitative relationships of LDL particles with CVD risk. Data from seven recently published or presented outcome studies provide evidence that NMR-measured LDL-P was a significantly stronger predictor of incident CVD events or disease progression than LDL-C *(52–58)*. In all of these studies, CVD associations with LDL-P and small LDL-P were independent of the standard lipid variables. Although LDL particle size was associated in univariate analyses with CVD risk in four of seven studies, it failed to retain significant prediction following multivariate adjustment for lipids or LDL particle number.

Additional insights have been reported recently regarding the inverse correlation between small and large LDL particle concentrations and the confounding effect of this interrelationship with observed CHD risk. Historically, studies of LDL subclass association with CHD risk have examined the relationship of qualitative size distribution or phenotype associations rather than the particle concentrations of LDL subclasses. In the MESA, baseline blood samples were used to examine relations of lipoprotein subclass concentrations and particle size, measured by NMR, to carotid intima-media thickness (IMT). In analysis of 5538 subjects not on lipid-lowering medications, the relationship of lipoprotein variables were determined individually, as well as in models that included both small and large LDL subclass concentrations together to assess the independent association of each LDL subclass with IMT. This analysis demonstrated that, owing to a strong inverse correlation of small and large LDL particle concentrations, small LDL confounded the association of large LDL with IMT *(58)*. Both small and large LDL particle concentrations were significantly associated with subclinical atherosclerosis, independent of each other, traditional lipids, and established risk factors. Furthermore, no significant association was demonstrated between LDL particle size and atherosclerosis after accounting for the concentrations of the large and small LDL particles. Veterans Affairs HDL Intervention Trial (VA-HIT) trial, where both large and small LDL particle concentrations, but not LDL particle size, were significantly associated with coronary events once their correlation was taken into account *(56)*.

Relationship of HDL Particle Subclasses and CHD Risk

In contrast to LDL, HDL particles are widely regarded as atheroprotective. A number of prospective epidemiologic investigations have demonstrated an inverse, independent relationship between serum levels of HDL-C and risk for CHD *(59–61)*. In addition to facilitating reverse cholesterol transport, several additional physiologic mechanisms have been elucidated which appear to contribute to the atheroprotective benefits attributed to HDL particles. At the arterial wall level, HDL particles stimulate endothelial nitric oxide production and vasodilatation *(62)*, reduce endothelial cell adhesion molecule expression *(63)*, inhibit platelet aggregation *(64)*, and inhibit endothelial cell apoptosis *(65)*. Additionally, through the actions of paraoxonase and acetylhydrolase enzymes, HDL particles reduce oxidized fatty acids within oxidized LDL particles resulting in decreased substrate for macrophage ingestion and foam cell formation *(66,67)*.

Over a dozen case–control and prospective epidemiologic studies using ultracentrifugation, electrophoresis, or differential precipitation to measure larger HDL_2 and smaller HDL_3 subfractions have yielded conflicting results regarding the association of these subclasses with CHD risk. Collectively these data demonstrate that HDL_2 levels are inversely associated with CHD risk *(68–75)*. By contrast, increasing CHD risk has been associated with increased *(69,72)* and decreased *(71,73,76,77)* HDL_3 levels. In two trials, CHD risk was significantly associated with a simultaneous reduction of HDL_2 and HDL_3 levels *(71,73)*. Similarly confounding data have been reported in clinical intervention trials where significant CHD risk reduction was associated with increased HDL_2 in niacin trials *(78–80)* and increased HDL_3 in fibrate trials *(81,82)*.

NMR spectroscopy has also been used in epidemiologic and clinical intervention trials to evaluate the association of small and large HDL particle numbers with CHD risk. In agreement with the results of other studies with fibrates, evaluation of HDL particles in the VA-HIT trial *(56)* demonstrated that the number of small HDL particles accounted for the increase in HDL-C *(83,84)* brought about by drug treatment and that on-trial numbers of the small HDL subclass particles, but not large-size HDL particles, predicted CHD events *(81,82)*. In apparent disagreement, previous studies of NMR-measured HDL subclasses suggested large HDL particles, not small, were primarily responsible for the cardioprotection associated with elevated HDL-C levels *(54,55,85,86)*.

Important insights into the reasons for these conflicting results can be gained from an appreciation of the quantitative interrelationships between concentrations of HDL particle subclasses over the wide range of HDL-C levels encountered in the general population, as provided by NMR lipoprotein analyses in the Framingham Offspring Study *(87)*. Figure 6 presents a graphical representation of relations of HDL subclass and total HDL particle numbers as a function of HDL-C in 3467 Framingham men and women. For HDL-C levels below 40 mg/dL (representative of the VA-HIT study population), increasing HDL-C is accounted for by increases in levels of all three HDL subclasses, but primarily small HDL. Increases in HDL-C beyond 40 mg/dL are due primarily not to increasing numbers of HDL particles, but to changes in HDL particle composition, with large HDL increasing steadily at the expense of decreasing numbers of the smaller HDL subclasses due to a product–precursor relationship between these species. The subclass relations in Figure 6 provide an explanation for the inconsistency of previous study results, which we believe are due mainly to variations in the proportion of study participants with HDL-C levels above and below 40 mg/dL. Those studies with predominantly low HDL-C subjects, like VA-HIT, show cardioprotection arising mainly from increases in small HDL, reflecting significant increases in total numbers of HDL particles and a presumed related increase in effectiveness of reverse cholesterol transport, antioxidation, and other processes by which HDL is believed to exert its protective effects *(84,88)*. In studies of populations with higher HDL-C or a wider range of HDL-C values, the large HDL subclass typically shows a strong inverse relation with CHD whereas the association with small HDL is weaker or even positive in some studies *(54,55,69,75,85,86)*.

The results of previous studies and NMR analysis in VA-HIT lead us to the conclusion that both large and small HDL subclasses are cardioprotective. Determining whether one subclass is more cardioprotective than the other and whether therapies that primarily affect levels of one or the other subclass are more or less beneficial are questions that await further investigation, employing multivariate modeling to account

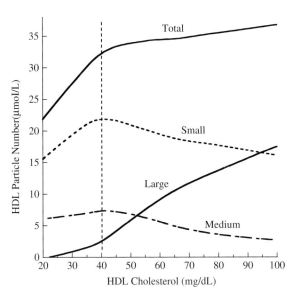

Fig. 6. Relations in the Framingham Offspring Study ($n = 3437$) of nuclear magnetic resonance (NMR)-measured total high-density lipoprotein (HDL) particle concentrations and levels of small, medium, and large HDL subclasses to HLD cholesterol (*x* axis). Smoothed lines were fitted using locally weighted scatterplot smoother (Lowess; SAS Institute, Cary, NC).

properly for the interrelationships not only among individual HDL subclasses but also those between HDL, LDL, and VLDL particle subclasses.

Relationship of TRL Subclasses and CHD Risk

Although the association of elevated plasma TG with increased CHD risk has been noted in numerous epidemiologic studies, the basis of this association has been a subject of brisk debate. As previously noted, elevated TG levels are commonly present as a component of the broader atherogenic lipoprotein phenotype characterized by high TGs, low HDL-C, small-sized LDL particles, and increased LDL particle number despite normal to minimally elevated LDL-C values. Additionally, given the high prevalence of this dyslipoproteinemia among patients with obesity, insulin resistance, type 2 diabetes, and the metabolic syndrome, it is unclear whether the increased risk associated with elevated TG is attributable to atherogenicity of TRL particles (chylomicrons, VLDL, and their remnants), other lipoprotein abnormalities concomitantly present in elevated TG states (increased LDL-P and small LDL-P, decreased HDL-P), or simply a consequence of the broader pathophysiology of which elevated TG is a part. Several individual trials demonstrate that the CHD risk associated with elevated TG levels does not remain significantly predictive following multivariate adjustment for these co-segregating factors. By contrast, meta-analysis by Austin et al. *(89,90)* supports the independent association of TG with CHD risk.

In addition to these data, numerous trials have focused attention on the physiologic origin of TRLs and the CHD risk associated with TRLs. As discussed earlier in this chapter, newly formed chylomicron and VLDL particles undergo a series of intravascular modifications resulting in the formation of TGL remnant particles varying in size, density, and composition. Owing to the lack of chemical characteristics that easily differentiate remnants from their larger TGL precursors, several methods have been used to quantify remnant lipoproteins based on density, density and charge, density and lipid content, size, size and apolipoprotein content, and apolipoprotein immunospecificity *(91)*. Use of the latter method has become increasingly common in recent years and entails the separation of remnant-like particles (RLPs) from plasma by immunoaffinity chromatography in which a gel matrix selectively retains apo AI and certain apo B-100 (those without enrichment in apo-E) particles. This results in HDL, LDL, large chylomicrons, and the majority of VLDL particles being removed by the gel. Cholesterol and TG testing of the unbound RLP lipoproteins (made up of remnant-like VLDL and TRL) are reported as RLP-C and RLP-TG, respectively.

Epidemiologic studies have shown univariate associations between RLP concentrations and atherosclerosis, CVD, and CHD *(92–103)*. However, RLP levels are highly correlated with plasma TG, VLDL TG, and VLDL-C, but not with LDL-C *(104–106)*. Thus, the degree to which RLP provides significant CHD risk prediction beyond plasma TG and other highly intercorrelated lipid variables remains unclear. Recently, results from the Honolulu Heart Study reported the association of RLP with incident CHD events in a cohort of 1156 initially healthy Japanese American men aged 45–68 years followed over a 17-year period *(106)*. Both univariate and multivariate regression analyses (adjusted for lipids age, BMI, smoking, alcohol intake, physical activity, systolic blood pressure, fasting glucose, history of diabetes, use of antihypertensive medication, and lipid variables) were performed. Univariately, RLP-C and RLP-TG were significantly associated with incident CHD events. In multivariate regression analyses, baseline RLP-C and RLP-TG associations with CHD events were independent of non-lipid risk factors, total cholesterol, HDL-C, and LDL-C. However, in models including RLP and total TG simultaneously, neither variable was significant when adjusted for the other. This finding can be attributed to the strong correlation between RLP-C and RLP-TG levels and total TG, suggesting that RLP levels did not provide additional information regarding incident CHD events over and above total TG levels. From these data, it is unclear whether a unique clinical role exists for remnant lipoprotein testing if measures of fasting TGs are available.

UTILIZATION OF LIPOPROTEIN INFORMATION FOR INDIVIDUAL PATIENT MANAGEMENT

Current National Cholesterol Education Panel Adult Treatment Panel III (ATP III) guidelines advise clinicians to mange LDL-C as the primary target for lipid-lowering therapy. The basic principal guiding LDL-lowering intervention is that the intensity of treatment is directly related to the degree of CVD risk. Individuals with known CHD, type 2 diabetes mellitus, or a CHD risk equivalent are designated as high-risk patients with an initial LDL-C target of therapy < 100 mg/dL *(107)*. A recent guideline update called into question whether the target of < 100 mg/dL is low enough and suggested an optional goal of < 70 mg/dL for those patients who, according to physician judgment, are considered to be at very high risk *(108)*. In a similar fashion, ATP III guidelines call for moderately high risk individuals (two or more risk factors and a 10-year Framingham Risk Score of 10–20%) to be managed to < 130 mg/dL as their initial LDL-C goal, with an optional target of < 100 mg/dL for those at deemed higher clinical risk based on clinical judgment advised in the update. ATP III also acknowledges that individuals with high TG have CHD risk that may be inadequately accounted for by LDL-C due to the presence of increased numbers of atherogenic lipoproteins in this setting. Because of various concerns regarding the use of apo B, and the presence of a high correlation between non-HDL-C and apo B, the report advocated use of non-HDL-C levels less than 30 mg/dL above the patient's LDL-C goal as a secondary treatment target for patients with TG > 200 mg/dL.

The impetus for the "lower is better" LDL-C guidance comes, in part, from statin intervention trials in which increasing intensity of therapy was associated with an approximate 20% reduction in relative CHD risk. Thus, even with LDL-C < 70 mg/dL, there is no guarantee that high-risk patients do not still have a substantial residual risk of suffering a CVD event. Does this residual risk come from LDL that has still not been lowered enough? Or does it come from other contributors to risk such as low HDL and/or high TG? Would lipoprotein testing provide unique information that could potentially change the clinical management of these patients? The following case study is designed to address these and other considerations regarding the potential clinical utilization of lipoprotein information.

Case HC

A 46-year-old male patient presents to your CVD prevention center for initial evaluation.

Six months prior to presentation, the patient was seen by his primary care physician for a routine annual physical examination. At that time, he was asymptomatic and able to engage in activities of choice without complaints of chest pain, pressure, heaviness, or shortness of breath. He followed no specific diet and engaged in no regular physical activity. Pertinent history and physical findings included the following:

Past medical history: unremarkable
Current medications: none
Medication allergies: none
Social history: denied use of tobacco products or alcohol
Family history: father – died of myocardial infarction (age 49 years)
Review of systems: unremarkable
Physical examination: blood pressure 128/80 mm Hg, pulse 60, weight 200 pounds, height 6 feet, and waist 38 inches
Otherwise unremarkable

Fasting laboratory studies at that time showed the following:

Total cholesterol: 185 mg/dL
TGs: 160 mg/dL
HDL-C: 41 mg/dL
LDL-C: 112 mg/dL
Glucose: 98 mg/dL
Additional laboratory studies including electrolytes, Blood Urea Nitrogen (BUN), creatinine, liver function studies, TSH, and urinalysis were Within Normal Limits (WNL).
10-year Framingham Risk: 3%

Given a history of two major risk factors (family history and low HDL-C), insufficient criteria for the diagnosis of metabolic syndrome, and a 10-year Framingham Risk score of 3%, the patient was found to be at ATP III LDL-C goal (< 130 mg/dL). Following Electrocardiogram (EKG) and exercise stress testing (both of which were unremarkable), the patient was advised to begin walking 45–60 min daily, as well as adopt a step I American Heart Association diet.

Two months prior to presentation, the patient's 50-year-old brother was hospitalized with unstable angina and found to have 20–30% lesions in his mid Left anterior descending (LAD) and circumflex coronary arteries. No additional information is known to the patient. He now presents to your office for a second opinion regarding his cardiovascular risk status. He has been compliant with increasing physical activity and now walks 60 min daily, which he tolerates without symptoms. However, he has been noncompliant with dietary recommendations. Additional history and review of systems is otherwise unremarkable. Significant findings now include the following:

Physical examination: blood pressure 122/74 mm Hg, pulse 60, weight 195 pounds, height 6 feet, and waist 38 inches.
Otherwise unremarkable

Fasting lipid profile now shows:

Total cholesterol: 171 mg/dL
TGs: 140 mg/dL
HDL-C: 45 mg/dL
LDL-C: 98 mg/dL ($< $ 20th percentile)

To determine whether significant lipoprotein-related risk is present at these lipid values, additional laboratory studies were performed, which demonstrated:

LDL particle number (LDL-P): 1860 nmol/L (> 80th percentile)
Small LDL particle number: 1466 nmol/L (> 75th percentile)
LDL particle size: 19.2 nm (pattern B)
Large VLDL particle number: > 75th percentile
Large HDL particle number: < 25th percentile

Clinical Considerations

From a clinical perspective, three questions are significant in approaching the management of this patient. First, is continued lipoprotein-related risk present at this time? Second, what is the source of the lipoprotein-related risk present? Third, has the lipoprotein-related risk been effectively managed? Despite being at ATP III LDL-C target of < 130 mg/dL (< 50th percentile), as well as the optional target of < 100 mg/dL (< 20th percentile) for intermediate risk patients clinically judged to be at higher cardiovascular risk, the patient manifests significant LDL-related risk due to extremely high LDL particle number (> 80th percentile) due to elevated small LDL particle number (> 75th percentile). As expected, LDL particle size is small (pattern B) in this setting. In addition, the finding of markedly elevated large VLDL (> 75th percentile) and decreased large HDL (< 25th percentile) are strongly associated with insulin resistance and increased risk for incident type 2 diabetes mellitus as previously noted. As a result, the patient harbors significant lipoprotein-related CHD risk, driven by the elevated LDL particle number that is not discernable from the fasting lipid profile.

Given the significant heterogeneity of LDL particle number among individuals at LDL-C goal levels, as well as clinical outcome data demonstrating that LDL-related CHD risk is most significantly associated with LDL particle number (not the amount of cholesterol carried by a patient's LDL particles), the use of LDL particle number as a clinical tool has been advocated *(6,7,13)*. While we await the results of clinical trials that may help to determine specific target levels of LDL particle numbers (with different targets possibly needed for large- and small-size particles), for patients in different CHD risk categories, utilization of LDL-P goals that mirror current NCEP ATP III targets at a population equivalent level has been suggested. Thus, for high-risk patients (CHD or CHD risk equivalents), the population equivalent LDL-P goal is LDL-P < 1000 nmol/L (20th percentile). For moderately high risk patients (multiple risk factors and a Framingham 10-year risk score < 20%), the goal is LDL-P < 1300 nmol/L (50th percentile). Additionally, many authors also emphasize treatment strategies targeted at reducing increased small LDL particles *(109,110)*. Once LDL values have been managed, secondary targets of therapy include (if indicated) increasing HDL-C and lowering TGs.

As seen in Table 2, medications used routinely for lipid optimization *(111)* have well-documented effects on LDL particle number *(112–126)*. In patients with elevated numbers of small LDL particles, diet therapy, as well as the combination of niacin or fibrates with statins, usually decreases TG, raises HDL-C, and increases LDL size – causing LDL-P to be decreased more than LDL-C.

Intervention

Options were discussed with the patient. The patient agreed to see a dietician for formal instruction in a Step II American Heart Association diet. To further assess the presence of subclinical atherosclerotic CVD, the patient was scheduled for a computed tomography (CT) coronary calcium score. Additionally, simvastatin 40 mg daily was started.

Table 2
Low-Density Lipoprotein (LDL) Particle Number *(112–126)* and Lipid *(111)* Altering Efficacy of Common Lipid-Altering Agents

Lipid-altering agent	Change in LDL particle number (%)	Change in LDL-C (%)	Change in triglyceride (%)	Change in HDL-C (%)
Statins	↓ 18–55[c]	↓ 18–55	↓ 7–30	↑ 5–15
Nicotinic acid (niacin)[a]	↓ 10–25	↓ 5–25	↓ 20–50	↑ 15–35
Fibric acids (fibrates)[a]	↓ 5–20[c]	↓ 5–20[b]	↓ 20–50	↑ 10–20
Ezetimibe	↓ 15–25[c]	↓ 17–22	↓ 4–11	↑ 2–5
Bile acid sequestrants	↓ 15–30[c]	↓ 15–30	No change to increased	↑ 3–5
Fish oils[d]	Trials in progress	No change to increased	↓ 20–50	No change to increased

[a] In patients with elevated numbers of small LDL particles, combination with statins usually decreases triglycerides, raises HDL cholesterol, and increases LDL size – causing LDL-P to be decreased more than LDL-C.

[b] Fibrates may increase LDL-C blood levels in some patients with hypertriglyceridemia. This is the so-called "beta-effect" of fibrates and can occur secondary to a large increase in the conversion of VLDL to LDL as lipoprotein lipase is activated.

[c] Combination of NMR and apo B data.

[d] The lipid-altering effects of oil listed are with administration of 5–9 g of omega-3 fatty acids per day.

Originally published in the following: Cromwell WC, Bays HE, Toth PP. Lipoprotein subfraction analysis using nuclear magnetic resonance spectroscopy. In: Adams JE, Jaffe AS, Apple F eds. *Markers in Cardiology: A Case-Oriented Approach.* Oxford, UK: Blackwell Publishing; 2007 in press.

Three months later, the patient returned for follow-up. He has been compliant with all therapeutic lifestyle measures and medication without complaints. Coronary calcium score returned elevated at 150 (LAD score 75, Circumflex score 50, Right Coronary Artery (RCA) score 25). Additional history and review of systems are otherwise unremarkable. Significant findings now include:

Physical examination: blood pressure 118/68 mm Hg, pulse 60, weight 185 pounds, height 6 feet, and waist 36 inches
Otherwise unremarkable

Fasting lipid profile now shows:

Total cholesterol: 153 mg/dL
TGs: 100 mg/dL
HDL-C: 48 mg/dL
LDL-C: 85 mg/dL (< 20th percentile)

To determine whether significant lipoprotein related risk is present at these lipid values, additional laboratory studies were performed, which demonstrated:

LDL particle number (LDL-P): 960 nmol/L (< 20th percentile)
Small LDL particle number: 450 nmol/L (< 20th percentile)
LDL particle size: 20.7 nm (pattern A)

Clinical Discussion

Growing evidence supports the placement of patients with an elevated coronary calcium score into a CHD-risk equivalent category with an appropriate LDL-C target of < 100 mg/dL (20th percentile). An optional LDL-C target of < 70 mg/dL has been increasingly discussed as a consideration for those judged to be at very high CHD risk. Much debate continues regarding when adequate LDL lowering has been achieved. Despite significant morbidity and mortality improvements noted in recent trials utilizing aggressive LDL-lowering therapy, LDL-C levels have not been identified, which confer optimal CHD risk reduction. Because CHD risk is more significantly associated with particles rather than traditional lipid values, the determination of whether lipoprotein-related risk continues to be present at these lipid values requires additional laboratory testing. In this case, only a modest LDL-C response (13 mg/dL) was noted. However, a much more robust lipoprotein response was noted with treated LDL particle number now < 1000 nmol/L (< 20th percentile) driven by a substantial reduction in small LDL particles. Having achieved concordant reductions in LDL-C and LDL-P (< 20th percentile), attention should now be paid to optimizing other sources of modifiable risk.

CONCLUSION

Substantial variability in CHD risk is present among patients across a wide range of lipid values, including the residual CHD risk harbored by those near or at recommended lipid targets on lipid-lowering therapy. The degree to which lipoprotein information may have clinical utility in daily practice continues to be an area of robust research. Although early studies focused on the ability of lipoprotein data to add incremental value in the assessment of CHD risk, focus is now shifting to the question of how lipoprotein information can be used to affect management of CHD risk. An increasing amount of research supports the fact that several on-therapy lipoprotein values (especially LDL particle number) are independently and significantly predictive of CHD risk after multivariate adjustment for major risk factors, lipids, and a variety of co-segregating variables. Because different lipoprotein methods report dissimilar information, CHD outcome associations for each commercially available lipoprotein test are needed to determine the value of individual parameters versus standard lipid values. Monitoring the therapeutic response of lipoprotein parameters meeting these expectations allows clinicians to tailor use of therapies and dosages used to ensure an optimal individual response to therapy.

REFERENCES

1. Fredrickson DS, Levy RI, Lees RS. Fat transport in lipoproteins – an integrated approach to mechanisms and disorders. *N Engl J Med* 1967;276:148–156.
2. Kannel WB, Castelli WP, Gordon T. Cholesterol in the prediction of atherosclerotic disease: new perspectives based on the Framingham study. *Ann Intern Med* 1979;90:85–91.
3. Kannel WB. Range of serum cholesterol values in the population developing coronary artery disease. *Am J Cardiol* 1995;76(9):69C–77C.
4. Phillips NR, Waters D, Havel RJ. Plasma lipoproteins and progression of coronary artery disease evaluated by angiography and clinical events. *Circulation* 1993;88:2762–2770.
5. Haskell WL, Alderman EL, Fair JM, et al. Effects of intensive multiple risk factor reduction on coronary atherosclerosis and clinical cardiac events in men and women with coronary artery disease. The Stanford Coronary Risk Intervention Project (SCRIP). *Circulation* 1994;89:975–990.
6. Sniderman AD, Furberg CD, Keech A, et al. Apolipoproteins versus lipids as indices of coronary risk and as targets for statin therapy treatment. *Lancet* 2003;361:777–780.

7. Barter PJ, Ballantyne CM, Carmena R, et al. Apo B versus cholesterol in estimating cardiovascular risk and in guiding therapy: report of the thirty-person/ten-country panel. *J Int Med Res* 2006;259:247–258.
8. Gotto AM, Whitney E, Stein EA, Shapiro DR, Clearfield M, Weis S. Relation between baseline and on-treatment lipid parameters and first acute major coronary events in the Air force/Texas Coronary Atherosclerosis Prevention Study (AFCAPS/TexCAPS). *Circulation* 2000;101:477–484.
9. Simes RJ, Marschner IC, Hunt D, Colquhoun D, Sullivan D, Stewart RAH. Relationship between lipid levels and clinical outcomes in the long-term intervention with pravastatin in the ischemic disease (LIPID) trial. To what extent is the reduction in coronary events with pravastatin explained by on-study lipid levels? *Circulation* 2002;105:1162–1169.
10. van Lennep JE, Westerveld HT, van Lennep HW, Zwinderman AH, Erkelens DW, van der Wall EE. Apolipoprotein concentrations during treatment and recurrent coronary artery disease events. *Arterioscler Thromb Vasc Biol* 2000;20:2408–2413.
11. Otvos JD, Jeyarajah EJ, Cromwell WC. Measurement issues related to lipoprotein heterogeneity. *Am J Cardiol* 2002;90(suppl):22i–29i.
12. Otvos JD. Why cholesterol measurements may be misleading about lipoprotein levels and cardiovascular disease risk – clinical implications of lipoprotein quantification using NMR spectroscopy. *J Lab Med* 2002;26:544–550.
13. Cromwell WC, Otvos JD. Low-density lipoprotein particle number and risk for cardiovascular disease. *Curr Atheroscler Rep* 2004;6:381–387.
14. Dominiczak MH. Apolipoproteins and lipoproteins in human plasma. In: *Handbook of Lipoprotein Testing* (Rifai N, Warnick GR, Dominiczak MH, eds), AACC Press, Washington DC, 2000, pp. 1–29.
15. Hussain MM, Fatma S, Pan X, Iqbal J. Intestinal lipoprotein assembly. *Curr Opin Lipidol* 2005;16:281–285.
16. Havel RJ. Origin, metabolic fate, and metabolic function of plasma lipoproteins. In: *Contemporary Issues in Endocrinology and Metabolism, Volume 3* (Steinberg D, Olefsky JM, eds), Churchill Livingstone, New York, 1986, pp. 117–141.
17. Eisenberg S. Metabolism of apolipoproteins and lipoproteins. *Curr Opin Lipidol* 1990;1:205–215.
18. Blum CB. Dynamics of apolipoprotein E metabolism in humans. *J Lipid Res* 1982;23:1308–1316.
19. Huff MW, Breckenridge WC, Strong WLP, Wolfe BM. Metabolism of apolipoproteins C-II, C-III and B in hypertriglyceridemic men: changes after heparin-induced lipolysis. *Arteriosclerosis* 1988;8:471–479.
20. Thuren T, Wilcox RW, Sisson P, Waite M. Hepatic lipase hydrolysis of lipid monolayers. Regulation by apolipoproteins. *J Biol Chem* 1991;266:4853–4861.
21. Herz J. The LDL-receptor related protein: portrait of a multifunctional receptor. *Curr Opin Lipidol* 1993;4:107–113.
22. Shachter NS. Apolipoproteins C-I and C-III as important modulators of lipoprotein metabolism. *Curr Opin Lipidol* 2001;12:297–304.
23. Kang S, Davis RA. Cholesterol and hepatic lipoprotein assembly and secretion. *Biochim Biophys Acta* 2000;1529:223–230.
24. Shelness GS, Sellers JA. Very-low-density lipoprotein assembly and secretion. *Curr Opin Lipidol* 2001;12:151–157.
25. Packard CJ, Shepherd J. Lipoprotein heterogeneity and apolipoprotein B metabolism. *Arterioscler Thromb Vasc Biol* 1997;17:3542–3556.
26. Berneis KK, Krauss RM. Metabolic origins and clinical significance of LDL heterogeneity. *J Lipid Res* 2002;43:1363–1379.
27. Krauss RM. Lipids and lipoproteins in patients with type 2 diabetes. *Diabetes Care* 2004;27:1496–1504.
28. Rye KA, Barter PJ. Formation and metabolism of prebeta-migrating, lipid-poor apolipoprotein A-I. *Arterioscler Thromb Vasc Biol* 2004;24:421–428.
29. Lee JY, Parks JS. ATP-binding cassette transporter AI and its role in HDL formation. *Curr Opin Lipidol* 2005;16:19–25.
30. Gofman J, Lindgren F, Elliott H. Ultracentrifugal studies of lipoproteins of human serum. *J Biol Chem* 1949;179:973–978.
31. Gofman JW, Lindgren, F. The role of lipids and lipoproteins in atherosclerosis. *Science* 1950;111(2877):166–171.
32. Lindgren FT, Elliott HA, Gofman JW. The ultracentrifugal characterization and isolation of human blood lipids and lipoproteins, with application to the study of atherosclerosis. *J Phys Colloid Chem* 1951;55:80–93.

33. Fredrickson DS, Levy RI, Lees RS. Fat transport in lipoproteins: an integrated approach to mechanisms and disorders. *N Engl J Med* 1967;276:148–156.
34. Otvos JD. Measurement of lipoprotein subclass profiles by nuclear magnetic resonance spectroscopy. In: *Handbook of Lipoprotein Testing* (Rifai N, Warnick GR, Dominiczak MH, eds), AACC Press, Washington DC, 2000, pp. 609–623.
35. Lounila J, Ala-Korpela M, Jokisaari J. Effects of orientational order and particle size on the NMR line positions of lipoproteins. *Phys Rev* 1994;72:4049–4052.
36. Jeyarajah EJ, Cromwell WC, Otvos JD. Lipoprotein particle analysis by nuclear magnetic resonance spectroscopy. *Clin Lab Med* 2006;26(4), pp. 847–870.
37. Sniderman AD, Acantlebury T, Cianflone K. Hypertriglyceridemic hyperapoB: the unappreciated atherogenic dyslipoproteinemia in type 2 diabetes mellitus. *Ann Intern Med* 2001;135:447–459.
38. Austin MA, King MC, Vranizan KM, Krauss RM. Atherogenic lipoprotein phenotype. A proposed genetic marker for coronary heart disease risk. *Circulation* 1990;82(2):495–506.
39. Sniderman AD, Vu H, Cianflone K. The effect of moderate hypertriglyceridemia on the relation of plasma total and LDL apoB levels. *Atherosclerosis* 1991;89:109–116.
40. Durrington PN, Bolton CN, Hartog H. Serum and lipoprotein apolipoprotein B levels in normal subjects and patients with hyperlipoproteinaemia. *Clin Chim Acta* 1978;82:151–160.
41. Garvey WT, Kwon S, Zheng D, et al. The effects of insulin resistance and type 2 diabetes mellitus on lipoprotein subclass particle size and concentration determined by nuclear magnetic resonance. *Diabetes* 2003;52:453–462.
42. Goff DC, D'Agostino RB, Jr, Haffner SM, Otvos JD. Insulin resistance and adiposity influence lipoprotein size and subclass concentrations. Results from the Insulin Resistance Atherosclerosis Study. *Metabolism* 2005;54:264–270.
43. Festa A, Williams K, Hanley AJG, et al. Nuclear magnetic resonance lipoprotein abnormalities in prediabetic subjects in the Insulin Resistance Atherosclerosis Study (IRAS). *Circulation* 2005;111:3465–3472.
44. Kathiresan S, Otvos JD, Sullivan LM, et al. Increased small LDL particle number: a prominent feature of the metabolic syndrome in the Framingham Heart Study. *Circulation* 2006;113:20–29.
45. Cromwell WC, Otvos JD. Heterogeneity of low-density lipoprotein particle number in patients with type 2 diabetes mellitus and low-density lipoprotein cholesterol < 100 mg/dL. *Am J Cardiol* 2006; Dec 15;98(12):1599–602.
46. Krauss RM. Heterogeneity of plasma low-density lipoproteins and atherosclerosis risk. *Curr Opin Lipidol* 1994;5:339–349.
47. Austin MA. Triglyceride, small, dense low-density lipoprotein, and the atherogenic lipoprotein phenotype. *Curr Atheroscler Rep* 2000;2:200–207.
48. Lamarche B, Lemieux I, Despres JP. The small, dense LDL phenotype and the risk of coronary heart disease: epidemiology, pathophysiology, and therapeutic aspects. *Diabetes Metab* 1999;25;199–211.
49. McNamara JR, Campos H, Ordovas JM, Peterson J, Wilson PWF, Schaefer EJ. Effect of gender, age, and lipid status on low density lipoprotein subfraction distribution. Results of the Framingham Offspring Study. *Arteriosclerosis* 1987;7:483–490.
50. Austin MA, King MC, Vranizan KM, Krauss RM. Atherogenic lipoprotein phenotype: a proposed genetic marker for coronary heart disease risk. *Circulation* 1990;82:495–506.
51. Reaven GM, Chen YD, Jeppesen J, Maheux P, Krauss RM. Insulin resistance and hyperinsulinemia in individuals with small, dense low density lipoprotein particles. *J Clin Invest* 1993;92:141–146.
52. Blake GJ, Otvos JD, Rifai N, Ridker PM. LDL particle concentration and size as determined by NMR spectroscopy as predictors of cardiovascular disease in women. *Circulation* 2002;106:1930–1937.
53. Mackey RH, Kuller LH, Sutton-Tyrell K, Evans RW, Holubkov R, Matthews KA. Lipoprotein subclasses and coronary artery calcification in postmenopausal women from the Healthy Women Study. *Am J Cardiol* 2002;90(8A):71i–76i.
54. Rosenson RS, Freedman DS, Otvos JD. Relations of lipoprotein subclass levels and LDL size to progression of coronary artery disease in the PLAC I trial. *Am J Cardiol* 2002;90:89–94.
55. Kuller L, Arnold A, Tracy R, Otvos J, Burke G, Psaty B, Siscovick D, Freedman DS, Kronmal R. NMR spectroscopy of lipoproteins and risk of CHD in the Cardiovascular Health Study. *Arterioscler Thromb Vasc Biol* 2002;22:1175–1180.
56. Otvos JD, Collins D, Freedman DS, et al. Low-density lipoprotein and high-density lipoprotein particle subclasses predict coronary events and are favorably changed by gemfibrozil therapy in the Veterans Affairs High-Density Lipoprotein Intervention Trial. *Circulation* 2006;113(12): 1556–1563.

57. Schaefer E, Parise H, Otvos J, McNamara J, D'Agostino R, Wilson P. LDL particle number, size, and subspecies in assessing cardiovascular risk: results from the Framingham Offspring Study. *Circulation* 2004;110:III-777.
58. Mora S, Szklo M, Otvos JD, et al. LDL particle subclasses, LDL particle size, and carotid atherosclerosis in the Multi-Ethnic Study of Atherosclerosis (MESA). *Atherosclerosis* 2006; June 9 [epub ahead of print].
59. Castelli WP, Garrison RJ, Wilson PW, et al. Incidence of coronary heart disease and lipoprotein cholesterol levels: the Framingham Study. *JAMA* 1986;256:2835–2838.
60. Multiple Risk Factor Intervention Trial Research Group. Relationship between baseline risk factors and coronary heart disease and total mortality in the Multiple Risk Factor Intervention Trial. *Prev Med* 1986;15(3):254–273.
61. Assmann G, Schulte H, von Eckardstein A, Huang Y. High-density lipoprotein cholesterol as a predictor of coronary heart disease risk. The PROCAM experience and pathophysiological implications for reverse cholesterol transport. *Atherosclerosis* 1996;124 Suppl:S11–S20.
62. Li X-P, Zhao S-P, Zhang XY, Liu L, Gao M, Zhou Q-C. Protective effect of high density lipoprotein on endothelium-dependent vasodilatation. *Int J Cardiol* 2000;73:231–236.
63. Barter PJ. Inhibition of endothelial cell adhesion molecule expression by high density lipoproteins. *Clin Exp Pharmacol Physiol* 1997;24:286–287.
64. Nofer JR, Walter M, Kehrel B, et al. HDL_3-mediated inhibition of thrombin-induced platelet aggregation and fibrinogen binding occurs via decreased production of phosphoinositide-derived second messengers 1,2-diacylglycerol and inositol 1,4,5-tris-phosphate. *Arterioscler Thromb Vasc Biol* 1998;18:861–869.
65. Nofer JR, Levkau B, Wolinska I, et al. Suppression of endothelial cell apoptosis by high density lipoproteins (HDL) and HDL-associated lysosphingolipids. *J Biol Chem* 2001;276:34480–34485.
66. Aviram M, Hardak E, Vaya J, et al. Human serum paraoxonases (PON1) Q and R selectively decrease lipid peroxides in human coronary and carotid atherosclerotic lesions. *Circulation* 2000;101:2510–2517.
67. Toikka J, Ahotupa M, Viikari J, et al. Constantly low HDL-cholesterol concentration relates to endothelial dysfunction and increased in vivo LDL oxidation in healthy young men. *Atherosclerosis* 1999;147:133–138.
68. Miller NE. Associations of high-density lipoprotein subclasses and apolipoproteins with ischemic heart disease and coronary atherosclerosis. *Am Heart J* 1987;113(2 Pt 2):589–597.
69. Johansson J, Carlson LA, Landou C, Hamsten A. High density lipoproteins and coronary atherosclerosis. A strong inverse relation with the largest particles is confined to normotriglyceridemic patients. *Arterioscler Thromb* 1991;11(1):174–182.
70. Silverman DI, Ginsburg GS, Pasternak RC. High-density lipoprotein subfractions. *Am J Med* 1993;94(6):636–645.
71. Gofman JW, Young W, Tandy R. Ischemic heart disease, atherosclerosis, and longevity. *Circulation* 1966;34(4):679–697.
72. Cheung MC, Brown BG, Wolf AC, Albers JJ. Altered particle size distribution of apolipoprotein A-I-containing lipoproteins in subjects with coronary artery disease. *J Lipid Res* 1991;32(3):383–394.
73. Stampfer MJ, Sacks FM, Salvini S, Willett WC, Hennekens CH. A prospective study of cholesterol, apolipoproteins, and the risk of myocardial infarction. *N Engl J Med* 1991;325(6):373–381.
74. Salonen JT, Salonen R, Seppanen K, Rauramaa R, Tuomilehto J. HDL, HDL2, and HDL3 subfractions, and the risk of acute myocardial infarction. A prospective population study in eastern Finnish men. *Circulation* 1991;84(1):129–139.
75. Lamarche B, Moorjani S, Cantin B, Dagenais GR, Lupien PJ, Despres JP. Associations of HDL2 and HDL3 subfractions with ischemic heart disease in men. Prospective results from the Quebec Cardiovascular Study. *Arterioscler Thromb Vasc Biol* 1997;17(6):1098–1105.
76. Hattori H, Kujiraoka T, Egashira T, et al. Association of coronary heart disease with pre-beta-HDL concentrations in Japanese men. *Clin Chem* 2004;50(3):589–595.
77. Yu S, Yarnell JW, Sweetnam P, Bolton CH. High density lipoprotein subfractions and the risk of coronary heart disease: 9-years follow-up in the Caerphilly Study. *Atherosclerosis* 2003;166(2):331–338.
78. Brown G, Albers JJ, Fisher LD, et al. Regression of coronary artery disease as a result of intensive lipid-lowering therapy in men with high levels of apolipoprotein B. *N Engl J Med* 1990;323:1289–1298.

79. Brown BG, Zhao XQ, Chait A, et al. Simvastatin and niacin, antioxidant vitamins, or the combination for the prevention of coronary disease. *N Engl J Med* 2001;345:1583–1592.
80. Asztalos BF, Batista M, Horvath KV, et al. Change in alpha1 HDL concentration predicts progression in coronary artery stenosis. *Arterioscler Thromb Vasc Biol* 2003;23:847–852.
81. Ruotolo G, Ericsson C-G, Tettamanti C, et al. Treatment effects on serum lipoprotein lipids, apolipoproteins and low density lipoprotein particle size and relationships of lipoprotein variables to progression of coronary artery disease in the Bezafibrate Coronary Atherosclerosis Intervention Trial (BECAIT). *J Am Coll Cardiol* 1998;32:1648–1656.
82. Syvänne M, Nieminen MS, Frick MH, et al. Associations between lipoproteins and the progression of coronary and vein-graft atherosclerosis in a controlled trial with gemfibrozil in men with low baseline levels of HDL cholesterol. *Circulation* 1998;98:1993–1999.
83. Mänttäri M, Koskinen P, Manninen V, Huttunen JK, Frick MH, Nikkilä EA. Effect of gemfibrozil on the concentration and composition of serum lipoproteins. *Atherosclerosis* 1990;81:11–17.
84. Guerin M, Le Goff W, Frisdal E, et al. Action of ciprofibrate in Type IIB hyperlipoproteinemia: modulation of the atherogenic lipoprotein phenotype and stimulation of high-density lipoprotein-mediated cellular cholesterol efflux. *J Clin Endocrinol Metab* 2003;88:3738–3746.
85. Freedman DS, Otvos JD, Jeyarajah EJ, Barboriak JJ, Anderson AJ, Walker J. Relation of lipoprotein subclasses as measured by proton nuclear magnetic resonance spectroscopy to coronary artery disease. *Arterioscler Thromb Vasc Biol* 1998;18:1046–1053.
86. Soedamah-Muthu SS, Chang Y-F, Otvos J, Evans RW, Orchard TJ. Lipoprotein subclass measurements by nuclear magnetic resonance spectroscopy improve the prediction of coronary artery disease in type 1 diabetes. A prospective report from the Pittsburgh Epidemiology of Diabetes Complications Study. *Diabetologia* 2003;46:674–682.
87. Freedman DS, Otvos JD, Jeyarajah EJ, et al. Sex and age differences in lipoprotein subclasses measured by nuclear magnetic resonance spectroscopy: the Framingham Study. *Clin Chem* 2004;50:1189–1200.
88. Brewer HB, Remaley AT, Neufeld EB, Baso F, Joyce C. Regulation of plasma high-density lipoprotein levels by the ABCA1 transporter and the emerging role of high-density lipoprotein in the treatment of cardiovascular disease. *Arterioscler Thromb Vasc Biol* 2004;24:1755–1760.
89. Austin MA, McKnight B, Edwards KL, et al. Cardiovascular disease mortality in familial forms of hypertriglyceridemia: a 20-year prospective study. *Circulation* 2000;101:2777–2782.
90. Hokanson JE, Austin MA. Plasma triglyceride level as a risk factor for cardiovascular disease independent of high-density lipoprotein level: a meta-analysis of population-based prospective studies. *J Cardiovasc Risk* 1996;3:213–219.
91. Cohnn JS, Marcoux C, Davignon J. Detection, quantification, and characterization of potentially atherogenic triglyceride-rich remnant lipoproteins. *Arterioscler Thromb Vasc Biol* 1999;19:2474–2486.
92. Inoue T, Uchida T, Kamishirado H, et al. Remnant-like lipoprotein particles as risk factors for coronary artery disease in elderly patients. *Horm Metab Res* 2004;36:298–302.
93. Hamano M, Saito M, Eto M, et al. Serum amyloid A, C-reactive protein and remnant-like lipoprotein particle cholesterol in type 2 diabetic patients with coronary heart disease. *Ann Clin Biochem* 2004;41(pt 2):125–129.
94. Schaefer EJ, McNamara JR, Tayler T, et al. Effects of atorvastatin on fasting and postprandial lipoprotein subclasses in coronary heart disease patients versus control subjects. *Am J Cardiol* 2002;90:689–696.
95. Kawakami A, Tanaka A, Nakajima K, Shimokado K, Yoshida M. Atorvastatin attenuates remnant lipoprotein-induced monocyte adhesion to vascular endothelium under flow conditions. *Circ Res* 2002;91:263–271.
96. Inoue T, Uchida T, Kamishirado H, et al. Possible relationship between insulin resistance and remnant-like lipoprotein particles in coronary endothelial dysfunction. *Clin Cardiol* 2002;25:532–536.
97. Fukushima H, Kugiyama K, Sugiyama S, et al. Comparison of remnant-like lipoprotein particles in postmenopausal women with and without coronary artery disease and in men with coronary artery disease. *Am J Cardiol* 2001;88:1370–1373.
98. McNamara JR, Shah PK, Nakajima K, et al. Remnant-like particle (RLP) cholesterol is an independent cardiovascular disease risk factor in women: results from the Framingham Heart Study. *Atherosclerosis* 2001;154:229–236.

99. Karpe F, Boquist S, Tang R, Bond GM, de Faire U, Hamsten A. Remnant lipoproteins are related to intima-media thickness of the carotid artery independently of LDL cholesterol and plasma triglycerides. *J Lipid Res* 2001;42:17–21.
100. Takeichi S, Yukawa N, Nakajima Y, et al. Association of plasma triglyceride-rich lipoprotein remnants with coronary atherosclerosis in cases of sudden cardiac death. *Atherosclerosis* 1999;142:309–315.
101. Takeichi S, Nakajima Y, Osawa M, et al. The possible role of remnant-like particles as a risk factor for sudden cardiac death. *Int J Legal Med* 1997;110:213–219.
102. Kugiyama K, Doi H, Takazoe K, et al. Remnant lipoprotein levels in fasting serum predict coronary events in patients with coronary artery disease. *Circulation* 1999;99:2858–2860.
103. Seman LJ, McNamara JR, Schaefer EJ. Lipoprotein (a), homocysteine, and remnant like particles: emerging risk factors. *Curr Opin Cardiol* 1999;14:186–191.
104. Twickler TB, Dallinga-Thie GM, Cohn JS, Chapman MJ. Elevated remnant-like particle cholesterol concentration: a characteristic feature of the atherogenic lipoprotein phenotype. *Circulation* 2004;109:1918–1925.
105. Satoh A, Adachi H, Tsuruta M, et al. High plasma level of remnant-like particle cholesterol in the metabolic syndrome. *Diabetes Care* 2005;28(10):2514–2518.
106. Imke C, Rodriguez BL, Grove JS, et al. Are remnant-like particles independent predictors of coronary heart disease incidence? The Honolulu Heart Study. *Arterioscler Thromb Vasc Biol* 2005;25:1718–1722.
107. Executive Summary of the Third Report of the National Cholesterol Education Program (NCEP) Expert Panel on Detection, Evaluation, and Treatment of High Blood Cholesterol in Adults (Adult Treatment Panel III). *JAMA* 2001;285:2486–2497.
108. Grundy SM, Cleeman JI, Merz CNB, et al. Implications of recent clinical trials for the National Cholesterol Education Program Adult Treatment Panel III guidelines. *Circulation* 2004;110:227–239.
109. Rizzo M, Berneis K, Corrado E, Novo S. The significance of low-density-lipoproteins size in vascular diseases. *Int Angiol* 2006;25(1):4–9.
110. Rizzo M, Berneis K. Should we measure routinely the LDL peak particle size? *Int J Cardiol* 2006 Feb 15;107(2):166–170.
111. Bays H, Stein EA. Pharmacotherapy for dyslipidaemia – current therapies and future agents. *Expert Opin Pharmacother* 2003;4(11):1901–1938.
112. McKenney JM, McCormick LS, Schaefer EJ, Black DM, Watkins ML. Effect of niacin and atorvastatin on lipoprotein subclasses in patients with atherogenic dyslipidemia. *Am J Cardiol* 2001;88(3):270–274.
113. Soedamah-Muthu SS, Colhoun HM, Thomason MJ, et al.; CARDS Investigators. The effect of atorvastatin on serum lipids, lipoproteins and NMR spectroscopy defined lipoprotein subclasses in type 2 diabetic patients with ischaemic heart disease. *Atherosclerosis* 2003;167(2):243–255.
114. Schaefer EJ, McNamara JR, Tayler T, et al. Effects of atorvastatin on fasting and postprandial lipoprotein subclasses in coronary heart disease patients versus control subjects. *Am J Cardiol* 2002;90(7):689–696.
115. Schaefer EJ, McNamara JR, Tayler T, et al. Comparisons of effects of statins (atorvastatin, fluvastatin, lovastatin, pravastatin, and simvastatin) on fasting and postprandial lipoproteins in patients with coronary heart disease versus control subjects. *Am J Cardiol* 2004;93(1):31–39.
116. Miller M, Dolinar C, Cromwell W, Otvos JD. Effectiveness of high doses of simvastatin as monotherapy in mixed hyperlipidemia. *Am J Cardiol* 2001;87(2):232–234.
117. Rosenson RS, Otvos JD, Freedman DS. Relations of lipoprotein subclass levels and low-density lipoprotein size to progression of coronary artery disease in the Pravastatin Limitation of Atherosclerosis in the Coronary Arteries (PLAC-I) trial. *Am J Cardiol* 2002;90(2):89–94.
118. Rosuvastatin Dose Ranging Study. FDA PI Aug 2003.
119. Morgan JM, Capuzzi DM, Baksh RI, et al. Effects of extended-release niacin on lipoprotein subclass distribution. *Am J Cardiol* 2003;91(12):1432–1436.
120. Al-Shaer MH. The effects of ezetimibe on the LDL-cholesterol particle number. *Cardiovasc Drugs Ther* 2004;18(4):327–328.
121. Pearson TA, Denke MA, McBride PE, Battisti WP, Brady WE, Palmisano J. A community-based, randomized trial of ezetimibe added to statin therapy to attain NCEP ATP III goals for LDL cholesterol in hypercholesterolemic patients: the Ezetimibe Add-On to Statin for Effectiveness (EASE) Trial. *Mayo Clin Proc* 2005;80(5):587–595.

122. Ikewaki K, Tohyama J, Nakata Y, Wakikawa T, Kido T, Mochizuki S. Fenofibrate effectively reduces remnants, and small dense LDL, and increases HDL particle number in hypertriglyceridemic men – a nuclear magnetic resonance study. *J Atheroscler Thromb* 2004;11:278–285.
123. Steinmetz A, Schwartz T, Hehnke U, Kaffarnik H. Multicenter comparison of micronized fenofibrate and simvastatin in patients with primary type IIA or IIB hyperlipoproteinemia. *J Cardiovasc Pharmacol* 1996;27(4):563–570.
124. Bilz S, Wagner S, Schmitz M, Bedynek A, Keller U, Demant T. Effects of atorvastatin versus fenofibrate on apoB-100 and apoA-I kinetics in mixed hyperlipidemia. *J Lipid Res* 2004;45(1):174–185.
125. Superko HR, Greenland P, Manchester RA, et al. Effectiveness of low-dose colestipol therapy in patients with moderate hypercholesterolemia. *Am J Cardiol* 1992;70(2):135–140.
126. Rosenson RS. Colesevelam HCl reduces LDL particle number and increases LDL size in hypercholesterolemia. *Atherosclerosis* 2006;185(2):327–330.

16 Cardiovascular Disease in Women
The Management of Dyslipidemia

Emma A. Meagher, MD

CONTENTS

INTRODUCTION
IDENTIFYING CARDIOVASCULAR RISK
CARDIOVASCULAR RISK ASSESSMENT
TREATMENT
LIFESTYLE AND BEHAVIORAL CHANGES
DRUG THERAPY
COMBINATION THERAPY FOR THE TREATMENT
   OF DYSLIPIDEMIA
REFERENCES

INTRODUCTION

Cardiovascular mortality amongst women has reached close to epidemic proportions. Approximately 500,000 women die of cardiovascular diseases (CVDs) per annum *(1)*. This accounts for close to half of all female mortality and is more than the next seven causes of death combined. For at least 20 years, the number of women dying from CVD has exceeded the number of men dying from CVD (Fig. 1). These data underscore the need to actively identify risk factors that contribute to CVD in women and aggressively implement CVD risk management strategies.

IDENTIFYING CARDIOVASCULAR RISK

Identification of at risk patients requires acknowledgement of current risk factors that go beyond the traditional, well-known risk factors of increasing age, post-menopausal status smoking status, and family history (Table 1).

Awareness

A significant barrier to positively affecting the CVD epidemic in women is a lack of awareness of the disease as a problem by at least 50% of the population. Lack of awareness of CVD risk warrants significant attention and needs to be the key initial focus in successfully addressing adequate risk management strategies in women.

From: *Contemporary Cardiology: Therapeutic Lipidology*
Edited by: M. H. Davidson, P. P. Toth, and K. C. Maki © Humana Press Inc., Totowa, NJ

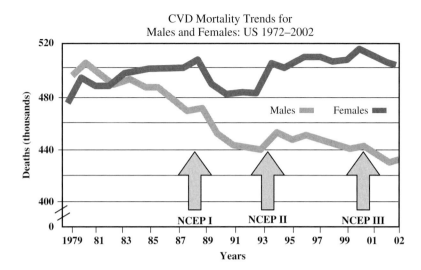

Fig. 1. Cardiovascular disease mortality trends for males and females: US 1979–2002, National Cholesterol Program *(1)*.

Data from an American Heart Association (AHA)-sponsored national survey of 1024 women aged ≥ 25 years demonstrated that awareness of CVD risk is generally low amongst women but particularly so in women of black and Hispanic origin *(2)*. This is particularly relevant because minority women, in particular black and Hispanic women, have a higher rate of death due to both coronary heart disease (CHD) and stroke compared with white women. The Hispanic community is the most rapidly expanding community in the US and therefore likely to carry an even greater burden of CVD over the next several decades. The lack of awareness also extends to the health care community. The high rate of undiagnosed and/or untreated cardiovascular risk in women is in part attributed to physicians' relatively low index of suspicion for CVD in women. In one survey, 70% of female patients stated that their physician had never discussed CVD with them *(3)*. However, women are more likely than men to have multiple risk factors for CHD *(4)*. Both health professionals and patients need to recognize the existence of risk factors and the potential for developing future cardiovascular risk.

Table 1
Identifying Risk Factors for
Cardiovascular Disease
in Women

Diabetes
Family history
Increasing age
Lack of awareness
Metabolic syndrome
Obesity
Post-menopausal status
Smoking history

Obesity

Obesity is a rapidly growing epidemic in the US and first world nations in general. An estimated 300,000 US adults die of causes related to obesity each year. The obesity epidemic has now extended into the pediatric population with the latest figures suggesting that 31% of American children in the age group of 6–19 years are at risk of becoming overweight *(1)*. The increased incidence of obesity has important implications for the incidence and type of dyslipidemia, hypertension, diabetes, metabolic syndrome, and ensuing vascular disease. Specifically related to dyslipidemia, weight gain has been reported to be associated with a reduction in high-density lipoprotein cholesterol (HDL-C), an important predictor of risk in women *(5,6)*.

Metabolic Syndrome

The increased incidence of metabolic syndrome is tightly related to the increased incidence of obesity. The diagnosis of metabolic syndrome in women is established when three of the following criteria are met: waist circumference > 35 inches, triglycerides $\geq 150\,\text{mg/dL}$, HDL-C $< 50\,\text{mg/dL}$, blood pressure $\geq 130/85\,\text{mmHg}$, and a fasting glucose $\geq 110\,\text{mg/dL}$ (Table 2) *(6)*. It is equally prevalent in both genders across the age spectrum with an increased prevalence among African-Americans and Mexican-American women *(5)*. The important implication of this diagnosis is its association with an increased incidence of cardiovascular mortality *(7)*.

Diabetes

Much like cardiovascular death, diabetes prevalence has increased by 60% over a 10-year period of time *(8)*. Diabetes is more common in women and is considered to essentially negate any protective effect of female gender against CVD *(9)*. Twenty-year follow-up data from the Framingham Heart Study of 5209 men and women aged 45–57 years demonstrated that the relative impact of diabetes on cardiovascular mortality

Table 2
Criteria for Diagnosis of the Metabolic Syndrome

Abdominal obesity (waist circumference)	
Men	$> 102\,\text{cm}$ (> 40 inches)
Women	$> 88\,\text{cm}$ (> 35 inches)
Triglycerides	$\geq 150\,\text{mg/dL}$
HDL-C	
Men	$< 40\,\text{mg/dL}$
Women	$< 50\,\text{mg/dL}$
Blood pressure	$\geq 130/ \geq 85\,\text{mmHg}$
Fasting glucose	$\geq 110\,\text{mg/dL}$

HDL-C, high-density lipoprotein cholesterol. Three of five elements must be present to meet the criteria for a diagnosis of metabolic syndrome.

National Cholesterol Education Program Expert Panel on Detection, Evaluation, and Treatment of High Blood Cholesterol in Adults (Adult Treatment Panel III). *JAMA* 2001;285:2486–2497.

appears to be greater in women than in men *(10,11)*. It is an established major risk factor for CVD, predominately through its association with obesity, hypertension, and dyslipidemia *(9–14)*.

CARDIOVASCULAR RISK ASSESSMENT

The first step in CVD risk management is to determine what the level of risk is for our patients. There are numerous assessment tools available. One of them is the Framingham risk assessment tool shown in Fig. 2, which helps to determine the absolute risk of CHD over 10 years *(15)*. This tool, which can be accessed at http://www.nhlbi.nih.gov/guidelines/cholesterol/index.htm, is incorporated into three sets of national guidelines which delineate a set of recommendations for CVD risk management, the National Cholesterol Education Program (NCEP), the AHA Evidenced Based Guidelines for Cardiovascular Risk Reduction, and the American Diabetes Association (Table 3). All recommend a comprehensive assessment of cardiovascular risk irrespective of gender *(6,16–18)*. The evaluation should include a complete medical history to identify the presence of a known history of CVD or CVD risk equivalent (diabetes, other vascular disease, and chronic kidney disease), hyperlipidemia, hypertension, metabolic syndrome, diabetes, thyroid disease, and obesity (Fig. 2). A family history for each of these should also be determined, as should a family history of premature CVD. A laboratory workup should initially include a complete lipid panel and fasting glucose level. In patients who have known hyperlipidemia, a thyroid-stimulating hormone level should be obtained to rule out secondary hypercholesterolemia. There is significant focus in all guidelines on primary prevention in persons with multiple risk factors. They use Framingham projections of 10-year risk

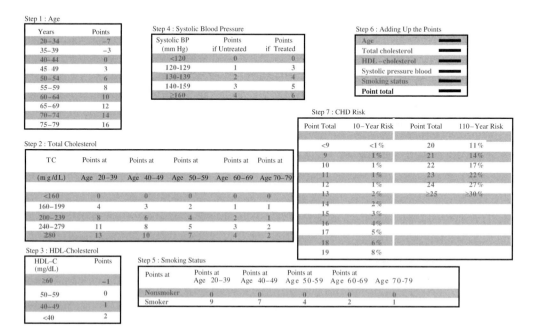

Fig. 2. Assessing cardiovascular disease risk in women. Risk estimates were derived from the experience of the Framingham Heart Study, Expert Panel on Detection, Evaluation, and Treatment of High Blood Cholesterol in Adults. *JAMA* 2001;285:2486–2497.

Table 3
Cardiovascular Disease (CVD) Risk Assessment

History
    Age
    Postmenopausal status
    Hypertension
    Diabetes
    Hypothyroidism
    Renal disease
    Other vascular disease
    Family history of premature CVD
        Female member < 60 years
        Male member < 50 years
Examination
    Height, weight, waist circumference
    Blood pressure
    Thyroid exam
    Examination of eyes with fundoscopy
    Cardiovascular exam
    Stigmata of dyslipidemia, xanthelasma, and thickened tendons
Laboratory workup
    Comprehensive metabolic panel
        Fasting glucose, creatinine level and Liver Function Test (LFTs)
    Fasting lipid panel
        Total cholesterol, LDL, HDL, and non-HDL cholesterol
    TSH level
    Additional optional labs
        HS-CRP level
        Lipoprotein (a) level

HS-CRP, high-sensitivity C-reactive protein; HDL, high-density lipoprotein; LDL, low-density lipoprotein; TSH, thyroid-stimulating hormone.

to identify certain individuals with multiple risk factors for more intensive treatment, including those with the metabolic syndrome. For this reason, using both historical data and results of the physical examination and laboratory workup, patients should be evaluated for the presence of the metabolic syndrome. It is recommended that an evaluation of a fasting total lipid profile is performed in adults aged ≥ 20 years at least once every 5 years (6). The utility of measuring lipoprotein (a) or high-sensitivity C-reactive protein (HS-CRP) as part of routine cardiovascular risk assessment is hotly debated.

The AHA evidence-based guideline for CVD prevention in women describes risk groups based on the Framingham global risk and other clinical characteristics that help to determine the aggressiveness of our preventive strategy (Table 4). For example, a high-risk group, based on the Framingham global risk score, corresponds to an absolute risk of CHD in the next 10 years of ≥ 20%, an intermediate risk would be classified as 10–20% absolute risk, and a lower risk is ≤ 10%. Typical clinical examples of intermediate risk women are summarized in Table 4. Importantly, subclinical CVD

Table 4
American Heart Association Evidence-Based Guidelines for CVD Prevention in Women

Stratify women by risk
  High (> 20%) – CHD, CVA, PAD, AAA, DM, CKD
  Intermediate (10–20%) – Sub-clinical CHD, metabolic syndrome, multiple risk factors, marked elevation of single risk factor, family history of premature CVD
  Lower (< 10%) – multiple risk factors, metabolic syndrome, 0–1 risk factors
  Optimal (< 10%) – no risk factors and healthy lifestyle

CHD, coronary heart disease; CVD, cardiovascular disease; DM, diabetes mellitus; CKD, chronic kidney disease; AAA, abdominal aoztic Aneveysm; PAD, peripheral arterial disease.

is being identified more frequently with the availability of tools such as the ultrafast computed tomography (CT) scan. Women who are identified as having atherosclerosis but have not yet had a clinical event fall into this category. The metabolic syndrome, a first-degree relative with premature heart disease, and multiple risk factors are other examples of intermediate risk. Finally, despite a Framingham risk score of low or intermediate risk, a markedly abnormal levels of a single risk factor may move women into the high-risk category (16).

Lipoproteins and CVD

Increased LDL-C has been definitively related to the development of CVD in both men and women. Data from the Lipid Research Clinics Program Follow-up Study, a mortality study with baseline data gathered from 1972 through 1976 from 2406 men and 2056 women aged 40–64 years, clearly delineated the association between increased LDL-C and CVD incidence and CVD mortality over a 19-year time frame (19). There is abundant evidence showing a reduction in clinical events in both men and women when LDL-C levels are lowered (4,20–26). The current guidelines from the NCEP Adult Treatment Panel (ATP) III, as well as the more recent AHA evidence-based guidelines for CVD prevention in women, reinforce LDL-C as the primary target of therapy (6,16). LDL-C levels arc generally lower in women than in men until menopause when levels increase and LDL particles become smaller and more dense and, therefore, more atherogenic (27,28).

The guidelines also recognize non-HDL-C is a secondary target for therapy (6,16). Despite the fact that non-HDL-C, which includes all atherogenic lipoproteins, has been shown to be a strong predictor of cardiovascular mortality in women, it is significantly underutilized as test in CVD risk assessment (29).

The Framingham Heart Study established both HDL-C and triglycerides as important predictors for coronary events (15). Castelli et al. obtained fasting lipid profiles on 1025 men and 1445 women aged 49–82 years between 1969 and 1971. During the 4-year follow-up period, CVD developed in 79 of the men and 63 of the women. This association was noted to be independent of total and LDL-C levels and applied to both genders (30). This and a subsequent analysis by the same author were the first data to suggest that triglyceride and HDL levels may have greater predictive potential in women when compared with men (30,31). More recently, the Lipid Research Clinics' Follow-Up Study also demonstrated that both HDL-C and triglycerides were better predictors of coronary risk and cardiovascular mortality in women than total cholesterol or LDL-C (19). Importantly, this study showed that when HDL-C level is < 50 mg/dL in a women compared with ≥ 50 mg/dL, there is a threefold to fourfold increase of CVD mortality irrespective of the baseline LDL-C level. In 1995, a meta-analysis performed by Hokanson and colleagues supported this data by showing a 1 mmol/L increase in triglycerides was associated with a 76% increased risk of CVD in women versus 32% in men *(32)*. The clinical relevance of the synergy between low HDL-C and elevated triglyceride level in women described in the Framingham data set is increasingly important given the increasing prevalence of combined dyslipidemia and its association with excess CVD morbidity and mortality.

There are two significant differences in lipoprotein levels between genders. First, women have on average HDL-C levels 10 mg/dL greater than men *(6,16)*. Second is the change that occurs in these levels throughout a woman's lifecycle, particularly during the pre-, peri-, and post-menopausal periods *(33)*. The former is well described in the ATP III and the AHA women's guidelines. The latter is thought to contribute to the sharp increase in cardiovascular mortality that occurs in the early post-menopausal period. A study of the influence of the menopause on serum lipids and lipoproteins was undertaken to examine the serum lipid profiles at 6-week intervals for 2–3 years in 1360 pre-menopausal women undergoing the menopause. Results from the study characterize the increase in total and LDL-C and triglycerides levels and the decrease in HDL-C levels that occur throughout the peri-menopausal period (Fig. 3) *(33)*.

Lipoprotein (a) is now emerging as a risk factor for CHD *(6)*. Data from the Heart and Estrogen/progestin Replacement Study indicate that lipoprotein (a) is an independent predictor of the risk of recurrent CHD in post-menopausal women *(34)*. This may have important implications for therapy, because lipoprotein (a) is unaffected by diet, exercise, and most lipid-modifying medications, with the exception of niacin, which decreases it. According to ATP III lipoprotein (a), measurement can be considered in

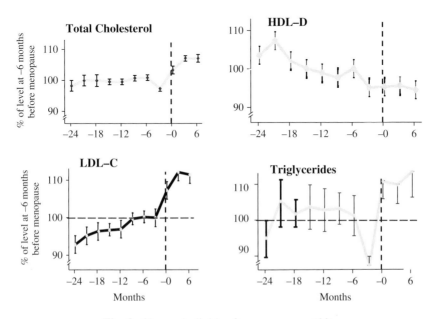

Fig. 3. Change in lipids after menopause *(33)*.

patients with less obvious risk but who may warrant more aggressive evaluation based on the presence of one significantly abnormal risk factor *(6)*.

The role of inflammation in the development and progression of atherosclerosis has received increased attention in recent years *(35)*. CRP, an acute-phase marker of systemic inflammation, has been identified as an independent risk factor for cardiovascular events, adding predictive value to that of individual lipoprotein fractions. Data from the Women's Health Study indicate that HS-CRP is related to several cardiovascular risk factors in women, including age, body mass index, blood pressure, cigarette smoking, and, to a lesser extent, HDL-C *(35)*. In one prospective follow-up of 28,263 women over 3 years, CRP was found to be the strongest predictor of CVD, superior to several other markers of inflammation and to homocysteine and lipoprotein levels *(36)*. However, this marker also correlates with other risk factors and may lose predictive value when adjustment for other factors occurs *(37)*. It has been suggested that screening patients for CRP elevations may help clinicians identify less obvious patients who are candidates for more aggressive primary prevention strategies *(38)*.

TREATMENT

The lipid goals are presented in Table 5. Elevated LDL-C is the primary target of therapy with elevated non-HDL-C as a secondary target of therapy for patients with triglyceride levels ≥ 200 mg/dL. An optimal LDL-C is defined as < 100 mg/dL. Statin therapy is indicated in all high-risk women regardless of the LDL-C level. An optimal level of HDL-C is defined as > 50 mg/dL for women. An optimal non-HDL-C is 130 mg/dL. Niacin or fibrate therapy is indicated for women who are classified under high or intermediate risk and have a low HDL or an elevated non-HDL after LDL-C goal is reached *(16)*. Both genders respond equally to risk management. However, even when women are identified as having risk factors for CVD, there is lower utilization by physicians of standard accepted therapy compared with men *(39–41)*. In comparisons of women and men with similar cardiovascular risk profiles, women have been found to be significantly less likely than men to undergo additional coronary evaluation

Table 5
Lipid Goals

Parameter	*ATP III + Update[a]*	*Women (16)*	*ADA position (17)*
Optimal LDL-C	< 100 mg/dL	< 100 mg/dL	< 100 mg/dL
Very high risk (2004 Update) *(18)*	< 70 mg/dL		
Optimal TG	< 150 mg/dL	< 150 mg/dL	< 150 mg/dL
Optimal HDL-C	> 40 mg/dL	> 50 mg/dL	> 40 mg/dL men > 50 mg/dL women
LDL-C goal for CHD or equivalents	< 100 mg/dL	< 100 mg/dL	< 100 mg/dL
Non-HDL goal	< 130 mg/dL	< 130 mg/dL	

[a] Expert Panel. *JAMA* 2001;285:2486–2497.
CHD, cardiovascular disease; HDL, high-density lipoprotein; LDL, low-density lipoprotein; TG, triglyceride.

(38 versus 62%; $P = 0.002$) or coronary revascularization (2 versus 5%; $P = 0.03$) *(40)*. Less aggressive lipid-modifying strategies are employed when treating women when compared with men with similar risk profiles. In the Heart and Estrogen/progestin Replacement Study, approximately half of the women with established CVD were not receiving lipid-modifying medications *(41)*. In a study of 825 men and women with CHD, over 3 years, use of lipid-modifying therapy increased and low-density lipoprotein cholesterol (LDL-C) decreased in men, but utilization of therapy and, not surprisingly, LDL-C levels remained the same in women despite LDL-C levels above goal in the women under study *(39)*.

This may partially derive from the lack of relevant data. Dyslipidemia is one of the most important modifiable risk factors for CHD *(6,16)*. However, historically, women have been largely excluded from the original primary and secondary prevention trials; most available data on lipid modification in women come from subanalyses of the relatively small female populations enrolled in clinical trials that clearly show that women benefit from such therapy *(4,20–22)*. An additional explanation for laxity in the approach to CVD risk in women may be the widely held belief that cardiovascular risk is more time dependent in women, increasing markedly only after menopause *(42)*. The results of various large-scale observational studies documenting an increase in CVD after menopause and a decline in risk with the use of estrogen laid the foundation for starting cardiovascular risk assessment after menopause and initiating hormone replacement therapy (HRT) as a preventive strategy *(1,43,44)*. Given our present understanding of the progressive nature of atherosclerosis *(45)*, it now seems that the time demarcation at menopause may be an artificial distinction and that the presence of other factors in women may warrant the initiation of risk intervention strategies much earlier.

LIFESTYLE AND BEHAVIORAL CHANGES

The first step to reducing overall CHD risk involves therapeutic lifestyle changes. To this end, clinicians need to encourage patients to adopt a healthy lifestyle, including smoking cessation, low fat and possibly reduced carbohydrate diet, weight control, and regular physical activity. Specific therapeutic lifestyle recommendations are presented in detail in ATP III *(6)*. As a general recommendation, patients should reduce their dietary intake of saturated fats ($< 7\%$ of total calories) and cholesterol (< 200 mg/day), increase their intake of foods that lower LDL-C (plant stanols/sterols and soluble fiber), reduce weight, and incorporate regular physical activity into their daily routine (30 min or more on most days of the week) *(16)*. Patients with the metabolic syndrome should also consider reducing their carbohydrate intake. The Mediterranean Diet Study randomized 180 individuals with the metabolic syndrome to one of two diets for 2 years. The intervention diet included daily consumption of 250–300 g of fruits, 125–150 g of vegetables, 25–50 g of walnuts, and 400 g of whole grains, and subjects were encouraged to increase use of olive oil. The Control diet, with less-specific heart-healthy recommendations of 50–60% carbohydrate, 15–20% protein, and $< 30\%$ total fat, is consistent with the AHA step 1 diet recommended in the ATP III guidelines. Subjects in the intervention diet demonstrated a significant reduction in blood pressure, weight, waist circumference, triglyceride level, LDL-C level, and plasma glucose and CRP levels *(46)*.

DRUG THERAPY

Pharmacological treatment options exist for the management of dyslipidemia in both genders. It is recognized however that women and, importantly, minority women are less likely than men to receive to receive optimal lipid management, despite the fact that they receive equal benefit from lipid management *(47)*. Furthermore, giving attention to appropriate lipid management in the post-menopausal years is particularly relevant given that the Women's Health Initiative (WHI) demonstrated that HRT did not prevent cardiovascular events in women despite some beneficial lipid effects. Part of this study randomized 16,608 primary prevention post-menopausal women between the ages of 50 and 79 years to receive estrogen plus progesterone or placebo *(48)*. The primary efficacy outcome of the trial was CHD [nonfatal myocardial infarction (MI) or death due to CHD]. After a mean follow-up of 5.2 years, the Data Safety Monitoring Board (DSMB) recommended terminating this part of the study because of an increased incidence of coronary CHD, stroke, breast cancer, and thrombo-embolic events. Despite the fact that there may be some benefit for colorectal cancer, hip fractures, and total fractures, and the known benefit of HRT on lipoproteins, it is recommended that combined HRT is not used for CVD prevention in women *(48)*. More recently, the estrogen alone arm of the WHI, also failed to prevent coronary events and was associated with an increase in stroke *(49)*. A decrease in the risk of hip fracture was reported. These data have important clinical implications. As women deemed to be at risk for CVD discontinue HRT, their lipoprotein profiles deteriorate with increases in LDL-C and decreases in HDL-C becoming evident. Also up until the publication of these data, HRT was commonly prescribed for cardiovascular risk modification in female patients. Fortunately, several lipid-lowering trials have shown an unequivocal benefit of statin therapy in both men and women. Meta analysis of data from the five trials in which 30,817 participants were randomized to statin or control therapy for at least 4 years demonstrated that total cholesterol was reduced by 20%, LDL-C by 28%, and triglycerides by 13%, and HDL-C was increased by 5% *(49)*. Overall, statin treatment reduced risk of major coronary events by 31%, fatal CHD by 29%, and all-cause mortality by 21%. The Heart Protection Study randomized 5082 women (25%) and 15,454 men (75%) with known CHD between ages 40 and 80 years to receive simvastatin 40 mg or placebo in a 2×2 factorial design and were followed for 5 years. The study demonstrated that overall reductions in major vascular events with statin therapy were similar in both men and women (25 and 20%, respectively) and were unaffected by age *(23)*. Among the 3421 patients with entry LDL-C levels below 100 mg/dL, a similar reduction in risk of major events is seen when compared with those participants with higher baseline LDL levels. This effect was irrespective of gender. Some of this data is shown in Fig. 4. Two recent trials, the Prospective Study of Pravastatin in Elderly at Risk *(50)* and the Anglo-Scandinavian Cardiac Outcomes Trial – Lipid Lowering Arm *(51)*, had contradictory findings, which showed no benefit with statin therapy in women. These trials were of shorter duration, however, accruing a smaller number of events in the women studied.

Statins have proved to be extremely safe and well tolerated in the majority of patients. Their most common serious adverse effects, hepatotoxicity and myopathy, occur at very low rates *(52)*. The risk of myopathy increases with advanced age, especially in women, in patients with multisystem disease, and in patients taking specific concomitant medications *(52)*. With appropriate care, statins can be used safely in these patients. Statins, however, should not be used in pregnant women, as they have

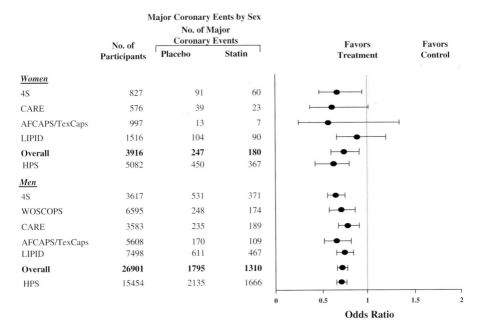

Fig. 4. Effects of statin therapy on cardiovascular disease based on gender.

not been tested in this population. Women of childbearing potential who use statins should be counseled about the need for adequate contraception and prenatal planning.

Whilst the benefit of aggressive LDL-C lowering is clear, cardiovascular events continue to occur. This should not be surprising given the known risks associated with other lipid abnormalities and other co-morbidities such as hypertension and diabetes. Many of our high-risk patients have multiple lipid abnormalities. Both the ATP III and the AHA women's guidelines recommend addressing risk beyond LDL-C. Specifically, the AHA women's guideline recommends treatment of low HDL-C with fibrate or Niacin therapy.

Fibrates are peroxisome proliferator-activated receptor (PPAR)-α agonists, a nuclear transcription factor involved in lipid metabolism and activated by fatty acids *(53)*. Fibrates increase fatty acid oxidation reducing the formation of very-low-density lipoprotein (VLDL) triglycerides, thereby reducing serum triglyceride levels. Fibrates raise HDL-C by increasing the synthesis of apolipoprotein A-I and A-II and reducing HDL catabolism. Fibrates are primarily effective in lowering triglyceride levels but also have moderate HDL-C-raising and LDL-C-lowering effects, and they increase LDL particle size *(16,53)*.

Reductions in cardiovascular events associated with fibrate therapy have been documented mainly in men. In the Helsinki Heart Study, gemfibrozil reduced the incidence of CHD events by 34% ($P < 0.02$) in 4081 men with no symptoms of CHD *(54)*. Data supporting intervention beyond LDL in high-risk patients with the metabolic syndrome come in part from the VA-HIT study *(55)*. This study demonstrated that treatment with a fibrate in men with established CVD was associated with a substantial reduction in both nonfatal MI and CHD death (primary endpoint) and also the individual endpoint of stroke. The 22% reduction in the primary endpoint was associated with a 6% increase in HDL-C and a 31% reduction in triglyceride levels with no change in LDL-C – demonstrating for the first time that raising HDL-C, with no change in LDL-C, was associated with a reduction in cardiovascular events *(56)*. A post hoc analysis of the data reported that the vast majority of the benefit was seen in subjects with hyperinsulinemia and/or diabetes *(57)*. The limited outcome data pertaining to women have generally shown benefit from treatment with fibrates. In the Stockholm Ischaemic Heart Disease Secondary Prevention Study, which included 113 women with a history of MI, the combination of clofibrate and niacin reduced total mortality in the study population overall by 26% ($P < 0.05$) *(58)*. Men and women showed equivalent benefits in the Diabetes Atherosclerosis Intervention Study, which included 113 women with diabetes and CHD *(59)*. Treatment with fenofibrate in this study reduced angiographic progression compared with placebo. Although the study was not powered for clinical outcomes, the fenofibrate group also experienced fewer cardiac events (38 versus 50). By contrast, the Coronary Drug Project study with clofibrate failed to demonstrate any short- or long-term risk reduction, an outcome that is presumed likely attributable to the study drug *(60)*. The Bezafibrate Infarction Prevention Study, in which 9% of the study population was female, also showed no significant reduction in the risk of MI *(61)*.

Conflicting data from the Field Trial have recently been reported *(62)*. This study was undertaken to examine the effect of fenofibrate in diabetic patient. Over 9000 patients were randomized to receive 200 mg of fenofibrate or placebo. Seventy three percent of these subjects had no known history of CVD. The study failed to demonstrate a statistical difference in the primary out come of combined fatal and nonfatal MI. There was a slight increase in cardiovascular death and a reduction in retinopathy and nephropathy. Efforts to explain these somewhat disappointing data include the following: the patients did not have dyslipidemia at study entry, and although patients could not be on a statin at the time of randomization, the caring physician could add statin therapy at his or her discretion throughout the study. This occurred in 18% of the patients in the placebo group and 7% of the patients in the fenofibrate group. It has been suggested that this caused a blunting of the actual difference in response in the two groups. Furthermore, the number of events seen in both groups was smaller that had been anticipated for this diabetic population at the time of study design, and finally, HDL-C was increased by only 2%. In reality, however, these data are unlikely to affect our clinical practice because diabetic patients are treated with statin therapy irregardless of LDL-C levels and consideration of adjunctive therapy for optimizing lipid management would include the addition of a fibrate or niacin. Importantly, the Action to Control CardioVascular Risk in Diabetes (ACCORD) trial is currently underway to examine the clinical effects of combining statin and fibrate therapy on cardiovascular outcomes.

Fibrates are generally well tolerated. Gastrointestinal complaints are the most common side effects; they may also increase the risk of cholesterol gallstones because they enhance the lithogenicity of bile. In addition, fibrates are strongly protein bound and may interact with other protein-binding drugs such as warfarin. The main concern is an increased risk of myopathy when fibrates are used in combination with statins, particularly in patients with renal impairment. In 871 reports of statin-associated rhabdomyolysis, concomitant use of fibrates was listed in 80 cases (9.2%) *(63)*. However, this combination can be used safely with appropriate monitoring *(6,52)*. Fenofibrate may carry a lower safety risk than gemfibrozil in combination therapy with a statin *(64)*.

Niacin reduces the hepatic synthesis of VLDL and triglycerides by inhibiting mobilization of free fatty acids from peripheral adipose tissue to the liver. Because less VLDL is available to be used as a substrate, LDL-C levels fall. By blocking hepatic uptake of apolipoprotein A-1, niacin also decreases HDL-C clearance, thereby increasing the amount of HDL available for reverse cholesterol transport. Among available lipid-modifying drugs, niacin has the greatest effect on HDL-C; it also effectively decreases triglycerides, LDL-C, and lipoprotein (a) *(6)*. Niacin has also been shown to increase LDL particle size, which may make it a particularly good choice when the atherogenic lipid triad is present *(65)*.

The Coronary Drug Project, which recruited 8341 men from 1966–1969, demonstrated after 5 years of follow-up that treatment with niacin was associated with a significant reduction of nonfatal MI (27%), and CHD death or nonfatal MI (14%). Whilst this study is often overshadowed by the modern-day statin trials, these results are similar both in absolute and relative magnitude *(60)*. Importantly, even after 15 years, there was a reduction in total mortality associated with the use of niacin compared with placebo *(66)*. There have been no long-term clinical trials of niacin monotherapy in women. The majority of the data regarding niacin's effect on CVD derives from studies in which it has been used in combination with other therapies, as will be discussed in a later section. Current data suggest that women tend to achieve greater LDL-C reductions with niacin than do men *(67,68)*. It is possible, therefore, that given the relative prognostic significance of HDL-C and triglycerides in women added to the gender-based difference in LDL-C lowering, niacin may be especially important in women because of its effects on HDL-C and lipoprotein (a). The more substantial LDL-C response to niacin in women may allow use of lower doses which will be clinically relevant given the tolerability profile for this drug.

While niacin has a beneficial effect on the overall lipid profile, its use has been severely limited by side effects, namely, prostaglandin-mediated facial and truncal flushing. This effect occurs equally in men and women. Both the incidence and the severity of this side effect can be reduced by starting with a low dose of extended release niacin and slowly titrating up and by avoiding compounds that are known to increase flushing such as caffeine-containing beverages, alcohol, and spicy foods. Flushing can also be dramatically reduced by taking nonenteric coated aspirin or ibuprofen 30–60 minutes prior to taking the first niacin dose *(69)*. With continued therapy, the incidence and severity of flushing frequently attenuate. Tolerability and safety of extended release niacin has improved our ability to utilize this medication in clinical practice. Other dose-dependent side effects related to niacin use include hyperuricemia and gout, and hyperglycemia. Sustained release formulations have been associated with hepatotoxicity *(70)*. Finally, niacin has been associated, although rarely, with statin-induced myopathy *(52)*. Whether niacin actually increases this risk is unclear; in one phase IV post-marketing surveillance study, only 4 of the 871 reports of statin-induced rhabdomyolysis to the Food and Drug Administration involved concomitant niacin *(63)*.

Bile Acid Sequestrants/Cholesterol Absorption Inhibitors

Bile acid sequestrants (BASs) are anion exchange resins that bind bile acids in the ileum preventing their reabsorption and reducing enterohepatic reabsorption of bile salts. By decreasing circulating levels of bile acids, through feedback regulation, there is an increase in the conversion of hepatic cholesterol into bile acids. Decreased intrahepatic cholesterol results in up regulation of the LDL-C receptor on heptocytes with resultant increased uptake of LDL-C from the plasma. BASs produce minimal effects on HDL-C and triglycerides. However, because they can potentially raise triglycerides, their use is contraindicated in individuals with significant elevations in triglyceride levels (> 400 mg/dL) or familial dysbetalipoproteinemia *(6)*. BASs have been shown to reduce cardiovascular events, total mortality, and atherosclerotic progression in clinical trials involving men *(71,72)*. In addition to binding cholesterol, BASs can also reduce the absorption of numerous drugs through the same mechanism. Thus, concomitant medications should be administered at least 1 h before or 4 h after dosing. In this class, colesevelam is the least likely to bind with other drugs in the gut *(73)*. BASs are limited by gastrointestinal side effects; however, because they lack systemic effects, they are a useful therapeutic option in women of childbearing potential and in patients who have demonstrated hepatotoxicity in response to statin therapy and have been unable to tolerate niacin *(6,74)*.

Cholesterol absorption inhibitors inhibit the intestinal absorption of cholesterol. The reduction in delivery of intestinal cholesterol to the liver results in a reduction of hepatic cholesterol stores and therefore an increase in the hepatic uptake of cholesterol from the blood leading to lower serum levels. Ezetimibe, the only available selective cholesterol absorption inhibitor, is thought to involve selective inhibition of the putative sterol transporter on the brush border surface of intestinal epithelial cells *(75)*. Ezetimibe has been shown in clinical studies of men and women to significantly reduce LDL-C although having minimal effects on HDL-C and triglycerides *(75,76)*. Safety data have been principally derived from short-term clinical studies, in which adverse effects and discontinuation rates were similar to those with placebo *(75,76)*. Clinical outcome studies have yet to be completed. Ezetimibe has not been adequately evaluated in either pregnant or nursing women. This is also a reasonable alternative for patients who are intolerant of statin therapy.

COMBINATION THERAPY FOR THE TREATMENT OF DYSLIPIDEMIA

Cardiovascular events continue to occur despite aggressive LDL-C lowering. There is an increasing body of evidence from numerous "proof of concept" studies in support of combination drug therapy in the management of dyslipidemia in high-risk patients. Two large randomized clinical endpoint studies of combination regimens are underway. Whilst we await these results, we must examine the available data that evaluate this type of approach. The majority of the evidence comes from studies of combining statin therapy together with niacin.

The combination of a statin plus niacin is perhaps one of the most useful combinations for treating dyslipidemia in women, as it adds the favorable effects of niacin on atherogenic dyslipidemia to the LDL-C-lowering action of statins. In numerous clinical studies, this combination has shown improvements across the lipid profile *(68,77–79)*. In a small study of fluvastatin and niacin, LDL-C levels were reduced by 54.6% in women versus 38.2% in men ($P < 0.0005$) *(68)*. The FATS trial combination therapy with niacin, colestipol, and/or lovastatin also resulted in a better reduction in coronary artery stenosis in women than men among individuals with familial hypercholesterolemia *(80)*. A 10-year evaluation of therapy with lovastatin, niacin, and colestipol in these same patients subsequently enrolled in the Familial Atherosclerosis 10-year follow-up showed that this triple therapy was associated with a significant reduction in LDL-C (−48%) and triglyceride levels (−36%), an increase in HDL-C (+23%), and a significantly lower rate of death and cardiovascular events ($P < 0.05$) *(81)*. The HDL-Atherosclerosis Treatment Study (HATS) was a placebo-controlled secondary prevention study of 160 patients with CHD designed to look at the impact of the combination of simvastatin and niacin with or without antioxidant vitamins on the progression of CHD *(82)*. The primary endpoint was the mean change in the percent stenosis caused by the most severe lesion from the initial arteriogram to the final arteriogram. The mean HDL was 31 mg/dL at baseline and was increased by 24%. The mean LDL was 125 mg/dL at baseline and was decreased by 42%. A significant reduction in events was was associated with the use of simvastatin and niacin in combination. Antioxidant vitamins appeared to attenuate the beneficial effects of simvastatin and niacin on cardiovascular outcome *(82)*. This may possibly be related to the fact that antioxidant vitamins may interfere with the HDL-raising impact of niacin therapy *(83)*. The University of California, San Francisco, Arteriosclerosis Specialized Center of Research trial demonstrated the benefit of combination therapy with an LDL-C-lowering drug and niacin on clinical outcomes as well as on regression of atherosclerotic lesions *(84)*. Considering that most of the patients in University of California, San Francisco, Arteriosclerosis Specialized Center of Research (92%) were asymptomatic, the findings suggest that combination lipid-modifying therapy can be beneficial to patients with subclinical atherosclerosis who are at high risk.

A large phase III randomized NIH sponsored study, the AIM HIGH study, is underway to evaluate the effect of simvastatin combined with extended release niacin on a wide range of vascular endpoints.

Statin/Fibrate

Combining statin therapy with a fibrate is another good option for patients with mixed dyslipidemia *(85–87)*. Although no outcome studies have been performed with this combination, one study showed a reduction in projected coronary risk *(86)*. As mentioned, myopathy is a potential concern when using this combination, but the risks can be attenuated by avoiding use in patients with renal impairment and appropriately monitoring patients and using moderate doses *(52,88)*.

Statin/BAS or Statin/Cholesterol Absorption Inhibitors

Adding a BAS or ezetimibe to statin therapy may be a good option for patients who need additional LDL-C lowering but do not have abnormalities of other lipoproteins. The combination of pravastatin plus cholestyramine was found in one study to be more effective than monotherapy with either agent alone in lowering LDL-C in both men and women *(89)*. Ezetimibe added to ongoing statin therapy resulted in significant reductions in LDL-C of 25% and triglycerides of 14% and increases in HDL-C of 2.7%; all changes were significantly greater than those seen with statin monotherapy *(90)*. Clinical endpoint data are not yet available for this approach.

Niacin/BAS

Niacin plus a BAS may be suitable for patients who are refractory to or intolerant of statins. This combination has been shown to favorably alter the lipid profile, reduce cardiovascular events, and promote regression of atherosclerotic plaque; these data are in an all-male population *(72,91)*.

REFERENCES

1. American Heart Association. Heart Disease and Stroke Statistics – 2005 Update. American Heart Association: 2005. Available at: http://www.americanheart.org/downloadable/heart/1040391091015 HDSStats 05.pdf.
2. Mosca L, Jones WK, King KB, Ouyang P, Redberg RF, Hill MN, for the American Heart Association Women's Heart Disease and Stroke Campaign Task Force. Awareness, perception, and knowledge of heart disease risk and prevention among women in the United States. *Arch Fam Med* 2000;9:506–515.
3. Mosca L, Ferris A, Fabunmi R, Robertson RM. Tracking women's awareness of heart disease: an American Heart Association National Study. *Circulation* 2004;109:573–579.
4. Lewis SJ, Sacks FM, Mitchell JS, et al., for The CARE Investigators. Effect of pravastatin on cardiovascular events in women after myocardial infarction: the cholesterol and recurrent events (CARE) trial. *J Am Coll Cardiol* 1998;32:140–146.
5. Ford ES, Giles WH, Dietz WH. Prevalence of the metabolic syndrome among US Adults: findings from the Third National Health and Nutrition Examination Survey. *JAMA* 2002;287:356–359.
6. Expert Panel on Detection, Evaluation, and Treatment of High Blood Cholesterol in Adults (Adult Treatment Panel III). Third Report of the National Cholesterol Education Program (NCEP) expert panel on detection, evaluation, and treatment of high blood cholesterol in adults (Adult Treatment Panel III) (final report). *Circulation* 2002;106:3143–3421.
7. Sattar N, Gaw A, Scherbakova O, et al. Metabolic syndrome with and without C-reactive protein as a predictor of coronary heart disease and diabetes in the West of Scotland Coronary Prevention Study. *Circulation* 2003;108:414–419.
8. Mokdad AH, Ford ES, Bowman BA, et al. Prevalence of obesity, diabetes, and obesity-related health risk factors, 2001. *JAMA* 2003;289:76–79.
9. Wenger NK. Hypertension and other cardiovascular risk factors in women. *Am J Hypertens* 1995;8:94s–99s.
10. Kannel WB, McGee DL. Diabetes and cardiovascular disease: the Framingham study. *JAMA* 1979;241:2035–2038.
11. Kannel WB, McGee DL. Diabetes and glucose tolerance as risk factors for cardiovascular disease: the Framingham study. *Diabetes Care* 1979;2:120–126.
12. UK Prospective Diabetes Study 27. Plasma lipids and lipoproteins at diagnosis of NIDDM by age and sex. *Diabetes Care* 1997;20:1683–1687.
13. Siegel RD, Cupples A, Schaefer EJ, Wilson PW. Lipoproteins, apolipoproteins, and low-density lipoprotein size among diabetics in the Framingham offspring study. *Metabolism* 1996;45(10):1267–1272.
14. Isomaa B, Almgren P, Tuomi T, et al. Cardiovascular morbidity and mortality associated with the metabolic syndrome. *Diabetes Care* 2001;24(4):683–689.
15. Castelli WP, Garrison RJ, Wilson PWF, Abbott RD, Kalousdian S, Kannel WB. Incidence of coronary heart disease and lipoprotein cholesterol levels. The Framingham Study. *JAMA* 1986;256:2835–2838.
16. Mosca L, Appel LJ, Benjamin EJ, et al. Evidence-based guidelines for cardiovascular disease prevention in women. *Circulation* 2004;109:672–693.
17. American Diabetes Association. Dyslipidemia management in Adults with Diabetes. *Diabetes Care* 2004;27(suppl 1):S68–S71.
18. Grundy SM, Cleeman JI, Merz CN, et al. Implications of recent clinical trials for the National Cholesterol Education Program Adult Treatment Panel III guidelines. *Circulation* 2004;110:227–239.
19. Bass KM, Newschaffer CJ, Klag MJ, Bush TL. Plasma lipoprotein levels as predictors of cardiovascular death in women. *Arch Intern Med* 1993;153:2209–2216.
20. Clearfield M, Downs JR, Weis S, et al. Air Force/Texas Coronary Atherosclerosis Prevention Study (AFCAPS/TexCAPS): efficacy and tolerability of long-term treatment with lovastatin in women. *J Womens Health Gend Based Med* 2001;10:971–981.
21. Miettinen TA, Pyörälä K, Olsson AG, et al., for the Scandinavian Simvastatin Study Group. Cholesterol-lowering therapy in women and elderly patients with myocardial infarction or angina pectoris: findings from the Scandinavian Simvastatin Survival Study (4S). *Circulation* 1997;96:4211–4218.
22. Long-Term Intervention with Pravastatin in Ischaemic Disease (LIPID) Study Group. Prevention of cardiovascular events and death with pravastatin in patients with coronary heart disease and a broad range of initial cholesterol levels. *N Engl J Med* 1998;339:1349–1357.

23. Heart Protection Study Collaborative Group. MRC/BHF Heart Protection Study of cholesterol lowering with simvastatin in 20536 high-risk individuals: a randomised placebo-controlled trial. *Lancet* 2002;360:7–22.
24. Downs JR, Clearfield M, Weis S, et al., for the Air Force/Texas Coronary Atherosclerosis Prevention Study. Primary prevention of acute coronary events with lovastatin in men and women with average cholesterol levels: results of AFCAPS/TexCAPS. *JAMA* 1998;279:1615–1622.
25. Scandinavian Simvastatin Survival Study Group. Randomised trial of cholesterol lowering in 4444 patients with coronary heart disease: the Scandinavian Simvastatin Survival Study (4S). *Lancet* 1994;344:1383–1389.
26. Shepherd J, Cobbe SM, Ford I, et al., for the West of Scotland Coronary Prevention Study Group. Prevention of coronary heart disease with pravastatin in men with hypercholesterolemia. *N Engl J Med* 1995;333:1301–1307.
27. Blake GJ, Otvos JD, Rifai N, Ridker PM. Low-density lipoprotein particle concentration and size as determined by nuclear magnetic resonance spectroscopy as predictors of cardiovascular disease in women. *Circulation* 2002;106:1930–1937.
28. Carr MC, Kim KH, Zambon A, et al. Changes in LDL density across the menopausal transition. *J Investig Med* 2000;48:245–250.
29. Cui Y, Blumenthal RS, Flaws JA, et al. Non-high-density lipoprotein cholesterol level as a predictor of cardiovascular disease mortality. *Arch Intern Med* 2001;161:1413–1419.
30. Castelli WP, Anderson K, Wilson PWF, Levy D. Lipids and risk of coronary heart disease: the Framingham Study. *Ann Epidemiol* 1992;2:23–28.
31. Castelli WP. Epidemiology of triglycerides: a view from Framingham. *Am J Cardiol* 1992;70:3H–9H.
32. Hokanson JE, Austin MA. Plasma triglyceride level is a risk factor for cardiovascular disease independent of high-density lipoprotein cholesterol level: a meta-analysis of population-based prospective studies. *J Cardiovasc Risk* 1996;3:213–219.
33. Jensen J, Nilas L, Christiansen C. Influence of menopause on serum lipids and lipoproteins. *Maturitas* 1990;12(4):321–331.
34. Shlipak MG, Simon JA, Vittinghoff E, et al. Estrogen and progestin, lipoprotein(a), and the risk of recurrent coronary heart disease events after menopause. *JAMA* 2000;283:1845–1852.
35. Bermudez EA, Rifai N, Buring J, Manson JE, Ridker PM. Interrelationships among circulating interleukin-6, C-reactive protein, and traditional cardiovascular risk factors in women. *Arterioscler Thromb Vasc Biol* 2002;22:1668–1673.
36. Ridker PM, Hennekens CH, Buring JE, Rifai N. C-reactive protein and other markers of inflammation in the prediction of cardiovascular disease in women. *N Engl J Med* 2000;342:836–843.
37. Mosca L. C-reactive protein – to screen or not to screen? *N Engl J Med* 2002;347:1615–1617.
38. Pearson TA, Mensah GA, Alexander RW, et al. Markers of inflammation and cardiovascular disease: application to clinical and public health practice: a statement for healthcare professionals from the Centers for Disease Control and Prevention and the American Heart Association. *Circulation* 2003;107:499–511.
39. Miller M, Byington R, Hunninghake D, Pitt B, Furberg CD, for the Prospective Randomized Evaluation of the Vascular Effects of Norvasc Trial (PREVENT) Investigators. Sex bias and underutilization of lipid-lowering therapy in patients with coronary artery disease at academic medical centers in the United States and Canada. *Arch Intern Med* 2000;160:343–347.
40. Shaw LJ, Miller DD, Romeis JC, Kargl D, Younis LT, Chaitman BR. Gender differences in the noninvasive evaluation and management of patients with suspected coronary artery disease. *Ann Intern Med* 1994;120:559–566.
41. Schrott HG, Bittner V, Vittinghoff E, Herrington DM, Hulley S, for the HERS Research Group. Adherence to National Cholesterol Education Program Treatment goals in postmenopausal women with heart disease: the Heart and Estrogen/Progestin Replacement Study (HERS). *JAMA* 1997;277:1281–1286.
42. Mosca L, Manson JE, Sutherland SE, Langer RD, Manolio T, Barrett-Connor E. Cardiovascular disease in women: a statement for healthcare professionals from the American Heart Association. *Circulation* 1997;96:2468–2482.
43. Grady D, Rubin SM, Petitti DB, et al. Hormone therapy to prevent disease and prolong life in postmenopausal women. *Ann Intern Med* 1992;117:1016–1037.
44. Stampfer MJ, Colditz GA. Estrogen replacement therapy and coronary heart disease: a quantitative assessment of the epidemiologic evidence. *Prev Med* 1991;20:47–63.
45. Lee RT, Libby P. The unstable atheroma. *Arterioscler Thromb Vasc Biol* 1997;17:1859–1867.

46. Esposito K, Esposito K, Marfella R, et al. Effect of a Mediterranean-style diet on endothelial dysfunction and markers of vascular inflammation in the metabolic syndrome: a randomized trial. *JAMA* 2004;292:1440–1446.
47. Basaria S, Dobs AS. Treatment of hyperlipidemia in women. *Int J Fertil Womens Med* 2000; 45:22–33.
48. Rossouw JE, Anderson GL, Prentice RL, et al., for the Writing Group for the Women's Health Initiative Investigators. Risks and benefits of estrogen plus progestin in healthy postmenopausal women: principal results from the Women's Health Initiative randomized controlled trial. *JAMA* 2002;288:321–333.
49. Anderson GL, Limacher M, Assaf AR, et al. Effects of conjugated equine estrogen in postmenopausal women with hysterectomy: the Women's Health Initiative randomized controlled trial. *JAMA* 2004;291:1701–1712.
50. Shepherd J, Blauw GJ, Murphy MB, et al., on behalf of the PROSPER study group. Pravastatin in elderly individuals at risk of vascular disease (PROSPER): a randomised controlled trial. *Lancet* 2002;360:1623–1630.
51. Sever PS, Dahlöf B, Poulter NR, et al., for the ASCOT investigators. Prevention of coronary and stroke events with atorvastatin in hypertensive patients who have average or lower-than-average cholesterol concentrations, in the Anglo-Scandinavian Cardiac Outcomes Trial – Lipid Lowering Arm (ASCOT-LLA): a multicentre randomised controlled trial. *Lancet* 2003;361:1149–1158.
52. Pasternak RC, Smith SC, Jr, Bairey-Merz CN, Grundy SM, Cleeman JI, Lenfant C. ACC/AHA/NHLBI clinical advisory on the use and safety of statins. *J Am Coll Cardiol* 2002;40: 567–572.
53. Fruchart JC, Duriez P, Staels B. Peroxisome proliferator-activated receptor-alpha activators regulate genes governing lipoprotein metabolism, vascular inflammation and atherosclerosis. *Curr Opin Lipidol* 1999;10:245–257.
54. Frick MH, Elo O, Haapa K, et al. Helsinki Heart Study: primary-prevention trial with gemfibrozil in middle-aged men with dyslipidemia. Safety of treatment, changes in risk factors, and incidence of coronary heart disease. *N Engl J Med* 1987;317:1237–1245.
55. Rubins HB, Robins SJ, Collins D, et al., for the Veterans Affairs High-Density Lipoprotein Cholesterol Intervention Trial Study Group. Gemfibrozil for the secondary prevention of coronary heart disease in men with low levels of high-density lipoprotein cholesterol. *N Engl J Med* 1999;341: 410–418.
56. Robins SJ, Collins D, Wittes JT, et al., for the VA-HIT study group. Relation of gemfibrozil treatment and lipid levels with major coronary events. VA-HIT: a randomized controlled trial. *JAMA* 2001;285:1585–1591.
57. Rubins HB, Robins SJ, Collins D, et al. Diabetes, plasma insulin, and cardiovascular disease: subgroup analysis from the Department of Veterans Affairs high-density lipoprotein intervention trial (VA-HIT). *Arch Intern Med* 2002;162:2597–2604.
58. Carlson LA, Rosenhamer G. Reduction of mortality in the Stockholm Ischaemic Heart Disease Secondary Prevention Study by combined treatment with clofibrate and nicotinic acid. *Acta Med Scand* 1988;223:405–418.
59. Diabetes Atherosclerosis Intervention Study Investigators. Effect of fenofibrate on progression of coronary-artery disease in type 2 diabetes: the Diabetes Atherosclerosis Intervention Study, a randomised study. *Lancet* 2001;357:905–910.
60. The Coronary Drug Project Research Group. Clofibrate and niacin in coronary heart disease. *JAMA* 1975;231:360–381.
61. BIP Study Group. Secondary prevention by raising HDL cholesterol and reducing triglycerides in patients with coronary artery disease: the Bezafibrate Infarction Prevention (BIP) study. *Circulation* 2000;102:21–27.
62. Keech A, Simes RJ, Barter P, et al., for FIELD study investigators. Effects of long-term fenofibrate therapy on cardiovascular events in 9795 people with type 2 diabetes mellitus (the FIELD study): randomised controlled trial. *Lancet* 2005;366:1849–1861.
63. Omar MA, Wilson JP. FDA adverse event reports on statin-associated rhabdomyolysis. *Ann Pharmacother* 2002;36:288–295.
64. Davidson MH. Combination therapy for dyslipidemia: safety and regulatory considerations. *Am J Cardiol* 2002;90(suppl):50K–60K.
65. Superko HR, Krauss RM. Differential effects of nicotinic acid in subjects with different LDL subclass patterns. *Atherosclerosis* 1992;95:69–76.

66. Canner PL, Berge KG, Wenger NK, et al., for the Coronary Drug Project Research Group. Fifteen year mortality in Coronary Drug Project patients: long-term benefit with niacin. *J Am Coll Cardiol* 1986;8:1245–1255.
67. Goldberg AC. Clinical trial experience with extended-release niacin (Niaspan): dose-escalation study. *Am J Cardiol* 1998;82:35U–38U.
68. Jacobson TA, Jokubaitis LA, Amorosa LF. Fluvastatin and niacin in hypercholesterolemia: a preliminary report on gender differences in efficacy. *Am J Med* 1994;96:64S–68S.
69. Whelan AM, Price SO, Fowler SF, Hainer BL. The effect of aspirin on niacin-induced cutaneous reactions. *J Fam Pract* 1992;34:165–168.
70. McKenney JM, Proctor JD, Harris S, Chinchili VM. A comparison of the efficacy and toxic effects of sustained- vs immediate-release niacin in hypercholesterolemic patients. *JAMA* 1994;271:672–677.
71. Lipid Research Clinics Program. The Lipid Research Clinics Coronary Primary Prevention Trial results. II. The relationship of reduction in incidence of coronary heart disease to cholesterol lowering. *JAMA* 1984;251:365–374.
72. Blankenhorn DH, Nessim SA, Johnson RL, Sanmarco ME, Azen SP, Cashin-Hemphill L. Beneficial effects of combined colestipol-niacin therapy on coronary atherosclerosis and coronary venous bypass grafts. *JAMA* 1987;257:3233–3240.
73. Donovan JM, Stypinski D, Stiles MR, Olson TA, Burke SK. Drug interactions with colesevelam hydrochloride, a novel, potent lipid-lowering agent. *Cardiovasc Drugs Ther* 2000;14:681–690.
74. Mosca LJ. Optimal management of cholesterol levels and the prevention of coronary heart disease in women. *Am Fam Physician* 2002;65:217–226.
75. Dujovne CA, Ettinger MP, McNeer JF, et al., for the Exetimibe Study Group. Efficacy and safety of a potent new selective cholesterol absorption inhibitor, *ezetimibe*, in patients with primary hypercholesterolemia. *Am J Cardiol* 2002;90:1092–1097.
76. Bays HE, Moore PB, Drehobl MA, et al., for the Ezetimibe Study Group. Effectiveness and tolerability of ezetimibe in patients with primary hypercholesterolemia: pooled analysis of two phase II studies. *Clin Ther* 2001;23:1209–1230.
77. Davignon J, Roederer G, Montigny M, et al. Comparative efficacy and safety of pravastatin, nicotinic acid and the two combined in patients with hypercholesterolemia. *Am J Cardiol* 1994;73:339–345.
78. Kashyap ML, McGovern ME, Berra K, et al. Long-term safety and efficacy of a once-daily niacin/lovastatin formulation in patients with dyslipidemia. *Am J Cardiol* 2002;89:672–678.
79. Bays HE, Dujovne CA, McGovern ME, et al. Comparison of once-daily, niacin extended-release/lovastatin with standard doses of atorvastatin and simvastatin (the Advicor Versus Other Cholesterol-Modulating Agents Trial Evaluation [ADVOCATE]). *Am J Cardiol* 2003;91:667–672.
80. Brown G, Albers JJ, Fisher LD, et al. Regression of coronary artery disease as a result of intensive lipid-lowering therapy in men with high levels of apolipoprotein B. *N Engl J Med* 1990;323:1289–1298.
81. Brown BG, Brockenbrough A, Zhao XQ, et al. Very intensive lipid therapy with lovastatin, niacin, and colestipol for prevention of death and myocardial infarction: a 10-year familial atherosclerosis treatment study (FATS) follow-up [abstract 3341]. *Circulation* 1998;98(suppl I):635.
82. Brown BG, Zhao XQ, Chait A, et al. Simvastatin and niacin, antioxidant vitamins, or the combination for the prevention of coronary disease. *N Engl J Med* 2001;345:1583–1592.
83. Cheung M, Zhao XQ, Chait A, Albers JJ, Brown BG. Antioxidant supplements block the response of HDL to simvastatin-niacin therapy in patients with coronary artery disease and low HDL. *Arterioscler Thromb Vasc Biol* 2001;21:1320–1326.
84. Kane JP, Malloy MJ, Ports TA, Phillips NR, Diehl JC, Havel RJ. Regression of coronary atherosclerosis during treatment of familial hypercholesterolemia with combined drug regimens. *JAMA* 1990;264:3007–3012.
85. Murdock DK, Murdock AK, Murdock RW, et al. Long-term safety and efficacy of combination gemfibrozil and HMG-CoA reductase inhibitors for the treatment of mixed lipid disorders. *Am Heart J* 1999;138:151–155.
86. Athyros VG, Papageorgiou AA, Hatzikonstandinou HA, et al. Safety and efficacy of long-term statin-fibrate combinations in patients with refractory familial combined hyperlipidemia. *Am J Cardiol* 1997;80:608–613.
87. Iliadis EA, Rosenson RS. Long-term safety of pravastatin-gemfibrozil therapy in mixed hyperlipidemia. *Clin Cardiol* 1999;22:25–28.
88. Shek A, Ferrill MJ. Statin-fibrate combination therapy. *Ann Pharmacother* 2001;35:908–917.

89. Eriksson M, Hadell K, Holme I, Walldius G, Kjellström T. Compliance with and efficacy of treatment with pravastatin and cholestyramine: a randomized study on lipid-lowering in primary care. *J Intern Med* 1998;243:373–380.
90. Gagne C, Bays HE, Weiss SR, et al., for the Ezetimibe Study Group. Efficacy and safety of ezetimibe added to ongoing statin therapy for treatment of patients with primary hypercholesterolemia. *Am J Cardiol* 2002;90:1084–1091.
91. Cashin-Hemphill L, Mack WJ, Pogoda JM, Sanmarco ME, Azen SP, Blankenhorn DH. Beneficial effects of colestipol-niacin on coronary atherosclerosis: a 4-year follow-up. *JAMA* 1990;264: 3013–3017.

17 Management of Lipids in the Elderly

Micah J. Eimer, MD, *and Neil J. Stone,* MD

CONTENTS

> INTRODUCTION
> REDUCING RISK IN THE ELDERLY THROUGH LIPID
>   TREATMENT
> REVIEW OF RELEVANT STUDIES PUBLISHED AFTER
>   ATP III
> SAFETY CONCERNS
> PREDICTING RISK IN THE ELDERLY
> MEASURES OF SUBCLINICAL CVD IN OLDER ADULTS
> KEY POINTS
> REFERENCES
> ELDERLY CASE HISTORIES

INTRODUCTION

Adult Treatment Panel (ATP) III defined those men aged 65 years and older and women aged 75 years and older as elderly *(1)*. Risk stratification for this group is similar to younger groups with one important difference: simply on the basis of age, the elderly, as defined above, would appear to have at least intermediate risk of coronary heart disease (CHD). For example, we noted that a 75-year-old man with no additional risk factors would have a 10-year risk of CHD of 12% *(2)*. This underscores the elevated CHD risk in the elderly even before considering other CHD risk factors present. Yet, elevated age alone is not enough to recommend intensive treatment for abnormal lipid/lipoprotein values. Physicians need to review the available data to identify those at risk who might, based on the evidence, benefit from a more intensive risk reduction approach.

REDUCING RISK IN THE ELDERLY THROUGH LIPID TREATMENT

LaRosa et al. *(3)* performed a rigorous meta-analysis of major clinical statin trials that preceded the recent ATP III report. It was especially helpful because it examined understudied subgroups such as women and the elderly. The authors used the following strict inclusion criteria: (1) placebo controlled and randomized to reduce bias;

From: *Contemporary Cardiology: Therapeutic Lipidology*
Edited by: M. H. Davidson, P. P. Toth, and K. C. Maki © Humana Press Inc., Totowa, NJ

(2) no other concomitant intervention; (3) at least 4 years of follow-up; and (4) clinical disease or death as endpoints.

The two primary prevention trials examined were the West of Scotland Coronary Prevention Study (WOSCOPS) *(4)* and the Air Force/Texas Coronary Atherosclerosis Prevention Study (AFCAPS/TexCAPS) *(5)*. The three secondary prevention trials were the Scandinavian Simvastatin Survival Study Group (4S) *(6)*, Long-Term Intervention with Pravastatin in Ischaemic Disease (LIPID) *(7)*, and the Cholesterol and Recurrent Events Trial (CARE) *(8)*.

These studies comprised a total of 30,817 subjects with a mean follow-up of 5.4 years. Relative risk reduction with statin therapy was striking for major coronary events in both primary and secondary prevention trials:

- Primary prevention: RRR = 34% (95% CI, 23–43%; $P < 0.001$)
- Secondary prevention: RRR = 30% (95% CI, 24–35%; $P < 0.001$)

Among treated subjects without known CHD, there were significant decreases in cardiovascular mortality, although decreases in CHD mortality and all-cause mortality were not significant. This contrasted with studies of secondary prevention, where there was a significant decrease in CHD mortality, cardiovascular mortality, and all-cause mortality. Subgroup analysis of this large data set showed that subjects over 65 years of age received a similar proportional risk reduction (32%; 95% CI, 23–39%; $P < 0.001$) compared with younger patients (31%; 95% CI, 24–36%; $P < 0.001$).

One of the studies in the LaRosa meta-analysis, the LIPID study *(9)*, looked specifically at relative and absolute effects of pravastatin on cardiovascular disease (CVD) outcomes in its oldest subjects. Older subjects (aged 65–75 years) were at greater risk than younger subjects (aged 31–64 years) for death (20.6 versus 9.8%), myocardial infarction (MI) (11.4 versus 9.5%), unstable angina (26.7 versus 23.2%), and stroke (6.7 versus 3.1%) (all $P < 0.001$). Pravastatin reduced the risk for all CVD events, and similar effects were observed in older and younger subjects in this trial. In older patients with CHD, pravastatin therapy reduced the risk of all major cardiovascular events and all-cause mortality. This trial emphasized that because older patients are at greater risk than younger patients for cardiovascular events, the absolute benefit of treatment is significantly greater in older patients.

REVIEW OF RELEVANT STUDIES PUBLISHED AFTER ATP III

The Heart Protection Study

The Heart Protection Study (HPS) *(10)* enrolled 20,536 subjects in the UK, of which 5806 (28%) were between the ages of 70 and 80 years at the time of enrollment. Patients eligible for HPS had total cholesterol values > 135 mg/dL and were felt to be at high risk of a cardiovascular endpoint during the 5-year study with one of the following: (1) history of CHD or peripheral vascular disease, (2) diabetes, or (3) men over the age of 65 years with hypertension. Subjects were randomized to 40 mg/day of simvastatin or placebo. Primary outcomes in this study were all-cause mortality and CHD.

During the study period, there was a highly significant difference in all-cause mortality among subjects assigned to the simvastatin group compared with placebo (12.9 versus 14.7%; $P < 0.01$). Not surprisingly, there was also an 18% relative risk reduction in CHD (5.7 versus 6.9%; $P < 0.001$). Those assigned to simvastatin had reduced nonfatal MI compared with placebo (3.5 versus 5.6%; $P < 0.001$). With regard to stroke, subjects assigned to simvastatin enjoyed a 30% relative reduction in the incidence of first ischemic stroke (2.8 versus 4%; $P < 0.001$), although there was no difference in hemorrhagic strokes (0.5% in each group). When the study population was analyzed by age, the decrease in first vascular event attributed to simvastatin was stronger among subjects younger than 65 years of age, but still significant among subjects older than 65 years of age. The investigators noted that benefit from simvastatin could not be attributed to the initial level of low-density lipoprotein cholesterol (LDL-C) or to the degree of "LDL responsiveness" as assessed during the trial's run-in period. In fact, the subjects with an initial baseline LDL-C $ < 100$ mg/dL still had a significant risk reduction in first major vascular event rates.

The Anglo-Scandinavian Cardiac Outcomes Trial—Lipid-Lowering Arm

The Anglo-Scandinavian Cardiac Outcomes Trial (ASCOT)—Lipid-Lowering Arm (ASCOT-LLA) was an arm of the larger ASCOT trial, which randomized subjects in Europe to two antihypertensive medications (11). Among the 19,342 subjects entered into this primary prevention trial, 10,305 were further randomized to 10 mg/day of atorvastatin or placebo. Of those, 6570 (64%) were above the age of 60 years as investigators recruited subjects from ages 40 to 79 years. This was a primary prevention trial, but subjects were required to have (1) treated blood pressure $>$ 140/90 mg/dL or untreated blood pressure $ < 160/100$ mm Hg and (2) total cholesterol $ < 252$ mm Hg (6.5 mmol/L). In addition, ASCOT-LLA subjects were required to demonstrate additional cardiovascular risk by having three additional risk factors.

Primary endpoints in the ASCOT-LLA study were nonfatal MI and fatal CHD. The Data and Safety Monitoring Board stopped the study prematurely (median of 3.3 years of follow-up) due to a significant 37% relative risk reduction in the primary endpoint in subjects receiving atorvastatin compared with placebo (1.9 versus 3.0%; $P < 0.001$). In the subgroup analysis, this benefit was maintained in patients older than 60 years of age. One consequence of premature termination of the trial was that it was not possible to show a difference in total mortality that might have been found had the study been allowed to go to completion. Similar to HPS, ASCOT-LLA showed a significant 27% reduction in fatal and nonfatal stroke (1.7 versus 2.4%; $P = 0.02$), although specific data on ischemic versus hemorrhagic strokes are not provided. In a post hoc analysis, this finding was maintained for patients older than 70 years of age.

The Prospective Study of Pravastatin in the Elderly at Risk

The Prospective Study of Pravastatin in the Elderly at Risk (PROSPER) study (12) looked specifically at the value of statin use in an elderly population. Investigators recruited 5804 men and women from Scotland, Ireland, and the Netherlands. The age of the subjects ranged from 70 to 82 years. This was a mixed primary and secondary prevention trial. Subjects had serum cholesterol between 155 and 350 mg/dL (4–9 mmol/L) and serum triglycerides $ < 533$ mg/dL (6.0 mmol/L). After a 4-week lead-in period, eligible subjects were randomized to 40 mg/day of pravastatin or placebo.

The primary outcome in the PROSPER study was a combined endpoint of death from CHD, nonfatal MI, and fatal or nonfatal stroke. The trial duration was a mean of 3.2 years. Treatment with pravastatin resulted in a 15% relative decrease in the primary endpoint (14.1 versus 16.2%; $P = 0.014$) as well as a 17% relative reduction in the secondary endpoint of MI (10.1 versus 12.2%; $P = 0.006$). The endpoints of fatal or nonfatal stroke did not appear to be influenced by treatment. When compared with the findings of HPS, the placebo group in PROSPER had a much lower rate of stroke than that seen in the HPS, which may explain the differences seen. However, the incidence of transient ischemic episodes in PROSPER was very close to statistical significance (2.7 versus 3.5%; $P = 0.051$). There was no effect on all-cause mortality, although a significant 21% relative reduction in CHD mortality was observed (3.3 versus 4.2%; $P = 0.043$).

The Antihypertensive and Lipid-Lowering Treatment to Prevent Heart Attack Trial—Lipid-Lowering Trial

Like ASCOT-LLA, this is a substudy of the larger Antihypertensive and Lipid-Lowering Treatment to Prevent Heart Attack Trial (ALLHAT) *(13)*, which was a four-armed antihypertensive study. ALLHAT randomized 10,355 North American subjects older than 55 years to either 40 mg/day of pravastatin or usual care. Subjects either had known CHD or had an additional CHD risk factor. The lipid criteria were an LDL-C of 100–129 mg/dL in those known to have CHD and 120–189 mg/dL in those greatly at risk. The mean study duration was 4.8 years.

The primary outcome was all-cause mortality. Investigators noted that the LDL-C difference between treatment and control subjects was less than expected. This was due to a greater than anticipated "drop-out" among the subjects assigned to pravastatin and "drop-in" statin use in those assigned to usual care. After 4 years of follow-up, pravastatin lowered LDL 27% versus an 11.0% reduction seen in the usual-care group. This reduced differential effect on LDL-C was considered an explanation why this was a "negative" result.

Pravastatin or Atorvastatin Evaluation and Infection Therapy

The Pravastatin or Atorvastatin Evaluation and Infection Therapy (PROVE-IT) trial *(14)* enrolled 4162 subjects from North America and Europe within 10 days after hospital admission for an acute coronary syndrome. The mean age for subjects in this trial was 58 ± 11 years. Subjects were randomized in a double-blind fashion to either 40 mg/day of pravastatin or 80 mg/day of atorvastatin and followed for 24 months. The primary composite outcome in this study included all-cause mortality, MI, unstable angina, stroke, and need for revascularization occurring at least 30 days after enrollment.

The median baseline LDL-C in both groups was 106 mg/dL. By study's end, the pravastatin subjects had a median LDL-C of 95 mg/dL (interquartile range 79–113 mg/dL), and the atorvastatin subjects had a median LDL-C of 62 mg/dL (interquartile range 50–79 mg/dL). Among the 75% of subjects who were statin naïve, the maximum changes in serum cholesterol had occurred by 30 days and were significantly higher in the atorvastatin arm (51 versus 22%; $P < 0.001$). Among the remainder of subjects who were already on statin upon enrollment, pravastatin did not lower the cholesterol further, but atorvastatin resulted in an additional 32% decrease. Importantly, there was a 15% relative reduction in the primary endpoint (26.3 versus 22.4%; $P = 0.005$) among those taking pravastatin versus those on atorvastatin. This trial raises the important point of the benefit/risk of therapy that would occur if the same drugs at the same dosages were used in those in the 70–85 range due to the real concern of potentially more side effects from high-dose statin therapy in the elderly.

Treating to New Targets

The Treating to New Targets (TNT) trial *(15)* enrolled 1003 subjects with stable CHD from 256 sites in 14 countries. Eligible subjects were 30–75 years of age (mean age ± SD in both groups was 60.9 ± 8.8 and 61.2 ± 8.8) and had an LDL-C < 130 mg/dL. After a run-in phase of atorvastatin 10 mg/day, subjects were then randomized in a double-blind fashion to either 10 or 80 mg/day of atorvastatin. The subjects were followed for a median of 4.9 years. The primary endpoint was the occurrence of a first major cardiovascular event, defined as death from CHD, nonfatal nonprocedure-related MI, resuscitation after cardiac arrest, or fatal or nonfatal stroke. The primary endpoint occurred in 8.7% of subjects receiving 80 mg atorvastatin as compared with 10.9% of subjects receiving 10 mg atorvastatin. This gave an absolute risk reduction in the rate of major cardiovascular events of 2.2 and a 22% relative reduction in risk [hazard ratio (HR) 0.78; 95% CI 0.69–0.89; $P < 0.001$]. There was no difference between the two treatment groups in overall mortality, although the study was not adequately powered to detect changes in the risk of death from any cause. These trials are summarized in Table 1.

Cholesterol Treatment Trialists' Collaborators

A recent meta-analysis by the Cholesterol Treatment Trialists' Collaborators included data from 14 randomized statin trials and over 90,000 patients *(16)*. In that analysis, which included all the studies mentioned above except PROVE-IT and TNT, patients over the age of 75 randomized to statin therapy had a substantial reduction in major coronary events [10.6 versus 12.8% (RR 0.82, 99% CI 0.70–0.96; $P = 0.002$)]. In addition, statin therapy resulted in a significant decrease in the incidence of major vascular events among those over 75 years of age [16.8 versus 19.7% (RR 0.82, 99% CI 0.72–0.93; $P = 0.0001$)].

SAFETY CONCERNS

The two most concerning side effects seen with statins are myopathy and hepatic transaminase elevations *(17,18)*. Of the two, myopathy is relatively more common, although some confusion may exist regarding the terminology of muscle toxicity (i.e., myalgia versus myositis versus rhabdomyolysis). Myalgia, when defined as muscle aches, appears with similar frequency in patients treated with statins and placebo. Objective evidence of muscle inflammation such as myalgia with elevated creatine kinase (CK) less than 10 times the upper limit of normal (ULN) also occurs infrequently and with similar incidence in blinded statin trials. Statin-induced rhabdomyolysis is defined as CK greater than 10 times ULN and acute renal failure from myoglobinuria. The incidence of fatal rhabdomyolysis is approximately one in 1 million prescriptions. Statin-induced rhabdomyolysis was strikingly more likely with cerivastatin, a statin withdrawn from the market in August 2001. This feared drug complication was especially likely in combination with gemfibrozil. A possible clue to myopathy in the elderly should be a complaint of muscle weakness without muscle pain. Indeed, the elderly are less likely to complain of muscle discomfort than are younger patients.

As renal insufficiency is a significant predictor of all-cause and cardiovascular mortality, patients with renal dysfunction are frequently considered for statin therapy *(19)*. Although post hoc analyses have to be viewed with caution, results of two such

Table 1
Clinical Trials with Significant Numbers of Elderly

Name of clinical trial	Age range	Patients enrolled	Entry criteria	Intervention	Net mean change in lipids	Follow-up	Selected outcomes (all reach statistical significance unless otherwise indicated)
HPS	40–80 years % Older: 28% > 70 years old	20,536	Primary and secondary prevention	Simvastatin 40 mg versus placebo	TC: ↓ 20% LDL: ↓ 29% HDL: ↑ 3% TG: ↓ 14%	5 years	All-cause mortality: ↓ OR (0.87) CHD: ↓ 18% RRR of CHD death Stroke: ↓ 30% RRR of ischemic stroke
ASCOT-LLA	40–79 years % Older: 64% > 60 years old	10,305	Primary prevention	Atorvastatin 10 mg versus placebo	TC: ↓ 18% LDL: ↓ 28% HDL: No Δ TG: ↓ 12%	3.3 years (planned for 5)	All-cause mortality: Not significant CHD: ↓ 30% RRR in total coronary events Stroke: ↓ 29% RRR in fatal and nonfatal stroke
PROSPER	70–82 years	5804	Primary and secondary prevention	Pravastatin 40 mg versus placebo	TC: No data LDL: ↓ 32% HDL: ↑ 5% TG: ↓ 12%	3.2 years	All-cause mortality: Not significant CHD: ↓ 17% RRR in CHD death or non-fatal MI Stroke: No difference
ALLHAT-LLT	> 55 years % Older: 55% > 65 years old	10,355	Primary and secondary prevention	Pravastatin 40 mg versus placebo	TC: ↓ 14% LDL: ↓ 16% HDL: No Δ TG: ↓ 3.5%	4.8 years	All-cause mortality: Not significant CHD: Not significant Stroke: Not significant

Study	Population	N	Indication	Intervention	Lipid changes	Duration	Outcomes
PROVE-IT	> 18 years old % Older: Unknown	4162	Secondary prevention after acute coronary syndrome	Pravastatin 40 mg versus Atorvastatin 80 mg	TC: No data LDL: ↓ 10% HDL: ↑ 8.1% TG: No data TC: No data LDL: ↓ 42% HDL: ↑ 6.5% TG: No data	2 years	All-cause mortality: Not significant CHD: ↓ 12% RRR in MI, revascularization or CHD death Stroke: Not significant The above compares two statins, there was no placebo arm.
TNT	35–75 years % Older: Unknown	10,001	Secondary prevention in stable CHD	Atorvastatin 10 mg versus Atorvastatin 80 mg	TC: No Δ LDL: ↓ 33% HDL: No Δ TG: No Δ TC: Sig ↓ LDL: ↓ 49% HDL: No Δ TG: Sig ↓	4.9 years	All-cause mortality: Not significant CHD: HR 0.78 (0.69–0.89) for MACE Stroke: HR 0.75 (0.59–0.96)
Meta Analysis (LaRosa, et al.)	Mean = 59 % Older: 23% > 65 years old	30,817	Primary and secondary prevention	Simvastatin or pravastatin or lovastatin versus placebo	TC: ↓ 20% LDL: ↓ 28% HDL: ↑ 5% TG: ↓ 13%	5.4 years	All-cause mortality: ↓ OR (0.77) in secondary prevention CHD: ↓ 32% in major coronary events (> 65 years old) Stroke: No data provided

LDL, low-density lipoprotein; HDL, high-density lipoprotein; TC, total cholesterol; TG, triglyceride; CHD, coronary heart disease; HR, hazard ratio; HPS, The Heart Protection Study; ASCOT-LLA, The Anglo-Scandinavian Cardiac Outcomes Trial—Lipid-Lowering Arm; PROSPER, The Prospective Study of Pravastatin in the Elderly at Risk; ALLHAT-LLT, The Antihypertensive and Lipid-Lowering Treatment to Prevent Heart Attack Trial; PROVE-IT, The Pravastatin or Atorvastatin Evaluation and Infection Therapy; TNT, Treating to New Targets.

analyses of statin use in this setting are reassuring. First, in the placebo-controlled HPS *(20)*, allocation to simvastatin was associated with a smaller increase in mean plasma creatinine concentration than seen with allocation to placebo. Diabetic subjects appeared to benefit even more. Participation in HPS, however, was limited to those whose creatinine values were < 2.4 mg/dL (200 µmol/L). Second, in CARE *(21)*, there were 1711 subjects with chronic renal insufficiency defined as creatinine clearance ≤ 75 mL/min using the Cockcroft-Gault equation. Pravastatin 40 mg was effective in reducing cardiac events as contrasted with placebo and was found to be safe despite the renal status.

Statin-induced hepatic toxicity is infrequent. It is more likely at higher dosages and may occur more with some drugs than others. In most trials, however, hepatic enzyme changes more than three to four times the ULN are similar to placebo. One exception is high dose (80 mg) atorvastatin therapy. In the PROVE-IT trial, the 80-mg dose of atorvastatin had an increased incidence of transaminase elevation compared with pravastatin. The transaminase elevation was also higher than that seen in those with chronic CHD who received low-dose (10 mg) atorvastatin therapy in TNT. It is not clear whether this transaminase elevation, particularly of milder degree, represents drug toxicity. Indeed, statin-induced liver failure is extremely rare. The differences in statin metabolism may help explain those at increased risk of side effects. For example, statins such as atorvastatin, lovastatin, and simvastatin utilize the hepatic P450 3A4 system. Thus, use of drugs that inhibit this system such as erythromycin and clarithromycin can lead to toxicity. Certain drugs such as fibrates and niacin, amiodarone, cyclosporine, or warfarin when used with statins may increase the risk of toxicity. Subjects who should be considered at higher risk of potential problems include the elderly (especially women), multisystem medical disease, and alcohol abusers. See Table 2 for a suggested checklist to review before starting statins in the elderly.

It should be noted that low rates of toxicity in trials such as the HPS, PROVE-IT, and TNT were in part due to the pretrial screening, run-in periods, and the exclusion of those subjects on strong inhibitors of P450 3A4 system in PROVE-IT. Thus, actual clinical experience may demonstrate higher than expected rates of muscle and liver toxicity unless the clinician uses the same degree of caution used in these trials.

The data on malignancy in these trials must be reviewed carefully. In the PROSPER study *(11)*, there was a significant 25% excess risk of new malignancy in the pravastatin group. Because significant differences can be seen due to change when many comparisons are made, it is reassuring to note that older subjects have not shown similar increases in malignancy when other statin trials were reviewed (Table 3). Taken as a whole, the experience with statins in large-scale clinical trials shows no systematic increased risk of malignancy over placebo. A recent meta-analysis from the Cholesterol Treatment Trialists' Collaborators, which included over 90,000 patients, found no excess incidence of malignancy among patients treated with statins, although nonmelanoma skin cancers were not included *(16)*.

PREDICTING RISK IN THE ELDERLY

Although age is a strong predictor of atherosclerotic disease burden, it is not specific enough to guide decisions about the need for drug therapy in the elderly. Indeed, traditional global risk assessment strategies based on the Framingham data are not available for patients aged 80 years and older. We believe the following short-list of factors may be useful when additional data are needed to determine if statin therapy is

Table 2
A Suggested Checklist of Factors to Consider Before Prescribing Statins in the Elderly (2)

1. Exogenous factors:
    Dehydration
    Heavy alcohol usage
    Drugs of abuse (cocaine, amphetamines, phencyclidine, heroin)
    Medications that have an interaction with statins
        Cyclosporine or tacrolimus therapy
        Gemfibrozil (consider fenofibrate if combination therapy required)
        Niacin (long-acting forms associated with the most reports of toxicity; for all niacin therapy, toxicity with statins more likely if niacin-induced hepatic damage occurs)
        Coumadin [watch the prothrombin time carefully with all statins including those metabolized by P450 2C9 (fluvastatin and rosuvastatin]
        For statins primarily metabolized by P450 3A4 (atorvastatin, lovastatin, simvastatin), eryhromnycin, clarithromycin, azole antifungal therapy, large quantities of grapefruit juice, for example
        If renal failure, consider atorvastatin and fluvastatin as they require no dosage modification.
2. Systemic factors:
    Impaired liver function (important to define nature of liver dysfunction)
    Impaired renal function (creatinine levels improve with statins in some clinical trials; nonetheless, use with caution with creatinine over 2.4 mg/dL (200 μmol/L)
    Hypothyroidism (it is prudent to check a TSH level before starting statin therapy in the elderly; impaired thyroid function could explain enhanced tendency to myalgias, muscle toxicity)
3. Overall personal descriptors:
    frail
    elderly females
    low body weight
    poor compliance
    prior history of rhabdomyolysis
    have conditions that require multiple medications prescribed by different physicians

The authors acknowledge that although not a complete listing of all factors, this table is designed as a guide to aid in best drug selection.

appropriate. A more detailed description of dyslipidemia as a risk factor in the elderly is found in a review by Wenger (22) (Table 4).

High-Density Lipoprotein Cholesterol in the Elderly

Prospective cohort data showed that low high-density lipoprotein cholesterol (HDL-C) predicts CHD mortality and occurrence of new CHD events in persons older than 70 years of age (23). Moreover, several statin trials confirm that those with low HDL-C levels at baseline enjoy the largest on-trial benefit from statin therapy (5,12). Low HDL-C levels also predict the metabolic syndrome. In a recent trial of women with angiographic coronary disease, those women with the metabolic syndrome were older and had a significantly lower 4-year survival rate as contrasted with those without the metabolic syndrome (24). Beacuse HDL-C levels are available and standardized,

Table 3
Major Adverse Events in Clinical Trials with Significant Numbers of Elderly

Name of clinical trial	Myalgias/ myositis/ CK elevations	Rhabdomyolysis	Liver function test (LFT) Abnl ($> 4 \times ULN$)	Any malignancy
HPS	No significant difference	0.05% versus 0.03% (nonsignificant)	0.42% versus 0.31% (nonsignificant)	7.9% versus 7.8% (nonsignificant)
ASCOT-LLA	No significant difference	One nonfatal case in the atorvastatin group	No difference	No data
PROSPER	No significant difference	None in either group	One patient in each group For LFT $> 3 \times$ ULN	25% increased risk ($P = 0.02$)
ALLHAT-LLT	Not reported	Not reported	Not reported	4.1% versus 3.7% (nonsignificant)
PROVE-IT	No significant difference	None in either group	1.1% pravastatin versus 3.3% Atorvastatin ($P < 0.001$) For LFT $> 3 \times$ ULN	Not provided
TNT	No significant difference	3 cases in the 10 mg group and 2 in the 80 mg group.	1.2% in the 80 mg group v. 0.2% in the 10 mg group ($P < 0.001$)	No difference in death from malignancy (1.5 versus 1.7%)
Meta Analysis (LaRosa, et al.)	No significant difference	Not provided	No difference	No significant difference

HPS, The Heart Protection Study; ASCOT-LLA, The Anglo-Scandinavian Cardiae Outcomes Trial—Lipid-Lowering Arm; PROSPER, The Prospective Study of Pravastatin in the Elderly at Risk; ALLHAT-LLT, The Antihypertensive and Lipid-Lowering Treatment to Prevent Heart Attack Trial Arm; PROVE-IT, The Pravastatin or Atorvastatin Evaluation and Infection Therapy; TNT, Treating to New Targets; ULN, upper limit of normal.

a strong case can be made for using HDL-C levels to influence therapy decisions in those with an intermediate risk of CHD.

C-Reactive Protein and the Elderly

Measurement of high-sensitivity C-reactive protein (hs-CRP) has been shown in multiple studies to stratify cardiovascular event risk among healthy middle-aged

Table 4
Additional Tests to Risk Stratify the Elderly for Cholesterol-Lowering Therapy

Tests	Availability	Standardization	Low cost	Comments
HDL-C	Yes	Yes	Yes	Low HDL-C indicative of increased risk and the metabolic syndrome. Low HDL-C predicts response to statins in several clinical trials
Lp(a)	Yes	No	Variable	Further studies are required as well as standardization of the methods before it can be recommended routinely
Lp–PLA$_2$	Yes	No	Yes	Available studies have used both Lp-PLA$_2$ mass and Lp-PLA$_2$ activity
hs-CRP	Yes	Yes	Yes	Elevated levels only "bump up" risk category in moderate risk men
Cystatin C	No	No	Yes	Not universally available. Not clear if low GFR is an acceptable substitute
Ankle-brachial index (ABI)	Yes	Yes	Yes	Has good sensitivity and specificity and may be especially useful in older populations for detecting asymptomatic peripheral vascular disease in older populations
Carotid imaging and intimal-media thickness (IMT) measurements	Yes	Yes	No	Carotid IMT requires special training; in elderly, carotid imaging can detect increased stroke risk. Simple presence of plaque may be equally predictive
Coronary artery calcium	Yes	No	No	Standardization needed so that results obtained with electron beam tomography and multi-slice scanners are comparable

HDL, high-density lipoprotein; hs-CRP, high-sensitivity C-reactive protein; GMR, glomerular filtration rate; Lp(a), lipoprotein(a).

patients *(25–28)*. In a study of 3971 elderly subjects without vascular disease, a single baseline CRP level was predictive of CHD risk at 10 years of follow-up *(29)*. Among those with elevated CRP (> 3 mg/L), the relative risk of CHD, after adjustment for traditional risk factors, was 1.45 (1.14–1.86, $P < 0.004$). Although CRP seemed to further stratify risk within Framingham risk groups, only among moderate risk men with CRP > 3mg/L would reassingment to a higher risk group occur.

Lipoprotein(a) and the Elderly

Lipoprotein(a) (Lp(a)), a complex of LDL with a glycoprotein called apoprotein (a), is considered an emerging risk factor. It is homologous (sequence similarity) with plasminogen and competes with it for binding sites. In those with symptomatic intracranial atherosclerotic disease, high levels predict silent coronary ischemia *(30)*. Although extensively studied as a predictor of cardiovascular morbidity and mortality, primarily in middle-aged men, results are not consistent. The use of this marker in nearly 4000 subjects over age 65 and without known CHD and followed for a median of 7 years was recently described *(31)*. A significant result was seen in men, but not in women. Among men, there was a threefold increase in the risk of stroke from the lowest quintile to the highest (HR 3.00; 95% CI 1.59–5.65). Similarly, there was a more than doubled increased risk of death from vascular causes (HR 2.54; 95% CI 1.59–4.08) and a nearly doubled increased risk of death from all causes (HR 1.76; 95% CI 1.31–2.36). In a small, but carefully done angiographic trial, it was found that Lp(a) lost its predictive value when LDL-C values were lowered below 100 mg/dL *(32)*. A workshop on Lp(a) convened by the National Heart, Lung, and Blood Institute examined the structural complexity and size heterogeneity of Lp(a) as well as the large inter-method variation in Lp(a) values *(33)*. It concluded with a cautionary note that general use of Lp(a) as a risk marker was premature.

Lipoprotein-Associated Phospholipase A_2

Lipoprotein-associated Phospholipase A_2 (Lp-PLA2) is an enzyme that hydrolyzes oxidized LDL, producing highly atherogenic byproducts. Given its role in atherogenesis, the association of plasma Lp-PLA2 mass or activity with CHD has been studied in several large cohorts *(34–36)*. In a study of elderly patients *(37)*, relative to those in the lowest quartile, those in the highest quartile of LP-PLA2 activity had an increased risk of CHD, MI, and stroke [HR = 1.97 (1.28–3.02), HR = 1.92 (1.01–3.62), HR = 1.97 (1.03–3.79), respectively]. The increased risk was independent of traditional risk factors and hs-CRP. In addition, the increased risk was seen across the spectrum of lipid levels. This is another emerging risk factor.

Impaired Renal Function

As mentioned previously, the presence of renal dysfunction has been linked to adverse cardiovascular outcomes across a spectrum of age as well as clinical scenarios. The Kaiser Permanente Renal registry included data on glomerular filtration rate (GFR) in over 1 million patients followed for nearly 3 years *(38)*. The mean age of patients in this cohort was 52.2 years. A strong, nonlinear relationship was found between decreased GFR and the risk of death, cardiovascular events, and hospitalization. For example, the adjusted HR for any cardiovascular event was 3.4 (95% CI 3.1–3.8) among patients with the lowest GFR ($< 15 \, mL/min/1.73 \, m^2$) relative to those with preserved GFR. An increase in cardiovascular events was also seen in those with more modest degrees of renal impairment.

A recent study conducted exclusively in the elderly reached similar conclusions. The Cardiovascular Health Study followed 4637 patients all over the age of 65 for a median of 7.4 years *(39)*. Study patients had measurement of creatinine and cystatin C from frozen plasma samples. Cystatin C is a ubiquitous protease inhibitor that is freely filtered and reflects GFR without the need for age and muscle mass correction. Among patients in the highest quintile of cystatin C levels ($\geq 1.60\,\text{mg/L}$), the adjusted HRs for cardiovascular death, MI, and stroke were 2.83 (95% CI 1.85–4.31), 1.65 (95% CI 1.03–2.64), and 1.80 (95% CI 1.16–2.79), respectively. Creatinine and estimated GFR performed poorly as predictors.

Therefore, the elderly with renal impairment likely constitute a higher risk subgroup than those with preserved renal function. Measurement of cystatin C may be a more sensitive discriminator of renal dysfunction than creatinine in the elderly. To date, there is little data regarding the use of statins in primary prevention for this group of patients, nonetheless the theoretical benefits are attractive. Some caution should be exercised in the dosing of certain statins (rosuvastatin in particular) as a higher incidence of side effects has been noted in those with impaired renal function. Indeed, when renal insufficiency is severe (creatinine clearance $< 30\,\text{mL/min}$), atorvastatin and fluvastatin are preferred as their dosage does not need to be modified in renal failure *(40)*.

MEASURES OF SUBCLINICAL CVD IN OLDER ADULTS

The Cardiovascular Health Study *(41)* examined subclinical atherosclerosis in older adults. Indeed, investigators found that subclinical atherosclerosis was as common as clinically overt CVD in this population and seemed to predict not only incident CVD, stroke, and mortality, but also frailty as well as physical and cognitive decline.

Coronary Artery Calcification and the Elderly

The coronary artery calcium (CAC) score as assessed by computed tomography correlates well with the presence of coronary atherosclerosis *(42)*. Some have referred to this as the "burden of atherosclerosis." An important distinction, however, is that the presence of CAC does not necessarily indicate the near-term risk for CHD *(43)*. The degree of CAC does, it appears, have prognostic value. In three studies *(44)* of prospectively followed cohorts, CAC scores ranging from 67th percentile to > 80th percentile had annualized rates of MI ranging from 1.8 to 4.5%. In 1795 elderly patients without known CHD, a CAC score was strongly predictive of cardiovascular events and mortality *(45)*. In a multivariate model adjusted for traditional risk factors, the CAC score in the highest strata (> 1000) conferred a relative risk of 7.6 (2.3–25.0) for MI and CHD mortality relative to the reference value of CAC (0–100). More modest elevations in the CAC (401–1000) were also associated with an increased risk of MI and CHD death. Although a barrier to more widespread use is cost, the CAC score is especially useful in those patients judged to be at intermediate risk as a high CAC score supports a clinical decision to treat more intensively.

Carotid Imaging and the Elderly

Measurement of carotid intima-media thickness (IMT) by ultrasound is a noninvasive method to assess subclinical atherosclerosis in the neck *(46)*. Carotid IMT measurements correlate with cardiovascular outcomes both for primary and secondary prevention. The Rotterdam study *(47)*, a population-based cohort study of 6389 subjects 55 years of age and older, found that noninvasive measures of subclinical atherosclerosis were strong predictors of MI. The detection of carotid plaques (HR 1.83; 95% CI 1.27–2.62, severe versus no atherosclerosis), increased carotid IMT (HR 1.95; 95% CI 1.19–3.19), presence of abdominal aortic atherosclerosis by X-ray (HR 1.94; 95%

CI 1.30–2.90), and low ankle-arm index (HR 1.59; 95% CI 1.05–2.39) all predicted events. In the Cardiovascular Health study *(48)*, in 5858 subjects 65 years of age or older, the incidence of cardiovascular events correlated with measurements of carotid-artery IMT. On the basis of IMT quintiles, the adjusted relative risk for MI or stroke in the highest quintile was over threefold that in the lowest quintile. Ultrasound studies have the advantage of being noninvasive and potentially point of care. Unlike ECGs, however, these studies are operator dependent and great care must be taken to ensure the laboratory provides reproducible and accurate results.

Ankle-Brachial Index and the Elderly

The ankle-brachial index (ABI) is especially valuable in older subjects in whom the pre-test likelihood of a positive test indicating disease is greater than among younger patients. The wide availability, ease of standardization, predictive value, and low cost of an ABI make it an ideal technique for this population. In the Peripheral Arterial Disease (PAD) Awareness, Risk, and Treatment—New Resources for Survival (PARTNERS) program *(49)*, a multicenter, cross-sectional study with a total of 6979 subjects aged 70 years or older or aged 50–69 years with history of cigarette smoking or diabetes, ABI was found to be very useful. PAD was detected in 1865 patients (29%), and 825 of these (44%) had PAD only. A simple ABI resulted in new diagnoses of PAD in 55% and PAD and CVD in 35%. An important insight was that patient's were more aware of their PAD diagnosis than their doctors were. This study found that classic claudication was distinctly uncommon in those with PAD. Indeed, the ABI was more associated with leg function in persons with PAD than is intermittent claudication or other leg symptoms *(50)*.

KEY POINTS

- The elderly have a high burden of CVD.
- Elderly patients with CHD or CHD risk equivalents (> 20% CHD risk over 10 years) should be strongly considered for intensive lipid-lowering therapy to an LDL-C goal < 100 mg/dL.
- The LDL-C goal of < 70 mg/dL is optional for those with CVD + diabetes, metabolic syndrome, or persistent major risk factors or acute coronary syndrome *(51)*.
- Elderly patients with intermediate risk require clinical judgment to determine intensity of therapy.
  - Multiplicity and severity of risk factors must be considered.
  - Factors such as low HDL-C, high CRP, elevated LpPLA2 indicate a pro-atherogenic milleau.
  - Measurement of ABI, CAC, or carotid IMT may prove useful.
- High priority to effective Rx of major risk factors:
  - elevated blood pressure;
  - type II diabetes;
  - cigarette smoking.
- Safety considerations important in the elderly: watch for
  - renal insufficiency;
  - poly-pharmacy with greater chance for drug interactions;
  - small body size;
  - poor communications causing delays in reporting adverse drug experience.

- Finally, consider recommending a healthy lifestyle (as risk reduction Rx may be unnecessary)
  - for elderly either at low risk
  - or with significant co-morbidity.
- For those 83 years of age and older, for whom major clinical trial data are lacking,
  - encourage a careful and frank discussion of presumed benefits and risks with the patient *(51)*
  - and make a joint decision about drug therapy based on this.

REFERENCES

1. Third Report of the National Cholesterol Education Program (NCEP) Expert Panel on Detection, Evaluation, and Treatment of High Blood Cholesterol in Adults. (Adult Treatment Panel III) final report. *Circulation* 2002;106(25):3143–3421.
2. Eimer MJ, Stone NJ. Evidence based treatment of lipids in the elderly. *Curr Atheroscler Rep* 2004;6:388–397.
3. LaRosa JC, He J, Vupputuri S. Effect of statins on risk of coronary disease a meta analysis of randomized controlled trials. *JAMA* 1999;282:2340–2346.
4. Shepherd J, Cobbe SM, Ford I, et al. Prevention of coronary heart disease with pravastatin in men with hypercholesterolemia. West of Scotland Coronary Prevention Study Group. *N Engl J Med* 1995;333:1301–1307.
5. Downs JR, Clearfield M, Weis S, et al. Primary prevention of acute coronary events with lovastatin in men and women with average cholesterol levels: results of AFCAPS/TexCAPS. *JAMA* 1998;279:1615–1622.
6. Scandinavian Simvastatin Survival Study Group. Randomized trial of cholesterol lowering in 4444 patients with coronary heart disease. *Lancet* 1994;344:1383–1389.
7. Long-term Intervention with Pravastatin in Ischaemic Disease (LIPID) Study Group. Prevention of cardiovascular events and death with pravastatin in patients with coronary heart disease and a broad range of initial cholesterol levels. *N Engl J Med* 1998;339:1349–1357.
8. Sacks FM, Pfeffer MA, Moye LA, et al. The effect of pravastatin on coronary events after myocardial infarction in patients with average cholesterol levels. Cholesterol and Recurrent Events Trial investigators. *N Engl J Med* 1996;335:1001–1009.
9. Hunt D, Young P. Simes J, et al., for the LIPID Investigators. Benefits of pravastatin on cardiovascular events and mortality in older patients with coronary heart disease are equal to or exceed those seen in younger patients: results from the LIPID Trial. *Ann Intern Med* 2001;134:931–940.
10. Collins R, Armitage J, Parish S, et al. MRC/BHF Heart Protection Study of cholesterol lowering with simvastatin in 20 536 high-risk individuals: a randomised placebo-controlled trial. *Lancet* 2002;360:7–22.
11. Sever PS, Dalhof B, Poulter NR, et al. Prevention of coronary and stroke events with atorvastatin in hypertensive patients who have average or lower-than-average cholesterol concentrations in the Anglo-Scandinavian Cardiac Outcomes Trial-Lipid Lowering Arm (ASCOT-LLA): a multicentre randomised controlled trial. *Lancet* 2003;361:1149–1158.
12. Shepherd J, Blauw GJ, Murphy MB, et al. Pravastatin in elderly individuals at risk of vascular disease (PROSPER): a randomised controlled trial. *Lancet* 2002;360:1623–1630.
13. ALLHAT Investigators. Major outcomes in moderately hypercholesterolemic, hypertensive patients randomized to pravastatin vs usual care. The Antihypertensive and Lipid-Lowering Treatment to Prevent Heart Attack Trial (ALLHAT-LLA). *JAMA* 2002;288:2998–3007.
14. Cannon CP, Braunwald E, McCabe CH, et al. Comparison of intensive and moderate lipid lowering with statins after acute coronary syndromes. *N Engl J Med* 2004;350:1495–1504.
15. LaRosa JC, Grundy SM, Waters DD, et al., for the Treating to New Targets (TNT) Investigators. Intensive lipid lowering with atorvastatin in patients with stable coronary disease. *N Engl J Med* 2005;352(14):1425–1435.
16. Cholesterol Treatment Trialists Collaborators. Efficacy and safety of cholesterol-lowering treatment: prospective meta-analysis of data from 90056 participants in 14 randomised trials of statins. *Lancet* 2005;366:1267–1278.
17. Pasternak RC, Smith SC, Bairey-Merz CN, et al. ACC/AHA/NHLBI clinical advisory on the use and safety of statins. *J Am Coll Cardiol* 2002;40:567–572.

18. Ballantyne CM, Corsini A, Davidson MH, et al. Risk for myopathy with statin therapy in high-risk patients. *Arch Intern Med* 2003;163:553–564.
19. Fried LF, Shlipak MG, Crump C, et al. Renal insufficiency as a predictor of cardiovascular outcomes and mortality in elderly individuals. *J Am Coll Cardiol* 2003;41:1364–1372.
20. Collins R, Armitage J, Parish S, et al. MRC/BHF Heart Protection Study of cholesterol-lowering with simvastatin in 5963 people with diabetes: a randomised placebo-controlled trial. *Lancet* 2003;361:2005–2016.
21. Tonelli M, Moye L, Sacks FM, et al. Cholesterol and Recurrent Events (CARE) Trial Investigators. Pravastatin for secondary prevention of cardiovascular events in persons with mild chronic renal insufficiency. *Ann Intern Med* 2003;138:98–104.
22. Wenger NK. Dyslipidemia as a risk factor at elderly age. *Am J Geriatr Cardiol* 2004;13(suppl 1):4–9.
23. Corti MC, Guralnik JM, Salive ME, et al. HDL cholesterol predicts coronary heart disease mortality in older persons. *JAMA* 1995;274:539–544.
24. Marroquin OC, Kip KE, Kelley DE, et al., for the Women's Ischemia Syndrome Evaluation Investigators. Metabolic syndrome modifies the cardiovascular risk associated with angiographic coronary artery disease in women: a report from the Women's Ischemia Syndrome Evaluation. *Circulation* 2004;109:714–721.
25. Ridker PM, Glynn RJ, Hennekens CH. C-reactive protein adds to the predictive value of total and HDL cholesterol in determining risk of first myocardial infarction. *Circulation* 1998;97:2007–2011.
26. Ridker PM, Hennekens CH, Buring JE, Rifai N. C-reactive protein and other markers of inflammation in the prediction of cardiovascular disease in women. *N Engl J Med* 2000;342:836–843.
27. Kervinen H, Palosuo T, Manninen V, et al. Joint effects of C-reactive protein and other risk factors on acute coronary events. *Am Heart J* 2001;141:580–585.
28. Ridker PM, Rifai N, Rose L, et al. Comparison of C-reactive protein and low-density lipoprotein cholesterol levels in the prediction of first cardiovascular events. *N Engl J Med* 2002;347:1557–1565.
29. Cushman M, Arnold AM, Psaty BM, et al. C-reactive protein and the 10-year incidence of coronary heart disease in older men and women. The cardiovascular health study. *Circulation* 2005;112:25–31.
30. Arenillas JF, Candell-Riera F, Romero-Farina G, et al. Silent myocardial ischemia in patients with symptomatic intracranial atherosclerosis: associated factors. *Stroke* 2005;36(6):1201–1206.
31. Ariyo AA, Thach C, Tracy R. Lp(a) lipoprotein, vascular disease, and mortality in the elderly. *N Engl J Med* 2003;349:2108–2115.
32. Maher VM, Brown BG, Marcovina SM, Hillger LA, Zhao XQ, Albers JJ. Effects of lowering elevated LDL cholesterol on the cardiovascular risk of lipoprotein(a). *JAMA* 1995;274:1771–1774.
33. Marcovina SM, Koschinsky ML, Albers JJ, Skarlatos S. Report of the National Heart, Lung, and Blood Institute Workshop on lipoprotein(a) and cardiovascular disease: recent advances and future directions. *Clin Chem* 2003;49:1785–1796.
34. Packard CJ, O'Reilly DS, Caslake MJ, et al. Lipoprotein-associated phospholipase A2 as an independent predictor of coronary heart disease. West of Scotland Coronary Prevention Study Group. *N Engl J Med* 2000;343:1148–1155.
35. Ballantyne C, Hoogeveen R, Bang H, et al. Lipoprotein-associated phospholipase A2, high sensitivity C-reactive protein, and risk for incident coronary heart disease in middle aged men and women in the Atherosclerosis Risk in Communities (ARIC) study. *Circulation* 2004;109:837–842.
36. Koenig W, Meisinger C, Trischler G, et al. Lipoprotein-associated phospholipase A2 and risk of coronary events in apparently healthy middle-aged men with moderately elevated cholesterol. Results from the 14 year follow up of the MONICA-Augsburg cohort. *Circulation* 2004;110:1903–1908.
37. Oei HS, van der Meer IM, Hofman A, et al. Lipoprotein-associated phospholipase A2 activity is associated with risk of coronary heart disease and ischemic stroke. The Rotterdam study. *Circulation* 2005;111:570–575.
38. Go AS, Chertow GM, Fan D, McCulloch CE, Hsu C. Chronic kidney disease and the risks of death, cardiovascular events, and hospitalization. *N Engl J Med* 2004;351:1296–1305.
39. Shilpak MG, Sarnak MJ, Katz R, et al. Cystatin C and the risk of death and cardiovascular events among elderly persons. *N Engl J Med* 2005;352:2049–2060.
40. Launay-Vacher V, Izzedine H, Deray G. Statins' dosage in patients with renal failure and cyclosporine drug-drug interactions in transplant recipient patients. *Int J Cardiol* 2005;101:9–17.
41. Chaves PH, Kuller LH, O'Leary DH, et al. Subclinical cardiovascular disease in older adults: insights from the cardiovascular health study. *Am J Geriatr Cardiol* 2004;13:137–151.

42. Rumberger JA, Simons DB, Fitzpatrick LA, et al. Coronary artery calcium area by electron-beam computed tomography and coronary atherosclerotic plaque area: a histopathologic correlative study. *Circulation* 1995;92:2157–2162.
43. Detrano RC, Wong ND, Doherty TM, et al. Coronary calcium does not accurately predict near-term future coronary events in high-risk adults. *Circulation* 1999;99:2633–2638.
44. Grundy SM. Coronary calcium as a risk factor: role in global risk assessment. *J Am Coll Cardiol* 2001;37:1512–1515.
45. Vliegenthart R, Oudkerk M, Hofman A, et al. Coronary calcification improves cardiovascular risk prediction in the elderly. *Circulation* 2005;112:572–577.
46. Espeland MA, Evans GW, Wagenknecht LE, et al. Site-specific progression of carotid artery intimal-medial thickness. *Atherosclerosis* 2003;171:137–143.
47. van der Meer IM, Bots ML, Hofman A, et al. Predictive value of noninvasive measures of atherosclerosis for incident myocardial infarction: the Rotterdam Study. *Circulation* 2004;109:1089–1094.
48. O'Leary DH, Polak JF, Kronmal RA, et al. Carotid artery intima and media thickness as a risk factor for myocardial infarction and stroke in older adults. *N Engl J Med* 1999;340:14–22.
49. Hirsch AT, Criqui MH, Treat-Jacobson D, et al. Peripheral arterial disease detection, awareness, and treatment in primary care. *JAMA* 2001;286:1317–1324.
50. McDermott MM, Greenland P, Liu K, et al. The ankle brachial index is associated with leg function and physical activity: the Walking and Leg Circulation Study. *Ann Intern Med* 2002;136:873–883.
51. Grundy SM, Cleeman JI, Merz CN, et al. Implications of recent clinical trials for the National Cholesterol Education Program Adult Treatment Panel III guidelines. *Circulation* 2004;110:227–239.

ELDERLY CASE HISTORIES

Patient 1

Patient 1 is an asymptomatic 72-year-old woman with a history of hypertension. She has a positive family history of CHD; her mother died at the age of 63 years because of complications following coronary artery bypass surgery. Her only medication is a thiazide diuretic.

The waist circumference is 36 inches. The blood pressure is 150/90 mm Hg.
Fasting laboratory results are given below:

Total cholesterol	Triglyceride	HDL-C	LDL-C	Non-HDL-C	Glucose
210	185	45	128	165	105

Her 12 lead ECG shows sinus rhythm and no evidence for LVH.
She said she would prefer not to take a statin drug.

DISCUSSION

She has the five criteria for the ATP III definition of the metabolic syndrome (low HDL-C, increased waist circumference, high triglyceride, elevated blood pressure, elevated blood sugar). Her estimated 10-year risk of CHD is 12% (Framingham risk calculator, http://hin.nhlbi.nih.gov/atoiii/riskcalc.htm).

The following treatment recommendations were made. Because she has criteria for ATP III metabolic syndrome, she had counseling directed at better diet, increased activity, and the need for approximately a 7% weight loss over the next year. She was given the Dietary Alternatives to Stop Hypertension (DASH) diet, a low saturated fat, dietary cholesterol diet with increased fruits and vegetables, nonfat dairy products, and low sodium. She was told to use a pedometer to help her achieve 10,000 steps daily. She was prescribed an angiotensin-converting enzyme inhibitor drug to lower her blood pressure < 130/85 mmHg. She was started on an 81 mg aspirin to reduce her risk of heart attack and stroke. She was not given a statin but encouraged to follow-up to monitor her blood pressure and her weight.

At the end of 6 months, she had lost 5 lbs and her lipids, blood pressure, and blood sugar had improved. Follow-up fasting laboratory results are given below:

Total Chol	Triglyceride	HDL-C	LDL-C	Non-HDL-C	Glucose
195	145	46	120	149	99

Patient 2

An 80-year-old man with a history of hypertension wants to know if he should take a statin. He has no known history of coronary disease or diabetes mellitus. He is a nonsmoking retired plumber with the following fasting laboratory results:

Total Chol	Triglyceride	HDL-C	LDL-C	Non-HDL-C	Glucose
220	210	40	138	180	90

Discussion

Accurate estimation of this patient's CHD risk is difficult as the Framingham risk equation does not address patients aged 80 years and older. Therefore, an approach to risk stratification might include a test for subclinical atherosclerosis. It seems especially appropriate because he has a low HDL-C, an important predictor of risk in his age group. Two possible choices are a carotid duplex or an ABI test. A low ABI or the presence of carotid plaque and/or an increased IMT would suggest an increased risk of cardiovascular events and justify an LDL-C goal of < 100 mg/dL.

Owing to its availability and low cost, an ABI was performed, which showed an ABI of < 0.90 in both lower extremities. A statin was begun with the initial dosage designed to lower LDL-C by at least 35% to allow his LDL-C to fall to goal levels.

The ATP III implications paper from July 2004 noted that in older persons, clinical judgment was required to determine when to begin intensive LDL-C-lowering therapy without definite CVD. The ATP III panel indicated that efficacy alone was not the key issue and that factors such as safety, tolerability, and patient preference must be determined before a final treatment decision was made. They cited the PROSPER and ASCOT-LLA trials as evidence of benefit of statin therapy in older, high-risk persons without established CVD.

Patient 3

A 76-year-old woman with noninsulin-dependent diabetes was started on 10 mg atorvastatin 2 years ago for the following laboratory values:

Total cholesterol	Triglyceride	HDL-C	LDL-C	Non-HDL-C	Glucose	TSH
212	200	38	134	174	129	2.58

After 1 year of therapy her lipids were as follows:

Total cholesterol	Triglyceride	HDL-C	LDL-C	Non-HDL-C	Glucose
174	185	39	98	135	118

After 2 years on atorvastatin, she complains of muscle aches which she attributes to the statin. On examination, it was found that there is no evidence of muscle tenderness or weakness. She had gained 5 lbs over the past year. Labs drawn 1 day before today's office visit show:

Total cholesterol	Triglyceride	HDL-C	LDL-C	Non-HDL-C
194	190	38	118	156

A CK was added to the blood from the day before and was found to be 284 mg/dL. The patient was instructed to hold the statin for 1 week, but the muscle symptoms persisted. The clinician talked to a lipid specialist who suggested two additional laboratory tests.

Erythrocyte sedimentation rate	TSH
24	12[a]

[a] Abnormal value (TSH normal range 0.550 uU/ml–4.5 uU/ml).

DISCUSSION

Side effects from statins are relatively rare, but the incidence increases with age. In the patient with complaints of muscle pain or weakness, it is reasonable to try a "drug holiday" to determine if the statin is the cause. If the symptoms persist, other causes of myopathy should be sought, specifically polymyalgia rheumatica and hypothyroidism. An increase in lipid values despite a stable dose of medication is a clue that the patient may have thyroid hormone deficiency. Other clues were weight gain and a CK that was mildly elevated. Hypothyroidism is a cause of an elevated CK level.

18 The Clinical Use of Noninvasive Modalities in the Assessment of Atherosclerosis

Atul R. Chugh, MD, Samir N. Patel, MD, Venkataraman Rajaram, MD, Rachel Neems, MD, Matt Feinstein, BA, Marshall Goldin, MD, and Steven B. Feinstein, MD

CONTENTS

INTRODUCTION
REVIEW OF NONINVASIVE CARDIOVASCULAR IMAGING
  TECHNOLOGIES
CONCLUSIONS
REFERENCES

INTRODUCTION

The maturing of the Baby Boomer generation is associated with dramatic social and economic changes worldwide, and importantly, this population is linked to a triad of malevolent clinical syndromes: obesity, diabetes, and the metabolic syndrome. These associated clinical findings inexorably lead to an increased prevalence of cardiovascular diseases. Therefore, it is incumbent upon health care providers to provide the public with education and technologies in order to provide an "early warning system." Several of these technologies have been clinically tested, while newer technologies await widespread clinical validation. Today, it is possible to identify the presence of early disease, thus permit aggressive preventive treatment for atherosclerosis. Furthermore, many of these imaging techniques may be used to monitor the progression and regression of cardiovascular diseases. This chapter will describe the variety of imaging tests that are clinically available for the early detection of atherosclerosis in the Boomer population.

From: *Contemporary Cardiology: Therapeutic Lipidology*
Edited by: M. H. Davidson, P. P. Toth, and K. C. Maki © Humana Press Inc., Totowa, NJ

REVIEW OF NONINVASIVE CARDIOVASCULAR IMAGING TECHNOLOGIES

The role of invasive angiography in gauging clinically significant atherosclerosis is well documented. However, in the era of efficacious medical treatment, it has become increasingly important to identify subclinical disease for the purposes of early prevention. The technical development of newer and more sophisticated noninvasive modalities for use in assessment of clinical and subclinical disease has created a new paradigm. These newer noninvasive imaging developments highlight the limitations of invasive coronary angiography [with the exception of intravascular ultrasound (IVUS)]. Interestingly, the use of invasive angiography has been limited for the prediction of future coronary events *(1)*. The inherent risks associated with invasive coronary angiography include the cumulative effects of radiation and the use of potentially nephrotoxic contrast agents. The newer, noninvasive imaging methods afford patients and physicians alike the opportunity detect atherosclerosis in a relatively cost-effective and increasingly safe manner, enabling clinicians to opt for aggressive medical management in those patients found to be at an increased risk for events. The following noninvasive cardiovascular imaging techniques are described in this section: (1) Ultrasound Carotid Intima-Media Thickness (c-IMT); (2) Two-Dimensional Echocardiographic for determination of left ventricular mass; (3) Coronary Artery Calcium Imaging using Electron Beam Computed Tomography (EBCT) and Multi-Detector Row Computed Tomography (MDCT); (4) Cardiac Magnetic Resonance Imaging (CMRI); (5) Ankle-Brachial Indices (ABI); (6) Brachial Artery Reactivity Testing; and (7) Doppler-Based, Epicardial Coronary Flow Reserve (CFR) determinations. As with most newly developed technologies, additional confirmatory clinical studies are required prior to the acceptance of widespread clinical use. Current clinical practices reveal a trend toward a prolific use of these noninvasive technologies for the early detection of preclinical disease in the baby boomer population.

Carotid Artery Intima-Media Thickness

INTRODUCTION

The genesis of c-IMT as a surrogate marker for atherosclerosis was reported in 1986, when Pignoli et al. *(2)* published data regarding B-mode ultrasound measurement of c-IMT with comparison to postmortem gross and microscopic examination. They reported a measurement error of less than 20% in 77% of all subjects studied *(2)*. In their subsequent studies, they noted that hypercholestrolemic patients had increased c-IMT augmented by including traditional risk factors of cigarette smoking, age, and male sex *(3)*. Fourteen years later, ultrasound-based measurement was deemed the only acceptable noninvasive method for assessment of c-IMT for the use of cardiovascular risk correlation by the American Heart Association *(4)*. From a pathophysiology point of view, any increase in IMT is thought to be a reactive process secondary to shear stress and transluminal pressure from processes such as hypertension and plaque formation *(5)*. However, not any amount of c-IMT increase is thought to be indicative of atherosclerosis, specifically, a measured c-IMT thickness greater than 900 μg was indicative of atherosclerosis *(6)*. Briefly, the details of the methodology require that the ultrasound probe measurements are used to calculate distance between the lumen-intima and the media-adventitia interfaces which correspond to the inner and outer echogenic lines seen on the B-mode image *(2)*. The changes of the intima are of great importance based on the presence of closer luminal proximity and profuse cell proliferation. However, due to the limited aspect of the intima cellular layers, the combined intima-media is measured as a complex *(7)*. The standardization of site measurements and analysis techniques has not been standardized. Each expert laboratory has established their guidelines. Subsequently, the literature on this subject reveals a variety of measurement locations including the far and near walls of the common carotid artery, the bifurcation, and internal carotid artery. Each data set of c-IMT measurements are generally averaged in order to account for anatomic variability. Variations in the IMT are not confined to anatomy but can be observed throughout the cardiac cycle; therefore, it is important to perform ECG-gated c-IMT measurements *(8)*. Notwithstanding the apparent dynamic nature of performing the c-IMT measurements, ultrasound quantification of c-IMT has been established as a reproducible method for the noninvasive assessment of cardiovascular disease. Clearly, the advantage of using ultrasound imaging for c-IMT measurements is the presence of a well-established clinical base, the cost-efficiency, and the clinical safety.

OBSERVATIONAL AND CLINICAL STUDIES

The Atherosclerosis Risk in Communities (ARIC) study *(9)* is the largest c-IMT observational cardiovascular endpoint study to date. In 13,870 middle-aged adults, c-IMT measurements were increased in participants with coronary artery disease (CAD). Quantification studies showed that a 0.2 mm increase in c-IMT yielded a 33% relative risk increase for myocardial infarction and 28% increase for stroke. Similarly, in the case-control Rotterdam study *(11)*, increased c-IMT significantly correlated with coronary and cerebrovascular disease in patients greater than the age of 55. Findings of the Cardiovascular Health Study (CHS) *(10)* revealed that coronary artery calcium scores correlated with c-IMT as markers of subclinical disease *(11)*. In the Suita study, strong correlations were noted between the aggregate of traditional coronary risk factors (e.g., hypertension, hyperlipidemia, and smoking) and c-IMT measurements. In their study, the progression rate of mean c-IMT was found to be higher in men than in women (0.0075 mm/year versus 0.0069 mm/year) *(12)*.

As an established marker of cardiovascular risk, c-IMT is also used as a primary endpoint in clinical trials. Several prospective, controlled clinical trials have been performed assessing the efficacy of lipid-lowering therapy on c-IMT. HMG Co-A reductase inhibitors have been published along with postulates as to the mechanism statin action on vascular reactivity *(13)*. The Arterial Biology for the Investigation of the Treatment Effects of Reducing Cholesterol (ARBITER) trial *(14)* randomized 161 patients meeting National Cholesterol Education Program (NCEP) II criteria for lipid-lowering therapy to receive either 40 mg/day of pravastatin or 80 mg/day of atorvastatin. One-year follow-up revealed that the Atorvastatin group showed regression of c-IMT with continued progression in the Pravastatin arm. The study by Smilde et al. *(15)* showed that aggressive low-density lipoprotein (LDL) reduction was more effective at causing regression of carotid IMT than conventional therapy. Similar results were seen in patients with familial hyperlipidemia who were enrolled in the Atorvastatin Versus Simvastatin on Atherosclerosis Progression (ASAP) study *(16)*. In another statin trial, the Asymptomatic Carotid Artery Progression Study (ACAPS), 919 patients with carotid atherosclerosis and elevated LDL cholesterol (LDL-C) were treated with either lovastatin or placebo. Over a 3-year-period follow-up, progression rate of c-IMT was found to be -0.009 ± 0.003 mm/year in those treated with lovastatin as compared with a progression rate of 0.006 ± 0.003 mm/year in those that received placebo *(17)*. Beneficial effects on c-IMT progression with lipid-lowering therapy have been shown in other studies *(18–23)*. Of significant interest will be the data yielded by the ezetimibe and simvastatin in Hypercholesterolemia Enhances Atherosclerosis Regression (ENHANCE) trial which will show whether lipid lowering by two separate pharmacologic mechanisms will confer superior c-IMT regression *(24)*. Both β-blockers and calcium antagonists have been studied in reference to c-IMT. The Effect of Long-Term Treatment of Metoprolol CR/XL on Surrogate Variables for Atherosclerotic Disease (ELVA) study *(25)* and the Beta-Blocker Cholesterol-Lowering Asymptomatic Plaque Study (BCAPS) *(26)* attributed metoprolol with a reduction in the rate of c-IMT progression of c-IMT in 3-year follow-up. Calcium antagonists isradipine and verapamil in the Multicenter Isradipine Diuretic Atherosclerosis Study (MIDAS) *(27)* and the verapamil in Hypertension and Atherosclerosis Study (VHAS) were also shown to decrease progression *(28)*, as well as in the International Nifedipine GITS Study: Intervention as a Goal in Hypertension Treatment-IMT (INSIGHT-IMT) *(29)*. In a head-to-head trial, the European Lacidipine Study on Atherosclerosis (ELSA) compared the effects of treatment with either the calcium antagonist lacidipine or the β-blocker atenolol on carotid atherosclerosis in approximately 2300 patients with diagnosed hypertension. Lacidipine was found to be more beneficial in regards to c-IMT progression and plaque formation following a 4-year treatment period *(30)*.

RISK GROUP IDENTIFICATION

A use of c-IMT measurements in patient subsets with less traditional risk factors is gaining considerable attention. For example, patients between the ages of 20 and 38 with World Health Organization (WHO)-defined or NCEP-defined metabolic syndrome revealed significant c-IMT measurements *(31)*. In a 3-year follow-up of 316 middle-aged men, those who met the WHO criteria for metabolic syndrome had significant increases in their c-IMT over time *(32)*. Some reports suggest that women meeting the NCEP ATP III criteria for metabolic syndrome have a greater increase in c-IMT and more pronounced atherosclerotic changes than observed in men *(33)*. An interesting association showed that women with polycystic ovarian syndrome have increased c-IMT correlated to increased abdominal adiposity, higher levels of androgens, insulin resistance, and total cholesterol and LDL-C *(34)*. Separately, the ARIC and Rotterdam studies noted the correlation with mild homocysteinemia and increased c-IMT. While obesity in youth has been attributed with an increased risk for cardiovascular disease, it has not shown to be an independent predictor of c-IMT *(35)*. Other potential risk groups have been studied in relation to c-IMT *(36)*. These studies not only confirm the acceptance of c-IMT as a marker of cardiovascular risk but demonstrate its use in translational studies to identify populations at risk for a cardiovascular event.

VARIATIONS IN METHOD AND TECHNOLOGY

Technical advancements in B-mode ultrasonography have reduced interobserver variability and have simplified the analysis techniques. The current investigator use of B-mode studies result in a description of mean c-IMT values over a relatively large surface area (1 cm length, far/posterior wall of the common carotid artery). Additionally, it is possible and important to analyze serial c-IMT measurements using an ECG-gated technique. These methods, including serial c-IMT measurement, have been shown to reduce measurement variability *(37,38)*. Newly developed computer software programs have been used to identify specific targets within an ECG-gated period of the cardiac cycle (i.e., far-wall c-IMT of the internal carotid artery in diastole; Tables 1 and 2 show the results of comparing manual and automated measurement techniques) *(9,39)*. Using a semi-automated c-IMT measurement technique (IoDP; SynArc, San Francisco, California), Macioch et al. *(40)* demonstrated that there was no significant difference between the manual and the automated c-IMT measurements (Table 1). The results of another published study that the interobserver correlation was improved with the automated measurements *(41)*. Traditionally, due to ease of acquisition and acoustic clarity, researchers have used measurements of the far-wall carotid IMT. Interestingly, Wong et al. *(42)* in 1993 reported a histopathology study with the conclusion that the near wall of the carotid IMT was thicker than the posterior wall by approximately 20%. Additionally, due to the inherent acoustic properties of ultrasound, the near wall of the carotid artery lacks the resolution and acoustic clarity of the far wall of the carotid artery *(43)*. However, this significant difference has been questioned, most notably in the ELSA study.

A recent report described the use of ultrasound contrast agents to enhance the near wall of the carotid artery c-IMT. Macioch et al. *(40)* showed that the additional use of a US Food and Drug Administration approved ultrasound contrast agent enhanced the detection and delineation of the carotid artery near wall, resulting in improved measurement precision. Furthermore, the use of these intravenous ultrasound contrast agents permitted clearer identification of luminal and plaque morphology. The effect of the ultrasound contrast enhancement on the near-wall c-IMT was statistically significant, resulting in a measured contrast-enhanced c-IMT of 19% larger than its corresponding noncontrast-enhanced carotid image of the near wall (Table 2). This is supported by previous studies that have shown that noncontrast ultrasound imaging underestimated the near-wall c-IMT by as much as 20% when compared with histologic measure *(42)*. The addition of three-dimensional ultrasound may permit a quantitative volumetric analysis of the carotid IMT and plaque.

Table 1
Manual Versus Automated Measurement of Carotid Artery Intima-Media Thickness

	Manual	Automated	P value
Near-wall c-IMT (mm)	0.74 ± 0.18	0.78 ± 0.20	0.02
Far wall c-IMT (mm)	0.65 ± 0.20	0.69 ± 0.21	0.13
P value	< 0.0005	< 0.0001	

c-IMT, carotid intima-media thickness.

Table 2
Effect of Contrast-Enhanced Ultrasound Imaging on Measurement of Carotid Artery Intima-Media Thickness

	Noncontrast	Contrast	P value
Near-wall c-IMT (mm)	0.62 ± 0.13	0.74 ± 0.18	< 0.0001
Far wall c-IMT (mm)	0.66 ± 0.14	0.66 ± 0.20	0.9
P value	< 0.02	< 0.001	

c-IMT, carotid intima-media thickness.

Extra-Carotid Sites

Researchers have reported on the IMT of other vascular locations. The Bogalusa heart study used data from the femoral arteries [femoral IMT (f-IMT)] and subsequently reported that the odds ratio for patients with three or more risk factors versus no risk factors having increased f-IMT which were located in the top fifth percentile was 4.7 ($P = 0.01$) *(44)*. While the APSIS study showed differences in the type of prognostic value between common f-IMT and carotid IMT measurements *(45)* the REGRESS study has shown significant regression in both c-IMT and f-IMT in 255 patients, while no significant changes were seen in 885 patients undergoing coronary angiography in men with documented CAD on 40 mg of pravastatin. This study illustrates the importance of IMT as a marker followed over time in comparison to invasive angiography *(43)*.

Left Ventricular Echocardiography

Assessed by echocardiography, the presence of left ventricular hypertrophy (LVH) is identified as an independent marker of sudden cardiac death, CAD, and myocardial ischemia. Echocardiographically proven LVH is a useful predictor of cardiovascular events independent of age, diabetes, smoking, and lipid profile. For every 50 g/month increase in left ventricular mass the relative risk for CAD was 1.49 in men and 1.57 in women *(46)*. The combination of ECG and echocardiographic data was complementary for prognostic stratification *(47)*. The use of ultrasound contrasts has been shown to enhance the left ventricular endocardial border, thereby increasing ease of use and reproducibility *(48)*. Additionally, LVH can be followed over time for regression and has been shown to be a favorable prognostic factor. A meta-analysis has shown that in hypertensive patients, a regression of LVH predicts a greater than 50% reduction in cardiovascular events *(49)*.

Coronary Computed Tomography Imaging

Coronary Computed Tomography (CT) Imaging has undergone dramatic changes over the last 10–15 years, largely due to advances in technology related to multi-slice computed tomography (MSCT). Much of the earlier work related to CT imaging for calcium scoring of coronary arteries. Clinical outcome studies correlated the presence and degree of calcium in order to provide novel risk stratification. However, with improved MSCT, in addition to identifying calcium in the coronary arteries, there is considerable interest in assessing the coronary artery lumen, noninvasively. This shift in focus has yielded newer data and will be discussed subsequently.

EBCT

The majority of clinical data pertaining to the use of EBCT relates to the detection and quantification of arterial calcium deposits using a standardized system. The scientific basis for these studies is that clinically significant atherosclerosis is detected by the presence of calcium notwithstanding the presence of a luminal stenosis. Therefore, the hallmark of atherosclerosis shifted from solely identifying a luminal stenosis pattern to a pattern of identifying calcium in the vessel wall. To perform these studies, EBCT is performed using a stationary tungsten target bombarded by electron beams. For analysis purposes, those regions imaged which exceed a radiographic density of 130 Hounsfield units are included in the quantification for calcified coronary atherosclerosis *(50)*. The original method of measurement, Agatston method, had been shown to provide greater inter-scan variability than that of the newer calcium volume scoring method *(51)*. Several clinical correlates have been correlated to EBCT results; however, the association of

EBCT with coronary angiography and cardiovascular risk stratification has had great relevance. The EBCT calcium score has been shown to correlate strongly with the number of segments showing 20% or greater stenosis by angiography *(52)*, though, a correlation to the benchmark of 50% stenosis has been mixed. In a study focusing on the detection of high-grade stenoses, Achenbach et al. *(53)* reported 92% sensitivity and 94% specificity when confirmed by angiography. The data reported in their study have not been as reproducibly reported. In another study, the reported sensitivity of comparing EBCT results to coronary angiography was 74% and specificity 14% *(54,55)*. The reported differences among researchers in the field may have resulted from the use of different radiographic and scoring protocols, thus limiting direct comparison of their data. As a screening tool, treadmill ECG, technetium stress, and calcium scoring were compared in a study by Shavelle et al. where a "positive" EBCT was defined as an Agatston score >0. A more significant negative predictive value was yielded by EBCT than technetium stress testing (82 versus 57%) *(56)*. Analysis methodology aside, EBCT appears to have an increasingly relevant role in the noninvasive assessment of CAD.

Large-scale clinical trials have been conducted to assess the use of calcium scoring as a noninvasive method for the determination of risk stratification. Subsequently, EBCT calcium scores have been developed for both men and women. Although the presence of any detectable calcium may be considered abnormal, the likelihood of having detectable calcium is roughly equal to the age in men and 10–15 points below the age in women. In addition, the extent of calcification increases with age, with women lagging approximately 10 years behind men *(57,58)*. Newly acquired data from the Multi-Ethnic Study of Atherosclerosis (MESA) shows disparate baseline calcium scores in Caucasians, Blacks, Hispanics, and Chinese subpopulations. In patients with no prior documented cardiovascular disease, 70.9% of white men and 34.9% of Hispanic women had an Agatston score >0 *(59)*. Several studies have suggested different thresholds for abnormal calcium, with scores ranging from the 50th to over the 90th percentile. Others have suggested an absolute threshold, with specific cut-points ranging from 80 to 640 Hounsfield units *(60)*. EBCT coronary calcium scores have been shown to be superior to the Framingham risk factors in predicting the measured proximal plaque burden, defined as the sum of percent stenosis for the worst lesion found in each of nine standard proximal coronary segments determined by quantitative coronary angiography (QCA) *(61)*. Although studies have shown that EBCT coronary calcium scores are highly predictive of coronary lesions, there is a high interobserver and intraobserver variability. One study demonstrated an intraobserver variability of 49 ±45% when the scan was read on 2 consecutive days *(62)*, with another study reported an interobserver variability of 24% *(61)*. In order to reduce measurement variability, electrocardiogram triggering, averaging studies, and controlled respiration efforts may improve variability by as much as 10% *(63)*. Recent data suggest calcium scoring is an additive predictor to the established Framingham risk score (FRS). A study by Shaw et al. *(64)* demonstrated 5-year mortality rates ranging from 2.0% in patients with a calcium score of <10 and 12.2% with a score of >1000 in the high risk FRS subset. Similarly, asymptomatic patients were risk stratified using FRS and calcium scoring. The study concluded that calcium scoring was predictive of risk only in patients with FRS >10% *(65)*. Correlation of EBCT coronary calcium scores to clinical endpoints remains to be established. Another study was conducted to assess the efficacy of coronary calcification screening in order to motivate patients to make changes in risk factors. This study showed no significant improvement during the 1-year study period *(66)*. Additionally, therapeutic approaches using calcium scoring as a guideline is being explored. In a hallmark study, moderate versus aggressive lipid-lowering therapy was administered in hyperlipidemic, postmenopausal women. Lipid profiles were favorably improved in the aggressive arm, with no substantial difference in the calcium score noted between the two subgroups (intensive versus moderate: median 15.1% and 14.3%, respectively) *(67)*. Current guidelines have not been developed defining which patient cohort benefits from EBCT imaging. Also, the variability of measurement may preclude current recommendations for the serial use of EBCT scans to determine therapeutic responses to interventions designed to cause regression of disease *(68)*. The ongoing prospective MESA project *(59)*, currently being conducted by the National Heart, Lung, and Blood Institute/National Institutes of Health, may clarify this issue.

MULTI-DETECTOR ROW COMPUTED TOMOGRAPHY

MDCT or MSCT produces multiple, simultaneous images reducing patient exposure and resulting in significantly higher spatial resolution. The current number of "slices" that can be taken at once is 64, whereas just 1 year ago, the total number was 16, with increases in technology likely to continue. In its current state, MDCT requires careful attention to ECG gating during diastole. Additionally, the use of β-blockers may be indicated to maintain the heart rate in a range of 60 beats/min *(69)*. The spatial resolution of MDCT exceeds that of EBCT, hence making it a logical, noninvasive analog to coronary angiography. However, the use of MDCT as a means of calcium scoring has shown inconsistent results. A study from Japan yielded low variability of coronary artery calcium measurement on two sequential scans using 16-slice MDCT with overlapping reconstruction and retrograde ECG gating *(70)*. Becker et al. examined calcium scores in EBCT versus MDCT which resulted in 32% variability between the two modalities. The variability widened in calcium scores <100 which was attributed to the Agatston score *(71)*. MDCT, 64-slice, calcium scoring data (see figure below) are yet to be released which may further delineate its use for this modality.

The use of MDCT for the detection of luminal stenosis has been studied in 4-slice and 16-slice MDCT models. Neiman et al. compared >50% stenoses with 16-slice MDCT and invasive angiography yielding very favorable numbers with a sensitivity of 95% and specificity of 86% ($n = 59$). Of note, vessels with a diameter >2 mm were included in this study *(72)*. Older, 4-slice MDCT studies resulted in lower sensitivities ranging from 72 to 93% *(73,74)*. While many 64-slice studies are ongoing, it is expected that this improved resolution may result in increased sensitivity and specificity. Recently, reports compared the use of a 64-slice scanner technology to QCA and IVUS. In proximal to midlesions, the MDCT yielded sensitivities of 80% in <50% stenosis, 75% in >50% stenosis, and 88% in >75% stenosis when placed head-to-head with QCA. Sensitivities were less impressive for distal and heavily calcified segments *(75)*. Plaque morphology has also been studied using MDCT. In the first study of its kind, proximal left anterior descending (LAD) lesions were studied using IVUS and MDCT which were then divided into soft, intermediate, and calcified plaques by IVUS criteria. The IVUS morphology of the plaques correlated significantly with the calcium score of each plaque *(76)*. Sparse literature is available regarding MDCT as an adjunct to echocardiography (i.e., assessment of wall motion). Evaluation of wall motion using processed animation with comparison to conventional left ventriculography was done which showed a 94% match of wall motion classification (normal, hypokinesis, and akinesis) between the two modalities *(77)*.

The anticipated clinical role of MDCT is not defined yet; however, it offers great promise as a screening tool or as prescursor to invasive angiography. The widespread

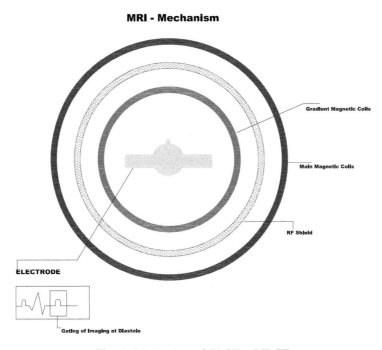

Fig. 1. Mechanism of 64-Slice MDCT.

use of the technology awaits further clinical trials. Additionally, patient accessibility, reimbursement, and cost-effectiveness of performing screening tests will need to be addressed.

Cardiac Magnetic Resonance Imaging

CMRI is a sophisticated powerful imaging system providing superb anatomic, functional, and tissue characterization images within a single imaging modality. The explosion of MRI has now led to a plethora of dedicated cardiovascular centers with MRI capabilities thus providing more access to the MR information *(78)*. At various points in time, cardiac MR has been viewed as an "all-in-one" tool; however, for purposes of this chapter, the scope of information will be restricted to a discussion of the uses of MRI in evaluating atherosclerosis.

Today, based on the current level of technology available, the MR-based evaluation of coronary anatomy remains challenging. An early study showed that 84% of proximal or middle coronary artery segments were interpretable on Magnetic resonance imaging (MRA). In these segments, 83% of lesions were also detected by MRA. The sensitivity, specificity, and accuracy of MRA for diagnosing any coronary disease (>50% stenosis by x-ray angiography) were 100, 85, and 87%, respectively, in patients with left main or three-vessel disease *(79)*. Atherosclerotic plaque has been studied using cardiac MR in both coronary and peripheral vasculature. Coronary vessel wall imaging using black blood MRI showed that proximal right coronary artery wall thickness was significantly greater in patients who were clinically healthy versus patients who had 10–50% stenosis of the vessel as noted by x-ray angiography (see figure below). Of note, lumen diameter and lumen area were similar in both arms, hence suggestive that CMRI may be useful in detecting subclinical disease *(80)*. As an adjunct to this study, a trial with patients scheduled for carotid endarterectomy underwent preoperative carotid

MRI. Plaque fibrous cap was then categorized as intact-thick, intact-thin, or ruptured for each carotid plaque. A higher percentage of symptomatic patients had ruptured caps (70%) compared with thick caps (9%), and patients with ruptured caps were 23 times more likely to have had a recent transient ischemic attack TIA or stroke *(81)*. In a comparative study of MRI versus 16-slice MDCT in rabbits, sensitivity and specificity, respectively, to detect noncalcified, atherosclerotic plaques were 89% and 77% for MDCT and 97% and 94% for MRI. The more favorable numbers for MRI were attributed to the better spatial resolution and tissue attenuation associated with MR *(82)*. Plaque studies with concurrent lipid-lowering therapy are of clinical value. Based on a study evaluating the effect of lipid-lowering therapy on aortic and carotid plaques, reproducibility of repeated vessel wall area measurements was found to be significant, with the error in measurement being 2.6% for aortic and 3.5% for the carotid plaques, respectively *(83)*. Plaque area and volume measurements were also associated with a similar low error rate of 4 and 6%, respectively *(84,85)*. This high reliability has led to the use of this technique in serial measurements and as a surrogate endpoint in lipid lowering trials.

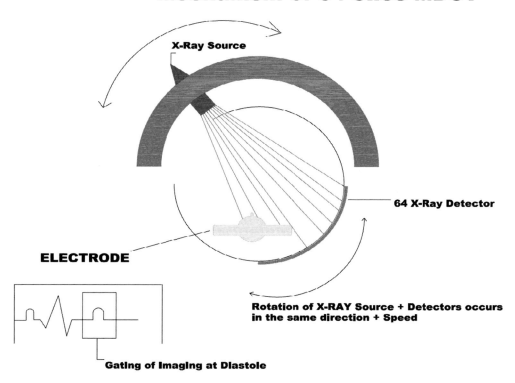

Fig. 2. Mechanism of 64 Slice MDCT.

Ankle-Brachial Index

ABI was initially used to identify the presence and extent of peripheral vascular disease. The ABI is measured as the ratio of the systolic blood pressure in the posterior tibial and dorsalis pedis arteries over the systolic blood pressure in the brachial artery. Differences in the two values are demonstrated due to reductions in the perfusion pressure distal to a critical stenosis. It is a "low-tech" modality because the only requirements include a sphynomanometer and a Doppler probe. The brachial pressures are generally measured in both arms, and the higher value is chosen due to the possibility of a falsely low brachial pressure secondary to unilateral subclavian artery stenosis. Similarly, the higher measurement is recorded from the lower extremities *(61)*. An early study used ABI as a marker of lipid-lowering therapy efficacy in which adverse lipid panels were represented by a decreased ABI. These concepts have applied to the theory that similar findings could be used for the detection of coronary heart disease based on the belief that the systemic affects of atherosclerosis can be detected in any or all vascular beds. In a definitive paper, after adjustments were made for age, LDL-C, and carotid and f-IMT, an ABI <0.9 was shown to be an independent predictor of cardiovascular events (i.e., cardiac death, nonfatal myocardial infarction, and unstable angina) *(86)*. An ABI <0.9 yielded a 90% sensitivity and a 98% specificity for moderate-to-severe obstructive peripheral arterial disease as determined by confirmatory angiography *(87,88)*. While ABI alone is not an ideal modality to screen for mild disease, it was estimated that 40% of patients with positive ABI were asymptomatic, rendering the method an effective means of finding subclinical disease *(89,90)*. Repeated measurements have been shown to be of benefit as the difference in values was only ±10% in comparison to ±16% in single testing *(91)*. However, the use of serial ABI measurements have not been found to be sensitive enough for serially monitoring milder forms of peripheral vascular atherosclerosis.

Brachial Artery Reactivity Testing

Endothelial mediators have achieved renewed interest because they may be used for provocative testing for the detection of clinical atherosclerosis. The blunting of the vasodilatory reaction mediated by endothelial nitric oxide is considered to be an early sign of endothelial dysfunction. As a result, delay or decreased vasodilatation in the setting of increased shear force may indicate some degree of endothelial dysfunction. Therefore, brachial artery reactivity testing or flow-mediated dilatation (FMD) may predict development of atherosclerosis *(61)*. Factors such as smoking, ingestion of fat or caffeine, and temperature affect the response. Patients are therefore instructed appropriately as not to influence the study. In order to perform this study, relative ischemia is induced by placing a cuff over the subject's antecubital fossa for 5 min at a pressure 50 mm Hg above patient's systolic pressure. Ultrasound/Doppler images of the brachial artery using a high-frequency ultrasound transducer (7–12 MHz) are performed permitting an image of the brachial artery before cuff inflation and 1 min after cuff release in order to produce vasodilatation. Reactive hyperemia produces a nearly sixfold increase of blood through the brachial artery following the release of the occlusive blood pressure cuff. Subsequently, subjects are given sublingual nitroglycerin and tested again to document endothelium-independent vasodilatation in the setting of extrinsically administered nitric oxide. Either manual or automated processes are employed; the change in diameter of the brachial artery before and after the stimulus is measured as a percentage of baseline diameters. Cuff placement in the antecubital fossa has been shown to produce a vasodilator response greater than 8% *(61)*. Factors such as cuff placement, baseline artery diameter, and age (decreased response in men over 40 years of age and women over 50 years of age) have been shown to affect the response. A poor prognosis has been shown in patients with chest pain and impaired vasodilatory response of the brachial artery *(92)*. In relation to postprocedural outcomes, a recent prospective study involving 136 patients with single-vessel CAD bare-metal stents showed that impaired FMD was an independent risk factor of clinically significant in-stent restenosis *(93)*.

Noninvasive Assessment of Coronary Flow Reserve

CFR is a relatively new use of epi-cardiac Doppler imaging. This technique is used to describe morphology and functional status of coronary blood flow. Initially introduced by Gould and Lipscomb *(94)*, coronary blood flow reserve is measured as the ratio of hyperemic and basal blood flow. Their initial studies showed the link between stenosis secondary to atherosclerosis and blood flow. Chilian and Marcus *(95)* showed that intramural and epicardial blood flow are closely related in the distal coronary artery but varied significantly in more proximal segments secondary to changes in segmental capacitance. This concept has defined our understanding of the pathophysiology of coronary flow and forms the basis for this noninvasive measurement of CFR of the distal segment *(96)*. While transthoracic coronary Doppler echocardiography has been validated, imaging capability is variable with each coronary artery [LAD coronary artery (100% with or without contrast), 50% for the posterior descending artery *(97–103)*]. Measurement of flow velocity using pulsed wave Doppler ultrasound with color flow mapping is done during infusion of potent microvascular dilators adenosine and dipyridamole. With the knowledge that flow is the product of velocity and cross-sectional area of the vessel, any increase in flow can only be explained by an increase in the velocity as cross-sectional area remains the same. Classification by CFR is grouped by nonsignificant (<50%), intermediate (50–69%), and severe (>90%) LAD narrowing. A value of less than 2 has a specificity of 82% and a sensitivity of 92% to predict significant (>70%) LAD stenosis as described by Hozumi et al. *(97,98)* This cutoff value was confirmed by Ouriel et al. and Hirsch et al. *(88,89)*. However, 2.5 was shown as the threshold number by the Doppler Endpoints Balloon Angioplasty Trial Europe (DEBATE) study *(104,105)*, with correlation of fewer coronary events after angioplasty. This derived value may assist a clinician in deciding whether a patient is a candidate for PCI and whether an individual lesion will benefit from a stent procedure. While large number studies exist for native coronary studies, less was available for stented segments until recent data described the use of CFR studies for decision-making in in-stent restenosis *(106)*.

IMAGING VULNERABLE PLAQUE

Conceptually, the identification of "vulnerable" or "culprit" plaque is an extremely attractive and critically important issue in cardiology today. The a priori knowledge of an unstable plaque would be paramount in preventing a potential untoward clinical event. Radiolabeled monoclonal antibodies targeted at specific molecular components of vulnerable plaque in a necrotic core has been described. Additionally, markers of local plaque inflammation such as 18F-fluorodeoxyglucose and 19F-fluorodeoxyglucose have been studied *(61)*. However, unstable plaque imaging with MRI continues to be highly visible in the published literature. These studies have been alluded to in the section *Cardiac Magnetic Resonance Imaging*.

PLAQUE NEOVASCULARIZATION: IN VITRO AND IN VIVO

Several studies have ascribed to the presence of proliferative neovascularization to culprit atherosclerotic plaque. These neovascularization changes are represented as increased adventitial, media, and intima vasa vasorum and have been associated with plaque instability. (Fig. 1A, B) *(107)*. This coronary artery vasa vasorum neovascularization noted in cholesterol-fed animals was reduced following oral injections of statins *(108)*. These data were further supported in a study which showed that the monoclonal endothelial cell marker CD34 staining revealed increased neovessel density in the base of ruptured plaque in the human aorta *(109)*. At the present time, contrast-enhanced, ultrasound imaging of the neovascular changes observed within the carotid artery is a promising noninvasive technique for the noninvasive detection of the unstable plaque. This ultrasound imaging technique is readily available to clinicians due to a large installed base and vascular trained technicians. Furthermore, this modality has a favorable economic basis based on the low threshold for entry into the imaging field. These practical considerations along with the remarkable spatial and temporal resolution of contrast ultrasound imaging contribute to the allure of using ultrasound contrast imaging for detection of unstable plaques (Figs. 2, 3, and 4). Following the use of contrast agents, the neovascularization (vasa vasorum) appears to be originating

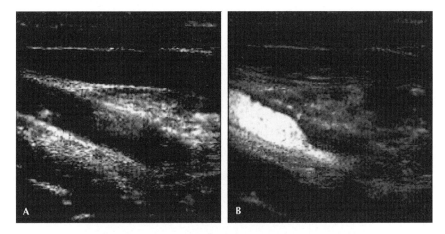

Fig. 1. (A) Carotid atery imaging without contrast enhancement. (B) Carotid artery imaging with contrast enhancement.

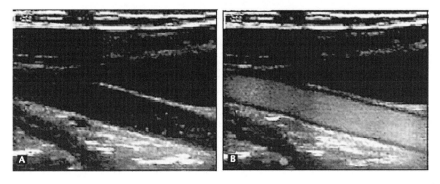

Fig. 2. (A) Vizualization of carotid plaque without enhancement. (B) Vizualization of carotid plaque with contrast enhancement.

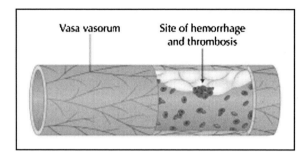

Fig. 3. Schematic representation of the contribution of the vasa vasorum to the genesis of the atherosclerotic plaque.

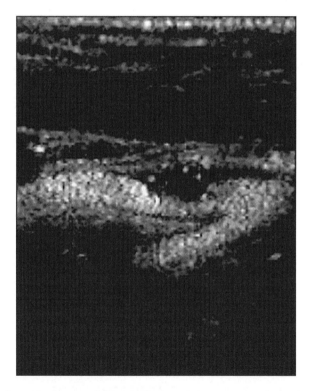

Fig. 4. Contrast-enhanced ultrasound image demonstrating neovascularization within a carotid plaque.

from the "outside-in," adventitia to the intima *(110)*. While the uniqueness of this imaging modality is only recently described, the pathophysiology observed within the atherosclerotic plaque is consistent with the prior pathology studies of Barger *(111)* in 1984, who also references the work of Winternitz in 1938. These authors aptly described the "presence of rich vascular channels surrounding and penetrating the sclerotic lesions" *(112)*.

CONCLUSIONS

Currently, the unprecedented technological advancements in the field of cardiovascular diagnostics may result in earlier identification and treatment of high-risk patients. These noninvasive, diagnostic imaging tests may play a pivotal role in developing new diagnostic and treatment paradigms.

REFERENCES

1. Kuller LH, Shemanski L, Psaty BM, et al. Subclinical disease as an independent risk factor for cardiovascular disease. Circulation 1995; 92:720–726.
2. Pignoli P, Tremoli E, Poli A, et al. Intimal plus medial thickness of the arterial wall: a direct measurement with ultrasound imaging. Circulation 1986, 74:1399–1406.
3. Poli A, Tremoli E, Colombo A, et al. Ultrasonographic measurement of the common carotid artery wall thickness in hypercholesterolemic patients. A new model for the quantitation and follow-up of preclinical atherosclerosis in living human subjects. Atherosclerosis 1988; Apr, 70(3):253–261.
4. Smith SC Jr, Amsterdam E, Balady GJ, et al. Prevention Conference V: beyond secondary prevention: identifying the high risk patient for primary prevention: tests for silent and inducible ischemia: Writing Group II. Circulation 2000; 101:E12–E16.
5. Glagov S, Zarins CK, Masawa N, et al. Mechanical functional role of non-atherosclerotic intimal thickening. Front Med Biol Eng 1993; 5(1):37–43.
6. Guidelines committee. 2003 European Society of Hypertension–European Society of Cardiology guidelines for the management of arterial hypertension. J Hypertens 2003; 21:1011–1053.
7. Grobbee DE, Bots ML. Carotid artery intima-media thickness as an indicator of generalized atherosclerosis. J Intern Med 1994; 236:567–573.
8. Meinders JM, Kornet L, Hoeks APG. Assessment of spatial inhomogeneities in intima media thickness along an arterial segment using its dynamic behavior. Am J Physiol Heart Circ Physiol 2003; 285:H384–H391.
9. Burke GL, Evans GW, Riley WA, et al. Arterial wall thickness is associated with prevalent cardiovascular disease in middle aged adults. The Atherosclerosis Risk in Communities (ARIC) Study. Stroke 1995; 26:386–391.
10. Newman AB, Naydeck B, Sutton-Tyrrell K, et al. Coronary artery calcification in older adults with minimal clinical or subclinical cardiovascular disease. J Am Geriatr Soc 2000; 48:256–263.
11. Bots ML, Hoes AW, Koudstaal PJ, et al. Common carotid intima-media thickness and risk of stroke and myocardial infarction: the Rotterdam Study. Circulation 1997; 96:1432–1437.
12. Mannami T, Baba S, Ogata J: Strong and significant relationships between aggregation of major coronary risk factors and the acceleration of carotid atherosclerosis in the general population of a Japanese city: the Suita Study. Arch Intern Med 2000; 160:2297–2303.
13. Hernandez-Perera O, Perez-Sala D, Navarro-Antolin J, et al. Effects of the 3-hydroxy-3-methylglutaryl-CoA reductase inhibitors, atorvastatin and simvastatin, on the expression of endothelin-1 and endothelial nitric oxide synthase in vascular endothelial cells. J Clin Invest 1998; 101:2711–2719.
14. Taylor AJ, Kent SM, Flaherty PJ, et al. ARBITER: Arterial Biology for the investigation of the treatment effects of reducing cholesterol: a randomized trial comparing the effects of atorvastatin and pravastatin on carotid intima medial thickness. Circulation 2002; 106:2055–2060.
15. Smilde TJ, van Wissen S, Wollersheim H, et al. Effect of aggressive versus conventional lipid lowering on atherosclerosis progression in familial hypercholesterolaemia (ASAP): a prospective, randomised, double-blind trial. Lancet 2001; 357:577–581.
16. Furberg CD, Adams HP Jr, Applegate WB, et al. Effect of lovastatin on early carotid atherosclerosis and cardiovascular events. Asymptomatic Carotid Artery Progression Study (ACAPS) Research Group. Circulation 1994; 90:1679–1687.
17. Hodis HN, Mack WJ, LaBree L, et al. Reduction in carotid arterial wall thickness using lovastatin and dietary therapy: a randomized controlled clinical trial. Ann Intern Med 1996; 124:548–556.
18. Mercuri M, Bond MG, Sirtori CR, et al. Pravastatin reduces carotid intima-media thickness progression in an asymptomatic hypercholesterolemic mediterranean population: the Carotid Atherosclerosis Italian Ultrasound Study. Am J Med 1996; 101:627–634.

19. Salonen R, Nyyssonen K, Porkkala E, et al. Kuopio Atherosclerosis Prevention Study (KAPS). A population-based primary preventive trial of the effect of LDL lowering on atherosclerotic progression in carotid and femoral arteries. Circulation 1995; 92:1758–1764.
20. Byington RP, Furberg CD, Crouse JR III, et al. Pravastatin, lipids, and atherosclerosis in the carotid arteries (PLAC–II). Am J Cardiol 1995; 76:54C–59C.
21. de Groot E, Jukema JW, van Boven AJ, et al. Effect of pravastatin on progression and regression of coronary atherosclerosis and vessel wall changes in carotid and femoral arteries: a report from the Regression Growth Evaluation Statin Study. Am J Cardiol 1995; 76:40C–46C.
22. Blankenhorn DH, Selzer RH, Crawford DW, et al. Beneficial effects of colestipol-niacin therapy on the common carotidartery. Two- and four-year reduction of intima-media thickness measured by ultrasound. Circulation 1993; 88:20–28.
23. Wiklund O, Hulthe J, Wikstrand J, et al. Effect of controlled release/extended release metoprolol on carotid intima-media thickness in patients with hypercholesterolemia: a 3-year randomized study. Stroke 2002; 33:572–577.
24. Kastelein JJ, Sager PT, de Groot E, et al. Comparison of ezetimibe plus simvastatin versus simvastatin monotherapy on atherosclerosis progression in familial hypercholesterolemia. Design and rationale of the Ezetimibe and Simvastatin in Hypercholesterolemia Enhances Atherosclerosis Regression (ENHANCE) trial. Am Heart J 2005; Feb, 149(2):234–239.
25. Wiklund O, Hulthe, J, Wikstrand, J, et al. Effect of controlled release/extended release metoprolol on carotid intima-media thickness in patients with hypercholesterolemia: a 3-year randomized study. Stroke 2002; 33:572–577.
26. Hedblad B, Wikstrand J, Janzon L, et al. Low-dose metoprolol CR/XL and fluvastatin slow progression of carotid intimamedia thickness: main results from the beta-blocker Cholesterol-Lowering Asymptomatic Plaque Study (BCAPS). Circulation 2001; 103:1721–1726.
27. Borhani NO, Mercuri M, Borhani PA, et al. Final outcome results of the Multicenter Isradipine Diuretic Atherosclerosis (MIDAS): a randomized controlled trial. JAMA 1996; 276:785–791.
28. Zanchetti A, Rosei E, Dal Palu C, et al. The Verapamil in Hypertension and Atherosclerosis Study (VHAS): results of long-term randomized treatment with either verapamil or chlorthalidone on carotid intima-media thickness. J Hypertens 1998; 16:1667–1676.
29. Simon A, Gariepy J, Moyse D, et al. Differential effects of nifedipine and co-amilozide on the progression of early carotid wall changes. Circulation 2001; 103:2949–2954.
30. Zanchetti A, Bond MG, Hennig M, et al. Calcium antagonist lacidipine slows down progression of asymptomatic carotid atherosclerosis: principal results of the European Lacidipine Study on Atherosclerosis (ELSA), a randomized, double blind, long-term trial. Circulation 2002; 106:2422–2427.
31. Tzou WS, Douglas PS, Srinivasan SR, et al. Increased subclinical atherosclerosis in young adults with metabolic syndrome: the Bogalusa Heart Study. J Am Coll Cardiol 2005; Aug 2, 46(3):457–463
32. Wallenfeldt K, Hulthe J, Fagerberg B. The metabolic syndrome in middle-aged men according to different definitions and related changes in carotid artery intima-media thickness (IMT) during 3 years of follow-up. J Intern Med 2005; 258:1–28.
33. Iglseder B, Cip P, Malaimare L, et al. The metabolic syndrome is a stronger risk factor for early carotid atherosclerosis in women than in men. Stroke 2005; Jun, 36(6):1212–1217.
34. Vryonidou A, Papatheodorou A, Tavridou A, et al. Association of hyperandrogenemic and metabolic phenotype with carotid intima-media thickness in young women with polycystic ovary syndrome. J Clin Endocrinol Metab 2005; May, 90(5):2740–2746.
35. Juonala M, Raitakari M, Viikari JSA, et al. Obesity in youth is not an independent predictor of carotid IMT in adulthood. The Cardiovascular Risk in Young Finns Study. Atherosclerosis 2006; Apr, 185(2):388–393.
36. Oren A, Vos LE, Uiterwaal CS, et al. Birth weight and carotid intima-media thickness: new perspectives from the atherosclerosis risk in young adults (ARYA) study. Ann Epidemiol 2004; Jan, 14(1):8–16.
37. Segers P, Rabben SI, De Backer J, et al. Functional analysis of the common carotid artery: relative distension differences over the vessel wall measured in vivo. J Hypertens 2004; 22:973–981.
38. Pitt B, Byington RP, Furberg CD, et al. Effect of amlodipine on the progression of atherosclerosis and the occurrence of clinical events. Prevent investigators. Circulation 2000; 102:1503–1510.
39. Touboul PJ, Prati P, Scarabin PY, et al. Use of monitoring software to improve the measurement of carotid wall thickness by B-mode imaging. J Hypertens Suppl 1992; 10:S37–S41.

40. Macioch JE, Patel SN, Bai CJ, et al. Contrast-enhanced ultrasound imaging measurement of carotid intimal medial thickness. J Am Coll Cardiol 2003; 41(suppl A):318A.
41. Secil M, Altay C, Gulcu A, et al. Automated measurement of intima-media thickness of carotid arteries in ultrasonography by computer software. Diagn Interv Radiol 2005; Jun, 11(2):105–108.
42. Wong M, Edelstein J, Wollman J, et al. Ultrasonic-pathological comparison of the human arterial wall: verification of intima-media thickness. Arterioscler Thromb 1993; 13:482–486.
43. de Groot E, Jukema JW, Montauban van Swijndregt AD, et al. B-mode ultrasound assessment of pravastatin treatment effect on carotid and femoral artery walls and its correlations with coronary arteriographic findings: a report of the Regression Growth Evaluation Statin Study (REGRESS). J Am Coll Cardiol 1998; Jun, 31(7):1561–1567.
44. Paul TK, Srinivasan SR, Wei C, et al. Cardiovascular risk profile of asymptomatic healthy young adults with increased femoral artery intima-media thickness: The Bogalusa Heart Study. Am J Med Sci 2005; Sep, 330(3):105–110.
45. Held C, Hjemdahl P, Eriksson SV, et al. Prognostic implications of intima-media thickness and plaques in the carotid and femoral arteries in patients with stable angina pectoris. Eur Heart J 2001; Jan, 22(1):62–72.
46. Levy D, Garrison RJ, Savage DD, et al. Prognostic implications of echocardiographically determined left ventricular mass in the Framingham Heart Study. N Engl J Med 1990; 322:1561–1566.
47. Sundstrom J, Lind L, Arnlov J, et al. Echocardiographic and electrocardiographic diagnoses of left ventricular hypertrophy predict mortality independently of each other in a population of elderly men. Circulation 2001; 103:2346–2351.
48. Al-Mansour HA, Mulvagh SL, Pumper GM, et al. Usefulness of harmonic imaging for left ventricular opacification and endocardial border delineation by optison. Am J Cardiol 2000; Mar 15, 85(6):795–799, A10.
49. Verdecchia P, Angeli F, Pittavini L, et al. Regression of left ventricular hypertrophy and cardiovascular risk changes in hypertensive patients. Ital Heart J 2004; July, 5:505–510.
50. O'Rourke RA, Brundage BH, Froelicher VF, et al. American College of Cardiology/American Heart Association Expert Consensus Document on electron-beam computed tomography for the diagnosis and prognosis of coronary artery disease. J Am Coll Cardiol 2000; 36:326–340.
51. Calister TQ, Cooil B, Raya BP, et al. Coronary artery disease: improved reproducibility of calcium scoring with electron beam CT volumetric method. Radiology 1998; 208:807–814.
52. Schmermund A, Denktas AE, Rumberger JA, et al. Independent and incremental value of coronary artery calcium for predicting the extent of angiographic coronary artery disease: comparison with cardiac risk factors and radionuclide perfusion imaging. J Am Coll Cardiol 1999; 34:777–786.
53. Achenbach S, Moshage W, Ropers D, et al. Value of electron-beam computed tomography for the detection of high-grade coronary artery stenoses and occlusions. N Engl J Med 1998; 339:1964–1971.
54. Nakinishi T, Ito K, Imazu M, et al. Evaluation of coronary artery stenoses using electron-beam CT and multiplanar reformation. J Comput Assist Tomogr 1997; 21:121–127.
55. Budoff MJ, Georgiou D, Brody A, et al. Ultrafast computed tomography as a diagnostic modality in the detection of coronary artery disease: a multicenter study. Circulation 1996; 93:898–904.
56. Shavelle DM, Budoff MJ, Lamont DH, et al. Exercise testing and electron beam computed tomography in the evaluation of coronary artery disease. J Am Coll Cardiol 2000; July, 36(1):32–38.
57. Hoff JA, Chomka EV, Krainik AJ, et al. Age and gender distributions of coronary calcium scores detected by electron beam tomography in 35,246 adults. Am J Cardiol 2001; 87:1335–1339.
58. Wong ND, Kouwabunpat D, Vo AN, et al. Coronary calcium and atherosclerosis by ultrafast computed tomography in asymptomatic men and women: relation to age and risk factors. Am Heart J 1994; 127:422–430.
59. Bild DE, Detrano R, Peterson D, et al. Ethnic differences in coronary calcification: the Multi-Ethnic Study of Atherosclerosis (MESA). Circulation 2005; Mar 15, 111(10):1313–1320.
60. Redberg RF, Vogel RA, Criqui MH, et al. 34th Bethesda Conference: Task Force #3-What is the spectrum of current and emerging techniques for the noninvasive measurement of atherosclerosis? J Am Coll Cardiol 2003; 41:1886–1898.
61. Brown BG, Morse J, Zhao XQ, et al. Electron-beam tomography coronary calcium scores are superior to Framingham risk variables for predicting the measured proximal stenosis burden. Am J Cardiol 2001; 88:23E–26E.
62. Devries S, Wolfkiel C, Shah V, et al. Reproducibility of the measurement of coronary calcium with ultrafast computed tomography. Am J Cardiol 1995; 75:973–975.

63. Mao S, Budoff MJ, Bakhsheshi H, et al. Improved reproducibility of coronary artery calcium scoring by electron beamtomography with a new electrocardiographic trigger method. Invest Radiol 2001; 36:363–367.
64. Shaw LJ, Raggi P, Schisterman E, et al. Prognostic value of cardiac risk factors and coronary artery calcium screening for all-cause mortality. Radiology 2003; 228:826–833.
65. Greenland P, LaBree L, Azen SP, et al. Coronary artery calcium score combined with Framingham score for risk prediction in asymptomatic individuals. [see comment] [erratum appears in JAMA 2004 Feb 4;291(5):563]. [Journal Article] JAMA 2004; Jan 14, 291(2):210–215.
66. O'Malley PG, Feuerstein IM, Taylor AJ. Impact of electron beam tomography, with or without case management, on motivation, behavioral change, and cardiovascular risk profile: a randomized controlled trial. JAMA 2003; 289:2215–2223.
67. Raggi P, Davidson M, Callister TQ, et al. Aggressive versus moderate lipid-lowering therapy in hypercholesterolemic postmenopausal women: beyond endorsed lipid lowering with EBT scanning (BELLES). Circulation 2005; Jul 26, 112(4):563–571 [Epub 11 Jul 2005].
68. Raggi P. Coronary calcium on electron beam tomography imaging as a surrogate marker of coronary artery disease. Am J Cardiol 2001; 87:27A–34A.
69. Auchenbach S, Hoffman U, Ferencik M, et al. Tomographic coronary angiography by EBCT and MDCT. Prog Cardiovasc Dis 2003; 46(2) (Septemeber/October):185–195.
70. Horiguchi J, Yamamoto H, Akiyama Y, et al. Variability of repeated coronary artery calcium measurements by 16-MDCT with retrospective reconstruction. AJR Am J Roentgenol 2005; Jun, 184(6):1917–1923.
71. Becker, C, Jakobs T, Aydemir S, et al. Helical and single slice conventional versus electron beam CT for quantification of coronary artery calcification. AJR Am J Roentgenol 2000; 174:543–547.
72. Neiman K, Cademartiri F, Lemos PA, et al. Reliable noninvasive coronary angiography with fast submillimeter multislice spiral computed tomography. Circulation 2002; 106:2051–2054.
73. Herzog C, Abolmaali N, Balzer JO, et al. Heart-rate-adapted image reconstruction of multidectector-row cardiac CT: influence of physiological and technical prequesite on image quality. Eur Radiol 2002; 12:1670–2678.
74. Kopp AF, Schroeder S, Knuetter A, et al. Non-invasive coronary angiography with high resolution multidetector-row computed tomography. Results in 102 patients. Eur Heart J 2002; 23:1714–1725.
75. Leber AW, Knez A, von Ziegler F, et al. Quantification of obstructive and nonobstructive coronary lesions by 64-slice computed tomography: a comparative study with quantitative coronary angiography and intravascular ultrasound. J Am Coll Cardiol 2005; 46:147–154.
76. Kopp AF, Schröder S, Baumbach A, et al. Non-invasive characterization of coronary lesion morphology and composition by multislice computed tomography: first results in comparison with intracoronary ultrasound. Eur Radiol 2000; 11:1607–1611.
77. Mochizuki T, Higashino H, Kayama Y, et al. Evaluation of wall motion using multi-detector-row CT: new application of post-processing interactive multi-planar animation of the heart. Radiology 2001; 221(P):413.
78. Lima JA, Desai MY. Cardiovascular magnetic resonance imaging: current and emerging applications. J Am Coll Cardiol 2004; September 15, 44(6):1164–1171.
79. Kim WY, Danias PG, Stuber M, et al. Coronary magnetic resonance angiography for the detection of coronary stenoses. N Engl J Med 2001; Dec 27, 345(26):1863–1869.
80. Kim WY, Stuber M, Bonert P, et al. Three-dimensional black-blood cardiac magnetic resonance coronary vessels wall imaging detects positive arterial remodeling in patients with non-significant coronary artery disease. Circulation 2002; 106:296–299.
81. Yuan C, Zhang SX, Polissar NL, et al. Identification of fibrous cap rupture with magnetic resonance imaging is highly associated with recent transient ischemic attack or stroke. Circulation 2002; 105;181–185.
82. Viles-Gonzalez JF, Poon M, Sanz J, et al. In vivo 16-slice, multidetector-row computed tomography for the assessment of experimental atherosclerosis: comparison with magnetic resonance imaging and histopathology. Circulation 2004; Sep 14, 110(11):1467–1472 [Epub 7 Sep 2004].
83. Corti R, Fayad ZA, Fuster V, et al. Effects of lipid-lowering by simvastatin on human atherosclerotic lesions: a longitudinal study by high-resolution, noninvasive magnetic resonance imaging. Circulation 2001; 104:249–252.
84. Chan SK, Jaffer FA, Botnar RM, et al. Scan reproducibility of magnetic resonance imaging assessment of aortic atherosclerosis burden. J Cardiovasc Magn Reson 2001; 3:331–338.

85. Kang X, Polissar NL, Han C, et al. Analysis of the measurement precision of arterial lumen and wall areas using high resolution MRI. Magn Reson Med 2000; 44:968–972.
86. Papamichael CM, Lekakis JP, Stamatelopoulos KS, et al. Ankle brachial index as a predictor of the extent of coronary atherosclerosis and cardiovascular events in patients with coronary artery disease. Am J Cardiol 2000; 86:615–618.
87. Yao ST, Hobbs JT, Irvine WT. Ankle systolic pressure measurements in arterial disease affecting the lower extremities. Br J Surg 1969; 56:676–679.
88. Ouriel K, McDonnell AE, Metz CE, et al. Critical evaluation of stress testing in the diagnosis of peripheral vascular disease. Surgery 1982; 91:686–693.
89. Hirsch AT, Criqui MH, Treat-Jacobson D, et al. Peripheral arterial disease detection, awareness, and treatment in primary care. JAMA 2001; 286:1317–1324.
90. Criqui MH, Fronek A, Klauber MR, et al. The sensitivity, specificity, and predictive value of traditional clinical evaluation of peripheral arterial disease: results from noninvasive testing in a defined population. Circulation 1985; 71:516–522.
91. Fowkes FG, Housley E, Macintyre CC, et al. Variability of ankle and brachial systolic pressures in the measurement of atherosclerotic peripheral arterial disease. J Epidemiol Community Health 1988; 42:128–133.
92. Neunteufl T, Heher S, Katzenschlager R, et al. Late prognostic value of flow-mediated dilation in the brachial artery of patients with chest pain. Am J Cardiol 2000; 86:207–210.
93. Patti G, Pasceri V, Melfi R, et al. Impaired flow-mediated dilation and risk of restenosis in patients undergoing coronary stent implantation. Circulation 2005; Jan 4, 111(1):8–10.
94. Gould KL, Lipscomb K. Effects of coronary stenoses on coronary flow reserve and resistance. Am J Cardiol 1974; 34:48–55.
95. Chilian WM, Marcus ML. Phasic coronary blood flow velocity in intramural and epicardial coronary arteries. Circ Res 1982; 50:775–781.
96. Feinstein SB, Voci P, Pizzuto F. Noninvasive surrogate markers of atherosclerosis. Am J Cardiol 2002; 89:31C–43C; discussion 43C–44C.
97. Hozumi T, Yoshida K, Ogata Y, et al. Noninvasive assessment of significant left anterior descending coronary artery stenosis coronary flow velocity reserve with transthoracic color Doppler echocardiography. Circulation 1998; 97:1557–1562.
98. Hozumi T, Yoshida K, Akasaka T, et al. Noninvasive assessment of coronary flow velocity and coronary flow velocity reserve in the left anterior descending coronary artery by Doppler echocardiography: comparison with invasive technique. J Am Coll Cardiol 1998; 32:1251–1259.
99. Caiati C, Montaldo C, Zedda N, et al. New noninvasive method for coronary flow reserve assessment: contrast–enhanced transthoracic second harmonic echo Doppler. Circulation 1999; 99:771–778.
100. Caiati C, Zedda N, Montaldo C, et al. Contrast-enhanced transthoracic second harmonic echo Doppler with adenosine: anoninvasive, rapid and effective method for coronary flowreserve assessment. J Am Coll Cardiol 1999; 34:122–130.
101. Pizzuto F, Voci P, Sinatra R, et al. Noninvasive assessment of coronary flow velocity reserve before and after angioplasty in a patient with mammary graft stenosis. Ital Heart J 2000; 1:636–639.
102. Pizzuto F, Voci P, Mariano E, et al. Assessment of flow velocity reserve by transthoracic Doppler echocardiography andvenous adenosine infusion before and after left anteriordescending coronary artery stenting. J Am Coll Cardiol 2001; 38:155–162.
103. Pizzuto F, Voci P, Mariano E, et al. Noninvasive coronary flow reserve assessed by transthoracic coronary Doppler ultrasound in patients with left anterior descending coronary artery stents. Am J Cardiol 2003; 91:522–526.
104. Piek JJ, Boersma E, Voskuil M, et al. The immediate and long term effect of optimal balloon angioplasty on the absolute coronary blood flow velocity reserve. A subanalysis of the DEBATE study. Doppler Endpoints Balloon Angioplasty Trial Europe. Eur Heart J 2001; 22:1725–1732.
105. Serruys PW, di Mario C, Piek J, et al. Prognostic value of intracoronary flow velocity and diameter stenosis in assessing the short- and long-term outcomes of coronary balloon angioplasty: the DEBATE Study (Doppler Endpoints Balloon Angioplasty Trial Europe). Circulation 1997; 96:3369–3377.
106. Kruger S, Koch KC, Kaumanns I, et al. Clinical significance of fractional flow reserve for evaluation of functional lesion severity in stent restenosis and native coronary arteries. Chest 2005; Sep, 128(3):1645–1649.

107. Kumamoto M, Nakashima Y, Sueishi K. Intimal neovascularization in human coronary atherosclerosis: its origin and pathophysiological significance. Hum Pathol 1995; 26:450–456.
108. Wilson SH, Heurmana J, Leuman L, et al. Simvastatin preserves the structure of coronary adventitial vasa vasorum in experimental hypercholesterolemia independent of lipid lowering. Circulation 2002; 105:415–418.
109. Moreno PR, Purushothaman KR, Fuster V, et al. Plaque neovascularization is increased in ruptured atherosclerotic lesions of human aorta: implications for plaque vulnerability. Circulation 2004; Oct 5, 110(14):2032–2038.
110. Martin RP, Lerakis S. Contrast for vascular imaging. Cardiol Clin 22(2):313–320.
111. Barger AC, Beeukes R, Lainey LL, et al. Hypothesis: vasa vasorum and neovascularization of human coronary arteries. A possible role in the pathophysiology of atherosclerosis. N Engl J Med 1984; 310:175–177.
112. Folkman T. Angiogenesis. In *Biology of Endothelial Cells*. Jaffe EA (ed.). Boston: Martinus Nijhoff Publishers; 1984:412–428.

19 Management of Dyslipidemia in Children

Stephen R. Daniels, MD, PhD

CONTENTS

    EARLY ATHEROSCLEROSIS
    CLINICAL APPROACH TO CHOLESTEROL ABNORMALITIES
      IN CHILDREN
    THE POPULATION APPROACH
    THE INDIVIDUAL APPROACH
    SUMMARY
    REFERENCES

EARLY ATHEROSCLEROSIS

The concern about the possible development of atherosclerosis in children has been longstanding, but clinically, it has been somewhat unfocused. However, it is the understanding of the natural history in the development of atherosclerosis which drives clinical decision making about identification and control of risk factors in order to prevent its development or slow its progression early in life. Clinical manifestations of atherosclerosis are generally seen in middle age. However, there is increasing evidence of early stages of atherosclerosis including the initiation of risk factors in childhood, and the process is progressive throughout life. Because prevention is thought to be important, it is clear that understanding the early stages of cardiovascular disease development and the role that risk factors play in that process is key.

The earliest pathological abnormality in atherosclerosis is thought to be the fatty streak. This can be characterized as an accumulation of lipid filled macrophages within the intima of an artery *(1)*. As the fatty streak progresses, lipid may continue to accumulate. This then results in the proliferation of macrophages and vascular smooth muscle cells, which migrate into the arterial media to form a fibrous plaque *(2)*. It is ultimately this lesion that can lead to a myocardial infarction or cerebrovascular accident.

Initially, the best understanding of this process came from pathology studies. Early pathology studies were performed during the Korean War *(3)* and the war in Vietnam *(4)*. These studies documented that atherosclerosis was present at a young age and was advanced early in life. What was less clear from these early studies was what factors were related to these early processes in young individuals. There was limited information on the status of lipids and lipoproteins, blood pressure, cigarette smoking, and other risk factors in the soldiers killed in combat who had autopsies.

More recent pathology studies have helped to fill in this gap. The Pathobiological Determinants Atherosclerosis in Youth (PDAY) study was a study of individuals aged 15–34 years who died of accidental causes unrelated to cardiovascular disease *(5)*. They used indicators of risk factor status present at autopsy to define the risk status. They found that traditional risk factors including elevation of cholesterol and blood pressure were associated with the presence of fatty streaks and fibrous plaques *(6)*.

The Bogalusa Heart Study improved on this understanding further by following individuals who had risk factor status measured during school age. When some of those individuals died of accidental causes, the investigators were able to obtain autopsies and evaluate for the presence of atherosclerosis *(7)*. They found that the extent of the arterial surface covered with fatty streaks and fibrous plaques increased with age. The prevalence of fibrous plaques increased from 8% in childhood to almost 70% in young adults. More importantly, they found that the percent of the surface of the artery covered with lesions was associated with elevation of total cholesterol and low-density lipoprotein cholesterol (LDL-C), triglycerides, blood pressure and body mass index, and lower high density lipoprotein cholesterol (HDL-C). The extent of atherosclerotic lesions increased with an increasing number of risk factors present *(8)*.

Since the publication of the autopsy studies, other noninvasive methods of evaluating the atherosclerotic process have become available. These include ultrasound evaluation of the carotid arteries and computed tomography of the coronary arteries. In a study of adolescents and young adults with heterozygous familial hypercholesterolemia, Gidding et al. found coronary calcium present in 7 of 29 subjects *(9)*. Because coronary artery calcium is thought to represent more advanced atherosclerotic lesions that are often not present until the fourth decade of life, these results underscore the potential impact of cholesterol elevation early in life.

Another important question is whether the atherosclerotic process can be ameliorated. Recent studies have demonstrated that treatment of children with familial hyperlipidemia can result in improvement of abnormal vascular function. This has been shown with treatment with statins *(10,11)*, use of antioxidant vitamins *(12)*, and oral supplementation with omega-3 fatty acids *(13)*. This is promising as an abnormality of endothelial function is thought to be an important early component of the atherosclerotic process *(14)*.

It is clear from this body of research that atherosclerosis begins in childhood, progresses through adolescence and young adulthood, and is governed by the traditional risk factors for cardiovascular disease already established for adults. These data establish the basis for the clinical approach to the evaluation and treatment of cholesterol abnormalities in childhood.

CLINICAL APPROACH TO CHOLESTEROL ABNORMALITIES IN CHILDREN

The clinical approach to lipid and lipoprotein abnormalities for adults and children has been presented by the National Cholesterol Education Program (NCEP) of the National Heart Lung and Blood Institute. However, although there have been three reports on adults from the NCEP with the most recent published in 2002, there has only been a single NCEP pediatric report that was published in 1992 *(15)*. Since the publication of that report, science in this area has made substantial progress. So although the overall framework of the approach outlined by the NCEP in 1992 remains valid, many of the details need to be updated and refined *(16)*. What follows is a synthesis of the original report and additional information published since that time.

The pediatric guidelines of the NCEP recommend a two-pronged approach to cholesterol elevation in children. The first approach is a population wide approach, which applies to all children and adolescents and their families. The second approach is an individualized approach, which focuses on the identification and more aggressive treatment of children at high risk of more rapid development of atherosclerosis.

THE POPULATION APPROACH

The rationale for the population approach to cholesterol modification is several fold. First, it was noted that population levels of total cholesterol and LDL-C in children in the US were higher than in many other countries around the world. Second, children in the US have a diet higher in saturated fat and cholesterol than in many other countries. Third, adults in the US also have higher blood cholesterol levels and higher rates of coronary heart disease than many other developed countries. This is important because epidemiologic studies have shown that cholesterol levels tend to track from childhood to adulthood. Thus, children who have high levels relative to their peers are likely to have higher levels as adults *(17)*. Finally, children and adolescents with elevated blood cholesterol often come from families in which other members also have high blood cholesterol. This is a reflection of both shared genes and shared environment. That environment may also be shared with others who come from families with less of a positive family history but, nevertheless, develop elevated blood cholesterol over time.

The goal of the population approach is to establish a healthier approach to eating and physical activity throughout the population. This means lowering the average intake of saturated fats and cholesterol in most children. The rationale is that these population-based changes in diet and activity would result in a lower average population level of blood cholesterol in children and adolescents. It has been suggested that even a relatively small population-based decrease in blood cholesterol, if continued throughout life, would reduce the incidence of coronary heart disease, as these children become adults, by reducing the number of adults with cholesterol elevation *(18)*. It was suggested that the recommended changes in childhood diet could eliminate the age-related increases in cholesterol observed in the US but not in other populations *(19)*.

Recommended Diet

The dietary recommendations are directed at children aged 2 years and older. The concern with younger children is that they need higher intakes of fat and cholesterol to support the rapid neural development that is occurring during this age range.

The general recommendation is that children should consume a diet that has the appropriate number of calories to support normal growth and match energy utilization to maintain a desirable body weight. When the pediatric guidelines were written, childhood obesity was less of a problem than it is currently. Today, fewer children are consuming an appropriate amount of calories to maintain a desirable weight. Table 1 presents information from the *Dietary Guidelines for Americans 2005*, published by the US Department of Agriculture *(20)*. This table outlines the number of "discretionary"

Table 1
Total and Discretionary Calories for Children and Adolescents

	Age (years)		
	4–8	*9–13*	*14–18*
Girls (moderately active)			
Total calories recommended	1600	1800	2000
Discretionary calories	132	195	267
Boys (moderately active)			
Total calories recommended	1600	2000	2600
Discretionary calories	132	267	410

Total and discretionary calories are lower for sedentary children and higher for very active children. Adapted from USDA Dietary Guidelines 2005.

calories that could be eaten by a moderately active child after the number of calories eaten to maintain a healthful diet has been taken into account. It is apparent that the number of available discretionary calories is low and is likely below what many children and adolescents are consuming on a daily basis. The recommendations for the population diet are presented in Table 2. The recommendations combine those of the NCEP pediatric panel *(15)* and the NCEP Adult Treatment Panel III *(21)*.

Recommendations for Children Aged 2 Years and Older

To achieve the goals, the NCEP pediatric panel recommended a variety of lifestyle strategies. These include eating a greater quantity and variety of fruits/vegetables, whole grains, and legumes. Trimmed, lean red meat, poultry without the skin, and fish were recommended in place of choices high in saturated fat. Low-fat dairy products including skim or low-fat milk were encouraged. It was recommended that children consume fewer processed and "fast" foods that tend to be higher in saturated fat. When eating out, it was recommended that children select items that are low in saturated fat and cholesterol.

Table 2
Daily Macronutrient Recommendations for Children Aged 2 Years and Older

Total Fat	25–35% of total calories
Saturated Fat	Less than 10% of total calories
Polyunsaturated Fat	Up to 10% of total calories
Monounsaturated Fat	Up to 20% of total calories
Cholesterol	Less than 200 mg per day
Carbohydrate	50–60% of total calories
Protein	Approximately 15% of total calories
Fiber	20–30 g per day

Adapted from National Cholesterol Education Program Adult and Pediatric Guidelines *(15,21)*.

Implementation of the population approach to diet and physical activity can be complex. There are numerous determinants of the environment in which children live that are important in the lifestyle choices they make. For example, children spend a substantial amount of time in school. Some children may eat two meals and two snacks each day in the school setting. Changing the school environment can be challenging. Children and adolescents are also subjected to intense direct marketing of food and sedentary pursuits. This may have an important impact on the choices they make and choices their parents make. The NCEP pediatric panel recommended that the government and schools, the food and entertainment industries, mass media, and health professionals should all be involved in promoting a healthier environment and implementing the population-based strategy to lower blood cholesterol in children.

THE INDIVIDUAL APPROACH

The individual approach to blood cholesterol in children and adolescents is of most concern to health professionals. It is the individual approach that is focused on the highest risk individuals. It is in this group of children and adolescents that more aggressive approach to lipid lowering is recommended.

Screening

The first component of the individualized approach is the identification of high-risk children and adolescents. This aspect of the individual approach has actually been quite controversial in terms of which children should have measurement of cholesterol and what levels of cholesterol should be considered abnormal.

WHO SHOULD HAVE CHOLESTEROL MEASUREMENTS?

The NCEP pediatric panel considered both a universal and a targeted approach to screening *(15)*. They concluded that a targeted approach was optimum. The recommended group for screening was children with a family history of premature atherosclerotic cardiovascular disease or elevated blood cholesterol. They estimated that approximately 25% of children would get cholesterol screening using their algorithm.

Since that time, investigators have evaluated the performance of this targeted screening approach. These studies to evaluate family history as a screening approach have generally used a spectrum of children from schools or in clinical settings. These investigations have found that from 35 to 46% of children and adolescents would currently be targeted for screening based on a positive family history *(22–26)*.

In addition, the use of the family history has repeatedly been demonstrated to miss children with elevated cholesterol. Estimates have ranged from 30 to 60% in most studies *(22,23,27,28)*. Thus, the approach using family history appears to have flaws. Some of the difficulty may arise from the fact that family history is often difficult to obtain. This may result in unavailable, incomplete, or even inaccurate information. The family history approach also relies on parents knowing their own cholesterol levels because they are often too young to have experienced a cardiac event. Unfortunately, many young adults do not have their cholesterol routinely measured and do not know their blood cholesterol values or are unable to remember or correctly interpret them when they have been tested. In reality, pediatric practices vary widely in their approach to cholesterol screening with some rarely screening, others using family history, and still others screening all children in their practice.

WHAT CUT POINTS FOR CHOLESTEROL SHOULD BE USED?

The NCEP pediatric panel recommended a single set of cut points for determining cholesterol elevation for all children aged 2–18 years. These values are generally based on percentile values when children and adolescents are considered as a group. These values are presented in Table 3. However, it has been demonstrated that these cut points do not take into account the differences in cholesterol seen with age or pubertal maturation *(29–31)*. They also do not take into account population differences by sex or ethnic group *(29–31)*. This means that the sensitivity specificity and predictive value will be different for boys and girls and at different ages. Total cholesterol and LDL-C levels tend to decline during puberty, meaning that sensitivity will be lower for pubertal adolescents. Future expert panels will need to consider whether the cut points should be based on percentiles by age and sex as is currently done for body mass index and for blood pressure.

Another concern about the NCEP pediatric panel is that it does not present percentile values for triglycerides or for HDL-C. In 2003, the American Heart Association published updated primary prevention guidelines for children and adolescents *(32)*. This report recommended that triglycerides above 150 mg/dL should be considered elevated and that HDL-C below 35 mg/dL should be considered too low.

Treatment of Elevated Cholesterol: Who Should Be Treated?

The NCEP pediatric panel *(15)* recommended that children and adolescents above the 95th percentile (Table 3) should be candidates for a therapeutic approach to changing lifestyle factors including diet and physical activity. They reserved more aggressive therapy with pharmacologic agents to children aged 10 years and over with higher levels of LDL-C as presented in Table 4. It is clear that this approach to treatment is based solely on LDL-C concentrations.

What is the Target for Treatment?

The NCEP pediatric panel recommended that the goal of treatment should be an LDL-C level of < 130 mg/dL *(15)*. This would place the patients' LDL-C at a value less than the 95th percentile. This has led to some confusion, however, as the range of values between 130 and 160 mg/dL (or 190 mg/dL with no risk factors) represents values at which more aggressive therapy with pharmacologic agents would not have

Table 3
Classifications of Total and LDL Cholesterol Levels in Children and Adolescents

Category	Percentile	Total cholesterol (mg/dL)	LDL-C (mg/dL)
Acceptable	>75th	<170	<110
Borderline	75th–95th	170–199	110–129
Elevated	>95th	>200	>130

LDL-C, Low-density lipoprotein Cholesterol.
Adapted from National Cholesterol Education Program Guidelines for Children and Adolescents.

Table 4
Recommended Values for Pharmacologic Treatment of Children
and Adolescents Aged 10 Years and older

Patient characteristics	Recommended cut points
1) No other risk factors for CVD	LDL-C is persistently >190 mg/dl despite diet therapy
2) Other risk factors present including obesity, hypertension, diabetes, or cigarette smoking, or positive family history of premature CVD	LDL-C is persistently >160 despite diet therapy

CVD, Cardiovascular disease; LDL-C, Low-density lipoprotein Cholesterol.

been recommended in the first place. This inconsistency remains a concern for pediatricians as they make the decision regarding the appropriate intensity of treatment. Unfortunately, there is no evidence base on which to determine the optimum level of lipid lowering to prevent further development of the atherosclerotic process in young patients could be determined.

How Should Patients Be Treated?

The recommended approach is to start therapy with lifestyle alterations, including changes in diet and physical activity. In this therapeutic setting, it is recommended that dietary changes be made with the assistance of a dietitian. A more aggressive approach is instituted when LDL-C values are quite high and after diet and lifestyle therapy has failed to sufficiently lower cholesterol levels. This more aggressive approach includes pharmacologic intervention.

Diet Therapy

The initial approach to diet therapy is to provide a more organized approach for families to comply with the population recommendation for diet presented in Table 2. In practice, even families with multiple members who have had a diagnosis of hyperlipidemia often have not achieved a diet with an appropriate content of saturated fat and cholesterol. This highlights how difficult this is in practice in our current environment. The population diet may normalize LDL concentrations that are in the borderline high range at baseline. The response of children and adolescents to diet therapy has been reported to be a drop of LDL-C from 3 to 10% *(33–37)*.

For children and adolescents who continue to have LDL-C greater than 130 mg/dL despite 3–6 months on the population diet, the NCEP pediatric panel recommends a more restrictive diet. Children and adolescents are advised to progressively reduce saturated fat to less than 7% of total calories and to reduce cholesterol intake to less than 200 mg per day. These nutrients are most important because they have been shown to have the largest impact on blood levels of LDL-C. This more aggressive dietary approach to lowering LDL-C has been reported to result in a reduction of LDL-C from 4 to 14% *(34–37)*.

A number of studies have demonstrated the safety of cholesterol-lowering diets. The Dietary Intervention Study in Children (DISC) *(38)* showed a significant lowering of LDL-C in the intervention compared with the control group with a mean difference of 3.23 mg/dL. A 7-year follow-up study also demonstrated that dietary changes could be maintained over time *(39)*. Furthermore, there were no abnormalities in growth, development, or other safety variables in the DISC intervention group. These results have been corroborated in other clinic-based studies of diet intervention *(40)*.

From a practical standpoint, making the recommended diet changes can be difficult for families. The NCEP pediatric report provides a guide to making these changes that can be used by health professionals *(15)*. The recommendations are presented in Table 5.

There are useful strategies for families to adopt to promote healthful eating for the children. Increased availability of low saturated fat foods and decreased accessibility of high saturated fat foods can be helpful. It is also important to remember that children may initially reject unfamiliar foods but may learn to like foods with repeat exposure. It has been shown that children may require a minimum of eight to ten exposures to new foods before their preference for the food will increase *(41–43)*. This means that parents should not assume that an initial rejection of a food means a long-term aversion. Another useful strategy is for the parents to model the desired eating behaviors. It has been shown that the preference of children and their acceptance of foods often follow the example of their parents. One study showed a correlation of the fat intake in children aged 3–5 years and that of their parents *(44)*. Children are more likely to eat a food when they observe an adult eating it *(45–46)*.

While there have been reports of children who have developed nutritional deficiencies and failure to grow appropriately on a fat restricted diet, it is important to note that these families were following parent imposed, medically unsupervised diets. These diets were greatly restricted in both total energy and fat and were far more restrictive than the dietary approaches that are currently recommended.

Other Dietary Approaches to Therapy

There has been emerging interest in the utility of dietary supplements to treat cholesterol elevation in children. Increased fiber has been proposed as a useful therapy.

Table 5
Diet Recommendations to Reduce Saturated Fat

Food	*Amount*	*Example*
Lean Meat	5–6 oz./day	Round or lean cuts/well trimmed
Poultry		Skin removed focus of white meat
Processed Meat		Less than 3 g of fat per ounce
Low fat Dairy	24–32 oz./day	Non-fat milk/yogurt or frozen yogurt
Cheese		Less than 2 g fat per ounce
Eggs	<2/week	Use egg white
Fruits/vegetables	Encourage	
Whole grains	Encourage	

The results of studies of dietary fiber have been variable. Davidson et al. found a 7% reduction in LDL-C with introduction of a psyllium-enriched cereal *(47)*. However, Dennison et al. were unable to demonstrate cholesterol reduction using a similar cereal *(48)*. Studies of soy-based protein have shown some reduction of LDL-C *(49)* and an increase in HDL-C *(50)*.

Plant stanol and sterol esters have also been an area of investigation. These compounds have been shown to reduce cholesterol absorption from the gastrointestinal tract *(51)*. A study by Tammi et al. showed a reduction of LDL-C of 7.5% *(52)*. Gyllin et al. showed that levels of LDL-C were reduced by a mean of 15% *(53)*. In general, plant stanols and sterols seem to be well tolerated and safe; however, absorption of some other dietary components such as β-carotene may also be reduced.

Other dietary supplements, such as garlic, have generally not been found to be effective in lowering cholesterol levels in children *(54)*. Other potential therapies that need systematic study in children include the use of antioxidant vitamins and omega-3 fatty acids in fish oil and flax seed. It is only with rigorous randomized clinical trials showing a benefit that such adjunctive treatments should be adopted in practice.

Pharmacologic Intervention

Pharmacologic intervention for hypercholesterolemia is recommended for children and adolescents aged 10 years and above with LDL-C >160 mg/dL when other risk factors or a family history of cardiovascular disease is present or >190 mg/dL when such factors are not present (Table 4). In general, pharmacologic intervention is reserved for patients who have had a 3- to 6-month trial of appropriate therapeutic lifestyle changes.

The approach to pharmacologic intervention in children and adolescents should be individualized. Each family may have a different level of concern about the risk of atherosclerosis, the importance of cholesterol lowering at a young age, and potential side effects of medication. Institution of pharmacologic intervention should occur only after a long discussion of these issues with the patient and their family.

As for adults, there are three main approaches to pharmacologic intervention for lowering LDL-C in children. These include bile acid binding agents, hydroxymethylglutaryl-coenzyme A (HMG-CoA) reductase inhibitors, and cholesterol absorption blockers. While other agents are potentially available, such as nicotinic acid, these are generally not used due to a high frequency of adverse effects, such as flushing.

Bile Acid Binding Agents

The NCEP pediatric panel included these agents as the primary pharmacologic approach to lipid lowering for children *(15)*. In part, this is because they had been available for some time and had the best evidence base at that time. The advantage of these agents is that they are not systemically absorbed. This is attractive because they have side effects limited to gastrointestinal concerns. These include bloating, abdominal pain, and constipation.

Bile acid binding resins have been shown to lower LDL-C from 13 to 20% *(55–57)*. Original studies of bile acid binding resins used the powder form *(55,56)*. This was generally found to be poorly tolerated in the pediatric population, thus resulting in poor compliance. There were few safety issues; however, there was a concern about absorption of fat-soluble vitamins; therefore, supplementation of folate and vitamin

D has been recommended. McCrindle et al. compared cholestyramine powder and a tablet form *(57)*. This randomized, crossover trial used a dose of 8 g per day. Children preferred the tablet form, but compliance was low and gastrointestinal complaints were common with both. The recommended approach to dosing of bile acid binding resins has been to start at a low dose of 1–4 g per day and to increase the dose over time. In general, doses above 8 g per day provide a diminishing cholesterol-lowering effect and increase the likelihood of side effects *(58)*.

Colesevelam, a bile acid binding resin, is approved by the Federal Drug Association (FDA) for use in adults. However, this agent has not been extensively tested in children and adolescents.

HMG-CoA Reductase Inhibitors

These medications, also known as statins, have been widely used in adults for cholesterol lowering. Some statin agents been shown to significantly reduce cardiovascular mortality in adults *(59)*.

There is increasing evidence of the safety and efficacy of these medications in children and adolescents *(60–66)*. The longest duration of these studies included 24 months of follow-up. Most studies have shown substantial LDL-C lowering with minimal adverse effects. Importantly, growth and maturation have not been shown to be adversely affected. Hormone levels have also been shown to be normal during treatment. Some studies have shown elevation in creatine kinase; however, these elevations have usually been transient and are often associated with vigorous physical activity.

De Jongh et al. studied the effects of statins on quality of life *(67)*. They found no concerns with quality of life; however, some children reported concerns about their disease and risk of future cardiovascular problems. Unfortunately, long-term effects of statin agents started in childhood remain unknown. In general, it is recommended that children and adolescents be started on a low dose with further titration of the dose upward to achieve increased cholesterol lowering. However, increasing the dose is generally associated with incrementally less lowering of LDL-C and an increase in the risk of side effects.

The side effects of statin agents include elevation of hepatic transaminases and elevation of creatine kinase. The most concerning side effect is the possibility of rhabdomyolysis. This is associated with muscle aches, cramping, and weakness. If such symptoms are present, the medication should be discontinued. Pediatric patients should be monitored for symptoms of muscle cramps and with periodic (every 3–4 months) measurement of creatine kinase and hepatic transaminases. Physicians and families should also be aware of possible drug interactions, which may increase the risk of rhabdomyolysis. These drugs include gemfibrozile, cyclosporine, and erythromycin. Drinking large quantities of grapefruit juice may also have a similar impact.

An important potential side effect for females is the concern about teratogenicity. Women taking statins who are sexually active should use a reliable form of birth control. Pregnancies should be planned and the statin agent should be discontinued prior to pregnancy. This may be problematic for adolescent females for whom planning sexual involvement and contraception may be less reliable. Despite the concern about potential side effects, statin agents are becoming an important part of the management of children and adolescents with dyslipidemia.

Cholesterol Absorption Inhibitors

Cholesterol absorption inhibitors are a relatively new class of agents that work by decreasing cholesterol absorption. Ezetimibe enters the enterohepatic circulation and works to decrease cholesterol absorption and bile acid reuptake. A single dose of 10 mg per day is used in all patients. Further increases in dose are not associated with additional cholesterol lowering. Further research on ezetimibe is needed in pediatric patients as there have not been any published studies of ezetimibe in children. In adults, it is often used in combination with a statin agent. Ezetimibe has been shown to reduce LDL-C levels by approximately 20% *(68)*. In studies of adults, this medication has been shown to have few side effects and is generally well tolerated. In children, it is possible that ezetimibe will be shown to be an effective medication alone.

Fibric Acid Derivatives

The fibrates are useful for lowering triglycerides and may also raise HDL-C. The NCEP pediatric panel had essentially no recommendations for treatment of non-LDL-C abnormalities. A very important aspect of management of triglyceride elevations is an aggressive approach to dietary modification with decreased simple carbohydrates as well as decreased fat intake. In addition, increased physical activity may be helpful in raising levels of HDL-C.

Recommendations for the level of triglycerides at which treatment should be instituted for children and adolescents have not been established. A concern about children with very high levels of triglycerides is the risk of developing pancreatitis. Patients with triglyceride levels of 400 mg/dL or greater while fasting may be candidates for pharmacologic treatment if aggressive diet therapy is not successful in lowering the triglyceride concentration.

Wheeler et al. studied 14 children with bezofibrate *(69)*. In general, the medication was well tolerated, but transient elevation of liver transaminases was observed in some patients. The known side effects of fibrates include gastrointestinal upset, a predisposition to cholelithiasis, and elevated hepatic transaminases and creatine kinase. The risk of rhabdomyolysis is increased in patients who take fibric acid derivatives with statins.

Further research on the optimum approach to the treatment of abnormalities of non-LDL-C in children is needed. This includes better understanding of whether the metabolic syndrome may also be a concern for children and adolescents. There have been some recent studies of the clustering of risk factors and obesity in children, but no consensus definition of the metabolic syndrome exists for pediatric patients *(70–72)*. This is another very important area for future investigation.

SUMMARY

- Pediatricians need to be aware of lipid and lipoprotein abnormalities in their young patients.
- The atherosclerotic process begins in childhood and is progressive.
- Elevated plasma cholesterol is an important risk factor for the early aspects of this process.
- Two combined strategies have been recommended to lower cholesterol and risk of cardiovascular disease: the population approach and the individual approach.

- The population approach focuses on improving diet and physical activity in all children and adolescents above the age of 2 years.
- The individual approach focuses on the identification and treatment of higher risk children and adolescents.
- The current recommendation for screening is that it should be targeted based on family history of CVD or hyperlipidemia. However, limitations of this strategy have been identified. Physicians may want to consider universal screening for pediatric patients.
- Once high-risk children are identified, treatment focuses on improved diet. When diet therapy is not successful, pharmacologic management is recommended for children aged 10 years and above with LDL >160 mg/dL (another risk factor present) or >190 mg/dL (no other risk factor present).
- Three types of medications can be used in children and adolescents: bile acid resins, HMG-CoA reductase inhibitors, and cholesterol absorption blockers.
- The initial goal for lowering cholesterol with diet therapy or with pharmacologic agents is < 130 mg/dL.
- Future pediatric research should focus on (1) additional studies of the safety and efficacy of pharmacologic treatment, (2) understanding of the metabolic syndrome and non-LDL-C, and (3) the ability of treatment of dyslipidemia to prevent the development of atherosclerosis.

REFERENCES

1. Stary HC, Chandler AB, Dinsmore RE, et al. A definition of advanced types of atherosclerotic lesions and a histological classification of atherosclerosis. A report from the Committee on Vascular Lesions of the Council on Arteriosclerosis, American Heart Association. *Arterioscler Thromb Vasc Biol* 1995;15:1512–1531.
2. Glagov S, Guyton JR, Insull W, Jr, et al. A definition of initial, fatty streak, and intermediate lesions of atherosclerosis. A report from the Committee on Vascular Lesions of the Council on Arteriosclerosis, American Heart Association. *Circulation* 1994;89:2462–2478.
3. Enos WF, Holmes RH, Beyer J. Coronary disease among United States soldiers killed in action in Korea; preliminary report. *J Am Med Assoc* 1953;152:1090–1093.
4. McNamara JJ, Molot MA, Stremple JF, Cutting RT. Coronary artery disease in combat casualties in Vietnam. *JAMA* 1971;216:1185–1887.
5. Pathobiological Determinants of Atherosclerosis in Youth (PDAY) Research Group. Relationship of atherosclerosis in young men to serum lipoprotein cholesterol concentrations and smoking (a preliminary report). *JAMA* 1990;264:3018–3024.
6. McHill HC, Jr, McMahan CA, Zieske AW, et al. Effects of nonlipid risk factors on atherosclerosis in youth with a favorable lipoprotein profile. *Circulation* 2001;11:1546–1550.
7. Newman WP, 3rd, Freedman DS, Voors AW, et al. Relation of serum lipoprotein levels and systolic blood pressure to early atherosclerosis. The Bogalusa Heart Study. *N Engl J Med* 1986;3124:138–144.
8. Berenson GS, Srinivasan SR, Bao W, et al. Association between multiple cardiovascular risk factors and atherosclerosis in children and young adults. The Bogalusa Heart Study. *N Engl J Med* 1998;338:1650–1656.
9. Gidding SS, Bookstein LC, Chomka EV. Usefulness of electron beam tomography in adolescents and young adults with heterozygous familial hypercholesterolemia. *Circulation* 1998;98:2580–2583.
10. de Jongh S, Lilien MR, op't Roodt J, Stroes ES, Bakker HD, Kastelein JJ. Early statin therapy restores endothelial function in children with familial hypercholesterolaemia. *J Am Coll Cardiol* 2002;40:2117–2121.
11. Wiegman A, Hutten BA, de Groot E, et al. Efficacy and safety of statin therapy in children with familial hypercholesterolaemia: a randomized controlled trial. *JAMA* 292;3:331–337.
12. Mietus-Snyder M, Malloy MJ. Endothelial dysfunction occurs in children with two genetic hyperlipidemias: improvement with antioxidant vitamin therapy. *J Pediatr* 1998;133:35–40.
13. Engler MM, Engler MB, Malloy M, et al. Docosahexaenoic acid restores endothelial function in children with Hyperlipidemia: results from the EARLY study. *Int J Clin Pharmacol Ther* 2004;42:672–679.

14. Gaeta G, De Michele M, Cuomo S, et al. Arterial abnormalities in the offspring of patients with premature myocardial infarction. *N Engl J Med* 2000;343:840–846.
15. American Academy of Pediatrics. National Cholesterol Education Program: Report of the Expert Panel on Blood Cholesterol Levels in Children and Adolescents. *Pediatrics* 1992;89:525–584.
16. Daniels SR. Cholesterol abnormalities in children and adolescents: time for an update of the 1992 National Cholesterol Education Program guidelines. *Prog Pediatr Cardiol* 2003;17:109–111.
17. Lauer RM, Lee J, Clarke WR. Factors affecting the relationship between childhood and adult cholesterol levels: the Muscatine Study. *Pediatrics* 1988;82:309–318.
18. Rose G. Sick individuals and sick populations. *Int J Epidemiol* 1985;14:32–38.
19. NCHS-NHLBI Collaborative Lipid Group. Trends in serum cholesterol levels among US adults aged 20 to 74 years: data from the National Health and Nutrition Examination Surveys, 1960 to 1980. *JAMA* 1987;257:937–942.
20. US Department of Health and Human Services and US Department of Agriculture. *Dietary Guidelines for Americans 2005*, 6th edition. Washington DC: US Government printing office January 2005.
21. Expert Panel on Detection, Evaluation, and Treatment of High Blood Cholesterol in Adults. Executive summary of the Third Report of The National Cholesterol Education Program (NCEP) Expert Panel on Detection, Evaluation and Treatment of High Cholesterol in Adults (Adult Treatment Panel III). *JAMA* 2001;286:2486–2497.
22. Dennison BA, Jenkins PL, Pearson TA. Challenges to implementing the current pediatric cholesterol screening guidelines into practice. *Pediatrics* 1994;94:296–302.
23. Rifai N, Neufeld E, Ahlstrom P, et al. Failure of current guidelines for cholesterol screening in urban African-American adolescents. *Pediatrics* 1996;98:383–388.
24. Rastam L, Hannan PJ, Luepker RV, et al. Seasonal variation in plasma cholesterol distributions: implications for screening and referral. *Am J Prev Med* 1992;8:360–366.
25. Williams RR, Hunt SC, Barlow GK, et al. Prevention of familial cardiovascular disease by screening for family history and lipids in youths. *Clin Chem* 1992;38:1555–1560.
26. Bachman RP, Scheon EJ, Sternbridge A, et al. Compliance with childhood cholesterol screening among member of a prepaid health plan. *Am J Dis Child* 1993;147:382–385.
27. Resnicow K, Cross D, Lacosse J, et al. Evaluation of a school-site cardiovascular risk factor screening intervention. *Prev Med* 1993;22(6):838–856.
28. Boulton TJ. The validity of screening for hypercholesterolaemia at different ages from 2 to 17 years. *Aust N Z J Med* 1979;9(5):542–546.
29. Dennison BA, Kikuchi DA, Srinivasan SR, et al. Serum total cholesterol screening for the detection of elevated low-density lipoprotein in children and adolescents: the Bogalusa Heart Study. *Pediatrics* 1990;85:472–479.
30. Labarthe DR, Dai S, Fulton J. Cholesterol screening in children: insights from Project HeartBeat! And NHANES III. *Prog Pediatr Cardiol* 2003;17(2):169–178.
31. Morrison J. A longitudinal evaluation of the NCEP-Peds guidelines for elevated total and LDL cholesterol in adolescent girls and boys. *Prog Pediatr Cardiol* 2003;17(2):159–168.
32. Kavey R-EW, Daniels SR, Lauer RM, et al. American heart association guidelines for primary prevention of atherosclerotic cardiovascular disease beginning in childhood. *J Pediatr* 2003;142(4):368–372.
33. Tershakovec AW, Shannon BM, Achterberg CL, et al. One-year follow-up of nutrition education for hypercholesterolemic children. *Am J Public Health* 1998;88:258–261.
34. Kuehl KS, Cockerham JT, Hitchings M, Slater D, Nixon G, Rifai N. Effective control of hypercholesterolaemia in children with dietary intervention based in pediatric practice. *Prev Med* 1993;22:154–166.
35. Sanchez-Bayle M, Gonzalez-Requejo A, Baeza J, et al. Diet therapy for hypercholesterolaemia in children and adolescents. *Arch Pediatr Adolesc Med* 1994;148:28–32.
36. Mietus-Snyder M, Baker AL, Neufeld D, et al. Effects of nutritional counseling on lipoprotein levels in pediatric lipid clinic. *Am J Dis Child* 1993;147:378–381.
37. Alexandrov AA, Maslennikova GY, Kulikov SM, Propirnij GA, Perova NV. Primary prevention of cardiovascular disease: 3-year intervention results in boys of 12 years of age. *Prev Med* 1992;21:53–62.
38. Lauer RM, Obarzanek E, Hunsberger SA, et al. Efficacy and safety of lowering dietary intake of total fat, saturated fat and cholesterol in children with elevated LDL-cholesterol: the Dietary Intervention Study in Children. *Am J Clin Nutr* 2000;72(suppl):1332S–1342S.

39. Obarzanek E, Kimm SY, Barton BA, et al. Long-term safety and efficacy of a cholesterol-lowering diet in children with elevated low-density lipoprotein cholesterol: seven-year results of the Dietary Intervention Study in Children. *Pediatrics* 2001;107:256–264.
40. Jacobson MS, Tomopoulus S, Williams Cl, et al. Normal growth at high-risk hyperlipidemic children and adolescents with dietary intervention. *Prev Med* 1998;27:775–780.
41. Birch LL. Development of food preferences. *Annu Rev Nutr* 1999;19:41–62.
42. Birch LL. Children: preferences for high fat foods. *Nutr Rev* 1992;50:249–255.
43. Birch LL, McPhee L, Shoba BC. What kind of exposure reduces children's food neophobia? *Appetite* 1987;9:171–178.
44. Oliveria SA, Ellison RC, Moore LL, Gillman HW, Garrhe EJ, Singer MR. Parent-child relationships in nutrition intake: the Framingham Children's Study. *Am J Clin Nutr* 1992;56:593–598.
45. Harper LV, Sander KM. The effect of adult's eating on young children's acceptance of unfamiliar foods. *J Exp Child Psychol* 1975;20:206–214.
46. Hertzler AA. Children's food patterns – a review II. Family and group behavior. *J Am Diet Assoc* 1983;83:555–560.
47. Davidson MH, Dugan LD, Burns JH, et al. A psyllium-enriched cereal for the treatment of hypercholesterolaemia in children: a controlled, double-blind, crossover study. *Am J Clin Nutr* 1996;63:96–102.
48. Dennison BA, Levine DM. Randomized, double-blind, placebo-controlled, two-period crossover clinical trial of psyllium fiber in children with hypercholesterolaemia. *J Pediatr* 1993;123:24–29.
49. Widhalm K, Brazda G, Schneider B, et al. Effect of soy protein diet versus standard low fat, low cholesterol diet on lipid and lipoprotein levels in children with familial or polygenic hypercholesterolaemia. *J Pediatr* 1993;123:30–34.
50. Laurin D, Jaccques H, Moorjani S, et al. Effects of a soy-protein beverage on plasma lipoproteins in children with familial hypercholesterolaemia. *Am J Clin Nutr* 1991;54:98–103.
51. Miettinen TA, Gylling H. Plant stanol and sterol esters in prevention of cardiovascular diseases. *Ann Med* 2004;36:126–134.
52. Tammi A, Ronnemaa T, Gylling H, et al. Plant stanol ester margarine lowers serum total and low-density lipoprotein cholesterol concentrations of healthy children: the STRIP project. Special Turku Coronary Risk Factors Intervention Project. *J Pediatr* 2000;136:503–510.
53. Gylling H, Siimes MA, Miettinen TA. Sitostanol ester margarine in dietary treatment of children with familial hypercholesterolaemia. *J Lipid Res* 1995;36:1807–1812.
54. McCrindle BW, Helden E, Conner WT. Garlic extract therapy in children with hypercholesterolaemia. *Arch Pediatr Adolesc Med* 1998;152:1089–1094.
55. Tonstand S, Sivertsen M, Aksnes L, et al. Low dose colestipol in adolescents with familial hypercholesterolaemia. *Arch Dis Child* 1996;74:157–160.
56. Tonstad S, Knudtzon J, Sivertsen M, et al. Efficacy and safety of cholestyramine therapy in peripubertal and prepubertal children with familial hypercholesterolaemia. *J Pediatr* 1996;129:42–49.
57. McCrindle BW, O'Neill MB, Cullen-Dean G, et al. Acceptability and compliance with two forms of cholestyramine in the treatment of hypercholesterolaemia in children: a randomized, crossover trial. *J Pediatr* 1997;130:266–273.
58. Superko HR, Greenland P, Manchester RA, et al. Effectiveness of low-dose colestipol therapy in patients with moderate hypercholesterolaemia. *Am J Cardiol* 1992;70:135–140.
59. Kwiterovich PO, Jr. State-of-the-art update and review: clinical trials of lipid-lowering agents. *Am J Cardiol* 1998;82:3U–17U.
60. Ducobu J, Brasseur D, Chaudron JM, et al. Simvastatin use in children. *Lancet* 1992;339:1488.
61. Knipscheer HC, Boelen CC, Kastelein JJ, et al. Short-term efficacy and safety of pravastatin in 72 children with familial hypercholesterolaemia. *Pediatr Res* 1996;39:867–871.
62. Lambert M, Lupien PJ, Gagne C, et al. Treatment of familial hypercholesterolaemia in children and adolescents: effect of lovastatin. Canadian Lovastatin in Children Study Group. *Pediatrics* 1996;97:619–628.
63. Stein EA, Illingworth DR, Kwiterovich PO, Jr, et al. Efficacy and safety of lovastatin in adolescent males with heterozygous familial hypercholesterolaemia: multicenter, randomized, placebo-controlled trial. *J Pediatr* 2003;143:74–80.
64. de Jongh S, Ose L, Szamosi T, et al. Efficacy and safety of statin therapy in children with familial hypercholesterolaemia: a randomized, double-blind, placebo-controlled trial with Simvastatin. *Circulation* 2002;106:2231–2237.

65. McCrindle BW, Ose L, Marais AD. Efficacy and safety of atorvastatin in children and adolescents with familial hypercholesterolaemia or severe hyperlipidemia: a multicenter, randomized, placebo-controlled trial. *J Pediatr* 2003;143:74–80.
66. Wiegam A, Hutten BA, de Groot E, et al. Efficacy and safety of statin therapy in children with familial hypercholesterolaemia: a randomized controlled trial. *JAMA* 2004;292:331–337.
67. de Jongh S, Kerckhoffs MC, Grootenhuis MA, et al. Quality of life, anxiety and concerns among statin-treated children with familial hypercholesterolaemia and their parents. *Acta Paediatr* 2003;92:1096–1101.
68. Gagne C, Gaudet D, Bruckert E. Efficacy and safety of ezetimibe coadministered with atorvastatin or Simvastatin in patients with homozygous familial hypercholesterolaemia. *Circulation* 2002;105:2469–2475.
69. Wheeler KA, West RJ, Lloyd JK, et al. Double blind trial of bezafibrate in familial hypercholesterolaemia. *Arch Dis Child* 1985;60:34–37.
70. Cook S, Weitzman M, Auinger P, Nguyen M, Dietz WH. Prevalence of a metabolic syndrome phenotype in adolescents: findings from the third National Health and Nutrition Examination Survey, 1988–1994. *Arch Pediatr Adolesc Med* 2003;157(8):821–827.
71. de Ferranti SD, Gauvreu K, Ludwig DS, Neufeld EJ, Newburger JW, Rifai N. Prevalence of the metabolic syndrome in American adolescents: findings from the Third National Health and Nutrition Examination Survey. *Circulation* 2004;19;110(16):2494–2497 (Epub 2004 Oct).
72. Goodman E, Dolan LM, Morrison JA, Daniels SR. Factor analysis of clustered cardiovascular risks in adolescence: obesity is the predominant correlate of risk among youth. *Circulation* 2005;111(15):1970–1977.

20 The Allied Health Professional's Role in the Management of Dyslipidemia

Lynne T. Braun, PhD, RN, CNP, and Joan E. Mathien, BSN

CONTENTS

    INTRODUCTION
    LIPID CLINIC/PREVENTION CLINIC MODELS
    NURSE'S ROLE
    DIETITIAN'S ROLE
    EXERCISE SPECIALIST'S ROLE
    CLINICAL PSYCHOLOGIST'S ROLE
    CONCLUSIONS
    REFERENCES

INTRODUCTION

An elevation in low-density lipoprotein cholesterol (LDL-C) is a major cause of coronary heart disease (CHD). In fact, the higher the LDL-C level, the greater the CHD risk *(1)*. Although national guidelines for cholesterol management have been written since 1988, under diagnosis and under treatment continue to be a major public health concern *(2)*. Data from the WHO Multinational MONItoring of trends and determinants of Cardiovascular disease (MONICA) Project *(3)*, conducted in adults aged 35–64 years in 19 countries, show that the prevalence of hypercholesterolemia varies across populations from 3 to 53% in men and from 4 to 40% in women. However, awareness of hypercholesterolemia was extremely low, ranging from 1 to 33% in men and from 0 to 31% in women. When hypercholesterolemia is identified and treated, studies *(4,5)* show that relatively few patients meet their guideline-suggested cholesterol target levels. In fact, patients with known CHD, who are in the highest risk category, rarely meet their cholesterol target level (Fig. 1).

The *Third Report of the National Cholesterol Education Program (NCEP) Expert Panel on Detection, Evaluation, and Treatment of High Blood Cholesterol in Adults* (Adult Treatment Panel [ATP] III) *(6)* asserts that LDL-C is the first target of therapy. Once LDL-C reaches the target level based on risk status, the metabolic syndrome, which includes low high-density lipoprotein cholesterol (HDL-C) and elevated triglyceride levels, is a secondary target of therapy. Dyslipidemia management poses a

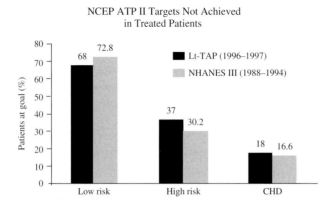

Fig. 1. National Cholesterol Education Program Adult Treatment Panel II targets not achieved in treated patients *(4,5)*.

challenge to both patients and health care providers. A multifaceted lifestyle approach is essential, often in combination with drug therapy. Patients require counseling on a heart healthy diet (referred to as the therapeutic lifestyle changes or TLC diet by ATP III), increased physical activity, weight reduction, and medication management. ATP III recommends multidisciplinary methods, including nurse case management and nutrition counseling, to help patients adhere to lifestyle and medication therapies. Therefore, this chapter will address the roles of allied health professionals, specifically the nurse/advanced practice nurse (APN), dietitian, exercise specialist, and clinical psychologist. These practitioners work in concert with physicians in the management of patients with dyslipidemia. Lipid clinics, which may be incorporated into broad-based cardiovascular risk reduction or preventive cardiology programs, model the team approach for improved outcomes in dyslipidemia management.

LIPID CLINIC/PREVENTION CLINIC MODELS

Midwest Heart Specialists (Naperville, Illinois) initiated a lipid clinic in 1985 in order to improve adherence to NCEP guidelines. The physician-directed, nurse-managed program demonstrated impressive results in guideline adherence compared with national averages. One hundred percent of lipid clinic patients were on lipid-lowering therapy, 97% of patients had a LDL-C level documented on the chart, and 71% of patients met their LDL goal compared with 11% nationally. These superior results inspired the development of a practice-wide lipid management program that has dramatically improved lipid outcomes *(7,8)*.

Ryan et al. *(9)* reported the outcomes of a collaborative-care clinic model designed to improve the care of high-risk patients with dyslipidemia. The clinic was led by a cardiologist with the assistance of an APN. Patients were evaluated at baseline and received an individualized treatment plan at 2–4 weeks, which included drug therapy and a specific diet and exercise plan. Patients were re-evaluated at 8–10 weeks with re-adjustment of drug therapy, review of diet and exercise plan, and assessment of adherence. Patients were seen every 8–10 weeks until lipid goals were achieved, with subsequent follow-up every 6 months. After 3.5 years, 62–74% of patients reached lipid goals and 35% achieved a combined lipid parameter goal. Of the patients without CHD at baseline, patients with Framingham 10-year CHD risk >20% were reduced from 6 to <1%.

A randomized clinical trial tested the effectiveness of a nurse case management program to lower blood lipids in patients with CHD *(10)*. Two hundred twenty-eight patients with hypercholesterolemia and CHD were recruited during hospitalization after revascularization and were randomized to receive either lipid management or usual care. The nurse case-management group was treated with pharmacotherapy and individualized lifestyle modification from a nurse practitioner for 1 year after discharge in addition to their usual care. The usual care group received feedback on their lipids from a nurse practitioner who then provided this report to their primary provider and/or cardiologist. More patients in the nurse case management program than in the usual care group achieved the LDL-C goal of <100 mg/dL (65 versus 35%, $P = 0.0001$). Improvement in diet and exercise regimen in the nurse case management group accompanied favorable changes in lipids and lipoproteins. After adjusting for age, sex, dietary fat, physical activity, and baseline cholesterol, multiple regression analysis showed that assignment to the nurse case management group ($P = 0.0001$) and lipid-lowering medication ($P = 0.001$) predicted LDL-C level *(10)*.

ATP III guidelines *(6)* call for more aggressive treatment of dyslipidemia. Based on recent clinical trial evidence, an update to the ATP guidelines *(11)* recommended that clinicians consider an LDL goal of <70 mg/dL in the highest risk patients. Those at high risk include patients with the following: established cardiovascular disease (CVD) plus multiple major risk factors (especially diabetes), severe and poorly controlled risk factors (especially continued cigarette smoking), and multiple risk factors of the metabolic syndrome, especially high triglycerides ≥ 200 mg/dL plus non-HDL-C ≥ 130 mg/dL with low HDL-C or patients with acute coronary syndromes. These guidelines significantly increase the number of individuals who require lifestyle and drug treatment for dyslipidemia. As demonstrated by the previously discussed clinical practice and research models, collaborative approaches that facilitate drug and lifestyle treatment strategies are highly effective in helping patients achieve target lipid levels, initiate and maintain healthy dietary and exercise habits, and reduce CHD risk *(12)*.

NURSE'S ROLE

Nurses have multiple roles in a lipid clinic model. Lipid clinics often employ both staff nurses and APNs (nurse practitioners and clinical nurse specialists). One of these nurses typically serves in the role of clinic manager or clinical coordinator and works collaboratively with the medical director to insure that patients receive coordinated care from all disciplines and that quality of care is maintained. The clinical coordinator typically has the responsibility to enrol new patients in the program, whether these are existing patients known to the practice or new patient referrals. A pre-appointment phone call with the patient is often useful to obtain some clinical background information about the patient, to explain the lipid clinic services, and to give the patient important instructions for the first visit, such as to fast the night before in order to obtain blood tests prior to the first appointment.

Nurses often work directly with the physician during a new patient's initial visit. Nurses are responsible for taking a comprehensive medical history and performing a global risk assessment. Computer software packages are available to facilitate risk assessment, generally using the Framingham scoring system. Palm Pilot™ CHD risk calculators (Fig. 2) may be used in the exam room to inform the patient of his/her 10-year CHD risk, to provide patient education, and to motivate patients to adhere to medications and lifestyle changes for CHD risk reduction *(13)*. The following case illustrates how the Framingham risk scoring system may be used as a motivational tool.

A male patient, age 52, has a total cholesterol of 242 mg/dL and a HDL-C of 40 mg/dL. Systolic blood pressure is treated and remains 144 mmHg; the patient smokes 1 ppd. His Framingham risk score is 16, which gives him a 10-year CHD risk of 25%. Following a discussion of individual risk factors and how they contribute to the development and progression of CHD, the nurse can illustrate how risk reduction will lower the patient's 10-year CHD risk. If total cholesterol was lowered to 150 mg/dL with drug and lifestyle therapies, the patient quit smoking, and systolic blood pressure was reduced to 120 mmHg by intensifying drug therapy and weight loss, the Framingham risk score is now 8, reducing the patient's 10-year CHD risk to 4%.

Nurses work collaboratively with the physician in defining patient-specific goals for cardiovascular risk reduction. Important aspects of the initial clinic visit include assessing the patient's level of understanding of the disease state and risk factors and the need for cardiovascular risk reduction. The patient is then involved in a comprehensive plan to achieve the desired goals *(14)*. The nursing role typically includes counseling on lifestyle approaches for cardiovascular risk reduction, such as a heart-healthy diet based on ATP III guidelines (Table 1), a regular exercise program, and weight loss goals. Nurses are often afforded a greater amount of time with patients than physicians in order to provide the necessary counseling. It is never sufficient to simply tell a patient to adhere to a particular diet or an exercise program. First, the nurse must assess the patient's readiness to make these lifestyle changes. Second, the nurse addresses

Fig. 2. Palm Pilot™ coronary heart disease risk calculator.

Table 1
Nutrient Components of the Therapeutic Lifestyle Changes Diet (6)

Nutrient	Recommended intake
Saturated fat[a]	Less than 7% of total calories
Polyunsaturated fat	Up to 10% of total calories
Monounsaturated fat	Up to 20% of total calories
Total fat	25–35% of total calories
Carbohydrate[b]	50–60% of total calories
Fiber	20–30 g/day
Protein	Approximately 15% of total calories
Cholesterol	Less than 200 mg/day
Total calories (energy)[c]	Balance energy intake and expenditure to maintain desirable body weight/prevent weight gain

[a] Trans fatty acids are another LDL-raising fat that should be kept at a low intake.
[b] Carbohydrates should be derived predominantly from foods rich in complex carbohydrates including grains, especially whole grains, fruits, and vegetables.
[c] Daily energy expenditure should include at least moderate physical activity (contributing approximately 200 kcal per day).

ways to incorporate changes into the patient's daily life. Brown-bagging a healthy lunch instead of eating out and waking up 30 min earlier to exercise are two common behavior changes often suggested. Establishing behavior change involves reviewing the patient's daily schedule with them in order to identify strategies for success. Finally, the nurse must provide tools for the patient to assist with making behavior changes, such as a pedometer and a walking program (Table 2). Referrals may be made, if necessary, to other health professionals, such as dietitians, exercise specialists, or clinical psychologists.

In addition to counseling patients on lifestyle changes, nurses will often, in collaboration with the physician, initiate and then titrate lipid-altering medications. An important aspect of medication management is interpreting baseline and follow-up laboratory studies to assess for medication safety and adverse effects. In addition, explaining laboratory studies to patients and educating patients about their prescribed medications, including indications, methods of administration (such as time taken, take with or without food), and potential side effects are part of the nurse's role. Well-established lipid clinics often have protocols for medication titration based on laboratory studies and patient response (Tables 3 and 4). Nurses are typically first to respond to patient phone calls and often must problem solve whether or not a patient's symptoms are related to a particular medication.

Nurses often employ a variety of strategies to improve adherence to lifestyle changes and medications. These include contracting with patients, self-monitoring, mail and telephone contact, prompts and reminders, and the use of spousal and family support (15). A combination of strategies is most effective for facilitating adherence, such as educational approaches, behavioral counseling, and supportive techniques (12).

Table 2
Pedometer-Guided Walking Program

	Sunday	Monday	Tuesday	Wednesday	Thursday	Friday	Saturday
Week 1							
Week 2							
Week 3							
Week 4							
Week 5							
Week 6							
Week 7							
Week 8							
Week 9							
Week 10							
Week 11							
Week 12							

Instructions:

1. Read pedometer instructions. Zero pedometer and wear it daily at your waist. Record number of steps you take each day.

2. Your goal is to reach 10,000 steps a day in 12 weeks. Compute your average steps per day during week 1.

Subtract this number from 10,000. (Example: If your average steps during week 1 is 3500, subtract 3500 from 10,000, which equals 6500 steps.)

3. Divide the difference (6500 steps) by 11. In this example, $6500 \div 11 = 591$ steps. Each week, your goal is to add 591 steps each day, for a 12-week goal of 10,000 total steps each day.

4. Add steps by: walk instead of drive, park farther away from your destination, take a 10-min walking break at work, walk the dog after work, walk yourself after work, etc. WALK, WALK, WALK!

5. Record your total steps each day and bring this walking diary to your next clinic visit.

Lipid clinics often employ a mix of nurses, including those with advanced practice degrees (clinical nurse specialists and nurse practitioners). APNs often have prescriptive authority and the practice may bill directly for their services. Therefore, patients may be scheduled for initial or follow-up appointments with APNs. The complexity of a patient's diagnosis may dictate if the patient is scheduled with the physician or APN. For example, patients with a recent cardiac event or severe hypertriglyceridemia may be scheduled with the physician, whereas primary prevention patients and patients with metabolic syndrome may be scheduled with an APN.

Nurses often collaborate with the physician in developing clinic policies and procedures. They take the lead role in producing assessment forms (Fig. 3) and patient education materials, as well as the utilization of computerized databases to track important outcomes, such as treatment goals and adherence to therapies. They also may be responsible for communicating a patient's progress with referring and primary care providers. In the lipid clinic setting, nurses provide consistent follow-up to ensure that a patient's lipid-lowering and other risk reduction goals are met *(14)*.

DIETITIAN'S ROLE

The ATP III guidelines emphasize the important role of the dietitian, especially during the early stages of making TLCs (Fig. 4). By obtaining a thorough diet history and reviewing food records, patient's diets are assessed for total daily calories,

Table 3
Midwest Heart Lipid Clinic Lab Protocol (Admission Lab: Comprehensive Metabolic Panel Complete Metabolic Panel (CMP), Creatine Phosphokinase (CPK), Thyroid Stimulating Hormone (TSH), Lipid Profile, Lp(a) and Apo B)

Medication	Lab[a] needed (prior to start)	Repeat (initially or dose change)	Maintenance monitoring[b] (if at goal)
Fibrates	CMP	Lipids, glucose, aspartate transaminase (AST), alanine transamine (ALT) and complete blood count (CBC) at 2 months	Lipids, AST, and ALT every 6 mos; CMP, CBC and TSH annually
Statins with or without Zetia	CMP and CPK	Lipids, AST, and ALT at 2 months	Lipids, AST, and ALT at 3 months, then every 6 months; TSH and glucose annually
Niacin Niaspan	CMP	Lipids, glucose, AST, ALT at 2 months	Lipids, glucose, AST, and ALT every 3 months for 1st year, then every 6 months, TSH annually
Advicor	CMP and CPK	Lipids, glucose, AST, ALT at 2 months	Lipids, glucose, AST, ALT at 3 months. Then every 6 months. TSH annually
Resins Zetia	CMP	Lipids at 2 months	Lipids every 6 months; Glucose, AST, ALT, and TSH annually
Fish oils	CMP	Lipids and glucose at 2 months	Lipids every 6 months. Glucose AST, ALT, and TSH annually
Niacin with any statin	CMP and CPK	Lipids, Glucose, AST, and ALT at 2 months	AST and ALT every 3 months; Lipids, glucose every 6 months; CMP and TSH annually
Niacin with fibrate	CMP and CPK	Lipids, AST, ALT at 2 months (If adding fibrate, CBC at 2 months)	AST and ALT every 3 months; Lipids, Glucose every 6 months; CMP, CBC, TSH annually
Fibrate with statin (see lipid MD annually to discuss)	CMP and CPK	Lipids, AST, ALT at 2 months (If adding fibrate, CBC at 2 months)	CPK, AST, ALT every 3 months. Add Lipids, glucose, blood urea nitrogen (BUN) and Creat. every 6 months. CBC, CMP, and TSH annually.

To rule out nephrotic syndrome, obstructive liver disease, diabetes, dysproteinemias, and hypothyroidism and assess CAD risks.

[a] For women capable of child bearing: pregnancy test (qualitative serum HCG ICD9 272.4) required prior to start of prescribed lipid agents.

[b] Patients with diabetes: glucose with each lipid panel.

Patients on coumadin: Instruct patient to have Protime/INR at 2 weeks thru MD initially or after any change in dose of lipid agent.

Patients with renal Insufficiency: Obtain BUN and Creat. annually from patient's Primary Care provider (PCP).

                    ☐Resistance training   ☐Calisthenics
                                    _____Limitations

- Stress sources _____
- Alcohol (1 drink = 12 oz. beer, 5 oz. wine, 1.5 oz. 80 proof spirits)
    ____No. drinks/week   ___Beer   ___Wine   ___Spirits
- Diet history _____Current _____Previous ___Ornish ___Pritikin ___Adkins ___Other
    ____No. serv. red meat/week   ___No. serv. fruits/veg./day (1/2 c.)   ____Eating out (no./week)
    Vitamins/minerals_____
    Dietary supplements_____
- Occupation _____

Physical examination
  General _____
  Skin ☐Xanthomas _____ ☐Xanthelasmas _____ ☐Other _____
  Eyes ☐ Arcus corneus _____ ☐Corneal opacities _____ ☐Other _____
  ENT Inspection of teeth, gums, palate, oral mucosa ☐Normal ☐Abnormal_____
  Neck Exam of jugular veins ☐Normal ☐Abnormal_____
    Exam of thyroid ☐Normal ☐Abnormal_____
  Resp/Lungs Assessment of respiratory effort ☐Normal ☐Abnormal Auscultation ☐Normal ☐Abnormal_____
  Heart   Palpation _____   Auscultation $S_1 S_2$ _____ ☐S_3 ☐S_4 ☐Murmur_____
  Vessels/Extremities
        Carotid arteries_____   Abdominal aorta_____ _____ Femoral arteries_____
        Pedal pulses _____Peripheral ed ema_____ Varicosities _____
  Abdomen ☐Tenderness_____   ☐ Masses _____Liver/spleen_____
  Musculoskeletal
        Exam of back (with notation of kyphosis or scoliosis) ☐Normal ☐Abnormal
        Exam of gait (with notation of ability to undergo exercise
            Testing and/or participation in exercise program   ☐Normal ☐Abnormal
        Assessment of muscle strength & tone (with notation of
            Any atrophy and abnormal movements)   ☐Normal ☐Abnormal
  Neuro Mood/affect _____ ☐Abnormal orientation to time/place/person_____
  Ext Inspection/palpation of digits, nails ☐Normal ☐Abnormal_____

Test results:
  ECG _____
  Stress/nuclear_____
  Echo _____
  Carotid doppler _____
  Peripheral vascular doppler _____
  Angiogram _____
  Fingerstick cholesterol values   TC____   LDL_____   HDL_____   TG_____
  EBCT_____

☐See dictated letter

Impression/Plan:_____
_____
_____
_____
_____
_____
_____
_____
_____

_____   _____
Signature   Date

Fig. 3. History and physical examination form for new patients entering the Preventive Cardiology Program.

CHD patients should undergo a symptom-limited exercise test prior to beginning an exercise program to assess maximal heart rate, as well as to determine the presence of ischemia or arrhythmias that might impact the safety of exercise. An exercise specialist can assess the exercise test results and formulate an exercise prescription tailored to the individual's functional capacity. Typically in a cardiac rehabilitation setting, patients exercise three times a week at an intensity of 70–85% of the measured peak heart rate while supervised and monitored *(17)*. In addition to aerobic exercise, calisthenics, flexibility, and strength training are often included.

Table 3
Midwest Heart Lipid Clinic Lab Protocol (Admission Lab: Comprehensive Metabolic Panel Complete Metabolic Panel (CMP), Creatine Phosphokinase (CPK), Thyroid Stimulating Hormone (TSH), Lipid Profile, Lp(a) and Apo B)

Medication	Lab[a] needed (prior to start)	Repeat (initially or dose change)	Maintenance monitoring[b] (if at goal)
Fibrates	CMP	Lipids, glucose, aspartate transaminase (AST), alanine transamine (ALT) and complete blood count (CBC) at 2 months	Lipids, AST, and ALT every 6 mos; CMP, CBC and TSH annually
Statins with or without Zetia	CMP and CPK	Lipids, AST, and ALT at 2 months	Lipids, AST, and ALT at 3 months, then every 6 months; TSH and glucose annually
Niacin Niaspan	CMP	Lipids, glucose, AST, ALT at 2 months	Lipids, glucose, AST, and ALT every 3 months for 1st year, then every 6 months, TSH annually
Advicor	CMP and CPK	Lipids, glucose, AST, ALT at 2 months	Lipids, glucose, AST, ALT at 3 months. Then every 6 months. TSH annually
Resins Zetia	CMP	Lipids at 2 months	Lipids every 6 months; Glucose, AST, ALT, and TSH annually
Fish oils	CMP	Lipids and glucose at 2 months	Lipids every 6 months. Glucose AST, ALT, and TSH annually
Niacin with any statin	CMP and CPK	Lipids, Glucose, AST, and ALT at 2 months	AST and ALT every 3 months; Lipids, glucose every 6 months; CMP and TSH annually
Niacin with fibrate	CMP and CPK	Lipids, AST, ALT at 2 months (If adding fibrate, CBC at 2 months)	AST and ALT every 3 months; Lipids, Glucose every 6 months; CMP, CBC, TSH annually
Fibrate with statin (see lipid MD annually to discuss)	CMP and CPK	Lipids, AST, ALT at 2 months (If adding fibrate, CBC at 2 months)	CPK, AST, ALT every 3 months. Add Lipids, glucose, blood urea nitrogen (BUN) and Creat. every 6 months. CBC, CMP, and TSH annually.

To rule out nephrotic syndrome, obstructive liver disease, diabetes, dysproteinemias, and hypothyroidism and assess CAD risks.

[a] For women capable of child bearing: pregnancy test (qualitative serum HCG ICD9 272.4) required prior to start of prescribed lipid agents.

[b] Patients with diabetes: glucose with each lipid panel.

Patients on coumadin: Instruct patient to have Protime/INR at 2 weeks thru MD initially or after any change in dose of lipid agent.

Patients with renal Insufficiency: Obtain BUN and Creat. annually from patient's Primary Care provider (PCP).

Table 4
Midwest Heart Lipid Clinic Statin Protocol (+ Zetia or Resin)

The lipid MD selects and prescribes the starting dose of the statin.

The lipid RN implements the order and instructs the patient regarding statin specifics and follow-up labs/appointment needed in 2 months.

Follow-up appointment: Patients who meet all of the following criteria may be treated according to this protocol.

Criteria:

    LDL is greater than 10% over LDL goal or greater than the LDL goal two times in a row.

    Liver enzymes are essentially normal. (Less than two times normal reference range for ALT and AST)

    Dose and compliance of lipid med previously ordered has been verified.

    Patient tolerates current statin dose without complaints.

Protocol:

    If all four of the above criteria are met, the lipid RN is to instruct the patient to increase the dose of the statin to the next standard dose up to 40 mg (Exception: patients on Amiodarone or Verapamil or their derivatives are not to take more than 20 mg of Zocor)

    Patients who are within 10% of the LDL or non-HDL goal are to enhance TLC and add viscous fiber and plant stanol/sterol esters for further LDL lowering.

    Patient's who do not meet their LDL goal on statin monotherapy at 40 mg[a] (or Zocor 20 mg with Amiodarone or Verapamil and deriviatives) are to add Zetia 10 mg daily.

    Patient's taking 40 mg[a] of a statin with 10 mg of Zetia who do not meet their LDL goal should be instructed to increase the statin dose to 80 mg with the Zetia.

    Patient's who are unable to tolerate Zetia should take the maximum tolerated statin dose and add Colesevelam to get to LDL goal.

[a] LDL, low-density lipoprotein; HDL, high-density lipoprotein; or on Altoprev 60 mg or Lescol 80 mg XL.

saturated, trans, and total fat, cholesterol content, as well as adequate nutrient and fiber content of their diet. Eating habits are evaluated, including the frequency of meals, meal, and snack times, portion sizes, food triggers, and sociocultural influences. Dietitians provide nutrition education, which promotes a healthier way of eating, one that is lower in sugar, fats, and cholesterol and higher in complex carbohydrates, fiber, and other nutrients. When weight loss is indicated, total daily caloric intake and percentage of fat calories are reduced *(16)*.

Dietitians who are employed in the lipid clinic setting are involved in the assessment and counseling of all patients; however, some clinics must refer selected patients to a nutrition clinic. Patients who require the expertise of a dietitian should receive an initial consultation as well as follow-up visits. Dietitians provide the patient with an individualized nutrition plan that will hopefully ease the burden of making long-term dietary changes (Tables 5–8).

EXERCISE SPECIALIST'S ROLE

The majority of patients with stable CHD should engage in regular physical activity. Exercise training improves all modifiable risk factors and specifically benefits the lipid panel by increasing HDL-C and reducing triglycerides. Patients who have had a recent cardiac event, percutaneous intervention, or coronary artery bypass grafting should be referred to an established cardiac rehabilitation program staffed by exercise specialists. Exercise specialists typically hold a master's degree in exercise physiology. In addition, many hold state licensure in nursing, physical therapy, or a related discipline. The nurses who also staff cardiac rehabilitation programs will often have specialized training in exercise science.

Date ___/___/___

PREVENTIVE CARDIOLOGY CENTER
(New Patient)

Patient name_____ ALL_____ ☐ No known allergies

Age ____ Gender: ☐ male ☐ female   Vital signs: BPR   BPL   Pulse   Weight   (lbs)
        Height (in)

Waist circumference (cm)   BMI (kg/m^2)

Meds: _____   _____   _____

_____   _____   _____

_____   _____   _____

HPI: Reason for visit _____

CV history   ☐ CAD _____ ☐ MI _____ ☐ Cardiac catheterization_____
               ☐ Coronary intervention_____ ☐ CABG _____ ☐ CVA_____
               ☐ Peripheral vascular disease _____ ☐ Carotid disease_____
               ☐ CHF_____ ☐ Arrhythmias_____ ☐ AICD_____
               ☐ Pacemaker_____ ☐ Other_____
      Symptoms
               General well-being_____
               ☐ Anginal symptoms_____ ☐ SOB_____
               ☐ PND_____ ☐ Cough_____ ☐ Edema_____
               ☐ Intermittent claudication_____ ☐ Dizziness_____ ☐ Palpitations_____

Medical/Surgical history:
    ☐ Pulmonary disease_____
    ☐ Thyroid disease_____
    ☐ Liver disease _____
    ☐ Gastrointestinal disease _____
    ☐ Renal disease _____
    ☐ Cancer _____
    ☐ Bleeding disorder _____
    ☐ Gout _____
    ☐ Past surgeries _____
    ☐ Other _____

Family history:
    Mother _____ Father _____ Sisters _____
    Brothers _____ Children ____ _____ Other _____

Cardiovascular risk factors:
    ☐ Dyslipidemia_____ Rx_____
    ☐ Hypertension_____ Rx_____
    ☐ Diabetes _____ Rx_____
    ☐ Smoking ___Never ___Previous ___Packs/day ___Years smoked ___Year quit
                       ___Current ___Packs/day ___Years smoked
    ☐ Fam. Hx. ___ Definite CAD ___Age of male(s) ___Age of female(s)
                       ___Dyslipidemia
    ☐ Menstrual Status ___ Premenopausal ___Post-menopausal (___Surgical?)
                       ___Age of onset ___ HRT

Lifestyle:
   • Weight history ___Weight at age 20 ___Recent change in weight
   • Physical Activity _____ Type ____Times/week ____Duration (min) _____Intensity

☐Resistance training ☐Calisthenics
_____Limitations

- Stress sources _____
- Alcohol (1 drink = 12 oz. beer, 5 oz . wine, 1.5 oz. 80 proof spirits)
  ___No. drinks/week ___Beer ___Wine ___Spirits
- Diet history _____ Current _____ Previous ___Ornish ___Pritikin ___Adkins ___Other
  ___No. serv. red meat/week ___No. serv. fruits/veg./day (1/2 c.) ___Eating out (no./week)
  Vitamins/minerals_____
  Dietary supplements_____
- Occupation _____

Physical examination
General _____
Skin ☐Xanthomas _____ ☐Xanthelasmas _____ ☐Other _____
Eyes ☐ Arcus corneus _____ ☐Corneal opacities _____ ☐Other _____
ENT Inspection of teeth, gums, palate, oral mucosa ☐Normal ☐Abnormal_____
Neck Exam of jugular veins ☐Normal ☐Abnormal_____
    Exam of thyroid ☐Normal ☐Abnormal_____
Resp/Lungs Assessment of respiratory effort ☐Normal ☐Abnormal Auscultation ☐Normal ☐Abnormal_____
Heart Palpation _____ Auscultation S_1 S_2 _____ ☐S_3 ☐S_4 ☐Murmur_____
Vessels/Extremities
    Carotid arteries_____ Abdominal aorta_____ _____ Femoral arteries_____
    Pedal pulses _____ Peripheral edema_____ Varicosities _____
Abdomen ☐Tenderness_____ ☐ Masses _____ Liver/spleen_____
Musculoskeletal
    Exam of back (with notation of kyphosis or scoliosis) ☐Normal ☐Abnormal
    Exam of gait (with notation of ability to undergo exercise
        Testing and/or participation in exercise program ☐Normal ☐Abnormal
    Assessment of muscle strength & tone (with notation of
        Any atrophy and abnormal movements) ☐Normal ☐Abnormal
Neuro Mood/affect _____ ☐Abnormal orientation to time/place/person_____
Ext Inspection/palpation of digits, nails ☐Normal ☐Abnormal_____

Test results:
    ECG _____
    Stress/nuclear_____
    Echo _____
    Carotid doppler _____
    Peripheral vascular doppler _____
    Angiogram _____
    Fingerstick cholesterol values TC____ LDL____ HDL____ TG____
    EBCT_____

☐See dictated letter

Impression/Plan:
_____
_____
_____
_____
_____
_____
_____
_____

_____   _____
Signature   Date

Fig. 3. History and physical examination form for new patients entering the Preventive Cardiology Program.

CHD patients should undergo a symptom-limited exercise test prior to beginning an exercise program to assess maximal heart rate, as well as to determine the presence of ischemia or arrhythmias that might impact the safety of exercise. An exercise specialist can assess the exercise test results and formulate an exercise prescription tailored to the individual's functional capacity. Typically in a cardiac rehabilitation setting, patients exercise three times a week at an intensity of 70–85% of the measured peak heart rate while supervised and monitored *(17)*. In addition to aerobic exercise, calisthenics, flexibility, and strength training are often included.

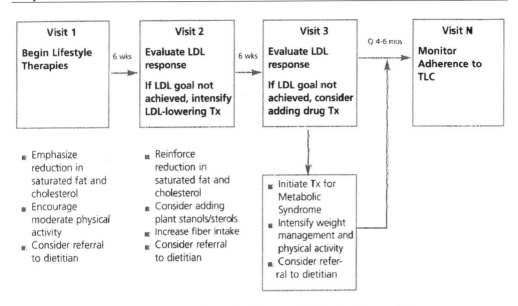

Fig. 4. A model of steps in therapeutic lifestyle changes *(6)*.

Most nurses in lipid clinics have the knowledge to counsel patients on the cardiovascular benefits of exercise and to provide low-risk patients with an appropriate exercise program. However, patients who have no exercise history may feel more comfortable doing so after consultation with an exercise specialist. An exercise specialist can meet with the patient in the cardiac rehabilitation setting to explain and demonstrate how to use the different types of equipment, thus increasing the patient's confidence in his/her ability to exercise.

Table 5
Limit LDL-Raising Nutrients – Fats and Nutrition Labeling

Nutrition facts	
Serving size 1 tbsp (14 g)	
Amount per serving	
Calories 70	Calories from fat 70
	% daily value
Total fat 8 g	13%
Saturated Fat 1.9 g	10%
Trans fat 0 g	
Polyunsaturated fat 4.5 g	
Monounsaturated fat 2 g	

In addition to the type of oil used, much of the fat consumed comes in commercially prepared products. Using the nutrition facts label fosters selection of products low in saturated and trans fatty acids. Totaling the amount of both fats can help to select products with the lowest total.

With permission from the Preventive Cardiovascular Nurses Association, 2005.

Table 6
Limit LDL-Raising Nutrients—Advice to Patients

To lower this fat	Instruct the patient to
Saturated	• Limit animal fat by selecting lean meats in moderate proportions • Limit animal fat by using fat-free or low-fat dairy products • Select products that do not contain significant amounts of coconut or palm oil
Trans-fatty acids	• Use tub or liquid margarines or spreads—ideally with less than 2 g of saturated and trans fat like Promise® Spreads or Country Crock • Limit baked goods and sweets made with partially hydrogenated fats, such as candy, pastry, pies, doughnuts, and cookies
Dietary cholesterol	• Limit egg yolks, organ meats, butterfat, and high-fat meats

Encourage the patient to use the nutrition label to identify hidden sources of saturated fat, trans fat, and cholesterol.

With permission from the Preventive Cardiovascular Nurses Association, 2005.

Table 7
Add Soluble Fiber and Plant Stanol/Sterol—Soluble Fiber and Advice to Patients

• Select high fiber foods at each meal to achieve a total of 10 to 25 g of soluble fiber per day
• Add fiber to foods—sprinkle oat bran on cereal; add almonds to salads; add dried beans to soups
• Eat whole fruit instead of drinking fruit juices
• Include vegetables rich in soluble fiber—lima beans, brussel sprouts, broccoli, eggplant, okra
• Eat whole grain products, especially oats and barley
• Add vegetables to sandwiches or have peanut butter on whole grain bread
• Gradually add more soluble fiber to the diet and make sure that plenty of liquids are consumed

With permission from the Preventive Cardiovascular Nurses Association, 2005.

CLINICAL PSYCHOLOGIST'S ROLE

Although a psychologist is usually not a staff member in a lipid clinic, he/she is often a valuable referral source. When obtaining a medical history, health care providers often identify extreme sources of stress in a patient's life or ineffective ways of managing stress which may interfere with necessary lifestyle changes. Psychologists are skilled in helping patients effectively manage their stress and learn specific stress reduction techniques. They can also provide in-depth counseling on behavior change strategies that will facilitate the therapeutic plan for patients.

Table 8
Maintain a Desirable Body Weight and Prevent Weight Gain – Advice to Patients

- Record everything eaten over a period of several days to a week to identify problems areas in caloric intake
- Be aware of portion sizes
- Be careful of calories consumed in beverages, especially soft drinks, fruit juices, and sweetened tea and coffee
- Increase fiber intake to reduce the caloric density of meals
- Eliminate unneeded fat from foods
- Partner with someone else with similar goals and share progress toward those goals
- Choose snacks that are healthy and low in calories
- Eat regularly; do not skip meals
- Eat out less frequently

With permission from the Preventive Cardiovascular Nurses Association, 2005.

There is a growing body of literature relating psychological factors to the development of CVD and to outcomes once a patient has established disease *(18–22)*. Lipid clinics may screen patients for anxiety disorders and depression.

Appropriate patients should be referred to a clinical psychologist for a thorough psychological assessment and therapeutic intervention.

An enormous benefit of having a clinical psychologist involved in a lipid clinic or preventive cardiology program is the expertise in smoking cessation strategies. When patients express that they are willing to quit smoking, they can be referred for individual or group sessions with the psychologist, because smoking cessation often requires both pharmacotherapy and counseling. In addition to smoking, patients may require the services of a clinical psychologist for other forms of substance abuse.

CONCLUSIONS

- Under diagnosis and under treatment of dyslipidemia are significant health problems. It is common for patients not to meet their guideline-suggested cholesterol target levels.
- Comprehensive and effective lipid management requires the collaboration of physicians, nurses, dietitians, and exercise specialists, among others.
- Lipid clinics that use a collaborative-care model show excellent outcomes in lipid management.
- Nurses have a multifaceted role in lipid clinics, including coordination of services from all disciplines, patient assessment (including assessment of global risk), patient education and motivation on the need for cardiovascular risk reduction, behavioral change counseling and the identification of specific strategies, medication management in collaboration with the physician, assessment of the patient's response to medications, assessment of laboratory parameters, and strategies to promote adherence to lifestyle and medication therapies.
- According to ATP III, dietitians have an integral role in lipid management. Through diet history and food records, they assess the detailed components of a patient's diet and ascertain specific eating habits, which allows them to tailor heart-healthy dietary recommendations to a patient's lifestyle, preferences, and weight loss goals.

- Exercise specialists evaluate the exercise test of high-risk or symptomatic patients and formulate an individualized exercise prescription. They may also provide a limited consultation for low-moderate risk patients who have never engaged in an exercise program.
- Clinical psychologists are important referral sources for psychological and behavior-related cardiovascular risk reduction. They provide counseling in the areas of stress management, depression and anxiety disorders, smoking cessation, substance abuse, and specific behavior change strategies that will facilitate a patient's therapeutic plan.

REFERENCES

1. Stamler J, Wentworth D, Neaton J. Is relationship between serum cholesterol and risk of premature death from coronary heart disease continuous and graded? Findings in 356,222 primary screens of the Multiple Risk Factor Intervention Trial (MRFIT). *JAMA* 1986;256:2823–2828.
2. Report of the National Cholesterol Education Program Expert Panel on Detection, Evaluation, and Treatment of High Blood Cholesterol in Adults. *Arch Intern Med* 1988;148:36–69.
3. Tolonen H, Keil U, Ferrario M, Evans A, for the WHO MONICA Project. Prevalence, awareness and treatment of hypercholesterolaemia in 32 populations: results from the WHO MONICA Project. *Int J Epidemiol* 2005;34:181–192.
4. Jacobson TA, Griffiths GG, Varas C, Gause D, Sung JCY, Ballatyne CM. Impact of evidence-based "clinical judgment" on the number of American adults requiring lipid-lowering therapy based on updated NHANES III data. *Arch Intern Med* 2000;160:1361–1369.
5. Pearson TA, Laurora I, Chu H, Kafonek S. The lipid treatment assessment project (L-TAP): a multi-center survey to evaluate the percentages of dyslipidemic patients receiving lipid-lowering therapy and achieving low-density lipoprotein cholesterol goals. *Arch Intern Med* 2000;160:459–467.
6. Expert Panel on Detection, Evaluation, and Treatment of High Blood Cholesterol in Adults. Executive summary of the third report of the National Cholesterol Education Program (NCEP) expert panel on the detection, evaluation, and treatment of high blood cholesterol in adults (Adult Treatment Panel III). *JAMA* 2001;285:2486–2497.
7. Brown AS, Cofer LA. Lipid management in a private cardiology practice (the Midwest Heart experience). *Am J Cardiol* 2000;85(3A):18A–22A.
8. Kinn JW, Brown AS. Cardiovascular risk management in clinical practice: the Midwest Heart Specialists experience. *Am J Cardiol* 2002;89(5A):23C–29C.
9. Ryan MJ, Gibson J, Simmons P, Stanek E. Effectiveness of aggressive management of dyslipidemia in a collaborative-care practice model. *Am J Cardiol* 2003;91:1427–1431.
10. Allen JK, Blumenthal RS, Margolis S, Young DR, Miller ER, Kelly K. Nurse case management of hypercholesterolemia in patients with coronary heart disease: results of a randomized clinical trial. *Am Heart J* 2002;144:678–686.
11. Grundy SM, Cleeman JI, Merz NB, et al. Implications of recent clinical trials for the National Cholesterol Education Program Adult Treatment Panel III Guidelines. *Circulation* 2004;110:227–239.
12. Fletcher B, Berra K, Ades P, et al. Managing abnormal blood lipids. A collaborative approach. *Circulation* 2005;112:3184–3209.
13. Davidson MH. Strategies to improve Adult Treatment Panel III guideline adherence and patient compliance. *Am J Cardiol* 2002;89(5A);8C–20C.
14. Thomas TS. Improving care with nurse case mangers: practical aspects of designing lipid clinics. *Am J Cardiol* 1997;80(8B):62H–65H.
15. Becker DM, Allen JK. Improving compliance in your dyslipidemic patient: an evidence-based approach. *J Am Acad Nurse Pract* 2001;13:200–207.
16. American Association of Cardiovascular and Pulmonary Rehabilitation. *Guidelines for Cardiac Rehabilitation and Secondary Prevention Programs* (4th edition). Champaign, IL: Human Kinetics, 2004.
17. Thompson PD. Exercise prescription and proscription for patients with coronary artery disease. *Circulation* 2005;112:2354–2363.
18. Rutledge T, Linden W, Davies RF. Psychological risk factors may moderate pharmacological treatment effects among ischemic heart disease patients. *Psychosom Med* 1999;61:834–841.
19. Kuper H, Marmot M, Hemingway H. Systematic review of prospective cohort studies of psychosocial factors in the etiology and prognosis of coronary heart disease. *Semin Vasc Med* 2002;2:267–314.

20. Sherwood A, Hinderliter AL, Watkins LL, Waugh RA, Blumenthal JA. Impaired endothelial function in coronary heart disease patients with depressive symptomatology. *J Am Coll Cardiol* 2005;46:656–659.
21. Barth J, Schumacher M, Herrmann-Lingen C. Depression as a risk factor for mortality in patients with coronary heart disease: a meta-analysis. *Psychosom Med* 2004;66:802–813.
22. Todaro JF, Shen B-J, Niaura R, Spiro III, A, Ward KD. Effect of negative emotions on frequency of coronary heart disease (The Normative Aging Study). *Am J Cardiol* 2003;92:901–906.

21 Development and Management of a Lipid Clinic

Carol M. Mason, ARNP, FAHA

CONTENTS

INTRODUCTION
RATIONALE FOR A LIPID CLINIC
THE LIPID CLINIC PLAN
LIPID CLINIC SERVICES
CLINICAL RESEARCH
BECOMING A LIPID SPECIALIST
LIPID CLINIC ORGANIZATION
MARKETING YOUR LIPID CLINIC
DOCUMENTATION, CODING, AND REIMBURSEMENT
COMMUNICATION
RESOURCES FOR LIPID CLINIC PROVIDERS
REFERENCES

INTRODUCTION

Current estimates are that 93 million adults in the US today (46% of the adult population) have low-density lipoprotein cholesterol (LDL-C) levels > 130 mg/dL, which would make them eligible for lipid-lowering drug therapy if they have coronary heart disease (CHD) or a 10-year risk of CHD of 10–20%. Yet, lipid disorders are not being aggressively managed, and patient adherence with prescribed therapy is often inadequate *(1)*. As the challenge to control the costly effects of CHD continues to grow, so too is the challenge to find ways for physicians and other health care professionals to manage patients with dyslipidemia as well as educate them for the risk factors for cardiovascular disease (CVD). Lipid clinics are "centers of excellence" for and about treating patients with lipid disorders. A lipid clinic is designed to employ accepted treatment pathways and is designed to manage all lipid disorders: general lipid disorders as well as the more complex dyslipidemias and dyslipoproteinemias. Lipid clinics have been around for the past 40 years paralleling national standards of care for the treatment of dyslipidemia as a major risk factor for atherosclerosis. Some were started as early as the 1960s, an example being the University of California, San Francisco, which began in 1961 and remains an active and well-respected lipid clinic

From: *Contemporary Cardiology: Therapeutic Lipidology*
Edited by: M. H. Davidson, P. P. Toth, and K. C. Maki © Humana Press Inc., Totowa, NJ

program today (2). Many universities in the US have research-based lipid clinics as well as many Veterans Administration (VA) medical centers, managed care organizations, private clinics, and hospitals.

Dyslipidemia is chronic and progressive, and for our purpose, we will use this term to describe all disorders that come under the heading of "high blood cholesterol" or "hypercholesterolemia," a major contributing risk factor for atherosclerosis (3). See section on International Classification of Diseases 9th Revision (ICD-9) codes in this chapter for a listing of associated diagnoses. The National Cholesterol Education Program, Adult Treatment Panel I (NCEP ATP I) published in 1988 recommends that all adults over the age of 20 years be screened for total cholesterol (TC) and if normal repeated every 5 years. In 2001, the NCEP ATP III recommended a complete lipid panel [TC, LDL-C, high-density lipoprotein (HDL)-C, and triglyceride (TG)] for all adults over the age of 20 years and if normal repeated in 5 years (4). A lipid clinic is a chronic disease management program providing clinical as well as research programs for patients with lipid disorders. Lipid clinics are staffed by physicians, nurse practitioners (NPs), physician assistants (PAs), clinical nurse specialists (CNSs), nurses, pharmacists, and nutritionists who have received specialized training in the science and management of lipid and lipoprotein disorders. Their goal is to diagnose, treat, and manage patients with lipid disorders.

RATIONALE FOR A LIPID CLINIC

A lipid clinic is designed to identify, manage, and treat individuals, children, adolescents, and adults, with lipid and lipoprotein disorders. Some lipid clinics chose to manage only adults, aged 18 years and older; others depending upon the training and experience of the medical director will treat children. Lipid clinics are also designed to educate patients on the risk factors for atherosclerotic vascular disease, although in most programs the management of hypertension and type 2 diabetes, two other very important risk factors for CVD, is referred to other specialists. The primary focus of management in a lipid clinic is to reduce the risk imposed by high levels of LDL-C, high levels of TGs, low levels of HDL-C, abnormal levels of Lp(a), homocysteine, high-sensitivity C-reactive protein (hsCRP), abnormal particle size, and familial abnormalities of lipoproteins. The minimum number of specialists needed to run a lipid clinic is one; usually a physician but one provider can also be an NP, a PA, or a pharmacist depending on the individual state licensure or "scope of practice." Many lipid clinics are "team managed," that is, the specialists who oversee and run the program are physicians, NPs, PAs, CNSs, nurses, pharmacists, and dietitians working together to manage patients with lipid disorders. There is extensive data supporting the case management model in risk reduction, and a lipid clinic is a good example of how a case management approach can be successful (5). The case management model will be discussed in further detail in Chap. 22. The lipid clinic program relies on national standards of care and practice guidelines, incorporates medical and lifestyle therapies, and relies on goal setting and outcome measurement. "Case managers" are often NPs, nurses, dietitians, or pharmacists who have received specialized training in lipid management and work with physicians following algorithms or guidelines. Other allied health professionals such as nurses often compliment the plan of care by providing the educational component to the program.

Lipid clinics can be found in many practice settings including university programs, hospitals, out-patient clinics both private and those affiliated with managed care organizations and public medical programs such as VA medical systems or other county or regional health care systems. Most often, lipid clinics are either found in university-based programs or are private out-patient clinics.

THE LIPID CLINIC PLAN

This section will cover the program committee, the mission statement, goals, and program objectives. It will also include information on how to develop a business plan (pro forma).

The Lipid Clinic Program Committee

The first step in designing a lipid clinic is to establish a lipid clinic program committee. The committee should include the primary providers: physicians and nurses involved in the program and all ancillary staff whose duties include assisting in the day-to-day running of the lipid clinic. Ancillary staff includes any nurse who will provide education to lipid clinic patients, the secretary who schedules patients for the lipid clinic, a member or members of the medical records department whose role involves lipid clinic chart preparation and responsibility for obtaining laboratory test results, a laboratory staff member with knowledge of specialized lipid and lipoprotein testing, an administrative support person, dietitians, and pharmacy or other allied health professional (cardiac rehab nurses, psychologists) who works directly with patients in this program. See section on *"Staffing"* for additional information on staffing and job descriptions. The committee's first job is to write the mission statement, goals, and objectives for the lipid clinic program. The committee should meet regularly, for example on a monthly basis, to discuss the clinic's progress and any issues that may arise. Having ancillary staff on the lipid clinic committee will improve its performance by enhancing a sense of ownership in the program by all involved. Ideas from all committee members are often a valuable resource in program planning and success.

Developing the Mission Statement

The mission statement describes the purpose and lays the foundation for development of goals and objectives. The mission statement has three main components: to identify the care providers who will service the program, to describe the type of patient population to be served, and to define the disease specialty or specialties to be treated. A sample mission statement might read:

> *John Heart, MD and Jane Smith, NP, of the Heartland Medical Group, are leaders in the diagnosis and management of patients with dyslipidemia and provide the community of Heartland with the newest and most advanced technology, research, and medical treatment in the area of lipid disorders.*

Establishing Lipid Clinic Goals

Goals are broad statements that describe the lipid clinic's expected short and long-term outcomes for patients with a specific diagnosis and course of therapy. Goals reflect nationally recognized standards of care in the field of cardiovascular medicine, e.g., the American Heart Association (AHA) and the National Institutes of Health (NIH) as well as the NCEP. Table 1 lists sample goals that are both clinically oriented and program specific.

Table 1
Sample Clinical Program and Operational Goals

Sample clinical program goals
  Screen all patients admitted to the lipid clinic for risk factors for cardiovascular disease (CVD)
  Diagnose, manage, and treat patients with dyslipidemia
  Identify all patients with established CVD and treat to NCEP III goals for LDL-C, triglycerides, and HDL-C
  Reduce clinical cardiac and vascular events
Sample operational goals
  Financially "break-even" or increase revenue
  Develop informational systems to meet the needs of patients and providers
  Continually explore new areas of preventive services and new methods of preventive care
  Implement a tracking system to evaluate the ongoing progress of the lipid clinic

HDL-C, high-density lipoprotein cholesterol; LDL-C, low-density lipoprotein cholesterol; NCEP, National Cholesterol Education Program.

When a lipid clinic is started in a family practice group or internal medicine practice, the patients are those who present with existing CVD (secondary prevention) and those who present with a moderate or high risk for acquiring it (primary prevention). In either case, appropriate goals need to be written based on the level of risk. When writing goals, it may be helpful to determine what specific areas within your current practice need improvement. One way is to determine by chart review how many of your patients with established CHD are at their NCEP III LDL-C goals. Having information as to areas that need improvement will give you a starting point from which goals can be set, measures can be derived, and outcomes can be determined.

Writing Program Objectives

Standards of care for patients with dyslipidemia are established by nationally recognized researchers [AHA, American College of Cardiology (ACC), National Heart, Lung and Blood Institute (NHLBI)] and leaders representing organizations in the field of lipidology, metabolism, and CVD and who can provide guidelines for writing your program objectives.

Operational or program objectives describe services within an existing practice that need improvement. This information is derived from current practice patterns within a medical group and the medical needs of patients who will be referred to the lipid clinic.

Program objectives should be specific, reflecting clinical guidelines, and targeted to specific time frames. Typically, objectives reflect conditions needing improvement, examples being too few specialists, time constraints, too great a variance in practice styles, too few patients at their LDL-C goal, or lack of patient satisfaction. Table 2 lists examples of program objectives for a lipid clinic that are time sensitive and measurable.

Developing the Business Plan

A business plan or pro forma is a budget that reflects an analysis of income and expenses over a given period of time and details the lipid clinic's operational plan.

Table 2
Examples of Dated Program Objectives

To provide CVD risk factor screening for all patients admitted to the clinic/hospital practice
To provide consultation for all patients at high or moderate risk for CVD and for all patients with a diagnosis of dyslipidemia or other diagnosis of lipid metabolism
To provide a minimum of three visits for management of patients with lipid disorders
To provide within 1 year a cost-effective, economically viable service with direct and indirect profit potential
To establish within 1 year a system whereby all providers use a risk factor assessment tool located in the front of each patient chart
To establish within 1 year a viable library of current literature and information for providers and patients on lipid disorders and CVD risk

CVD, cardiovascular disease.

Once the mission statement, objectives, and goals are established, the next step is to estimate revenue and expenses. The pro forma describes the economic plan for projecting patient volume, frequency of office visits, space needs, and overhead expenses. The following section reviews the information you will need to obtain before writing your business plan.

Revenue

Revenue includes estimating patient volume during a given period of time, average reimbursement, revenue from in-house diagnostics, and other revenue that might include research, laboratory services, etc. To reasonably estimate potential patient volume, you need to consider patient demographics, average reimbursement based on insurance carriers your clinic contracts with, potential referrals, potential competitive programs, and the interest and needs of the clinic and community. Table 3 lists sample questions that can be directed to your referral base to ascertain potential numbers of referrals.

Once you have an estimate of the number of insurance-covered, fee-for-service referrals, you can estimate the number of visits per month per patient. Patient volume equals the number of visits by both new and established patients (follow-up visits) within a given time period.

Table 3
Sample Demographic Questionnaire

Do you refer to other lipid clinics in your referral network?
Would you refer patients with complex dyslipidemia to a lipid clinic program in your area?
If so, what is the approximate number of patients that you would refer per month?
What percentage on the average of your patient population has coronary heart disease and/or hyperlipidemia?
What other medical specialties might you be interested in having patients referred to in addition to dyslipidemia, e.g. nutrition counseling, obesity, insulin resistance, and education on atherosclerosis risk factors?

Examples of types of patient care visits:

- Consultation: one visit to establish a diagnosis and formulate a treatment plan
  o Physician services, potential level 5 consultation
  o NP and PA services, potential level 5 consultation
- Follow-up: ongoing management, follow-up visits of patients for dyslipidemia
  o Physician services, potential level 4 "new" or "established visit"
  o NP and PA services, potential level 4 "new" or "established visit"
  o Registered nurse (RN) or dietitian services, level 1 "established visit"

The average time for an initial visit is 60 min whether it is a "new patient visit" or a "consultation." A follow-up visit is usually 30 min. You can use these average times as guides when estimating how many patients each provider can see during a given time period. Then the estimated number of patient visits per provider each month at an average level of service gives you the estimated expected income per provider.

Other Sources of Revenue

Most lipid clinics today are also are involved in some research, whether it is their own or sponsored by an outside organization. The expertise of the providers in the area of specialized laboratory testing and research methodology, and the availability of patients at risk provide lipid clinics the opportunities afforded by clinical trials. Clinical trials, both private and governmental, are important sources of income as well as provide a valuable resource to many patients who might otherwise not afford medication. A further discussion of research as a source of revenue will appear later in this chapter.

Other sources of revenue can be obtained from specialized laboratory testing services for patients with complex dyslipidemias. Other providers of the medical group may find this a beneficial service to them as well. Information obtained from specialized laboratory testing is an important adjunct to clinical decision making for patients with complex dyslipidemias. Examples of testing that is done in lipid clinics include "screening" as well as "advanced lipid testing," on-site point-of-service, finger-stick lipid panels, and testing for associated risk factors such as thyroid disease and diabetes.

Additional income can be obtained from educational grants, preceptor programs, and tutorials managed by lipid clinics. Tutorials can be supported by grants from both governmental agencies and pharmaceutical companies and can involve training of other medical care providers or pharmaceutical company representatives who have an interest in training their own staff or their customers. Preceptor programs or tutorials are generally 1- or 2-day programs developed and run by the lipid clinic staff providing continuing education in the area of lipid management to other providers in the community or region and are often sponsored by the pharmaceutical industry.

Expenses

SALARIES

To determine potential costs for personnel, first estimate the amount of time each provider will work within the lipid clinic program. Table 4 gives sample time estimates for staff. Use the minimum estimates of time initially, but review estimates frequently, such as every 3–6 months or at times when clinic population is increasing. Remember

Chapter 21 / Development and Management of a Lipid Clinic

Table 4
Sample Time Estimates for Staff

Sample	Percent of time	Individual
Medical Director	10–50	Physician or NP
Clinic Manager/Nurse Case Manager	25–100	NP, PA, RN, and Pharmacist
Nutrition counseling	10–50	Registered Dietitian or RN
Scheduler	5–25	
Medical Records	5	
Laboratory Staff	5–50	
Administrative Support	5–10	

NP, nurse practitioner; PA, physician assistant; RN, registered nurse.

the times estimated here are only an example of time spent in the lipid clinic. Once you have estimated the number of patients you expect to see in a given day, then that figure can be used to estimate the time spent in the program.

PHYSICAL SPACE AND SUPPLIES

Space, supplies, and other miscellaneous expenses are part of "overhead" costs. The cost for "space" is determined by estimating the cost per individual employee given a specific period of time, e.g., 1 day a week for such items as rent or mortgage, electricity, telephone, and insurance. This amount can be estimated for you by the "finance department" once they are given information as to number of exam rooms needed and days of operation.

MARKETING COSTS

Although many lipid clinics will not spend money on "direct-to-consumer" advertising, many devote some advertising space on medical group web sites and most if not all have a brochure that describes their program. The costs associated with adding information to an existing web site and printing a brochure describing the lipid clinic need to be included in the "miscellaneous costs" section of the pro forma.

A SAMPLE PRO FORMA

The pro forma provides an estimate of income and expenses over a given period of time. In addition, it estimates the lipid clinic's ability to attain predetermined goals, and it is used to evaluate the program's performance. Essentially, the pro forma includes three sections: (1) revenue, (2) expenses, and (3) summary, including total operating expense and net operating expenses.

In Table 5, you will see an example of a pro forma. In the lower last three columns, there are three areas that include "net operating income" (NOI), accounts receivable, and annual cash flow. The NOI is the difference between total operating income and total operating expense and is basically earned income. "Cumulative NOI" provides a running account of earned income over time. Accounts receivable shows the actual amount of money to be received from services that then can be compared with an

Table 5
Sample Pro Forma

	Year 1	Year 2	Year 3
Revenue			
Patient volume/year			
Average reimbursement			
In-office testing revenue			
Other revenue			
Total operating revenue			
Expenses			
Personnel costs			
In-office testing costs			
Supplies			
Space			
Miscellaneous costs			
Total operating expenses			
Net operating expenses			
Cumulative NOI			
Accounts receivable			
Annual cash flow			

NOI, net operating income.

accounting of actual cash based on what is "in the bank" at any given time. Cash flow is collected revenue minus total operating expenses. An amount for "annual cash flow" is usually provided by the finance office of your medical group.

During the initial process of determining costs, consider the need for capital expenditures for such items as new laboratory equipment and/or additional space requirements, e.g., additional "locked" cabinet space may be necessary for storing drugs during a clinical trial as well refrigerator space for stored medications. A discussion of space and equipment will be discussed later in this chapter.

"Break-Even Planning"

Providers often ask how to determine the number of patients they need to see in the lipid clinic to "break even" or exceed expenses. If lipid clinic staff is already employed and supplying revenue, you only need to estimate the additional number of visits needed to meet expenses. This can be estimated from the total operating expenses, total number of available days, and the average reimbursement income. To determine the "break-even amount" divide the "total operating expense" by 264 (the average number of patient days per year) and then divide that amount by the average reimbursement (in dollars) per patient visit. This amount will give you the required number of patients needed per day to be seen to "break even." For example, if the estimated cost of a provider is $52,000 (salary plus benefits), dividing this by 264 days equals $196.97. This amount is then divided by the average visit reimbursement amount, e.g., $30.00. You then have determined that 6.5 patients per day are needed to meet your expenses. This information gives you and your administrator a patient number to plan for when developing the lipid clinic.

Estimating Annual Clinic Revenue

A reasonable estimate of the annual revenue of a lipid clinic is important during planning stages, and the estimate should be reviewed and revised periodically. To estimate, you might want to use an average visit code: level 3 initial visit code, 99203, and the level 3 established patient visit code, follow-up visit, 99213. If you then multiply the number of new patients visits and follow-up visits for a given period of time by the average rate of reimbursement (based on the visit codes suggested), you will have an estimate of the provider's reimbursement income. Expenses per provider are such items as salary, benefits, and "overhead," and clerical support, supplies, phone/office equipment, educational material, and office space. Once you have the income based on 50 weeks, you can deduct from that the overhead, including another 25% for nonpayment and what is left is an estimate of your "net profit." Table 6 is an example of an annual physician's net profit.

LIPID CLINIC SERVICES

Lipid clinic services fall into three categories: patient care services, laboratory services, and research. Under the heading of *"patient care services,"* we will include a discussion on the various services a lipid clinic provides, and under *"patient selection,"* we have included examples of reasons for referral. In addition, there is a section entitled

Table 6
Example of Physician Reimbursement

10 new patients/week
  5 patients @ 99204 (Medicare allowable fee, $235.04)
  Allowable fee for physicians is 100% ($235.04)
  Total/week: $675.20

  5 patients @ 99205 (Medicare allowable fee, $171.52)
  Total/week: $857.60

30 established (follow-up) patients/week
  15 patients @ 99212 (Medicare allowable fee, $37.64)
  Total/week: $564.60
  15 patients @99213 (Medicare allowable fee, $52.23)
  Total/week: $783.45

Total revenue/1 week
  5 new patients @ 99204: $675.20
  5 new patients @ 99205: $857.60
  15 established patients @99212: $564.60
  15 established patients @99213: $783.45

Total weekly revenue/1 week: $2880.85
Yearly revenue (50 weeks): $144,042.25

Reprinted with permission, Long, J. *Evaluating Clinical Outcomes: Measuring Success, A Guide To Developing a Risk Reduction Guide*, Copyright, 2002, Preventive Cardiovascular Nurses Association.

"*Other patient care services,*" which includes a discussion of educational material, "offering classes to patients," and group visits.

Patient Care Services

Promoting a lipid clinic begins by selling its services to your own practice partners and community referral base. It is important to support all providers if they choose to manage their own patient's abnormal lipid values. In a specialized lipid clinic, it is important to demonstrate to colleagues the diagnostic and management skills of a specially trained lipid clinic staff, one that can consult and manage the more difficult and complex patients with dyslipidemia. In addition, it is also important that colleagues view the lipid clinic staff as a resource for current scientific information on lipid metabolism. Often lipid clinics develop newsletters or web-based programs that provide current scientific information pertaining to CHD prevention as well as lipid metabolism. Many patients with dyslipidemia are referred for the educational services of a lipid clinic. The services of a lipid clinic can be adjunctive to medical therapy and involve helping patients make nutrition and activity changes in their lives. Many lipid clinics develop their own patient education tools. For patients, the service most appreciated is a "newsletter" that can be mailed to homes, appear on a clinic web site, or just be used as a regular source on topics that pertain to CHD prevention such as nutrition and exercise. Specialized laboratory testing and clinical research can expand the services of a medical group. Providing the opportunity to be part of a clinical trial can provide an option for obtaining medications for patients who might otherwise have difficulty obtaining them, a service that is often appreciated by a medical group staff. Highlighting the specialized services that a lipid clinic can offer may help to overcome the often initial hard sell of a lipid clinic program.

Lipid clinics provide patient care that includes evaluation and management of patients with various disorders of lipid metabolism as well as family and genetic counseling for patients with premature heart disease, severe atherosclerosis, and new and emerging therapies. Examples of lipid clinic patient care services can be seen in Table 7.

Patient Selection for a Lipid Clinic

An important part of developing a lipid clinic and then "selling" it to a referral community is describing the type of patient population that a lipid clinic will consult and manage on. The following table lists potential patient problems appropriate for referral. They can be divided into several groups. The first are patients with an existing diagnosis of dyslipidemia who for various reasons are not at goal established by the clinic as well as national guidelines and standards of care. The second are those who do not have existing CHD but are at high risk and often fall into "gray" areas in the guidelines. The third area is those in the community in need of risk factor screening. These may include family members of established patients, the general public, and/or business groups interested in preventive services. Offering the screening services of a lipid clinic to one's own staff is a way to provide a clinic benefit as well as highlighting the lipid clinic program and the services it provides. Table 8 describes the patient population that is often referred to a lipid clinic program.

Table 7
Examples of Lipid Clinic Patient Care Services

Consultation for patients with lipid disorders
Management of patients with lipid disorders
Specialized laboratory testing and analysis for:
    Apolipoprotein (Apo) measurements
    Apo A1 and Apo B
    Homocysteine
    Particle testing
    Measurement of "direct LDL-cholesterol"
    hsCRP
    Lp(a)
    Genetic testing of Apo E
Screening for CVD risk factors
Educational services for community, patients, and staff, e.g. library and resource center
Research

CVD, cardiovascular disease; hsCRP, high-sensitivity C-reactive protein; LDL, low-density lipoprotein.

Table 8
Examples of Patients Referred to a Lipid Clinic

Patients with established CVD with known lipid abnormalities
    Patients not at goal on medication
    Patients intolerant to medication
    Patients preferring lifestyle therapy to medication
    Patients needing combination or multiple drug therapy
Patients with complex dyslipidemias (examples):
    Familial combined hyperlipidemia
    Polygenic primary elevation of LDL-C
    Familial dysbetalipoproteinemia
    Familial hypertriglyceridemia
    Familial hypoalphalipoproteinemia
    Elevated Lp(a)
Patients without established CVD
    Patients with known lipid disorders
    Patients with a family history of dyslipidemia or premature CAD
    Patients with type 2 diabetes or thyroid disease
    Patients with other vascular disorders
Patients in need of individual or group support, diet therapy, or education
Special populations: HIV, pediatric and adolescent cases, diabetic dyslipidemia
Family counseling for inherited lipoprotein disorders or high-risk patients not responsive to therapy

CAD, coronary artery disease; CVD, cardiovascular disease; LDL-C, low-density lipoprotein cholesterol.

Laboratory Services

The NCEP III guidelines recommend a complete fasting lipid panel: TC, LDL-C, HDL-C, and TG obtained once every 5 years. The LDL-C is generally calculated based on the Friedewald formula:

$$\text{LDL-C} = \text{TC} - \text{HDL-C} - \left(\frac{\text{TG}}{5}\right)$$

When the fasting TC levels are > 300 mg/dL, the majority of laboratories in this country will not calculate the LDL-C level. Therefore, to obtain an accurate LDL-C level in patients who present with TG levels > 300 mg/dL, a "direct LDL-C" may need to be ordered in addition to the standard lipid panel. These are normally available through all standard laboratories.

As part of the initial laboratory screening for lipid abnormalities, and to ascertain if secondary causes are present, the following additional tests are recommended: thyroid-stimulating hormone (TSH) level, a fasting glucose (FBG) level, and liver and kidney function. Additional tests such as uric acid may be indicated in certain patients. Table 9 lists some of the recommended laboratory tests as well as other tests that many lipid clinics will add.

Blood samples should be drawn after a 9–12 h fast. The patient needs to be in a steady state, which is the absence of any active weight loss, acute illness, recent trauma or surgery, pregnancy, or recent dietary changes. To ensure reliability, blood samples should be sent to an established standardization laboratory.

Creatinine phosphokinase (CK) levels are recommended as part of a baseline screening and follow-up in individuals taking a 3-hydroxy-3-methylglutaryl co-enzyme A (HMG-CoA) reductase inhibitor or "statin" medication or a combination of a "statin" with either niacin or a fibrate drug. CK levels should be obtained if the patient

Table 9
Recommended Laboratory and Other Tests of Lipid Clinic

Suggested lipid clinic laboratory tests:
    Complete lipid panel
        TC, LDL-C, HDL-C, TG
    Direct LDL-C
    TSH
    Bun, creatinine
    Fasting blood glucose
    Uric acid
Suggested other lipid clinic laboratory tests:
    hsCRP
    homocysteine
    Lp(a)
    Particle testing (see Chapter 17 on "advanced lipid testing")
    Apolipoprotein B-100
    Apolipoprotein A-1

LDL-C, low-density lipoprotein cholesterol; HDL, high-density lipoprotein; hsCRP, high-sensitivity C-reactive protein; TC, total cholesterol; TG, triglyceride; TSH, thyroid-stimulating hormone.

complains of "sore muscles" or "weakness"; complaints of myalgias and elevated CK levels suggest a diagnosis of myositis *(6)*. This usually resolves with discontinuation of the medication or elimination of one medication in a combination regimen. See Chap. 8 for a detailed discussion on side effects of the "statin" drugs. Many patients who are referred to a lipid clinic will require specialized laboratory testing, and although many laboratories today are capable of performing many of the specialized tests that you will need, you may want to try one or more of the specialty and reference laboratories that advertise testing the sub-fractions of LDL-C, HDL-C, very-low-density lipoprotein (VLDL), and Lp(a) primarily. Three of these are well known: the Berkeley Heart Lab, 839 Mitten Road, Burlingame, CA, 94010, telephone 877-454-7437, www.berkeleyheartlab.com; Atherotech, 201 London Pkwy, Birmingham, AL. 35211, telephone 800-719-9807, www.atherotech.com; and LipoScience, 2500 Sumner Blvd., Raleigh, NC, 27616, telephone 877-547-6837, web site; www.liposcience.com. All three companies have local representatives who can visit your clinic and describe the various advanced lipoprotein testing options they offer.

CLINICAL RESEARCH

The benefits of clinical research for a lipid clinic program are threefold: (1) it can enhance the clinic's reputation in the community, its physicians and staff as lipid specialists; (2) it is an excellent recruitment tool for patients, tying many patients and their families to the medical group; and (3) it is a source of additional revenue. The national networking that accompanies being part of a research trial can help providers remain informed of current scientific information. It may be simpler to begin with a phase IV clinical trial that is relatively easy to administer and can usually be done with existing staff. Phase IV studies are usually offered by pharmaceutical companies, do not require safety and monitoring board approval, and are often of shorter duration. Phase II and III trials that are more involved require clinical expertise in research implementation and administration to assure patient safety and compliance. National standards for human studies are associated with Phase II and III trials, and it is recommended that a medical director be knowledgeable in research methodology prior to undertaking them.

Group Visits

There are some lipid clinic programs that are using group visits as a means to educate and support large numbers of patients over a given period of time. Patients with lipid abnormalities often suffer from being overweight and inactive, are insulin resistant, have type 2 diabetes, or need additional education regarding either activity/lifestyle issues and/or medication adherence issues. Group visits been shown to improve adherence through group support and the positive influence of peers. A session usually lasts 1 h. Patients may be charged a nominal cash fee for attending. Depending on the insurance carriers, a 1-h group visit may be billed at a level 1 (99211), which is about a 10-min visit per patient with an RN or licensed practical nurse (LPN). The concept of a "support group" is well received in lipid clinic programs. You will need to seek guidance from your coding and billing specialists prior to using this type of billing.

Other Patient Care Services

Other lipid clinic services designed to assist patients with managing lifestyle changes and lowering their risk for CHD include providing educational materials on nutrition,

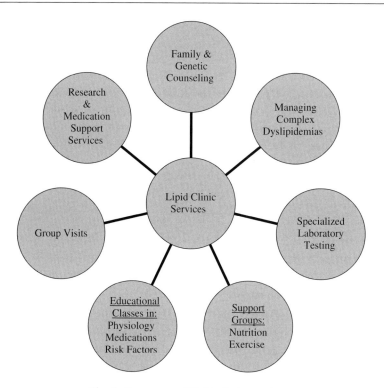

Fig. 1. Summary of lipid clinic services.

exercise, and nonpharmacologic agents. These can take the form of written materials, videos, CDs, posters, books, etc. Topics for patient education material can include "understanding atherosclerosis," its risk factors, and pharmacology of treatment, as well as topics that directly affect lifestyle, such as nutrition and exercise. Offering classes that patients can attend in the evening are well received. Family counseling is another option for those with a history of premature CHD. See Fig. 1 for a summary of these and other services.

BECOMING A LIPID SPECIALIST

It is imperative that lipid clinic providers be lipid specialists. It is certainly obvious that a physician, NP, or PA does not need to be an expert in lipid metabolism and genetics to treat a patient with high cholesterol or high TGs, and yet in order to manage a specialized lipid clinic program, one does need to have expert knowledge of lipidology and lipoprotein metabolism. In addition to a having a current and thorough knowledge of the biochemistry and pharmacology of lipid-lowering drugs, epidemiology, and genetic dyslipidemia, lipid specialists must be comfortable with such lifestyle issues as nutrition and exercise. A lipid expert needs to demonstrate they have received training and advanced education as well as have ready access to the latest information pertaining to lipid disorders. Becoming a lipid specialist involves lecturing, mentoring, consulting, and marketing the expertise of lipid management. A board certification process is now available in lipidology. The American Board of Clinical Lipidology (ABCL) is now offering credentialing and a board certification exam in lipidology.

For further information, check the ABCL website at http://www.lipidboard.org or call the ABCL office at 904-674-0752. In addition, there are a number of lipid conferences and tutorials available yearly throughout the US that offer classes in lipidology, lipid metabolism, genetics, pharmacology, and managing lifestyle issues. These courses often range from half-day programs to 1- to 3-day tutorials and are often affiliated with existing major university lipid clinic programs. The "Midwest Heart Specialists" Group in Oakbrook, Illinois, offer a yearly lipid tutorial as does the Duke University Lipid Clinic Preceptor Program in North Carolina. You can find out more by going to Midwest Heart Specialist's web site, http://www.midwestheart.com. Information regarding the Duke University Lipid Clinic Preceptor Program affiliated with Duke University's lipid clinic can be found on the National Lipid Association (NLA) web site, http://www.lipid.org. This and other information pertaining to several regional and national conferences can be found on the NLA web site or by mail at NLA, 8833 Perimeter Park Blvd., #301, Jacksonville, Fl 32216.

There are many published articles, web site resources, and books written today, from textbooks to pocket-sized guides, on lipidology. The NLA recently published (2005) a guidebook authored by Neil J. Stone, MD, and Conrad B. Blum, MD, titled *Management of Lipids in Clinical Practice*. This is a handy "pocket sized" text on lipids that can be obtained through the NLA, http://www.lipid.org. There is also an excellent guide published by the Preventive Cardiovascular Nurses Association (PCNA) entitled *A Guide to Developing a Successful Cardiovascular Risk Reduction Program*, published by PCNA and available on its website at http://www.PCNA.net.

Professional Associations

There are several excellent professional associations that lipid specialists ought to consider joining. Serving as a member in a professional association enhances one's credibility and suggests opportunities for leadership positions in lipidology and CVD prevention. For physicians, pharmacists, and nurses, the NLA not only offers a great deal of information in newsletters and on their web site pertaining to clinical management of dyslipidemia but now offers the opportunity for certification. The NLA has "Chapters" that offer regional programming, networking, and education. Information on the NLA can be found at http://www.lipid.org. For NPs, PAs, CNSs, and nurses, the PCNA organization is a leading nursing organization whose focus is nursing leadership in the arena of cardiovascular risk reduction. Its web site is http://www.pcna.net. There are two councils of the AHA that are noteworthy for their contribution to CHD prevention and worthy of joining; one is the Council of Cardiovascular Nursing (CVN) and the other is the Nutrition, Physical Activity, and Metabolism Council. The National Lipid Education Council (NLEC) is another national organization with a large steering committee composed of top lipid specialists in the US today. Its web site is a valuable resource for current information on lipidology and can be found at http://www.lipidhealth.org.

Areas of Expertise for the Lipid Specialist

GENETIC DYSLIPIDEMIA

Genetic dyslipidemia is a broad and complicated area of cardiovascular medicine and one that requires training and continuing education. Although rare in the population, patients with genetic dyslipidemias are frequently referred to lipid clinics as they are often the most difficult to manage. Further information regarding genetics of dyslipidemia can be found in Chap. 3.

Pharmacology

Pharmacology of lipid management is an important area of expertise of the lipid specialist and one that requires a current, thorough, and comfortable understanding with. Accurate knowledge and detailed understanding of the standard prescription lipid-lowering agents, their indications, side effects, etc. are essential for a lipid specialist. In addition, having a comfortable knowledge base of the many nonpharmacologic lipid-lowering agents available over-the-counter is important as well. A detailed description of the pharmacology of lipids can be found in Chap. 8.

Epidemiology

Epidemiology of lipid management is the science from which national guidelines are written and future clinical trials are based. Having a thorough knowledge and understanding of national guidelines is an essential component to developing a lipid clinic. As will be discussed elsewhere in this chapter, lipid clinic protocols and algorithms are based on nationally recognized guidelines and standards of care. These guidelines are developed from data collected in national and international clinical trials, and it is important for the lipid specialist to have a thorough knowledge of them. Information on international lipopoprotein treatment guidelines can be found in Chap. 4.

Biochemistry

Knowledge of biochemistry is essential to understanding the actions and mechanisms involved in the pharmacology of managing patients with lipid disorders. Patients with severe allergic reactions or intolerances to lipid-lowering agents are often a primary reason for referral to a specialized lipid clinic. It is important for the lipid specialist to have a thorough understanding of the physiology of atherosclerosis and the relationship of drug metabolism to the patient and the potential for disease.

Staying informed as to the latest therapies, clinical trials, and changes in treatment strategies and guidelines is imperative for a lipid specialist and his/her staff. There are a number of organizations that make available information that is current and reliable both in periodical form and on the Internet. A few of these include the NLA, (http://www.lipid.org), Advances in Prevention through Optimal Lipid Lowering (http://www.apollolipids.org), the NLEC (http://www.coronaryheartdisease.org), and the AHA (http://www.heart.org).

The NHLBI web site (http://www.nhlbi.nih.gov) provides a link to the NCEP, which publishes guidelines on managing lipid disorders.

Lipid Clinic Resources

Maintaining a "lipid library" is an important service that a lipid specialist and lipid clinic provide. Having available to other physicians, nurses, and other personnel information that is current and informative is a valuable resource to a medical group and one that sets a lipid clinic apart from usual practice. The library should contain information on pharmacology of lipids for the medical staff as well as educational material that is "patient friendly" for patients. The library should also have available information on new therapies such as electron beam tomography (EBCT) and "other emerging risk factors." It should include all new information regarding lipoprotein laboratory testing and all that is new in the arena of clinical trials that affect lipid management.

Lipid experts are "reference sources" on new and emerging risk factors that target atherosclerosis and dyslipidemia. As a lipid specialist, mentoring other health care professionals, offering preceptor programs, and speaking at local "grand rounds" and community service programs help to market your program and yourself. Accepting offers to guest lecture, developing in-house educational programs and brown-bag lunch programs, and writing and publishing in the community on news related to lipids help to support the role of the lipid specialist.

LIPID CLINIC ORGANIZATION

This section includes staffing and job descriptions, a description of a patient visit, specific lipid-related issues that affect the history and physical exam, visit schedules, physical space and equipment, and tracking programs.

An Overview of Staffing

Staff at some lipid clinics includes a physician lipid specialist and a nurse who provides education. The comfort level of the provider is what is important, and yet many lipid clinics operate with the "team approach" in which the physician, NP or PA, and nurse work collaboratively, often with existing support staff. Figure 2 depicts this model and is one that is currently used in many hospitals, private medical clinics, and university-based programs throughout the country. In this case, the patient is supported equally by a team that includes providers, educators, and support staff.

Examples of Lipid Clinic Staff Positions

The following is a list of potential staff:

- Medical Director: physician or NP with prescriptive-authority and specialized training in atherosclerosis, lipids, and metabolism.
- Clinic Manager: NP, PA, RN, or nonmedical administrator with expertise in CHD risk reduction, lipids, and preventive services.
- Nurse Case Manager: NP, RN, CNS, with specialized training in lipid disorders, risk reduction strategies, and clinic management.
- Dietitian, Nutritionist: registered dietitian, PhD nutritionist, trained in nutrition education, especially lipid metabolism and diabetes.
- Lipid Clinic Scheduler: a secretary or scheduler designated to schedule patients for the lipid clinic program.
- Medical Records: a health information specialist with knowledge regarding patient care information specific to the lipid clinic program.

Fig. 2. Team approach in lipid clinics.

- Laboratory Support: a laboratory technician with knowledge and training in handling specialized laboratory specimens.
- Pharmacy Support: a registered pharmacist available to consult on issues related to pharmacology of lipid management.
- Management support: management/supervisory staff knowledgeable and supportive of a lipid clinic program.

Job Descriptions

As previously discussed, staffing a specialized lipid clinic requires an experienced medical director skilled in diagnosis and management of lipid disorders. Although the minimum requirement for a successful lipid clinic is one staff member, more often a team (two or more health professionals) is involved. It is our recommendation that if a team of two, a physician and a NP for example, are working together to support a specialized lipid clinic, then both members of the team be qualified experts in the field of lipidology. Patients as well as colleagues need to feel comfortable that lipid clinic staff is knowledgeable in the field of lipid metabolism and pharmacology. See Chap. 22 for a detailed description of the role of the allied health provider. Table 10 lists job descriptions for the main providers in a lipid clinic setting.

Allied health professionals such as exercise therapists are frequently involved in lipid clinics, as a consultant, or by referral, or employed by the program. A valued service for a lipid clinic patient is often the exercise prescription they receive along with emotional support to sustain a beneficial level of activity. Clinical psychologists and counselors are often needed and contribute greatly when patients are suffering from depression or under a great deal of emotional stress. Other problem areas such as social isolation, the stress of chronic disease, or anxiety can make adherence to medical therapy difficult. Often these patients do well when seen and counseled by licensed therapists.

Scheduling Visits

Many lipid clinics today are part-time programs operating within existing clinic or hospital-based clinic practices. They usually begin with a half-day or full-day schedule, 1 day a week, and grow to either several days a week or integrate appointments for lipid clinic services into open clinic days as they become available. It often occurs that patients are not able to keep a visit on one specific day per week and many lipid clinic providers find that integrating them into their normal clinic hours works well also. This certainly depends on the schedule of the lipid specialists and his/her other commitments.

When setting up your clinic organization, it is helpful to plan a schedule that allows time for new patient visits and follow-up visits for each provider who will be seeing patients. Table 11 provides an example of a program time schedule for multiple providers: 60 min for new patients and 30 min for follow-up patients are only suggestions, these can be adjusted, and the coding department can assist you with making appropriate changes. We encourage working with your coding and billing department in this regard. In addition, you may want to encourage 50-min visits for new patients and 20-min for follow-up visits to allow for dictation and returning phone calls.

The lipid clinic scheduler needs to be aware that at the time the office visit is scheduled, any laboratory tests ordered for that visit need to be scheduled as well. It is

Table 10
Job Descriptions for Members of the Lipid Clinic Team

Medical Director:
  Can be MD, PhD, PharmD, NP
  Supervise other members of the lipid clinic team
  Develop lipid clinic protocols, pathways, and algorithms
  In conjunction with other team members, develop clinic mission, goals, and objectives
  Provide direct patient care based on scope of practice
  Mentor, educate, and provide a resource for lipid metabolism, pharmacology, and other issues related to patient care
Lipid Nurse Specialist:
  Can be NP, PA, CNS, or "advanced practice nurse"
  Coordinate patient, lab & research activities of lipid clinic program
  Screens all new patients to lipid clinic program
  Provide direct patient care based on scope of practice
  Coordinate patient referrals, compiles patient data base, and confers with Medical Director on patient care issues
  Implements lifestyle and medical interventions including pharmacotherapy
  Provides education for staff, patients, and family members
  Provides direct patient care to all "follow-up" patients
Nurse Educator:
  Can be NP, PA, RN, LPN
  Provides education to patients with lipid disorders
Dietitian/Nutritionist:
  Conducts initial dietary assessment
  Counsels patients in follow-up visits
  Provides dietary counseling as needed
  Develops nutrition education material
  Provides a resource to patients and providers

CNS, clinical nurse specialist; LPN, licensed practical nurse; NP, nurse practitioner; PA, physician assistant; RN, registered nurse.

also important that staff from medical records be aware that all current laboratory or diagnostic test results be obtained and placed on the patient chart prior to the patient visit. There is nothing more frustrating to a provider at the time of the lipid clinic visit than not having current laboratory results available. As mentioned in the section *"The Lipid Clinic Program Committee"* involving a scheduler and a medical records staff person at lipid clinic meetings enhances their cooperation and understanding of these processes.

The "Initial" or New Patient Visit

The initial visit, "new patient visit," or "initial consultation" is for a patient whom the provider has not seen in at least 3 years *(7)*. This visit usually takes 1 h and should be undertaken by a physician or NP, given state licensure. This visit generally follows the medical model and includes history, physical examination, laboratory tests, assessment, and plan. The aim of the initial clinical assessment has three main goals: (1) to establish the presence or absence of atherosclerosis and/or other risk factors,

Table 11
Sample Program Time Schedule for Physician, Return Visit RTN, and Dietitian

New patient visits (NPVs) and return visits (RVs)

Times	MD provider	RN provider	RD provider
NPV 60/RTN 30	60 or 30 min	60 or 30 min	45 min
8:30 a.m.	1 NPV/2 RTNs	1 RTN	1 NPV
9:00 a.m.		1 RTN	
9:30 a.m.	1 NPV/2 RTNs	1 NPV/2 RTNs	1 NPV
10:00 a.m.			
11:00 a.m.		1 RTN	
11:30 a.m.	1 RTN	1 RTN	1 RTN
12:00 noon	Lunch	Lunch	Lunch
1:00 p.m.	Catch up	Phone calls	Meetings
2:00 p.m.	1 NPV/2 RTNs	1 RTN	1 NPV
2:30 p.m.		1 RTN	
3:00 p.m.	1 NPV/2 RTNs	1 NPV/2 RTNs	1 RTN
3:30 p.m.			
4:00 p.m.	1 RTN	1 RTN	1 NPV
4:30 p.m.	1 RTN	1 RTN	1 RTN
5:00 p.m.	Catch up	Phone calls	Meetings

Reprinted with permission from M. Champaigne, *A Guide To Developing a Successful Cardiovascular Risk Reduction Clinic*, copyright © 2002, Preventive Cardiovascular Nurses Association.

(2) to establish any secondary causes of dyslipidemia, and (3) to discern if there are any genetic and/or lifestyle influences. Referrals can be made to other health care providers, some of whom may be members of the lipid clinic staff and others who are outside the clinic such as endocrinology or clinical psychology. Cardiac rehabilitation programs that are affiliated with hospitals may also be options for referral for one-time or longer-term exercise programs. Local community pharmacists as well as pharmacists affiliated with hospitals often serve as educators for patients and patient groups. Some lipid clinics are staffed by physicians and pharmacists working together in lipid clinic settings, e.g., VA medical centers. Referrals may need to be made for thyroid disorders or other complications that are discovered as part of initial screening. Other referrals for cardiology, cardiac surgery, or gastroenterology occasionally arise and require a referral as well.

The Medical History for a Lipid Clinic Patient

In addition to the "past medical history," a thorough family history taken during the "new patient visit," "consultation," or clinical evaluation is an important element in assessing for the possible presence of genetic factors in patients with dyslipidemia. Patients who have first-degree relatives with premature vascular disease are at increased risk for developing the disease. Familial combined hyperlipidemia (FCHL) is one of the most common inherited traits (1 in 300 persons) that predispose individuals to atherosclerosis. Increased levels of TC and LDL-C and/or TC, LDL-C, elevations in TGs, and (often) decreased levels of HDL-C characterize this phenotype. FCHL usually does not appear until an individual reaches adulthood. Individuals with this inherited defect usually present with TG levels that range from 200 to 500 mg/dL, with TC levels of < 240 mg/dL. Table 12 lists the more common genetic hyperlipidemias. Chapter 12 expands the discussion on familial hypercholesterolemia.

The patient's subjective complaints often begin the process of evaluating risk. Complaints of angina may lead to a diagnosis of atherosclerosis or determine progression of CHD. Complaints of leg or calf pain while walking that subsides with rest may indicate intermittent claudication, another possible risk for atherosclerosis. Taking a comprehensive medical history could reveal potential secondary causes of dyslipidemia as well as obtaining important information regarding the potential for allergies. This information could then help develop a medical treatment plan that is individualized, safe, and effective. A history of current medications, both prescription and over-the-counter as well as lifestyle issues, smoking, exercise, and successes or failures with previous diet programs will help all members of the health care team better plan a program individualized to that patient.

Physical Examination for a Lipid Clinic Patient

The physical examination done on a new lipid clinic patient includes recording vital signs, e.g., height, weight, blood pressure, respiration, heart rate, waist circumference, and body mass index (BMI). Many lipid clinics have available other diagnostic tools in addition to laboratory testing that can aid in assessing a patient's risk for atherosclerosis. These may include stress testing, endothelial function testing, carotid ultrasound, or EBCT. See Chap. 20 for further information regarding noninvasive measures for diagnosing atherosclerosis. The physical exam for dyslipidemia needs to include a search for evidence of lipid deposition and evaluate patients for clinical evidence of genetic dyslipidemias. Individuals with heterozygous familial hyperlipidemia (FH) are prone to premature coronary artery disease (CAD) and often have two important clinical findings: tendon xanthomas and corneal arcus *(8)*. Only rarely does a patient present with clinical manifestations of inherited dyslipidemia. Tendon xanthomas are an interesting clinical sign present in some patients with less common genetic hyperlipidemias. These cutaneous deposits of macrophage-derived foam cells are usually found in individuals with TC levels > 400 mg/dL, often secondary to familial hypercholesterolemia. These deposits are usually found in the tendons between the knuckles on the hands, Achilles tendons, and extensor surfaces of elbows, hands, and knees. Planar xanthomas are yellow-orange raised surface lesions found around eyelids or palms. Planar skin xanthomas occur with familial hypercholesterolemia,

Table 12
Common Genetic Dyslipidemias

Familial hypercholesterolemia
Polygenic hypercholesterolemia
Familial defective apolipoprotein B-100
Familial hypertriglyceridemia
Familial combined hyperlipidemia

Clinical Management of Dyslipidemia, NP Prescribing Reference 1998, Preventive Cardiovascular Nurses Association.

whereas plantar palmar xanthomas are more typical of dysbetalipoproteinemia. Another form of xanthomas, known as tuberous xanthomas, can be found on skin surfaces in areas of previous trauma. These soft, circular cutaneous nodules appear either singularly or in clusters. Premature corneal arcus is usually found in individuals under age 60 and is also a clinical manifestation of familial hypercholesterolemia *(9)*. Table 13 lists items found on a typical physical exam form that can be used at an initial or new patient visit.

Visit Schedules

Many patients are seen for an initial consultation and then returned to their primary care physician or referring medical group. Others are seen for several visits, e.g., three visits over a 6-month time period, to assess benefit from recommended therapy. Others are seen ongoing and are followed for many years in a lipid clinic program.

For patients on medication, diet, and/or an exercise plan, most lipid specialists recommend follow-up between 4 and 8 weeks. Often a 4-week follow-up visit for a patient needing to make lifestyle changes can positively reinforce their efforts. A 6- to 8-week follow-up laboratory test is closer to the average for assessing the response to medication. The follow-up to obtaining laboratory results may take the form of a telephone call, or post-card, or an actual office follow-up visit. One advantage in seeing the patient in the office is that seeing the patient reinforces the treatment plan and enhances adherence. If after two visits, the patient is not at their LDL-C goal, a third visit (or second follow-up visit) is scheduled in another 4–8 weeks. If the patient is on combination drug therapy and their lipid goals have been met, their visits move to a 3-month interval of time. In addition to the standard lipid panel, a "CK" is included as well as measurements of hepatic and renal function. For patients on single drug therapy and at goal, laboratory and/or office follow-up visits are recommended twice a year and can provide valuable reinforcement for overall risk reduction efforts.

Physical Space

Information regarding physical space is necessary for a lipid clinic pro forma to be developed and is necessary for determining the space needs of providers. As stated earlier, most lipid clinics develop within an existing medical practice, and therefore office space for providers often already exists. Then to determine the cost of lipid clinic office space for each provider, one merely needs to subtract the time spent in the lipid clinic from the total "overhead" time for that provider. Because exam rooms are often at a premium in medical groups, the next "space" question and the most important one is the cost of exam room space. Usually, a lipid clinic requires two exam rooms per provider, but because one exam room is primarily used for the patient who is waiting, this room can be shared with another provider, and when this is worked out, then only one exam room is needed per lipid clinic patient per provider per day. Some lipid clinics have tried using an "education" room to see patients in, especially 'follow-up' patients who may not need an exam with their visit. It is our recommendation that this not be done as it sends a message to patients that dyslipidemia is not a disease but a "lifestyle issue." It is recommended to only use real exam rooms for all patient appointments as this conveys the message that they have a legitimate medical condition that warrants serious medical treatment. Examination rooms provide the privacy for new patient physical examinations. Often in addition to weight and

Table 13
Physical Exam Assessment Sheet for Lipid Clinic Patients

Patient name: _____ DOB: _____ Date: _____
Height: _____ Weight: _____ Waist circumference: _____ BMI: _____

		Encircle one	
Eyes	Arcus:	Present	Absent
	Periorbital xanthomas:	Present	Absent
Tendinous xanthomas	Hands:	Present	Absent
	Elbows:	Present	Absent
	Achilles:	Present	Absent
Cardiac:	Rhythm: (describe)	_____	_____
	Murmurs:	Present	Absent
	Pulses: Carotid: Femoral: Dorsalis Pedis Posterior Tibial:	Present Present Present Present Present	Absent Absent Absent Absent Absent
Respiratory:	Lung Sounds: (describe)	_____	_____
Extremities:	Clubbing, cyanosis, or edema	Present Present	Absent Absent
Comments:			

blood pressure (B/P) measurements, waist circumference can be a difficult assessment for some patients to undergo, and the privacy of an examination room is appreciated.

Not all clinics have the luxury of having an education room, but certainly lipid clinic and other educational material related to reducing risk for CHD could be made available in whatever reception or patient waiting areas there are.

When assessing space needs remember that once you begin doing clinical trials, additional space may be needed for storage of medications, usually required to be stored in locked cupboards. Some study medication may require refrigeration and freezer space. Clinical trial coordinators will often require that medications be kept in separate storage areas away from the clinic's main medication supply. In addition, when undertaking clinical trials, it is important to consider space that storage of materials may require. In the section on "*Lipid clinic tracking programs*," you will see additional information regarding clinical trials.

Lipid Clinic Equipment

Similar to the issue of "space," lipid clinics are often able to use existing office and laboratory equipment. One item that many lipid clinics are using is "in-house testing equipment" for "point-of-service" lipid testing. This equipment would be listed in the pro forma under the heading "capital expenditure." Many lipid clinics have purchased specialized laboratory equipment for in-house-testing of lipids and lipoproteins as well as glucose and Lp(a). "Point-of-service lipid testing" is desirable when immediate feedback of test results is desired. The Cholestech LDX analyzer is one example of an in-house screening device that offers the lipid panel, glucose, alamine aminotransferase (ALT), aspartate aminotransferase (AST), and hsCRP in 5 min. You can find information about this machine at the Cholestech web site, http://www.cholestech.com. Other "tools" and equipment that are used in a lipid clinic include equipment to measure weight and waist circumferences, charts on BMI, computers, copiers, fax machines, and software programs for tracking patient progress.

Tracking Programs

A lipid clinic program provides medical care to patients with dyslipidemia and a treatment plan to reduce a major risk factor for CHD. Success can be measured in several ways, by tracking the progress of the patient's lipid values or measuring patient satisfaction with the program. Other measures include tracking changes in clinical outcomes within a patient population, e.g., number of angioplasties and bypass surgeries, or measuring the benefits a lipid clinic offers the medical group, e.g. increased referrals, enhanced visibility and income. The administration or management will reassess the progress or lack of progress a lipid clinic is making often after a specific amount of time (6 months or a year). Their purpose is to re-evaluate the business plan. Tracking the progress of the lipid clinic is important, and determining what to measure is an important first step in clinic planning. One important data set involves patient progress toward meeting nationally accepted standards of care. The following is a list of others:

- Percentage of lipid clinic patients at their LDL-C goal per year
- Percentage of lipid clinic patients at their TG goal
- Percent change in the Framingham risk screening score
- Percentage of patients with at least a 7% weight loss per year
- Percentage of patients who state they are "physically active"
- Percentage of patients who state they have "improved their quality of life (QOL)"

- Percentage of patients who state they are "satisfied" with their care
- Providers satisfaction with lipid clinic services
- Percent change in invasive procedures over a year
- Percent change in clinic referrals
- percent change in revenue

Data Tracking Programs

Computer tracking programs are specifically designed to measure and analyze outcomes. In the past, there have been several very good computer tracking programs that have been available to lipid clinic programs by the pharmaceutical industry. It will be necessary for you to inquire with your industry representatives if any of these programs are still available. Some tracking programs are available through commercial vendors, but today many programs are designed with the help of "in-house" programmers who develop their own computerized tracking programs.

MARKETING YOUR LIPID CLINIC

This section includes market assessment, market demographics, pricing, community wellness programs, and advertising.

Market Assessment

Table 14 lists the marketing issues a lipid specialist needs to address prior to establishing a lipid clinic. The marketing of specialized lipid clinic services although a new concept for health care providers is one that is necessary in order to develop a strategy for promoting referrals. Marketing assessment includes knowing who your referral base is, understanding the demographics of your referral base, pricing your services, understanding the potential in your community for wellness programs, and what media outlets are available and when to use them.

Marketing to Your Referral Base

The first step in market assessment is to know the area of referral and potential referring providers. This information is generally available from your business office as it relates to all potential patients within your practice area. The second step involves gathering information as to the potential for other lipid clinics in the same area. Competition may affect the number of referrals. Third, it is important to know who the insurers are and what services are covered. Knowledge of covered services is an important reimbursement issue when managing patients with dyslipidemia. Again, information pertaining to covered services can be provided by the documentation and coding specialists within your medical group.

Table 14
Marketing Patient Care Services

The referral base
Demographics of patient population
Pricing
Business/community wellness programs
Marketing through community media resources

If the lipid clinic is a division within a specialty practice, e.g., cardiology or endocrinology, and patients are covered under an health maintenance organization (HMO) (the referral for specialist follow-up is required), the marketing strategy should be directed to the primary care physicians who will likely send patients to your program. If the lipid clinic is in a primary care setting, e.g., internal medicine, family practice, or gynecology, a referral is usually not needed.

Advertising in the form of a brochure and/or person-to-person communication are ways to communicate that your lipid clinic is open for business. In whatever form of advertising you choose, be sure to describe the type of patient that makes for an appropriate referral. (See chapter section on *Patient Selection for a Lipid Clinic*, for examples of types of patients for referral.) Another important marketing strategy is to highlight being a lipid specialist and be available for "arm-chair" or "telephone-consults." This form of generosity and "good will" is often reciprocated in the form of referrals.

Patients covered under Medicare or Preferred Provider organization (PPO) providers can be targeted directly. For this group, direct mailing of a lipid clinic brochure often works well to inform potential patients. Inviting patients to an "open house" or to participate in a free introductory screening or class may be appropriate. Few lipid clinics actually advertise directly to the consumer via traditional newspaper advertising, although depending on your referral base this may be an option to consider.

Demographics and Patient Population

Patient demographics have a lot to do with how far individuals are willing to travel for lipid clinic services. In our experience, they are usually willing to travel any distance for a one-visit consult, which is fine when consults are a service your clinic is promoting. Yet, they are less likely to travel beyond a 50- to 60-mile radius of home for multiple visits. Marketing your lipid clinic beyond this distance is not usually warranted unless the areas are so remote that patients feel that they have no other options. It is also important to know if other lipid clinics are in your service area, what programs they are offering, and how many patients they are seeing. Many lipid clinics see only adults, aged 18 years and over, preferring to refer younger patients with dyslipidemia to pediatric cardiologists or lipid specialists trained to work with children. Whatever age group you decide to work with, it will be important to develop age, culture, and ethnic-specific resources. Many educational tools are available in languages other than English, and many reflect sensitivity to the various needs of different cultures.

PRICING

Pricing really develops from information obtained from billing and coding experts within your practice and should also include a market analysis of what others may be charging for similar services within your community. Some lipid clinic services such as nutrition counseling may need to be referred to dietitians employed by the hospital your clinic is affiliated with. In addition, activity and exercise therapists affiliated with outside programs may also be available by referral. For additional information on using the services of other allied health employees, see the section on "*Staffing*." Most lipid clinic patient visits will fall within the categories of either "consultation" or "follow-up" visits, may require "simple" or "complex" decision making, and involving single or multiple diagnoses, all of which will affect "pricing." Many lipid clinics today are basing their "pricing" on time spent with patients in face-to-face encounters. Your pricing options need to be discussed and approved with the clinic's billing and coding department.

Business/Community Wellness Programs

Many lipid clinics today market their services to businesses, state, and federal agencies and organizations. Many of these local-area companies are interested in screening employees for heart disease. Some have gone so far as to contract for a great many preventive services for their employees and/or members. These might include laboratory testing for lipid disorders as well as stress testing, EBCT, or more advanced testing such as nuclear imaging and endothelial function testing. Developing prevention programs targeting large groups of individuals for dyslipidemia is often a valuable marketing strategy for lipid clinics.

Advertising Through Community Media Resources

There are potentially large numbers of local sources for media coverage available to promote your lipid clinic program. Some choices might include television coverage within a "health and wellness" segment of local programming. The local newspapers with sections devoted to wellness programs in the area as well as free public service announcements about screening programs or classes. Local community newsletters can serve as an outlet for advertising as well as the Internet. Many large clinics today have their own web sites that could easily highlight the lipid clinic program. Marketing your lipid clinic can go from taking the bold form of direct-to-consumer advertising to a quieter approach by targeting only those patients within the clinic practice and referral base. If it is important that your lipid clinic offer programs that are educational and informative. How you choose to advertise them is important. Tagging onto existing programs that already appear in the public media, such as the AHA "Heart Month" in February and "Stroke Month" in May, is a way to promote your program. This form of advertising may offer an opportunity to let patients know about nutrition and exercise classes held during these months. Creative exercise programs such as Senior Stretch and Mall Walkers can be inexpensive options for patients.

When to Refer

Clinicians should refer patients when confronted with medical conditions that are outside their scope of practice or comfort level. Several examples of problems that are often diagnosed in a lipid clinic setting and often require a referral for management are type 2 diabetes and/or thyroid disease, to a diabetes nurse educator and endocrinologist, or to Internal Medicine for patients with elevations of liver function who do not respond to medication cessation.

DOCUMENTATION, CODING, AND REIMBURSEMENT

Thorough, efficient documentation and accurate coding are essential components to operating a lipid clinic. It is important that members of the lipid clinic staff adhere to the clinic's standard form of documentation. Today, most clinics employ billing and coding specialists, and it is recommended that prior to starting your lipid clinic, you consult with them on the method of documentation they prefer. Most clinics will code and bill based on the charting, enhancing the importance of accurate and appropriate documentation. The information pertained in this chapter is based on federal guidance. It is important to note that individual states may vary in their interpretation as well as individual payors may have rules as well that differ. Again, it is advisable to consult with your own billing and coding specialists to ascertain what is appropriate for your clinic program. In this chapter, I provide an overview of the terms and processes of billing and documentation that will guide you when starting a lipid clinic.

Providers

The Centers for Medicare and Medicaid Services (CMS), formerly called the Health Care Financing Administration (HFCA), is the federal government's managing agency for Medicare and Medicaid. Although for many years physicians were the only designated providers, rules updated in 1997 established new specifications for Medicare nonphysician providers, NPs, PAs, and CNSs. In 2002, provisions were established for dietitians. Many lipid clinics employ NPs, PAs, CNSs, RNs, and LPNs as well as dietitians to assist them in managing patients with lipid disorders. Nonphysician staff is usually employed by physicians or working under contract with them. When they are employed or under contract with a physician, "incident to" billing is applicable. NPs, PAs, and CNSs who obtain provider numbers and do not need the presence of a physician to perform their services may bill independently *(10)*.

Payors

MEDICARE

Medicare coverage is supplied in essentially two areas: Medicare Part A and Medicare Part B. Medicare is the federal health insurance program that is funded from Social Security payroll deductions and managed by CMS to provide benefits to individuals over age 65 who have contributed to Social Security as well as to individuals who are disabled. Medicare Part A reimburses hospital expenses and does not reimburse providers of individual services. Medicare Part B reimburses providers of services, supplies, and procedures, with individual providers submitting claims and receiving direct reimbursement *(11)*. Medicare rules and regulations set the precedent and are usually followed by other payors. NPs, PAs, or CNSs may receive their own provider numbers and then bill Medicare for their services. As stated previously "incident to" billing allows for RNs, LPNs, and other allied health employees who are employed by physicians to bill using the physician or nonphysician's (NP, PA, CNS) provider number. There are some states that do not allow RNs or LPNs to bill "incident to" a nonphysician provider (NP, PA, CNS), so again it is important to check with either state law or your own billing department.

MEDICAID

Individuals who qualify for care based on need may come under this federal program managed by CMS and funded through federal and state monies. Individual states control reimbursement fees with reimbursement for nonphysician providers (NPs, PAs, and CNSs) at 70–100% of allowable fees *(12)*.

OTHER INSURERS

Insurance companies often follow Medicare by using their rules and regulations as guides for their own fee schedules, but again it is important to consult with your billing specialists as to individual payor's coverage plans. Patients may be billed for differences between the insurer's standard reimbursement and the provider's fee for service.

MANAGED CARE ORGANIZATIONS

Managed care organizations (MCOs) provide both patient care services and payments to other providers for services. These organizations are called by various names including HMO, provider-sponsored organization (PSO), and physician-hospital organization (PHO). There are two types of models: the "Group Model" and the "Practice Model." The "Group Model" is salaried employees of the MCO and provides care for a group of patients. The "Practice Model" is an MCO that contracts for patient services with groups or individual providers who are reimbursed either by a "fee-for-service" method or a "capitated method." The "fee-for-service" method is based on a set fee schedule. The provider is aware of the fee schedule prior to signing a contractual agreement with the MCO but may negotiate for higher reimbursement. In the "capitated method," fees are based on contractual agreement. The HMO pays a provider to cover each patient in the plan for a set period of time, with a contractual agreement listing set fees applicable to the range of medical care delivered. The provider agrees to the set fees and must provide services as often as necessary based on the needs of each patient.

NEW PREVENTIVE SERVICE CODES

A revised informational edition of "Medlearn Matters" is available on the CMS web site, http://www.cms.hhs.gov/medlearn/qu_prevent_serv.pdf. This "special edition" is entitled "Informational and Educational Materials for New Preventive Services," effective October, 2005. This article covers *"Diabetes Screening Tests," "Cardiovascular Screening Blood Tests,"* and *"Initial Preventive Physical Examination* (IPPE)." All three of these are important services for a lipid clinic program to be aware of. The *"Cardiovascular Screening Blood Tests"* provides for coverage under Medicare Part B and tests "for early detection of CVD or abnormalities associated with an elevated risk for that disease," and is effective for services performed on or after January 1, 2005. Coverage includes TC, HDL-C, and TGs. Coverage for each of these tests is once every 5 years and must be ordered by the physician who is treating the patient for the purpose of early detection of CVD in patients without apparent signs or symptoms. On occasion, a patient will be referred to a lipid clinic because of a strong family history for premature CHD, and this coverage may be an option for these individuals. The IPPE examination is a "once-in-a-lifetime" benefit that must be performed within 6 months after the effective date of the beneficiary's first Part B coverage, on or after January 1, 2005. This service may be performed by either a physician or a qualified nonphysician practitioner (NPP), and coverage is provided for only one IPPE per beneficiary lifetime *(13)*. Table 15 lists the IPPE services.

In addition to the above services, the CMS has developed a variety of informational and educational products for health care professionals and ones that could be of value to those running a lipid clinic program. The list below summarizes some of these services:

- A preventive services educational resource web site: http://www.cms.hhs.gov /medlearn / preventiveservices.asp. on the CMS web site.

Table 15
Services for an "Initial Preventive Physical Exam" Visit

Review of individual's medical and social history
Review of individual's risk factors for depression
Review of individual's functional ability and level of safety
Physical examination
EKG and interpretation
Education, counseling, and referral
Education, counseling, and referral for other preventive services (see other services covered separately under Medicare Part B)
    "Under Other Services Covered":
        Lipid clinic related services:
            Medical nutrition therapy for individuals with diabetes
            Cardiovascular screening blood tests
            Diabetes screening blood test

- A guide to Medicare Preventive Services for Physicians, Providers, Suppliers, and other Health Care Professionals: http://www.cms.hhs.gov/medlearn/preventive/psguide.asp on the CMS web site.
- Brochures provide an overview of the coverage information for each preventive service. They may be downloaded, viewed, and printed at: http://www.cms.hhs.gov/medlearn/preventiveservices.asp on the CMS web site.
- Two-side laminated chart entitled, "Quick Reference Information": Medicare Preventive Services
  - Order copies at http://www.cms.hhs.gov/medlearn/preventiveservices.asp on the CMS web site.
- CD Rom describing the preventive services offered by Medicare
  - Order CD Rom at http://www.cms.meridianksi.com/kc/main/kc_fram.asp?kc_ident = kc0001&loc+5 on the CMS web site.

Additional information regarding preventive services from CMS can be found on its web site, http://www.medicare.gov, or can be obtained by calling 1-800-MEDICARE (1-800-633-4227).

Documentation

Key components of documentation include history, physical examination, decision-making, and time. In addition to working closely with your clinic's coding billing department, you may want to reference the *CPT Code Book* for the current year. Documentation of the visit will support the level of care and will support the Current Procedure Terminology (CPT) coding that is used *(14)*.

DOCUMENTING THE MEDICAL HISTORY

Often at the initial visit to a lipid specialist, the "chief complaint" is abnormal laboratory values. Either a referring provider has referred them because they have a "problem with high cholesterol" or they are concerned as to their own risk because a family member has been recently diagnosed with premature CHD. None of these referrals involve symptoms unless they exist alongside the dyslipidemia and portend to CHD. The "history of present illness" includes a history of dyslipidemia and all relevant information that may pertain to it. This would include any family history of CHD or diabetes. In addition, taking a complete history includes associated risk factors; hypertension, smoking, obesity, or insulin resistance is important for a dyslipidemia diagnosis and treatment plan. Additional information pertaining to clinical assessment of patients with dyslipidemia can be found in the section on *"The Medical History of a Lipid Clinic Patient"* as well as in Chaps. 9, 10, and 11.

MEDICAL DECISION MAKING

The assignment of risk involves four levels of medical decision making:

- Straightforward
- Low complexity
- Moderate complexity
- High complexity

In a lipid clinic setting, patient problems tend to favor more complex cases, in that the majority of referrals to a lipid clinic have multiple risk factors, e.g., dyslipidemia, hypertension, and insulin resistance. The complexity of the visit may increase based on various types of diagnostic testing and laboratory test results that require analyzing. These could include liver function testing for liver or gastrointestinal complications, fasting blood sugar or HgbA1c for diabetes, kidney function for potential drug interactions, and a urinalysis, noting protein, sediment, or microalbuminuria, as well as a TSH for testing thyroid function. Often electrocardiograms are added for screening purposes as well as more sophisticated noninvasive testing such as nuclear imaging or EBCT *(15)*. Table 16 lists the CPT codes. These codes determine the amount of reimbursement.

- Initial visit: In a lipid clinic, the initial visit is used for every new patient defined as anyone who has not received professional services form the same practice and specialty provider in 3 years.
- Consultation: This code is used when a patient is referred for evaluation and a request for a professional opinion is sought. A consulting provider seeks the opinion of the

Table 16
Current Procedure Terminology Codes
for Evaluation and Management Services

Codes used:

Level	Initial visit	Consultation	Established patient
1	99201	99241	99211
2	99202	99242	99212
3	99203	99243	99213
4	99204	99244	99214
5	99205	99245	99215

MD, NP, or PA in a different specialty. The consulting provider may threat the specific problem, but the referring physician continues to be involved with the care of the patient.
- Established visit: This code is used for follow-up visits with established patients by any provider in the lipid clinic program. As mentioned earlier, only an MD, NP, PA, or CNS can bill higher than a level 99211. This level is usually used by RNs and LPNs who may see a patient in follow-up for a blood pressure or weight check.

The ICD-9 incorporates diagnostic codes that describe the reason for the medical service. It is important to maximize reimbursement for the lipid clinic to use ICD-9 codes, often several, correctly. A common mistake is to use only one or two diagnostic codes, when in reality the patient's visit included other problems, treatments, and additional time *(16)*. Most lipid clinic patient visits are a level 3 or higher due to their complexity.

Table 17 lists the commonly used ICD-9 codes that are used in a lipid clinic.

TIME

Documentation and coding patient care can be based on time. The total amount of time (face-to-face) spent with a patient is factored into the code level. Often lipid clinic patients because of their need for managing lifestyle issues take more time than other patients, and for this reason, many lipid clinics find it more profitable to use time as a measure for coding. For example, if more than 50% of the time spent with patient was for counseling and/or coordination of care, then this time may determine the level of billing *(17)*. In this example, it is not necessary to meet the requirements of time by history, physical examination, and medical decision making. Documentation of the total time spent with the patient must be reflected in the notes (a place on the "office notes" sheet to write the actual "time in" and "time out" will suffice) or in the encounter form, with detailing of topics discussed or an outline for coordination of care.

PROLONGED SERVICE CODES

In addition to using the maximum time allotted per CPT code, there is another code called a "prolonged service code" that may be useful in a lipid clinic setting. This code is used in addition to the CPT code if you go beyond the time allotted for the usual visit code. There are two codes for this:

- 99354 is used for the first 30–74 min beyond the usual CPT code.
- 99355 is used for any additional time beyond that time (75–104 min).

Your visit must extend beyond the 30 min to use the additional code or codes. Again, we recommend documenting actual "time in" and "time out" to demonstrate actual face-to-face time spent with the patient *(18)*.

COMMUNICATION

Communication between the lipid specialist and the referring provider is critical to the success of the program. As mentioned earlier in the chapter, "arm-chair" consulting can assist to establish good rapport with practice colleagues; having the availability of a lipid specialist close at hand is an asset for a medical group. The referring provider needs to communicate to the lipid specialist the purpose of the referral. This is not only important to identify the purpose of the visit but is necessary if consultation codes are

Table 17
Examples of ICD-9 Codes Used in a Lipid Clinic

Symptoms, Signs, and Ill-Defined Conditions:
783.1 Abnormal weight gain
790.6 Hyperglycemia
791 Proteinuria
791.5 Glycosuria
729.1 Myalgia/Myositis
579.8 Intolerance of carbohydrates
359.4 Myopathy due to drugs
251.1 Hyperinsulinism
Disorders:
255.1 Hyperaldosteronism
256.4 Polycystic ovaries
272.0 Hypercholesterolemia
   272.1 Hypertriglyceridemia
   272.2 Mixed hyperlipidemia
   272.3 Hyperchylomicronemia
   272.4 Hyperlipidemia unspecified
277.7 Dysmetabolic syndrome
278.0 Obesity, unspecified
278.1 Morbid obesity
440.2 Atherosclerosis of the extremities (unspecified)
790.2 Impaired fasting glucose
790.9 Elevated CRP
Skin:
701.2 Acanthosis nigricans
272.4 Xanthelasma
272.2 Xanthoma (eruptive)
272.3 Xanthoma with hyperlipoproteinemia type 1
272.4 Xanthoma with hyperlipoproteinemia type 3
272.5 Xanthoma with hyperlipoproteinemia type 4
272.6 Xanthoma with hyperlipoproteinemia type 5
272.7 Xanthoma familial (tendon)
272.8 Xanthoma hyperlipidemic
Eyes:
371.41 Arcus senilis

CRP, high-sensitivity C-reactive protein.
Reprinted with Permission from J. Long, *Reducing CV Risk in the Insulin Resistant Patient*, copyright 2004, Preventive Cardiovascular Nurses Association.

used. In addition, the referring provider must state the method of consultation he or she prefers; either the specialist evaluates and recommends a course of treatment and refers the patient back to the referring provider who will implement the treatment or consultation where the patient is referred for evaluation and treatment by the specialist. Offering a variety of services provides options for referrals, consultations, one to three visit programs, or patient follow-up for an indefinite period of time. Making available other services in addition to medical therapy, laboratory and research, group visits, and educational classes affords options for referring providers. The more choice colleagues have, the more likely they are to use some or all of the services. Communication to the referral provider is a key component to maintaining a busy clinic. The visit report needs to include the full assessment and treatment plan in writing. A courtesy telephone call can enhance the communication but should not replace a letter. A copy of the letter to the referring provider can be used as a source of documentation for the patient file. In some clinics, either a copy of this letter is sent to the patient or the letter is addressed and sent to the patient with a copy to the referring provider.

RESOURCES FOR LIPID CLINIC PROVIDERS

There are a number of national organizations, most of whom have Internet web sites and can provide not only information pertaining to national guidelines but offer an abundance of educational material that can be used by lipid clinics. Many of these sites are linked. A few examples follow:

- The ACC at http://www.acc.org
- The AHA at http://www.meraicanheart.org
- The National Committee for Quality Assurance at http://www.ncqa.org
- The Center for Medicare and Medicaid Services at http://www.cms.hhs.gov
- The American Diabetes Association at http://www.americandiabetes.org
- The NIH at http://www.nih.gov
- The NHLBI at http://www.nhlbi.nih.gov

REFERENCES

1. American Heart Association. *Heart Disease and Stroke Statistics – 2005 Update*. Dallas, TX: American Heart Association, 2005.
2. UCSF Lipid Clinic, Operated by the Metabolic Faculty, University of California, San Francisco. Available online at http://www.uscf.edu/lipidclinic/lipidhome6.htm. Assessed October 2005.
3. Stone, NJ, Blum, CV. *Management of Lipids in Clinical Practice*. NY: Professional Communications, Inc., 2005.
4. Report of the National Cholesterol Education Program Expert Panel on Detection, Evaluation, and Treatment of High Blood Cholesterol in Adults. The Expert Panel. *Archives of Internal Medicine* 1988;148:36–69.
5. Thomas, TS. Improving care with nurse case managers: practical aspects of designing lipid clinics. *American Journal of Cardiology* 1997;80:62H–65H.
6. ACC/AHA/NHLBI Clinical Advisory on the Use and Safety of Statins. *Circulation* 2002;106:1024.
7. Fee Schedule for Physician Services. 15502 Payment for Office/Outpatient Visits Carriers Manual XV, Part 3, Center for Medicare and Medicaid Services. Available at http://www.cms.hhs.gov/manuals/14car3b15052.asp. (Assessed October, 2005).
8. Gotto, AM, Pownall, JH. *Manual of Lipid Disorders*, Baltimore, Hong Kong, London: Williams & Wilkins, Copyright, 1992.
9. Genest, J, Lippby, P, Gotto, AM. *Braunwald's Heart Disease, A Textbook of Cardiovascular Medicine*, Seventh Edition, Philadelphia, Elsevier, Copyright 2005.
10. Federal Register, Rules and Regulations Medicare Program. Revision to Payment Policies and Adjustments to the Relative Value Units. Listed Under the Physician Fee Schedule for Calendar Year 1999. Assessed at http://www.cms.hhs.gov/medicare. (Assessed October, 2005).
11. Physician Information Resource for Medicare. Available online at http://www.cms.hhs.gov. (Assessed October, 2005).
12. Center for Medicare & Medicaid Services (CMS). State & Territorial Government Information. Available online at http://www.cms.hhs.gov/states/default.asp. Assessed October 2005.
13. The Guide to Medical Preventive Services for Physicians, Providers, Suppliers and Other Health Care Professionals Coding and Diagnosis. Available online at http://www.cms.hhs.gov/medlearn/matters/mmarticles/2005/SE0556.pdf (Assessed October, 2005).

14. Current Procedural Terminology (CPT) 2002. American Medical Association Press, 2001.
15. Center for Medicare and Medicaid (CMS). 1997 Documentation Guidelines for Evaluation and Management Services, the Medicare Learning Network. Available online at http://www.cms.hhs.gov/medlearn.emdoc.asp. (Assessed October, 2005).
16. Center for Medicare and Medicaid Services (CMS), ICD-9-CM Official Coding Guidelines. Available online at http://www.cms.hhs.gov/paymentsystems/icd9/. Assessed October, 2005.
17. MedPAC. Medicare Payment Advisory Commission. Available online at http://medpac.gov/ publications/congressional_reports/Mar00%20/AcronymsTerm.pdf. Assessed October 2005.
18. MedPAC, Sources of Additional Coverage for Medicare Beneficiaries. Medicaid and Other State Programs. Available online at http://medpac.gov/.publications/ congressional_repots/Jun02_AppB.pdf. Assessed October, 2005.

22 The American Board of Clinical Lipidology Physician Certification Program

Nicola A. Sirdevan, MPA

CONTENTS

INTRODUCTION
WHY IS THERE A BOARD CERTIFICATION PROGRAM
  IN CLINICAL LIPIDOLOGY?
ELIGIBILITY CRITERIA
THE CERTIFYING EXAMINATION
CERTIFICATION FOR ALLIED HEALTH PROFESSIONALS
REFERENCES

INTRODUCTION

The American Board of Clinical Lipidology (ABCL) is an independent certifying organization offering the only certification program for physicians specializing in clinical lipidology. The ABCL mission is to reduce the morbidity and mortality from dyslipidemia and related diseases by assessing qualifications and certifying knowledge in clinical lipidology.

A certification program for lipid specialists was first envisioned by the leaders of the National Lipid Association (NLA)—the professional association for lipid specialists—to validate and recognize the specialized expertise of its members. Spearheaded by the NLA, the ABCL was incorporated as a nonprofit organization in 2003 to develop and administer the first certification examination of its type. Although the ABCL is a separate entity from the NLA, the two organizations work together to ensure that there are adequate educational opportunities for lipid professionals to prepare for certification. The Board receives no public funds or commercial support and is not a licensing authority, acting as a certifying body only.

Clinical lipidology has been defined by the NLA as a multidisciplinary branch of medicine focusing on lipid and lipoprotein metabolism and their associated disorders *(1)*. Because clinical lipidology is a young and evolving specialty, the ABCL Board of Directors first had to define the domain of knowledge and clinical expertise of the lipidologist. Thereupon, the Board approved a rigorous credentialing process

From: *Contemporary Cardiology: Therapeutic Lipidology*
Edited by: M. H. Davidson, P. P. Toth, and K. C. Maki © Humana Press Inc., Totowa, NJ

and a comprehensive written examination that assesses and validates the specialized knowledge and advanced training required to practice in this dynamic field.

Through the ABCL certification process, successful physician candidates are awarded "Diplomate" status and are certified as specialists in clinical lipidology. Physicians with ABCL Diplomate status have demonstrated that they have a board-certified knowledge base that is recognized and quantifiable.

WHY IS THERE A BOARD CERTIFICATION PROGRAM IN CLINICAL LIPIDOLOGY?

The specialty of lipidology is rapidly growing in importance for a number of reasons, including demographic trends that place an increasing number of Americans at risk for coronary heart disease and epidemic levels of obesity concurrent with a surge in new cases of diabetes. Lipid management has been demonstrated to be one of the most effective strategies for the prospective treatment of cardiovascular disease. Yet, the number of specialists with expertise in lipid management is inadequate to address this burgeoning population of patients. The ABCL certification program is designed to encourage the growth of the specialty of lipidology within medicine and to enhance physician practice behavior to improve the quality of patient care.

It is well established that physician board certification is associated with better clinical outcomes and higher rates of adherence to evidence-based guidelines. There is additional evidence that patients regard board certification as a measure of quality when selecting a physician (2). Achieving certification in clinical lipidology allows physicians to demonstrate that they have undergone a thorough training, self-assessment, and evaluation process that validates their expertise for patients, external organizations, and professional colleagues.

Unlike many subspecialties in medicine, lipidology has wide applicability across many different fields and types of patients. Lipid specialists represent a variety of backgrounds such as internal medicine, cardiology, endocrinology, family practice, and obstetrics and gynecology. Given this diverse candidate pool, the ABCL certification program establishes a consistent benchmark of proficiency in the field.

ELIGIBILITY CRITERIA

The ABCL certification program is open to licensed physicians in the United States or Canada. To become credentialed, candidates must meet the basic eligibility criteria and the training requirements. Although rigorous, the requirements have been designed to provide any physician with demonstrated knowledge and experience in lipid management an avenue to become certified as a clinical lipidologist (3).

Basic Requirements for All Candidates

1. Be a currently licensed physician (MD or DO) in the United States or Canada
2. Be certified by a primary care board (ABMS or equivalent, i.e., Internal Medicine, Family Practice, Preventive Medicine, Pediatrics, and Obstetrics and Gynecology)
   OR
   Have completed a 2-year minimum relevant accredited fellowship (e.g., cardiology and endocrinology)
   OR

Have 2 years of demonstrated appropriate experience and practice activity in the management of patients with lipid disorders

Training Requirements

Applicants are required to accumulate 200 credit hour equivalents (or "points"), which are earned through documented training and expertise in lipid management as delineated in Table 1.

THE CERTIFYING EXAMINATION

Applicants who meet the eligibility requirements must pass a comprehensive written examination, which is offered annually by the ABCL. The subject content of the certifying examination reflects the knowledge, skills, and attitudes deemed essential for the competent practice of lipidology. The content focuses on the biochemistry and metabolism of lipids; the genetics, diagnosis, and management of dyslipidemias and the metabolic syndrome; and the fundamentals of vascular biology. The examination is constructed at the upper level of difficulty and consists of approximately 250–275 multiple-choice questions.

The breakdown of question content by area of lipidology is listed in order of emphasis on the examination:

Pharmacological Therapy
Nutrition and Nonpharmacological Therapy
Risk Assessment and NCEP Guidelines
Genetic Disorders
Lipoprotein Metabolism

Table 1
Training Criteria

Training criteria	*Points*
1. Certification in a primary care board	50
2. Subspecialty certification in Endocrinology, Diabetes and Metabolism (EDM) or Cardiology	50
OR	
Other relevant advanced training and/or certification, including Certified Diabetes Educator, Clinical Nutrition, Gastroenterology, Nephrology, Specialist in Clinical Hypertension	10–50
3. Relevant academic practice and relevant faculty appointment at an ACGME recognized institution (i.e., relevant to lipid practice)	Up to 50
4. Clinical research and/or scholarly publications in the management of lipid disorders	Up to 50
5. Balance of points not obtained in 1–4 must be earned through lipid-focused continuing medical education credit obtained within the previous 3 years	1 per credit hour earned

Epidemiology and Clinical Trials
Metabolic Syndrome
Vascular Biology

The eligibility requirements and examination materials for the ABCL certification program have been developed based on substantial review and analysis of the current state of medical and scientific knowledge of the treatment of lipid disorders, as reflected in the medical literature. The ABCL Board of Directors, with the assistance and advice of professionals in relevant fields, has developed a certification examination that recognizes accepted levels of knowledge and expertise in the profession.

Attainment of Designation and Redesignation

Eligible candidates who successfully pass the examination are recognized as achieving status as a "Diplomate of the American Board of Clinical Lipidology" and receive certificates from the ABCL. A registry of designated board-certified individuals is maintained by the ABCL and is reported in various publications and on the ABCL website.

The "Diplomate of the American Board of Clinical Lipidology" designation is recognized for a period of 10 years at which time the candidate must retake and pass the current board examination to maintain the designation.

To Apply

For an application, an Applicant Handbook, complete eligibility requirements, and examination information, visit the ABCL website at http://www.lipidboard.org. Questions concerning the examination should be referred to the ABCL staff at (904) 674–0752.

Continuing Education and Lifelong Learning

The NLA is responsible for ensuring that there are adequate educational opportunities for lipid professionals to achieve and maintain certification to uphold the highest standards of care. To that end, the NLA provides a core curriculum and professional development track in lipid management to meet its members' lifelong learning and self-assessment needs. The ABCL certification program provides the capstone to the NLA professional development program. More information on lipid-focused continuing medical education activities is available on the NLA website at http://www.lipid.org.

CERTIFICATION FOR ALLIED HEALTH PROFESSIONALS

The Accrediation Council for Clinical Lipidology (ACCL) is an independent certifying organization that has developed standards and an examination for mid-level providers who manage patients with lipid and other related disorders. Eligibility requirements are available online at http://www.lipidspecialist.org. Questions should be referred to the ACCL staff at (904) 988–0356.

REFERENCES

1. McKenney JM. Who am I? Who are we? *The Lipid Spin.* Spring 2006:4(1):4.
2. Cassel CK. President's message: does certification make a difference? *ABIM Perspectives.* Spring 2006:2.
3. *ABCL Applicant Handbook.* American Board of Clinical Lipidology Website. Available at: http://lipidboard.org/handbook.pdf. Accessed October 2, 2006.

Index

A
ABCA1 deficiency, *see* Tangier disease
ABCG5 and *ABCG8* gene, 298–300, 312
ABCL, *see* American Board of Clinical Lipidology
ABCL physician certification program
  basic requirements for candidates, 478–479
  certifying examination, 479–480
  clinical lipidology and, 478
  training criteria, 479
Accreditation Council for Clinical Lipidology (ACCL), 480
Acetyl coenzyme A (CoA), 291
Action to Control CardioVascular Risk in Diabetes (ACCORD), 360
Acute coronary syndrome (ACS), 12, 70
Acyl cholesterol acyltransferase (ACAT), 6
Acyl-CoA cholesterol acyltransferase 2 (ACAT2), 297
  production, 300
Adiponectin, 93, 96
Adult Treatment Panel III (ATP III), 57, 222, 248, 357, 369, 426–427
Age-related macular degeneration (AMD), 282
American Board of Clinical Lipidology, 454–455, 477
American Heart Association (AHA), 37, 350, 443
  cardiovascular risk assessment and, 352–353
ε-aminocaproic acid, 242
Angina pectoris (AP), 70
Angiotensin converting enzyme (ACE), 273
Angiotensin-converting enzyme inhibitors, 52
Angiotensin II receptor blockers, 52
Anglo-Scandinavian Cardiac Outcomes Trial (ASCOT), 128
Anglo-Scandinavian Cardiac Outcomes Trial (ASCOT)—Lipid-Lowering Arm (ASCOT-LLA), 371
Ankle-Brachial Index (ABI), 382, 398–399
Annexin-2 (ANX2), 296
Antihypertensive and Lipid- Lowering Treatment to Prevent Heart Attack Trial (ALLHAT), 372
Antipsychotic medications, 212
ApoA-I on aortic atheromatous plaque, hepatic expression of, 167
ApoA-IV expression, in humans, 298
ApoB48, 297–298
Apolipoprotein(a) [apo(a)], 242, 259–261
  isoform size, cardiovascular disease risk and, 250–251
  statin effect and, 248
  structure and functional implication, 242–246
Apolipoprotein B, 41–42

Apolipoprotein C-II deficiency, 23
Apolipoprotein E (apoE) polymorphisms, 295
Apolipoprotein E3-Leiden (E3L) transgenic mice, 73
Apolipoproteins, 3–5
  classification, 5
Apoprotein C-II deficiency, 205
Aspirin therapy, 248
Assessment of Diabetes Control and Evaluation of Efficacy of Niaspan Trial (ADVENT), 137
Asymptomatic atherosclerosis, clinicopathologic correlation of, 11
Atherogenic dyslipidemia
  drug therapies for, 62
  features of, 57
  lifestyle management for, 61
  lipoprotein and hepatic lipases in, 61
Atherosclerosis, in children, 409–410
Atherosclerosis Risk in Communities (ARIC), 249, 391
Atherothrombosis, 10
  endothelial dysfunction, 12–13
  lipid efflux and plaque rupture, 15
  lipid influx, 13
  phases, 10–13
Atorvastatin, 63, 128
ATP-binding cassette transport A1 (ABCA1), 297, 300
Awareness, CVD in women and, 349–350

B
Behavioral changes and lifestyle, CVD in women and, 357
Bezafibrate infarction prevention, 86
Bile acid (BA), 296
  synthesis, 301
Bile acid binding agents, 417–418
Bile acid sequestrants (BASs), 134, 230, 361–362
  potential adverse affects and drug interactions with, 235
Blood viscosity, LDL apheresis and, 279–280
Body mass index (BMI), 461
Brachial artery reactivity testing, 399–400
"Break-even planning," 448
Bypass Angioplasty Revascularization Investigation 2D (BARI 2D), 155

C
Canadian Consensus Conference of Cholesterol (CCCC), 37
Cardiac Magnetic Resonance Imaging (CMRI), 397–398

Cardiovascular disease, 37, 441
  lipoproteins and, 354–356
  Lp(a) and, 248–250
  risk, APO(a) isoform size and, 250–251
  risk of developing, 50–51
Cardiovascular Health Study (CHS), 74, 391
Cardiovascular response and lipid changes, fibrate, statin, niacin and, 175–181
Cardiovascular risk assessment
  AHA and, 352–353
  lipoproteins and CVD, 354–356
Carotid Artery Intima-Media Thickness (c-IMT), 390
  extra-carotid sites, 394
  measurement, 183, 381–382
  method and technology, variations in, 392–393
  observational and clinical studies, 391–392
  risk group identification, 392
Caveolin-1 (CAV1), 296
CD40/CD40L signaling pathway, 89
c7E3 Anti-Platelet Therapy in Unstable REfractory angina (CAPTURE), 88
Cell adhesion molecules (CAMs), 72
Cerebral vascular disease (CVD), LDL apheresis and, 283–284
CHD, LDL-C lowering and, 221
  cholesterol absorption inhibitor, statin and, 234–238
  ezetimibe, 224–230
  lipid-altering drugs and, 222
  pathophysiology of, 222
  resins and polymers, 230–234
  statins, 224
  treatment goals, 222–224
Cholestech LDX analyzer, 464
Cholesterol, 1, 292–293; *see also* Sterols absorption
  absorption inhibitors, 234, 362–363, 419
  absorption, statins and, 307–309
  LDLr deficiency and, 297
Cholesterol abnormalities, in children, 410–411
  individual approach
    bile acid binding agents, 417–418
    cholesterol absorption inhibitors, 419
    diet therapy, 415–417
    fibric acid derivatives, 419
    HMG-CoA reductase inhibitors, 418
    pharmacologic intervention, 417
    screening, 413–414
    treatment, 414–415
  population approach, 411–413
Cholesterol 7α-hydroxylase (CYP7A1), 300
Cholesterol and Recurrent Events (CARE) trial, 79
Cholesterol ester (CE), 1–2, 296–300
Cholesterol ester transfer protein (CETP), 4, 9, 300
  deficiency, 32
Cholesterol-lowering therapy, 121
Cholesterol Treatment Trialists' Collaborators, 373
Cholesteryl ester transfer protein (CETP), 59, 142–143, 161
Cholestyramine, 114
Chylomicronemia, 204
Chylomicrons, 6

Clinical psychologist's role, dyslipidemia management and, 436–437
Coagulation and fibrinolysis cascade., 82
Collaborative Atorvastatin Diabetes Study (CARDS), 128
Combination therapy, dyslipidemia treatment and, 362–363
Combined hyperlipidemia, dietary therapy, 118
Compensatory hyperinsulinemia, 58
Concomitant TZD and statin therapy, 154
Coronary angiography, 84–85
Coronary artery calcification (CAC), and elderly, 381
Coronary artery disease (CAD), 269, 276
  LDL apheresis and, 281
Coronary Computed Tomography (CT) Imaging
  EBCT, 394–396
  MDCT, 396–397
Coronary Flow Reserve (CFR), noninvasive assessment of, 400–402
Coronary heart disease (CHD), 69, 221, 441
  and LDL apheresis, 275
  LDL-C and, 425
  Lp(a) level, as risk factor for, 248–249
  morbidity and mortality associated with, 121
  NCEP guidelines and, 242
Corticosteroids, 211
C-reactive protein (CRP), 43, 70–75, 276
  in atherogenesis, 73
  and elderly, 378–379
  transgenic mice, 73
CVD, *see* Cardiovascular disease
CVD, in women
  cardiovascular risk
    assessment, 352–356
    identification, 349–352
  combination therapy, for dyslipidemia, 362–363
  drug therapy, 358–362
  lifestyle and behavioral changes, 357
  treatment, 356–357
Cyclosporine, 130
CYP3A4 pathway, 129
Cytochrome CYP2C8-metabolized drugs, 140
Cytochrome P450 (CYP) 3A4 enzyme system, 231

D

Danazol and tibolone, Lp(a) levels reduction and, 247–248
Density Gradient Ultracentrifugation (DGU), 328–329
Dextran sulfate cellulose adsorption (DSA), 272–273
Diabetes, CVD in women and, 351–352
Diabetes Mellitus, 51–52
  Type 2, 56–57
Diabetes Outcome Progression Trial (ADPOT), 155
Diabetic dyslipidemia, 65
Diabetic nephropathy, LDL apheresis and, 280–281
Diacylglycerol acyltransferase (DGAT), 64
Dietary Intervention Study in Children (DISC), 415
Dietary prescriptions, dyslipidemia and, 113
Dietary therapy, 113

Diet, evolving role of
  early statin years, LDL lowering (1987–1994), 114
  pre-statin cholesterol lowering (1960–1987), 114
  statin Trials, LDL Lowering, (1994–2004), 115
  therapies beyond, (2004 to present), 115–116
Dietitian's role, dyslipidemia management and, 430–432
"Diplomate of the American Board of Clinical Lipidology," 480
Direct adsorption of lipoproteins blood perfusion (DALI) system, 275
Docosahexaenoic acid (DHA), 65
Drug therapy, CVD in women and, 358
  ACCORD trial and, 360
  BASs and, 361–362
  cholesterol absorption inhibitors and, 362
  PPAR-α agonists and, 359
Dysbetalipoproteinemia, 206, 215
Dyslipidemia, 221, 442
  lipid-altering drugs for, 222
  treatment, combination therapy for, 362–363
Dyslipidemia management
  allied health professional's role in
    clinical psychologist's role, 436–437
    dietitian's role, 430–432
    exercise specialist's role, 433–436
    lipid clinic/prevention clinic models, 426–427
    nurse's role, 427–430
  Canadian guidelines, 38
  in children
    cholesterol abnormalities, approaches to, 410–411
    early atherosclerosis, 409–410
    individual approach, 413–419
    population approach, 411–413
  European guidelines, 43
  guidelines for asymptomatic patients at high risk, 49
  in women (see CVD, in women)

E
Eicosapentaenoic acid (EPA), 65
Electron Beam Computed Tomography (EBCT), 394–396
Electrophoretic mobility, lipoprotein, 2–3
Elevated LDL-C, Inherited syndromes of, 27–28
  autosomal dominant hypercholesterolemia (ADH), 29
  autosomal recessive hypercholesterolemia (ARH), 28
  familial hypercholesterolemia (FH), 27–28
  sitosterolemia, 29
Endogenous pathway, 7–8
Endothelial cell adhesion molecules (ECAMs), 9
Endothelial dysfunction, 13
Endothelial nitric oxide synthase (eNOS), 13, 72
Enzyme immunoassay (ELISA) methods, 259–260
  LP(a) measurement and, 257–258
Exercise specialist's role, dyslipidemia management and, 432–436
EZE/simvastatin (EZE/SIMVA), 234
  potential adverse effects of, 237–238
Ezetimibe (EZE), 131, 133, 214, 221, 224–230, 304–307
  potential adverse effects of, 234

F
Familial Atherosclerosis Treatment Study (FATS), 249
Familial chylomicronemia syndrome, 23
Familial combined hyperlipidemia (FCHL), 205–207, 460
Familial dysbetalipoproteinemia (FD), 26
Familial hyperalp halipoproteinemia, 33
Familial hypercholesterolemia, 23, 267
  LDLR activity and, 268
  plasma LDL-C and, 268–269
  treatment guidelines, 270–276
Familial hypertriglyceridemia (FHTG), 26
Fasting glucose, 57
Fenofibrate, 60, 214
FH, see Familial hypercholesterolemia
Fibrate combination therapy, 232
Fibrates, 138–141, 182, 309–310, 359, 363, 419
Fibrin D-Dimer, 83–84
Fibrinogen and atherothrombosis, possible mechanisms, 80
Fibrinogen Studies Collaboration (FSC), 80
Fibrinolysis, 72
Flow-mediated dilatation (FMD), see Brachial artery reactivity testing
Fluvastatin, 127
  Lp(a) levels reduction and, 248
Framingham risk score (FRS), 74
Free cholesterol (FC), 1–2, 296–300
Free fatty acids (FFAs), 5, 57

G
Gemfibrozil, 214
Gene for apolipoprotein E *(APOE)*, 26
Gene therapy, 215
Genetic dyslipidemia, 455–456
Glomerular filtration rate (GFR), 380
Glucose and Lipid Assessment in Diabetes (GLAD), 156
Glucuronidation, 132
Gradient Gel Electrophoresis (GGE), 327–328

H
HDL:LDL ratio, 15–16
Head-to-head analysis, lipid effects of pioglitazone and rosiglitazone, 153
Health maintenance organization (HMO), 466
Heart Protection Study (HPS), 370–371
Hemolytic anemia, 31
Hemostasis, markers of, 79–81
  Fibrin D-dimer, 83–84
  plasminogen activator inhibitor-1, 81–82
  von Willebrand factor (vWF), 84–85
  WBC count, 85–87
Heparin-induced extracorporeal low-density lipoprotein precipitation (HELP) system, 274
Hepatic cholesterol, handling, 164
Hepatic lipase deficiency, 206
Hepatosplenomegaly, 26

High-density lipoprotein cholesterol (HDL-C), 226
  BAS and, 230
  dietary and lifestyle therapy, 118–119
  in elderly, 377–378
  FH and, 271
High-density lipoprotein (HDL), 9–10, 250
  antiatherogenic effects of
    endothelial cell function, 165–166
    net capacity to affect atherogenesis, 166–167
    oxidative phenomena, prevention of, 165
    pro-oxidative activity by HDL, 167
    reverse cholesterol transport, 160–165
    thrombotic phenomena, 166
  antiatherogenic functions of, 168
  apoprotein A-I$_{Milano}$, liver-x-receptor and CETP, 188
  compositional alterations in, 169
  D4F, 189
  effects of lifestyle modification on, 172–173
  inherited disorders of, 30
  metabolism, 9
  particle subclasses and CHD risk, relationship of, 334–336
  risk for cardiovascular disease, epidemiological evaluation, 167
  role in lipoprotein metabolism, 9
  subpopulations, 163
High-sensitivity C-reactive protein (hsCRP), 442
HMG-CoA reductase inhibitors, 29, 122, 270, 275, 292, 299, 312, 418
Hormone replacement therapy (HRT), 357–358
Hydroperoxyeicosatetraenoic acid (HPETE), 15
Hydroperoxyoctadecadienoic acid (HPODE), 15
3-hydroxy-3-methylglutaryl coenzyme A, 121
Hyperlipidemia augments endothelial dysfunction, 13
Hypertriglyceridemia, 26, 58, 201, 207–208, 208
  causes of, 204
  medications, 211
  pharmaceuticals, 213
  physiology of, 201–204
  secondary causes of, 208–211
  treatments/recommendations, 212–213

I

Ileal bile salt transporter (IBAT), 300
Inflammatory cytokines, 71
  interleukin-6, 75–77
  interleukin-18, 77–78
  tumor necrosis factor, 78–79
Insulin resistance (IR), 56, 58
Insulin resistance syndrome, 57
Insulin-resistant states, elevated FFA levels in, 58–59
β2-integrin Mac-1, 254
Intercellular adhesion molecule-1 (ICAM-1), 72
Interferon therapy, 211
Interleukin-6 (IL-6), 70
Intermediate-density lipoproteins (IDLs), 59
International Federation of Clinical Chemistry and Laboratory Medicine (IFCC) Working Group, 258
Intestinal-acting agents, mechanism of, 133–134
Isopentyl pyrophosphate, 292

K

Kringles (Ks), 245

L

Landmark statin trials, LDL lowering (1994-2004), 115
L-carnitine, Lp(a) levels reduction and, 247
LDL, see Low-density lipoprotein
LDL apheresis, applications of
  cardiac transplant, 281
  cerebral vascular disease, 283–284
  ocular microcirculatory disturbances, 281–283
  peripheral vascular disease, 283
  renal disease, 280–281
  sudden idiopathic hearing loss, 280
LDL Apheresis Atherosclerosis Regression Study (LAARS), 283
LDL apheresis methods, FH and, 271
  DALI system, 275
  double-membrane filtration, 275
  DSA system, 272–273
  HELP system, 274
  immunoadsorption, 275
  and lipid-lowering therapy, 275–276
LDL cholesterol (LDL-C), 26, 63, 249, 301–303, 441
  CHD and, see LDL cholesterol (LDL-C) lowering, CHD and
  ezetimibe and, 304–307
  familial hypercholesterolemia and, 268–269
  and HDL-C, CHD-related events and, 170–171
  homozygous FH treatment and, 271
  lipid-lowering therapy and, 267
LDL cholesterol (LDL-C) lowering, CHD and, 221, 425
  cholesterol absorption inhibitor, statin and, 234–238
  ezetimibe, 224–230
  lipid-altering drugs and, 222
  pathophysiology of, 222
  resins and polymers, 230–234
  statins, 224
  treatment goals, 222–224
LDL lowering and plaque regression, vascular effects in
  hemorheology, 279–280
  inflammation, 276
  Lp(a) and, 276–277
  thrombosis and fibrinolysis, 278–279
LDL-related protein receptor (LRPr), 8, 10
LDL subfractions and CHD, relationship of, 333–334
Lecithin cholesterol acyl transferase (LCAT), 32, 300
Left ventricular hypertrophy (LVH), 394
Leukotriene B4 (LTB4), 86
Liberalize fat intake, 118
Lifestyle and combination therapy, 39
Lipid clinic, development and management of, 441
  clinical research, 453–454
  documentation, coding and reimbursement, 465–467
  lipid specialists and, 454–457
  marketing of (see Lipid clinic, marketing)
  organization, 457–465
  plan for, 443–449
  rationale for, 442–443

resources for, 474
services, 449–453
Lipid clinic, marketing
business/community wellness programs, 467
community media resources and, 467
demographics, patient population and, 466–467
market assessment, 465–466
Lipid clinic/prevention clinic models, dyslipidemia management and, 426–427
Lipid clinic program committee, 443
Lipid-related biomarkers, 87–89
Lipid Research Clinics Coronary Primary Prevention Trial (LRC-CPPT), 114
Lipids, in vascular biology, 10
Lipids management, in elderly
predicting risk, 376–381
relevant studies, review of, 370–373
risk reduction, by lipid treatment, 369–370
safety concerns, 373–376
subclinical CVD measurements, 381–382
Lipid therapy, 115
"Lipid triad," 57, 60
Lipoprotein(a), 241
APO(a) and cardiovascular disease, relationship between, 250–251
and elderly, 380
and homocysteine levels, 42
measurement, 256–259
pathogenicity mechanism of, 251–256
plasma Lp(a) levels, 246–248
structure and functional implications, 242–246
uses of, 260–261
vascular disease, risk factor for, 248–250
Lipoprotein associated phospholipase A2 (Lp-PLA2), 91–93, 92–93, 380
Lipoprotein heterogeneity
analytic implications of, 329–337
in endogenous lipid transport, 324–325
in exogenous lipid transport, 323–234
in reverse cholesterol transport, 326
Lipoprotein lipase (LPL), 23, 202, 216
Lipoprotein(s)
classification and properties, 3
composition, 4
heterogeneity (*see* Lipoprotein heterogeneity)
lipase, 204
major classes of, 2
metabolism, 5–6, 10, 23, 24–25
structure, 3
Lipoprotein subfractions, utilization of, 321
assay methods, 326–329
heterogeneity, physiologic origins of, 322–326
analytic implications of, 329–337
patient management and, 337–341
Liposorber LA-15 apheresis system, *see* Dextran sulfate cellulose adsorption (DSA)
Liver X receptors (LXR), 300
L-lysine monohydrochloride, Lp(a) levels reduction and, 247

Lovastatin, 126
Low-density lipoprotein, 241, 247
basic manuvers to lower, 116
oxidized, and vascular motion, 276–277
plasma LDL concentrations, 248
Low-density lipoprotein receptor (LDLR), 267
gene, 268–269
Low HDL-C
guideline definitions and targets for therapy, 172
inherited syndromes of, 30
Low LDL-C, inherited syndromes of
abetalipoproteinemia, 29–30
ApoA-I deficiency and structural mutations, 30–31
familial hypobetalipoproteinemia, 30
Low serum HDL-C, pharmacologic therapy for, 173
Lp(a), *see* Lipoprotein(a)
LP(a) measurement, 256
ELISA methods, 257–258
non-immunologically based methods for, 259
standardization of, 257–259
LPL gene, 23
Lp-PLA2 and cardiovascular disease, overview of studies, 94–95
LXR/RXR complex, 10
Lysine-binding site (LBS), 245
Lysophosphatidylcholine (LysoPC), 91
Lysophosphatidylcholine, Pro-atherogenic activities of, 92

M
Macrophages, interaction of apoA-I and HDLs with, 162
Managed care organizations (MCOs), 469
Meta-analysis, lipid effects of pioglitazone and rosiglitazone, 152
Metabolic syndrome, 39–42, 42, 217
CVD in women and, 351
Meta-regression analysis CHD, 122
Microsomal TG transfer factor (MTF), 297, 300
Mixed dyslipidemia, 118
Monoclonal antibodies (MAbs), 257–258
Multi-Detector Row Computed Tomography (MDCT), *see* Multi-slice computed tomography (MSCT)
Multiple Risk Factor Intervention Trial (MRFIT), 114
Multi-slice computed tomography (MSCT), 396–397
Myeloperoxidase (MPO), 86, 87–89
Myocardial infarction (MI), 73, 81, 249–251

N
National Cholesterol Education Panel, Adult Treatment Panel III (NCEP ATP III) Guidelines, 38, 61
National Cholesterol Education Program (NCEP), 57, 128, 171, 222, 242, 267, 352
cholesterol abnormalities, in children, 410–411
lipid-lowering therapy and, 391
TLC diet, 270
National Institute of Diabetes and Digestive and Kidney, 56
National Lipid Association (NLA), 477, 480
Neutrophils, 86
Niacin, 65, 135–137, 174, 213

and BAS, 363
  Lp(a) levels reduction and, 247
Niemann–Pick C1-Like 1 protein (NPC1L1, 6), 296, 300, 304
Nonarteritic acute anterior ischemic optic neuropathy (NAION), 281
Noncholesterol sterols, 293–294; see also Sterols absorption
Noninvasive cardiovascular imaging technologies
  ankle-brachial index, 398–399
  brachial artery reactivity testing, 399–400
  cardiac magnetic resonance imaging, 397–398
  carotid artery intima-media thickness, 390–394
  coronary computed tomography imaging, 394–397
  coronary flow reserve, noninvasive assessment of, 400–402
  left ventricular echocardiography, 394
Nuclear factor-kappa-B (NFκB), 14, 254
Nuclear Magnetic Resonance (NMR) spectroscopy, 329
Nurse's role, dyslipidemia management and, 427–430

O
Obesity, 56
  CVD in women and, 351
  prevalence among US adult, 56
Omega-3 ethyl esters, 60
Omega-3 fatty acids, 65, 141–142
Oxidized LDL (Ox-LDL), 89–91, 277

P
Palm Pilot™ CHD risk calculators, 427–428
Pancreatic lipase inhibition, 215
Paraoxonase (PON), 15, 165
Pathobiological Determinants Atherosclerosis in Youth (PDAY) study, 410
Patient management, lipoprotein utilization and
  case HC, 337–339
  clinical considerations, 339
  clinical discussion, 341
  intervention, 339–240
Peripheral lipodystrophy, 58
Peripheral vascular disease (PVD), LDL apheresis and, 283
Peroxisome proliferator-activated receptor (PPAR)-δ, 296
Phospholipase, 88
Phytosterols, 293–294; see also Sterols absorption
  ACAT and, 297
Pioglitazone, 214
Plaque development., 99
Plasma LP(a) levels
  determination of, 246–247
  modulation of, 247–248
Plasmapheresis, 215
Plasminogen, 14
  KIV, 245
  tPA-mediated, 253
Plasminogen activator inhibitor-1 (PAI-1), 70
Platelet activating acetylhydrolase (PAFA), 165

PPAR-α agonists, see Fibrates
Pravastatin, 127
Pravastatin or Atorvastatin Evaluation and Infection Therapy (PROVE-IT), 372
Pre-statin cholesterol lowering (1960–1987), 114
Prognostic markers, of advanced disease, 98
Prospective Cardiovascular Munster (PROCAM) study, 250
Prospective Epidemiological Study of Myocardial Infarction (PRIME), 249
Prospective Study of Pravastatin in Elderly at Risk (PROSPER) study, 371–372
Protease inhibitors, 211

Q
Quebec Cardiovascular Study, 250

R
Ramipril, 155
Rapamycin, 211
Retinoic acid, 216
Reverse cholesterol transport (RCT), 161
Rhabdomyolysis, 128
Rosiglitazone, 155, 214
Rosiglitazone Evaluated for Cardiac Outcomes and Regulation of Glycemia in Diabetes (RECORD), 155
Rosiglitazone study, 155
Rosuvastatin, 63, 75, 122, 185, 229, 231, 381

S
Scavenger receptor-B type 1 (SR-B1), 296
Serum amyloid A (SAA), 70
Simvastatin, 63, 127
Sitosterol, 299
β-sitosterolemia, 310–313
Smoking cessation., 118
Stanols, 294, 301–303; see also Sterols absorption
Statins, 224, 234, 363
  cholesterol absorption and, 307–309
  fibrate combination therapy, 232
  potential adverse effects and potential drug interactions of, 231
Statin(s), 185, 214
  actions and comparative pharmacokinetics of, 126–127
  biosynthesis pathway and inhibitors, 124
  clinically relevant statin drug interactions, 132
  mechanism of action, 121–122
  outcomes studies of statin medications, 123–124
  pharmacophore, 125
  safety and drug interaction, 129–131
  structure and relative lipophilicity, 125
Statin Therapies for Elevated Lipid Levels Compared Across Doses to Rosuvastatin (STELLAR), 128–129
Statin therapy, 52, 63
Steroidogenic acute regulatory protein (StAR), 300
Sterol carrier protein-2 (SCP-2), 300

Sterol regulatory element-binding protein-2 (SREBP-2), 292
Sterols absorption
  intestinal, 294–301
    pharmacologic modulation of
      ezetimibe, 304–307
      fibrates, 309–310
      statins and cholesterol absorption, 307–309
      sterols and stanols, 301–303
Stockholm Heart Epidemiology Program (SHEEP), 85
Sudden idiopathic hearing loss, LDL apheresis and, 280
Symptomatic atherothrombosis, 11
Systematic Coronary Risk Evaluation (SCORE) system, 37

T
Tangier disease, 31
Taq1B polymorphism, 33
TC and LDL-C, score chart, 44–45
TC/HDL-C ratio, 91
Tendon xanthomas, FH and, 268–269
Tesaglitazar, 156
TG-rich lipoproteins (TRLs), 326
Therapeutic Lifestyle Changes (TLC) diet should, 270
Thiazides, 211
Thiazolidinediones (TZDs), 149, 187
  actions on serum lipoproteins, 149–150
  clinical evidence of lipid-lowering effects of, 150–151
  concomitant TZD and statin therapy, 154–155
  meta-analyses of pioglitazone and rosiglitazone trials, 152–153
  prospective comparative clinical studies, 153–154
Thrombogenicity, 16

Thrombomodulin, 72
Tissue factor pathway inhibitor (TFPI), 17
Tissue plasminogen activator (t-PA), 72, 245
Torcetrapib, 214
Transforming growth factor-β (TGF-β), 253, 255
Treating to New Targets (TNT) trial, 373
Triglycerides (TGs), 1–2, 2, 117, 201
Triglyceride (TG)-rich lipoproteins, catabolism of, 202
TRL subclasses and CHD risk, relationship of, 336–337
Tuberoeruptive xanthomas, 27
Tumor necrosis factor, 71
Type III hyperlipidemia, *see* Familial dysbetalipoproteinemia (FD)

U
Urokinase plasminogen activator (uPA), 245

V
Vascular biology, 1
Vascular cell adhesion molecule- 1 (VCAM-1), 72
Very low density lipoprotein (VLDL), 7, 272–273, 300, 311
  elevations of, 205
  overproduction of, 59
Virchow's triad of thrombogenicity, 17
Vitamins B12 and B6, 43

W
WBC, count and subtypes, 86, 87
Women's Health Initiative (WHI), 358
Women's Health Study (WHS), 76
World Heart and Stroke Forum (WHSF), 37

Editors Biography

MICHAEL H. DAVIDSON, MD, FACC, FACP, is Professor of Medicine and Director of Preventive Cardiology at University of Chicago Pritzker School of Medicine. He is also the founder and Executive Medical Director of Radiant Research Chicago formerly the Chicago Center for Clinical Research. Dr. Davidson has published more than 150 articles in peer reviewed journals and has conducted more than 1000 clinical trials in the area of preventive cardiology and nutrition over the past 20 years. Among his many honors he is president of the Midwest Lipid Association and was named one of the Best Doctor's in America 2004–2007.

PETER P. TOTH, MD, PhD, FAAP, FICA, FAHA, FCCP, FACC, is the Director of Preventive Cardiology at the Sterling Rock Falls Clinic as well as Chief of Medicine and Vice Chief of Cardiovascular Medicine at the CGH Medical Center in Sterling, Illinois. Dr. Toth is a Clinical Associate Professor of Family and Community Medicine at the University of Illinois College of Medicine in Peoria, Illinois, and Southern Illinois University School of Medicine in Springfield, Illinois. He is Editor-in-Chief of the Journal of Applied Research in Clinical and Experimental Therapeutics. Dr. Toth has authored and coauthored over 120 publications in journals as well as medical and scientific textbooks.

KEVIN C. MAKI, PhD, is the Founder, President and Chief Science Officer for Provident Clinical Research and Consulting, Inc., a private clinical research company with offices in Bloomington, Indiana and Glen Ellyn, Illinois. He received a PhD in Epidemiology from the University of Illinois School of Public Health in Chicago, Illinois. Dr. Maki has been an investigator, consultant or statistician for more than 200 clinical trials and has authored or co-authored more than 120 published manuscripts, books and book chapters.